Springer
Berlin
Heidelberg
New York
Barcelona
Budapest
Hong Kong
London
Milan
Paris
Santa Clara
Singapore
Tokyo

Joseph F. Zayas

Functionality of Proteins in Food

With 86 Figures and 21 Tables

Springer

Prof. Dr. Joseph F. Zayas †

Kansas State University
Dept. of Foods and Nutrition, USA

ISBN 3-540-60252-6 Springer-Verlag Berlin Heidelberg New York

Library of Congress Cataloging-in-Publication Data
Zayas, Joseph F., 1932 - Functionality of proteins in food / Joseph F. Zayas.
p. cm. Includes bibliographical references and index.
ISBN 0-387-60252-6
ISBN 3-540-60252-6
1. Proteins in human nutrition. 2. Proteins--Analysis.
I. Title. TX553.P7Z39 1996 664--dc20 96-28428 CIP

Typesetting: Fotosatz-Service Köhler OHG, Würzburg
SPIN: 10504072 52/3020 - 5 4 3 2 1 0 - Printed on acid-free paper

Preface

The book is devoted to expanding current views on the phenomena of protein functionality in food systems. Protein functionalities in foods have been the object of extensive research over the last thirty to forty years and significant progress has been made in understanding the mechanism and factors influencing the functionality of proteins. The functionality of proteins is one of the fastest developing fields in the studies of protein utilization in foods. Currently, a broad spectrum of data related to protein functionality in food systems has been collected, however, much more needs to be known. In this volume, the most important functional properties of food proteins are presented:

Protein solubility, water holding capacity and fat binding, emulsifying, foaming, and gelling properties as affected by protein source, environmental factors (pH, temperature, ionic strength) and protein concentration;
Relationships between protein conformation, physicochemical properties, and functional properties;
Protein functional properties as influenced by various food processing conditions, particularly heat treatment, dehydration, freezing and storage when frozen, extraction and other processes;
Effects of protein modification on the enhancement of protein functionality;
Utilization of various proteins in improving functional properties in food systems.

Those aspects of protein functionality are presented which the author believes to be interesting and most important for protein utilization in food systems. The book is recommended to students and food scientists engaged in food protein research and food industry research, and development scientists.

Table of Contents

Introduction

Proteins are the basic functional components of various high protein processed food products and thus determine textural, sensory and nutritional properties. Food products include various proteins with different structural, physical, chemical and functional properties, and sensitivity to heat and other treatments. The term „protein functional properties“ is of relatively recent origin. Functional properties of proteins are those physicochemical properties of proteins which affect their behavior in food systems during preparation, processing, storage, and consumption, and contribute to the quality and sensory attributes of food systems [1]. The most important functional properties of proteins in food applications are: – hydrophilic, i.e. protein solubility, swelling and water retention capacity, foaming properties, and gelling capacity; – hydrophilic-hydrophobic, i.e. emulsifying, foaming, and hydrophobic, i.e. fat binding properties. There is no generally accepted scheme of classification for the functionality of proteins with relation to specific physicochemical properties of the protein molecules. Attempts to classify functionality of soya and other proteins have been presented [1].

Because of their heterogeneous structure and interaction with other food components, proteins exhibit a broad spectrum of functional properties. The molecular basis for the functionality of proteins is related to their structure, and their ability to interact with other food ingredients. Functional properties of proteins as food components are affected by molecular weight and shape of protein molecules, structural diversity, structure and conformation, and charge distribution on the protein molecules. Functional properties are affected by the primary structure of proteins, i.e. thenumber of amino acids and their disposition in the polypeptide chain. The functional properties depend on the protein interactions with other proteins, lipids, carbohydrates, water, ions, and flavors. Functional properties are affected by hydrophobicity of proteins. The content of apolar amino acids (2.5–3.0% in most proteins) influences conformation of proteins, hydration, solubility and gelation properties. Charged amino acids in the protein molecule enhance electrostatic interactions which stabilize the globular proteins and influence water binding. The compactness of protein molecule structure and extent of bonding and interactions significantly influence the functional properties of proteins. The effect of conformation on protein functionality is demonstrated by globular proteins that have more polar charged amino acids toward the surface which accelerates solubility, swelling and hydration. High water retention may be obtained if there is a large proportion of hydrophilic residues on the surface of globular proteins. Functional properties of soya proteins are affected by covalent and non-covalent forces, i.e. covalent bonds as disulfide linkages play an important role in gel formation.

There are incomplete data on the effect of protein conformational structure on functionality and it remains a subject for future studies. The absence of progress is related to the mixed nature of food systems. The functional performance of proteins is determined by multiplicity of the reactions, mainly by interaction between different proteins and between proteins and nonprotein ingredients of food systems. In order to utilize all protein resources completely, it will be necessary to improve our understanding of protein structure, its modification, and to optimize functional properties of proteins in foods.

Functional properties of proteins are influenced by the source of protein (plant and animal origin), a processing history of the protein material, by the interaction with environmental factors. The following environmental factors influence functional properties of proteins: pH of medium, temperature of treatment (whether native or denatured), ionic strength, moisture content, oxidation/reduction potential, shear stress and others [2]. Processing temperature affects functional properties; denaturation of proteins may improve or impair functional properties. Functional properties of plant proteins are influenced by processing conditions, methods of defatting, the kind of solvent, temperature of extraction and drying. Proteins with desired functionalities can be obtained by designing selected protein recovery processes. Limited studies have been carried out to find correlations between processing conditions of protein flours, concentrates and isolates production and their functionality in foods. Knowledge of the relationship between the structure of proteins, their physical and chemical properties and functionality undervarious environmental conditions is critical for their proper food usage and can be utilized to control functional properties of proteins.

The preparation and processing of proteins for use as ingredients in different food formulations influences their functionality in foods. A significant effect on protein functionality is obtained by heating, dehydration, freezing, ultrafiltration, comminution, homogenization and other treatments. Desired functional properties of proteins are obtained as a result of the change of protein structure during baking, cheese ripening, drying of milk, and compatible combinations of proteins with other ingredients of food products.

Under commercial conditions, proteins exhibit several functional properties that are necessary in the production of food products. In comminuted meat products gelling and emulsifying properties of proteins, water retention and fat binding during comminution and heating are the determining characteristics for their incorporation in the formulations.

The properties of milk proteins which cause them to form curds under appropriate conditions leads to the development of cheese. The ability of soy proteins to coagulate and form tofu in the presence of calcium and heating facilitates their preservation, improves their quality characteristics and digestibility. The conformation and composition of the amino acids of egg white protein make it an excellent foaming agent in the production of bakery products and various foam-based foods. Numerous formulated foods have been developed in order to provide better high protein diets with limited animal protein content and to improve the functionality of proteins. In different additives of plant origin, proteins are the most important functional component. However functional properties of protein preparations apparently cannot be attributed to proteins alone, and other components of preparations, such as carbohydrates, lipids and minerals may also contribute to functionality.

Studying the functionality of food proteins is essential in order that their role may be fully understood and they may be used effectively in food products. Testing of protein functionality should be performed during designing processes of protein concentrates and isolates production, developing new sources of protein and incorporation of these proteins in food systems. This testing is necessary to demonstrate the applicability of new protein ingredients as replacements for conventional proteins.

Testing of protein functionality in model systems is effectively used in comparison functionality of various proteins in different food systems. A wide variety of model systems are used for studying protein functionality. The most commonly used method is studying the functionality of proteins in model systems as affected by conformational structure, environmental conditions pH, temperature, ionic strength, mechanical and heat treatment, protein interactions with other proteins and nonprotein components. Because of the differences in protein performance depending on the processing and environmental conditions, it may be necessary to use more than one model system. Requirements for model methods are: good repeatability and simplicity, and acceptable correlation with real food systems.

Functional properties of food proteins determined in model systems could be used as reliable indicators or predictors of protein functionality in real food systems. Molecular properties of proteins and the influence of environmental and processing factors should be utilized for predicting protein functional properties. Protein functionality may be predicted from solubility and hydrophobicity. Li-Chan et al. [3] used multiple regression analysis to predict the functional properties of salt-soluble proteins and meat minces from various physicochemical properties of proteins. At present, the evaluation of food proteins functionality is strictly empirical and the development of a standard methodology for predicting the functional properties of protein in various foods is necessary.

In the future, we will probably be able to predict and control protein functionality. Study of the protein functionality in model systems would help to predict functionality of proteins from their structural features and to improve their utilization in food products. In food processing, prediction of protein functionality could improve the quality of foods and reduce the cost of production. However, such predictions require knowledge not only of the structure, hydrophobic and hydrophilic properties, charge, flexibility, etc., but also the environmental conditions such as product pH, ionic strength, and thermal treatment.

Testing of protein functionality in model systems will not always generate data acceptable in real food systems. Difficulties in determining the functionality of proteins in foods are related to the diversity of their properties and functions, as well as the interactions between protein and nonprotein components. The value of data on protein functionality from model experiments is limited because of the complexity of foods and variation in the protein content and its physical properties. The additional factor is the absence of standard methods of testing protein functionality and the use of various empirical procedures. It is generally impossible to reproduce model systems with the specific conditions of the actual foods. Frequently, the only reliable way of determining functionality of proteins is to incorporate the protein ingredient into the formulation and produce the finished product.

It is highly recommended to develop model, standard methods for determining protein functionality under specific conditions and to extrapolate and utilize these methods for predicting the functionality of these proteins in commercial food systems. Using standardized tests, the results obtained by different scientists and laboratories can be compared. Standard testing procedures including conditions of sample storage, sample solubilization, testing temperature, pH, the protein concentration in solution, emulsion, foam, or gel should be established. The knowledge of protein functionality could be markedly enhanced if standard tests for determining protein functionality could be developed and implemented in food research.

In exploring and developing new sources of food proteins, the functional property is a main criterion in assessing their potential utilization in foods. The success of new protein applications is possible if, in addition to the pattern of essential amino acids, they possess desirable functional properties and acceptable sensory characteristics. Development of new processes for manufacturing protein concentrates and isolates should be carried out in order to provide protein products with improved functional and sensory properties. The knowledge of the functional property of food proteins provides information about performance of these additives in foods. Currently a variety of proteins with functional properties acceptable for many foods are commercially produced. Food manufacturers have utilized different sources of proteins to improve the quality of protein products and decrease cost. There is a need for the food manufacturers to understand the functionality of protein components in formulated foods. Intensive research and technological development must be carried out to increase protein quality and quantity. Food scientists and geneticists may modify food proteins to create new and to improve existing functional properties.

References

1. Kinsella, J. E. (1976). Functional properties of proteins in foods: a survey, Crit. Rev. Food Sci. Nutr., 7: 219.
2. Kinsella, J. E. (1982). Protein structure and functional properties: emulsification and flavor binding effects, In Food Protein Deterioration, Mechanisms and Functionality (J. P. Cherry, ed.), Amer. Chem. Soc., Washington, D.C., ACS Symp. Ser.: 206, p. 301.
3. Li-Chan, E., Nakai, S., and Wood, D. F., (1987). Muscle protein structure-function relationships and discrimination of functionality by multivariate analysis, J. Food Sci., 52: 31.

Chapter 1

Solubility of Proteins

1.1 Introduction

The solubility of proteins is considered as that proportion of nitrogen in a protein product which is in the soluble state under specific conditions. Solubility is the amount of protein in a sample that dissolves into solution. Proteins recommended as food additives can be partly or completely soluble or completely insoluble in water.

Protein solubility is the first functional property usually determined during development and testing of new protein ingredients. Protein solubility is a physicochemical property that is related to other functional properties. Knowledge of protein solubility can give useful information on the potential utilization of proteins and their functionality, especially in foams, emulsions and gels. Solubility is the main characteristic of proteins selected for use in liquid foods and beverages. High soluble proteins possess good dispersibility of protein molecules or particles, and lead to the formation of finely dispersed colloidal systems. Potential applications of proteins can be dramatically expanded if they possess high solubility. Kinsella [1] has suggested that solubility of proteins is the most important factor and excellent index for their functionality.

Protein solubility is influenced by amino acid composition and sequence, molecular weight, and conformation and content of polar and nonpolar groups in amino acids. It is important to utilize fully our knowledge of the amino acid composition and conformation of proteins, including hydrophobic and hydrophilic properties that influence protein solubility. Protein solubility is affected by environmental factors: ionic strength, type of solvent, pH, temperature, and processing conditions. A relationship between protein solubility and structure has not been demonstrated. McGowen and Mellors [2] reported relationships between the solubility of amino acids and hydrophobicity. Whey protein solubility is strongly influenced by exposed hydrophobic amino acids [3]. Protein-protein interaction in an aqueous medium is accelerated by hydrophobic interactions between the nonpolar groups on the protein.

1.1.1 Factors Affecting Solubility of Proteins

pH of Medium

The determining factor of protein solubility is the pH of the medium. The degree of protein solubility in an aqueous medium is the result of electrostatic and hydrophobic interactions between the protein molecules. Solubility is increased if electrostatic repulsion between the molecules is higher than hydrophobic interactions. Protein solubility is affected by a sensitive balance between repulsive and attractive intermolecular forces and proteins are soluble when electrostatic repulsion between proteins is greater than hydrophobic interactions (1).

Proteins to be soluble should be able to interact as much as possible with the solvent. At the isoelectric point (pI), proteins have a net zero charge, attractive forces predominate, and molecules tend to associate, resulting in insolubility. Above the pI, the net charge is negative and solubility is enhanced. Protein-water interactions increase at pH values higher or lower than the pI because protein carries a positive or negative charge. Minimal interactions with water were observed for protein molecules at pH values not far from pI. If the protein solubility is plotted as a function of the pH, a U-shaped curve is obtained and the minimum corresponds with the pI. Consequently, solubility and yield of extraction is greater at alkaline than acid pH and can be enhanced by increasing the net electrical charge of the proteins. Alkali treatment usually increases soy and other protein solubility by causing dissociation and disaggregation of the proteins. Solubility of proteins, especially 7S and 2S soy globulins is decreased by acid precipitation. In some protein beverages (i.e. carbonated) the solubility of protein in the acidic range is critical and such proteins are prepared in the form of hydrolysates. Protein solubility values could be misleading if they are obtained in pure water, because the protein solubility in real complex food systems in the presence of various ions could be different.

Ionic Strength

The functionality of proteins can be studied more effectively if a systematic study is first made of the protein solubility under various ionic conditions. The mechanism of the ionic strength effect on protein solubility is poorly understood and probably involves solvation, electrostatic and salting in and salting out phenomena [4]. Chloride ions increase solubility by electrostatic repulsion after binding to the positively charged protein groups. Low concentrations of neutral salts at molarities of the order of 0.5–1.0 M may increase the solubility of proteins, but above 0.15 M can reduce it. In salt solutions protein solubility initially increases (salting in) and after a maximum of solubility it starts to decrease (salting out). Increase in salt concentration greater than 1.0 M caused a decrease in protein solubility. Water molecules are strongly bound to the salt and there is competition between the salt ions and the protein molecules for the water molecules.

The Effect of Heating

Alteration in protein solubility during heating is evidence of conformational changes in the structure of protein. Solubility of various proteins decreases variously with temperature and time of heating. Conformational changes in protein structure result in diverse and complicated precipitation reactions. The mechanism of protein precipitation occurring after heat treatment is difficult to study because the process is irreversible. Moist heat has a more complex effect on the solubility of proteins than dry heat does, and is strongly affected by pH and ionic strength. The molecular structure of most proteins in aqueous medium is susceptible to changes in temperature. Solubility of proteins is influenced by temperature and increases with temperature between 0 and 40–50 °C. At temperatures higher than 40–50 °C, the solubility of proteins is less than that of native proteins. Heat treatment and protein denaturation causes a marked and irreversible reduction of protein solubility. The nitrogen solubility index (NSI) for commercial soy flours, concentrates and isolates is found to vary from 10–90% depending on the conditions of heat treatment. Protein solubility profiles are effective indicators of the degree of protein denaturation during processing and are used in commercial conditions to control emulsification, foaming, extraction, and gelation processes. Most food protein concentrates and isolates are prepared from minimum heat-treated flours and exhibit good solubility.

Processing Conditions

The solubility of proteins reflects the processing conditions which partially or completely insolubilize the protein. Processing conditions – such as pH of extraction, precipitation, and neutralization prior to drying – influenced protein solubility. The protein solubility of protein-containing products is dependent upon the physicochemical state of their protein molecules, which are either favorably or adversely affected by processing conditions particularly heating, mechanical treatment (emulsification, grinding, comminution), and drying during their manufacture. The degree of agitation or speed of blending influences protein solubility.

Solubility properties are limiting criteria in development of the optimal parameters for the protein extraction and purification, and for separation of protein fractions. During formation of emulsions and foams, diffusion of proteins of high solubility is facilitated at the air/water and oil/water interfaces. Aoki et al. [5] have reported that emulsifying properties and solubility were not well correlated. However, Li-Chan et al. [6] have reported that solubility properties influenced emulsifying capacity of proteins for proteins with low (<50%) solubility. The emulsifying, gelling and foaming capacity of whey proteins is enhanced by increased solubility of proteins. Caseinates with high solubility have better emulsifying and gelling properties than less soluble caseins. However, the functionality of proteins cannot always be enhanced by increased protein solubility. Protein solubility is the determining factor of the effectiveness of the protein extraction from raw materials and its subsequent precipitation. Solubility data give useful information for the optimization of the processing procedure.

Processes for the production of protein concentrates and isolates are investigated by studying the solubility of the proteins, according to the temperature and time of extraction, pH, ionic strength, and size of particles [7]. The protein extraction curves are obtained by dispersing flour in water, stirring the suspensions and adjusting the pH level. Centrifugation or filtration is used for extract separation and the amount of solubilized proteins is determined by the protein content of the extract. Protein solubility or extractability is an important factor, but not the only one governing efficiency of protein extraction. Other factors such as time of extraction, ratio meal/solvent, temperature of extraction, and the added salts are also important.

Protein solubility is measured as the concentration (%) of protein in aqueous solution or dispersion that is not sedimented by moderate centrifugal forces. Solubility of protein is utilized mostly as an indicator of other functional properties. Protein performance in emulsions, foams, and gels can be predicted from protein solubility. The solubility of food proteins has been designated in the following terms: water-soluble protein (WSP), water-dispersible protein (WDP), protein or nitrogen solubility index (PSI or NSI), and protein dispersibility index (PDI). American Oil Chemists Society approved PDI and NSI testing as official methods. The International Dairy Federation approved the standard procedure for determining milk protein solubility: the sample is dispersed in water under standardized conditions, followed by centrifugation at 20|000 g (or 3000 g) and determination of the nitrogen content in the supernatant [8].

NSI or PSI and PDI are used for measuring protein solubility. NSI and PSI are most widely used as quick tests for predicting protein functionality. In the PDI test, the protein is blended for mixing with water. NSI and PSI tests include more gentle mixing. Insoluble solids are separated by low-speed centrifugation. In these tests proteins may disperse without being soluble and solubility will be affected by the separation power of the centrifuge. A collaborative study was carried out to develop a simple, reliable and rapid procedure for determining the solubility of food protein and the procedure of NSI was modified [9].

1.2 Solubility of Meat and Fish Proteins

1.2.1 Solubility of Muscle Proteins

On the basis of solubility in aqueous solvents the muscle proteins may be divided into three major classes: sarcoplasmic (most soluble), myofibrillar (soluble in dilute salt solutions), and stroma proteins (least soluble) [10]. Some myofibrillar proteins are extractable with water.

Myosin, the major myofibrillar protein is soluble at high ionic strength (greater than 0.3 M), and is insoluble at low ionic strength. Myosin possesses three important functional properties: (1) it is an enzyme with ATP-ase activity; (2) myosin forms complexes with actin; and (3) myosin can aggregate with itself to form filaments. These properties are critically important for the functionality of myosin in muscle foods.

Actin is a globular protein made up of a single polypeptide chain which binds one molecule each of nucleotide (ATP of ADP). In rigor mortis the contractile proteins are considered to be present as the actomyosin complex. Myofibrillar protein solubility decreases in muscle with low pH. As SDS-polyacrylamide gel electrophoresis has shown, during rigor mortis, a strong fall of myofibrillar protein solubility is observed [11]. In unsalted muscle homogenates, a small amount of myosin is soluble. In the supernatant of salted pre-rigor mortis muscle homogenates (pH 6.1), however, relatively high amounts of myosin can be detected which become insoluble during development of rigor mortis. This change of protein solubility during rigor mortis is probably due to the strong association of myosin and actin and the decrease in pH. In NaCl-containing meat samples, the isoelectric point of the proteins will be moved toward a lower pH.

Muscle protein solubility is one of the major factors influencing the water retention properties of muscle. Solubility of sarcoplasmic proteins decreases during the first 24 hours postmortem. At the same time, myofibrillar proteins show no loss in solubility under conditions of slow pH decline and medium pH at the onset of rigor mortis as long as the temperature is low. However, significant decrease of myofibrillar protein solubility have been found under conditions of high temperature and a medium or low pH. The loss in myofibrillar protein solubility results from the precipitation of sarcoplasmic proteins or from a direct alteration of the myofibrillar proteins. There is correlation between sarcoplasmic protein solubility and pH at onset and completion of rigor mortis. Correlation is found between fibrillar protein solubility and initial temperature as well as muscle temperature at the onset of rigor mortis.

Aging of meat markedly influences protein solubility and meat quality. There is an increase in protein extractability with post-mortem aging during the first 14–16 days. This increase during aging results from weakening of the fibrous protein linkages, or from a disintegration of the insoluble stroma itself. It is apparent that pH values determine the rate of increase of protein extractability during aging.

Physical and chemical bonds in muscle proteins influence the structural properties and texture of muscle proteins [12]. The correlation was established between the structure of muscle proteins and their solubility and associated influence on tenderness or toughness during processing and storage. The solubility of the myofibrillar proteins for samples of Longissimus dorsi stored at 2 °C tended to increase throughout storage time (13). Increase in the solubility of the myofibrillar proteins is caused by the muscle enzyme activity; microbial growth has little effect on this solubility at 2 °C. The temperature of post-mortem storage is an important factor affecting the extractability of the myofibrillar proteins. The extractability at high storage temperature (25 °C and 37 °C) is slightly higher than at the lower temperature (2 °C). The temperature effect on protein solubility is probably due to the growth of natural enzymes or microbial growth.

During beef aging sarcoplasmic protein fractions remain constant in the longissimus dorsi muscle but decrease in the semitendinosus muscle [14]. The same trend in increase of the total extractable nitrogen, total soluble fibrillar protein nitrogen, and soluble actomyosin nitrogen during early post-mortem periods was found in meat of other species. The rate of post-mortem changes in chicken muscle is much more rapid than in either pork or beef muscle. In chicken muscle, there is significant decrease in extractability during 4–8 hours post-mortem, increasing to a maximum value after aging for 3–4 days.

1.2.2 Solubility of Stroma Proteins

Stroma proteins are usually measured as the insoluble proteins remaining after exhaustive extraction of all soluble muscle proteins. Stroma proteins influence meat quality directly[10]:

(1) they lower tenderness of meat and the effect depends on the amount of stroma proteins and the degree of cross-linking among stroma proteins;
(2) because of their insoluble nature, they decrease the emulsifying capacity of meat;
(3) because of their low content of charged and hydrophilic amino acids, stroma proteins lower water holding capacity of meat;
(4) they decrease the nutritive value of meat.

Collagen is a high-molecular weight, relatively insoluble, fibrous protein. The specific property of collagen disintegrated from its native state to repolymerize and form a fibrous structure is the basis for collagen commercial utilization including production of edible sausage casings. Miller et al. [15] observed decreasing levels of soluble collagen at elution time 4.3 min with increasing age. The soluble collagen fraction of the sample was calculated by reference to the standard. The acid soluble collagen from fetal corium (7.6%) was higher that that of the 18-month (3.4%) age group and higher than that of the 40-month (2.7%) group. The soluble collagen content of the 3 to 6-week age group was higher than that of the 18-month and the 40-month age groups. The general trend was observed, i.e. decreasing susceptibility of collagen to acid solubilization with increasing age. Collagen solubility decreased with the age of animals due to the formation of certain cross-linkages within the collagen macromolecule.

The differences in collagen solubility between sexes might be related to a lipid coating over the collagen molecules. Inter- and intramolecular cross-linkages in stromal proteins from muscles of young animals are fewer in number and more labile than in older animals. The stability and number of the cross-links increases with age while the solubility of collagen decreases [16]. Consequently, an increase in meat toughness from older animals is related more to stable cross-linkages than to an increase in collagen content. Solubility of collagen and elastin within bovine muscles are the major factors determining tenderness. An increase in extractable collagen during aging was reported. Wu et al. [17] found that the lysosomal enzymes, b-galactosidase and b-glucuronidase, increases the dissolution of collagen fibers by collagenase.

The solubility of fish collagen in 0.5 M NaCl solutions is not affected significantly by the age of the specimen. Because most of the connective tissue in fish is renewed annually, collagen is exhibiting its property of being a not highly cross-linked protein. The tenderization effect is achieved by utilizing proteolytic enzymes such as ficin, papain, and bromelain [18]. Solubilization of collagen fragments was measured by determining hydroxyproline concentrations in hydrolysates of both the soluble and insoluble fractions of the cooked meat. A small increase in collagen solubility was found with post-mortem injection of Achromobacter iophagus collagenase into beef muscle immediately after slaughter [19]. After enzyme treatment the collagen is more extractible. Different microbial collagenase preparations solubilize meat collagen at a level greater than that of the control (no added enzymes), and increase the fraction of total collagen in the soluble phase after cooking [18]. An increase in collagen solubility improves the meat tenderness and binding of structure, particularly in restructured beef products.

Elastin protein of connective tissue is unique because of its elastic properties, insolubility, content of lysine-derived cross-links and high resistance to digestive enzymes. Because of its low content, elastin is not so significant to meat tenderness. Content of elastin is less than 1% of the collagen content.

1.2.3 Protein Solubility in Processed Meats

In the processing of cooked meat products, especially comminuted and restructured meats, solubility of the contractile proteins with NaCl added is essential to the quality. Soluble proteins in comminuted and restructured meats bind together the insoluble components of the formulation in a stable protein matrix to form a stable system before and after heat treatment. Siegel and Schmidt [20] found that an increased level of extracted myosin distributed between meat surfaces significantly increased binding strength of the components. When myosin and actomyosin were heated in high ionic strength salt solutions they formed a strong, coherent three-dimensional network. In the absence of salts the same proteins formed a spongy structure.

Important processing factors influencing protein extractability were found to be mechanical treatments: mixing, tumbling, massaging, and mechanical tenderization. Protein extraction during blending, comminution, and tumbling is the main factor that determines binding strength of the sausage emulsions or restructured meats. Mechanical treatment caused cell disruption and breakage with a release of myofibrillar and other proteins [21]. Extracted proteins form more cohesive bonds between the protein matrix and the meat surface. There is an optimum mixing time, and if it is increased beyond a certain limit, the binding strength decreases.

Sodium chloride increases the binding capacity of proteins as a result of increased protein extractability and by its influence on the ionic strength of the medium. An important function of phosphates is acceleration of the extractability of myofibrillar proteins. Prusa and Bowers [22] reported that NaCl and sodium tripolyphosphate increased protein solubility and protein extractability in meat systems. The soluble myofibrillar proteins lead to a uniform, interwoven matrix with entrapment of water and fat during comminution and heat treatment. Incorporation of phosphates in raw meat emulsions increases protein solubility and improved emulsion stability and water holding capacity [23]. Tetrasodium and tetrapotassium pyrophosphates result in higher protein solubility than sodium or potassium tripolyphosphate. They improve sausage emulsion water holding capacity and stability. Myosin solubilization is accelerated by (1) depolymerization of the thick filament backbone promoted by increased ionic strength and by the presence of pyrophosphate; (2) dissociation of the myosin heads from actin, promoted by NaCl and pyrophosphate in the presence of Mg^{2+} [24]. These processes cause an increase in capacity to retain moisture. Extraction of titin, myosin and other myofibrillar proteins from muscular beef tissue has been increased by addition of NaCl and pyrophosphate [25]. An increase in extraction of these proteins, especially titin, causes an increase in myofibril swelling and water binding and improves meat tenderness.

The effect of added collagen on protein functionality in sausage batters has been reported [26]. Alkali-soluble and insoluble fractions of proteins increases with increasing levels of collagen in bologna. In cooked bologna, a considerable decrease in solubility of muscular tissue proteins has been observed: sarcoplasmic proteins, 5.4 to 84%, and myofibrillar proteins soluble in Guba-Straub buffer, 35 to 54%. During heating at 68 °C internal temperature, collagen was partially solubilized, and the amount of alkali-soluble fraction increased almost two-fold. The extent of collagen solubilization depends on the biological maturity of collagen, the degree of cross-linkages, heating temperature, and time. In summer sausages, contractile protein solubilization is limited by processing conditions, including low temperature. As a result contractile proteins in summer sausages are exposed to minimal alteration by heat. Extractability of muscle proteins is decreased if elevated temperature and low pH combinations are applied in the early post-mortem period.

1.2.4 Solubility of Blood Proteins

Proteins from blood represent an important source of protein for human consumption. The plasma from beef blood is utilized in bologna and wiener formulations in different countries. Beef plasma isolate is an excellent binding agent, emulsifier, and foaming agent. Different pH-solubility profiles of the phosphated plasma protein and ultrafiltered protein in the pH range 2.5–8.0 were observed [27]. Slight depression in solubility was found for utlrafiltered protein near the isoelectric point. The phosphated plasma protein exhibited complete solubility above the isoelectric pH range (pH 4.8–5.2) and very low solubility below pI range. Extremely high solubility of plasma proteins above pH 5.5–6.0 is of practical importance because most foods are produced at pH values 6.0–7.0.

Globin, the protein portion of hemoglobin separated by the acetic acid-acetone-calcium chloride method, may be utilized as a source of protein [28]. Globin protein showed a good essential amino acid content and can be considered as an excellent source of lysine, leucine, and valine. Threonine and phenylalanine content was higher than the FAO/WHO pattern. Globin protein has a wide minimum of NSI between pH 7.0 and 9.5, which is probably related to the interaction of the acetic acid and acetone with partial globin denaturation. Proteins of globin were almost completely soluble in the vicinity of pH~4 and 11.

There is an interest in the fractionation of whole blood and the upgrading of different fractions for utilization as ingredients in human foods. Purified red blood cell concentrates were prepared by ultrafiltration or diafiltration and spray drying, and they contained 95% protein [29]. The removal of nonprotein components from blood cell concentrate increased solubility compared to the control. Blood cell powders exhibited protein solubilities in water of 75–95% over the pH range 2–10. The minimum protein solubility was at pH 7.2. The ionic strength had little effect on solubility at pH 8.0 but had a marked effect at pH 3.0. At pH 3.0, solubility fell from 86% to 5% as the ionic strength increased from 0 to 4.0 M NaCl.

1.2.5 The Effect of Heating on Solubility of Proteins

The most drastic changes in muscle during heating involve muscle proteins. Shrinkage of tissues, release of meat juice, and change of color are the results of the denaturation of proteins. The level of solubility decrease is affected by time and temperature of heating. Development of the optimal heating methods and parameters is necessary when the toughening effect of heating on myofibrillar proteins will be offset by the tenderizing action of heat on collagen. Myofibrillar proteins start to coagulate at 30–40 °C, they are losing solubility and the coagulation is nearly completed at 55 °C. At temperatures of 40–50 °C or greater, proteins will denature with a decrease in solubility, unfolding of polypeptide chains, and the formation of new electrostatic and hydrogen cross-linkages.

Bovine longissimus muscles were heated at 45 °C, 50 °C, 55 °C, 60 °C, 70 °C, and 80 °C and myofibrillar protein solubility was studied [30]. The most heat-labile muscle protein was found to be a-actinin. It became insoluble at 50 °C; heavy and light chains of myosin became insoluble at 55 °C. Other myofibrillar proteins were significantly less sensitive to heating.Actin was insolubilized between 70 and 80 , tropomyosin and troponin above 80 °C.

Heating temperature affected protein dispersibility, content of sulfhydryl groups, and surface hydrophobicity determined as S<2,3>, ANS or S<2,3>, CPA (Fig. 1.1) [31]. CPA is composed of an unsaturated aliphatic hydrocarbon chain. ANS is composed of aromatic rings. It is possible that CPA and ANS are bound at protein sites differing in aliphatic and aromatic hydrophobicities. Surface hydro-

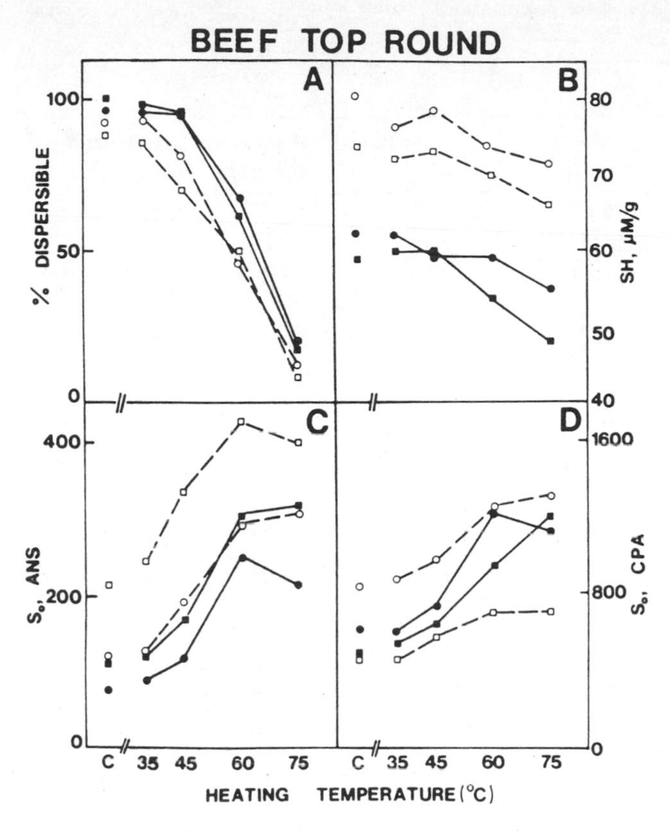

Fig. 1.1. Dispersibility **(A)**, sulfhydryl content **(B)** and hydrophobicity **(C,D)** of beef top around extracts as a function of heating temperature. (*Closed symbols*, pH 7.0: ● sample 1, ■ sample 2. *Open symbols*, pH 5.5: ○ sample 1, □ sample 2). From Ref. 31

phobicity of proteins increased upon heating to a temperature of 45 °C or lower. In this range of temperatures dispersibility remained fairly constant.

Moderate heating may improve functional properties of beef proteins by controlled denaturation without coagulation or insolubilization, and by partial unfolding and exposing of hydrophobic sites which positively influence functional properties. However, at heating temperatures in the 45 °C to 75 °C range, a significant decrease in dispersibility and an increase in hydrophobicity were observed. Functionality (emulsifying capacity and fat binding) of meat proteins heated in the temperature range from 45 °C to 75 °C generally decreased. Hydrophobicity increased gradually upon heating to 45 °C, then showed an increase from 50 °C to 70 °C and finally reached a plateau at temperatures above 70 °C. No additional exposure of hydrophobic groups was indicated at temperatures above 70 °C.

Table 1.1. Mean values obtained for bicep femoris muscle heated by microwave and conventional method

Method of heat treatment				
Measurement	Microwave 65 °C	Conventional 65 °C	Microwave 80 °C	Conventional 80 °C
Shear cohesiveness, kg (perpendicular to fibers)[c]	6.33[a]	6.10[a]	7.01[a]	6.70[a]
Shear cohesiveness, kg (parallel to fibers)[c]	5.09[a]	6.01[b]	5.36[a]	5.85[a]
Firmness, kg/min (perpendicular to fibers)[c]	25.87[a]	25.98[a]	29.36[a]	31.31[a]
Firmness, kg/min (parallel to fibers)[c]	17.77[a]	20.38[b]	16.46[a]	19.54[a]
% Solubilized hydroxyproline	27.61[a]	18.52[b]	28.32[a]	20.30[b]
% Moisture	61.28[a]	63.49[a]	58.15[a]	60.94[a]
% Heating losses	38.78[a]	27.01[b]	40.61[a]	30.64[b]
Water holding capacity	0.73[a]	0.76[a]	0.67[b]	0.73[a]
Heating time, min	7.53[a]	14.25[b]	7.94[a]	19.31[b]

[a,b] Means bearing the same letter in row are not significantly different ($P < 0.05$).
[c] Average from 4 muscles; measurements were done in triplicates. Determined from shear force deformation curve. From Ref. 69

The utilization of fibrous collagen in meat processing as a replacement for meat proteins is achieved by the dilution of myosin, a very functional meat protein. When minced muscle was replaced by collagen, a reduction in the quantity of salt soluble nitrogen was found in unheated samples [32]. When heated at a sufficient temperature for a sufficient time, solubilization of the collagen of the meat block was observed. When a meat slurry was heated, the amount of soluble nitrogen was markedly reduced during the initial 15 min of heating at 50 °C, 60 °C, or 70 °C. The initial decrease in solubility at 50 °C may be due to the release of tropocollagen molecules that have not undergone crosslinking.

During cooking, the temperature of collagen gelation is about 50 °C for young animals and about 60 °C for older animals [33]. Attempts have been made to relate collagen content, properties, and amount of collagen solubilized during heat treatment to tenderness of meat. Changes in contractile proteins and collagenous connective tissue during post-mortem storage and subsequent heat treatment predominantly influence textural properties of meat. The amount of collagen solubilized during heat treatment and the shear force for cooked meat were affected by heating temperature and time. The percent of solubilized collagen increased with increasing internal temperature of meat.

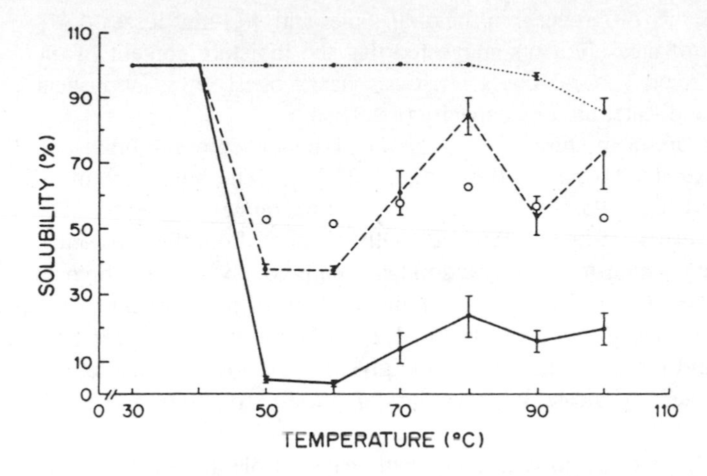

Fig. 1.2. Solubility curves of protein solutions upon heating. *Bars* denote S.E. Beta-conglycinin, -----Myosin,----Mixture of beta-conglycinin and myosin. From Ref. 35

Research has been focused on studying the influence of two methods of heat treatment, microwave heating and boiling, on the quality characteristics of meats (Table 1.1) [34]. Solubility of collagen and textural properties were reported as being dependent on the heating time and the end-point temperature of the product. Comparing microwave and conventionally heated samples has indicated differences in their characteristics. Rapid heating by microwave had complex effects on the contractile and connective tissue fibers. The tenderness of cooked meat is generally considered to be controlled by the heat-induced changes in collagenous connective tissue and in the contractile proteins. The number of crosslinkages between the collagen molecules within the connective tissue is associated with collagen solubility. Microwave heat treatment solubilized more collagen from the meat samples than did the conventional heating method [34]. Microwave heating resulted in a higher percentage of hydroxyproline solubilization: 27.65% for end-point temperature 65 °C and 28.53% for end-point temperature 80 °C. For control samples, percentage of hydroxyproline solubilization was 19.18% for end-point temperature 65 °C and 18.97% for end-point temperature 80 °C. The data on 65 °C and 80 °C internal temperatures showed variations among the samples. Collagen solubility should be considered when attempting to explain the toughness of meat biochemically. The amount of collagen solubilized was higher for microwave at the end-point temperature 65 °C and at the end-point temperature 80 °C.

The effect of microwave heating on the collagen of connective tissue of beef makes an important contribution to the quality of the finished product. The data show an increase in the percentage of solubilized collagen brought about by microwave heating and increasing internal temperature of the samples. Some of the textural properties of meat heated by microwave were different than those heated by boiling, despite the differences in heating time and heating losses [34]. Correlation analysis of measurements suggested that the moisture content of the cooked samples affected textural characteristics, shear cohesiveness and shear firmness (r = 0.52), and water holding capacity (r = 0.59).

Chicken myosin and a mixture of myosin with b-conglycinin exhibited two minima of solubility, at 60 °C and 90 °C (Fig. 1.2) [35]. The solubility of b-conglycinin remained high (95–90%) in the temperature range 30–100 °C. The solubility curve between 50 °C and 100 °C generally deviated from the theoretical values. Solubility data demonstrated interaction between b-conglycinin and myosin upon heating at 50–100 °C. The theoretical solubility values of the protein mixture if no interaction takes place (open circles, Fig. 1.2) are the means of the separate values for myosin and b-conglycinin. The b-conglycinin and myosin interactions were also demonstrated by measuring turbidity of protein solutions and SDS-PAGE analysis.

The solubility of blood globin was high when heated at pH 4.0 and pH 10 at 100 °C for 10 min [36]. The globin heated at pH 4 was highly soluble in the pH regions below 6 and above 10. No difference was found in the pH-solubility profile between heated and unheated globin. Heated and unheated globin showed low solubility at neutral pH which can be explained by instability at room temperature. The globin heated at temperature 50 °C to 100 °C, and pH below 5 remained almost 100% soluble. However, heating of globin at pH 5.2 and 100 °C caused a drastic decrease in solubility, with an increase in viscosity; globin showed solid-like textural properties. Spray-dried globin had a maximum solubility at the pH range ≤ pH 6.0.

During conventional drying the increased NaCl concentration in the nondehydrated part of the muscle promotes the denaturation of muscle proteins. Protein denaturation during freeze-drying is insignificant and it is accelerated when the freeze-dried meat is exposed to heat. Kim et al. [37] showed that myosin is involved in the textural changes in freeze-dried meat and is primarily responsible for the loss of protein extractability. The extractability of proteins decreases upon heating and the loss of solubility of myosin is evident. A decrease in extractability of myosin heavy chain is due to covalent cross-linking between molecules of myosin. The loss of myosin extractability significantly decreased by treatment of the meat with N-a-acetyl-L-lysine, even upon heating of the treated meat [37]. This inhibiting effect of lysine on the cross-linking of myosin is of practical significance, and can be utilized for the treatment of freeze-dried meat.

1.2.6 The Effect of Freezing and Storage When Frozen on Protein Solubility

Freezing and storage in the frozen state is an effective method for preventing microbial spoilage of foods. Microbial growth and some of the chemical reactions can be controlled by freezing and subsequent storage. However, changes in the functional properties of protein of animal origin have been observed [38]. Most of

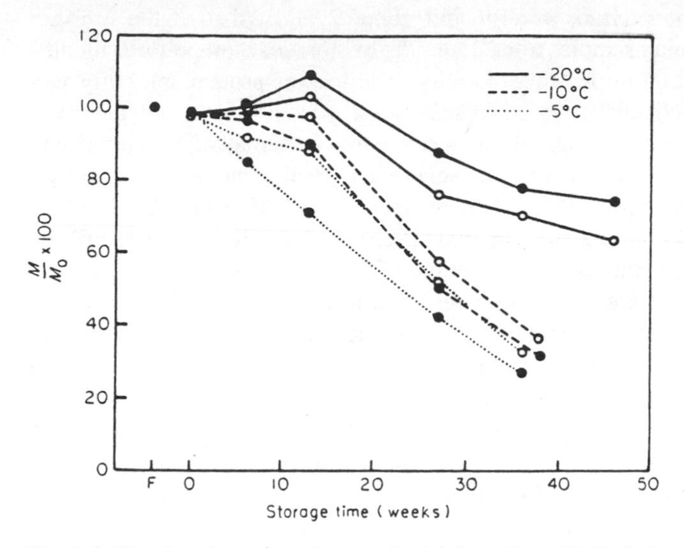

Fig. 1.3. The time-dependent change of solubility of myofibrils during frozen storage of muscle at -20, -10 and -5 °C; for fast freezing (●) and slow freezing (○). M_0 and M are respectively the solubility of myofibrils isolated from fresh and frozen stored muscle. From Ref. 38

the studies demonstrated that protein denaturation, in particular myofibrillar proteins, is the main factor of meat quality deterioration during frozen storage. The degree of protein denaturation and insolubilization during freezing and frozen storage is influenced by many factors such as pretreatment before freezing, the degree of the autolytic process before freezing, freezing rate, freezing and storage temperature, storage time, stability of storage conditions, especially temperature, and thawing methods and conditions.

The functionality of protein in frozen meat is better preserved by using fast freezing rates. Using fast frozen meat increased juiciness and tenderness of ground beef patties because of smaller ice crystals and less protein denaturation [39]. Sarcoplasmic proteins were more sensitive to denaturation than myofibrillar proteins. This was demonstrated by a decrease in their solubility and an increase in the difficulty of extracting actomyosin, a loss of myosin ATPase activity, and a decrease in titratable SH groups. A considerable decrease in solubility in a 0.6 M KCl solution of frozen beef muscle (with fast and slow freezing methods) stored at –20 °C, –10 °C, and –5 °C for 40–45 weeks is presented in Fig. 1.3 [38]. The loss of solubility of myofibrillar proteins was greater at the higher storage temperatures. At –5 °C the solubility of proteins decreased from the first week; at –10 °C the solubility decreased significantly after 6 weeks, while at –20 °C solubility increased slightly for about 13 weeks then decreased slowly.

Protein solubility of gels prepared from prerigor and postrigor meats and containing cryoprotectants (polydextrose, sucrose-sorbitol) decreased during 6 months of frozen storage [40]. A significant difference in protein solubility was found between prerigor and postrigor meat gels. The loss of protein extractability was about 30% in the prerigor control and about 20% of that in the prerigor cryoprotectant-containing samples after 2 months of storage. Storage for 6 months showed further reduction in protein solubility. The loss of protein solubility was prevented more effectively by a sucrose/sorbitol mixture than polydextrose. However, polydextrose is a lower calorie and nonsweet alternative to sucrose and sorbitol. Frozen muscle that underwent severe contraction at thawing showed greater protein extractability (35%) than muscles stored at 0 and 15 °C (27% extractability) [41]. Improved protein extraction in thaw-shortened myofibrils resulted from physical disruption and loosening of muscle fibers.

Fish muscle proteins are much more sensitive to freezing than meat proteins. Insolubilization of myofibrillar proteins was the main factor affecting functional properties of fish proteins after frozen storage [42]. Protein solubility in fish muscle stored in a frozen state has been used as a criterion for the alterations of proteins. It was difficult to test independently the different theories of fish protein denaturation that had been proposed: the „lipid hydrolysis“, „lipid oxidation“, „salt denaturation“, and „formaldehyde-protein interaction“.

Properties of fish myofibrillar proteins were affected more significantly at –25°C of frozen storage than at –35°C [43]. The activities of Ca-ATPase and Mg(Ca)-ATPase, Ca-sensitivity and actomyosin solubility in 0.6 M KCl decreased at a higher rate at–20 °C than at –35 °C. The amount of the salt-soluble protein fraction decreased with storage time and the decreasing rate was higher for –20 °C than –35 °C. The increase of the urea-soluble fraction of actomyosin indicated formation of disulfide, hydrogen, and hydrophobic bonds during storage. The level of reactive SH groups decreased considerably during freezing, and the rate of decrease was higher at –20 °C than –35 °C storage temperatures. This demonstrated that more disulfides were formed during freezing in fish muscle at –20 °C than at –35 °C. As a result, the tertiary structure of actomyosin changed during storage at –20 °C. The decrease in Ca-ATPase activity of actomyosin was found during freezing at –20 °C and –35 °C and storage at –20 °C, and no significant change was observed during storage at –35 °C. The decrease of Ca-ATPase activity was much higher at –20 °C than at –35 °C. Jiang et al. [44] reported that muscle proteins of amberfish and mullet were much more stable when stored at –40 °C that at –20 °C, and protein denaturation progressed during storage when the storage temperature was in the range –8 °C to –10 °C.

Different stabilities of various fish proteins to denaturation were established [45]. Actin was relatively more stable than tropomyosin. Generally, both salt- and water-soluble proteins of red hake muscle decreased in solubility during frozen storage. However, the rate of insolubilization was higher for salt-soluble proteins. Different susceptibilities to denaturation of fish proteins may be important in textural deterioration of red hake.

The influence of fish lipids on protein denaturation during frozen storage varies according to the state of the lipid oxidation and lipolysis. Insolubilization of actomyosin was observed in all fish species during frozen storage [46]. Extractable myosin was more affected by freezing and frozen storage than the salt-soluble proteins. Salt-soluble proteins extracted included myofibrillar and sarcoplasmic proteins, and the sarcoplasmic proteins were not affected markedly during frozen storage. Changes in actomyosin were reflected by considerable loss in ATPase activity of actomyosin. The actomyosin ATPase activity lost 66% of its original value after 6 weeks of storage [46]. Variations in denaturation patterns among fish species were demonstrated by changes in electrophoretic properties. Krivchenia and Fennema [47] found a significant decrease in protein extractability during the early stages of storage at –12 °C or –60 °C, exhibiting values of about 500–700 micrograms protein/0.1 ml at 4 weeks, about 300–400 micrograms protein/0.1 ml at 10 weeks, and remaining relatively constant thereafter. Considerably greater protein extractability was found for the control than for other samples. Protein extractability of control samples did not remain constant and followed the same pattern exhibited by other samples through a 28-week storage period.

Freeze-denaturation of actomyosin was mainly caused by formation of disulfide, hydrogen, and hydrophobic bonds [48]. Matsumoto [49] reported formation of aggregates during frozen storage of fish muscle. The formation of aggregates was caused by the progressive formation of hydrogen bonds, ionic bonds, hydrophobic interactions, and disulfide bridges. A decrease of protein solubility in the washed minced cod during the freeze-thaw cycle was found [50]. Exposure to hydroxyl free radicals caused a 10% loss of protein solubility. The combined exposure to hydroxyl radicals and to the freeze-thaw cycle caused a loss of 45% protein solubility.

The fish in which half of the protein has become unextractable in salt solutions is considered extremely tough. Protein extractability of frozen minced red hake muscle progressively decreased to about 40% of original values over 70 days of frozen storage [51]. High molecular weight fraction and actin were found in the insolubilized fraction that underwent the greatest reaction with formaldehyde produced from trimethylamine-oxide (TMAO). In certain fish (gadoids) TMAO is converted by enzymes to dimethylamine and formaldehyde. It has been suggested that formaldehyde causes crosslinking of the muscle proteins and toughening of the tissue during frozen storage [52]. Correlations were observed between formaldehyde formation and loss of protein solubility and an increase in toughness of frozen stored gadoid fish [53]. There is a tendency for formaldehyde to bind, reversibly or irreversibly, to single amino acid residues. Formaldehyde produced during frozen storage increased denaturation of cod myosin, decreased loss of solubility, and increased surface hydrophobicity. Formaldehyde might increase the rate of protein denaturation during frozen storage by interacting with the side chain groups on the fish proteins resulting in protein aggregation through noncovalent interactions. The loss in solubility increased to 70% at concentrations of formaldehyde of 20 mM and above. However, formation of crosslinking was not a prerequisite to loss of solubility because high molecular weight fractions were formed in frozen samples in the absence of formaldehyde.

The effect of fish preprocessing holding time in ice prior to freezing on protein solubility during frozen storage is shown in Fig. 1.4 [54]. Solubility of protein during frozen storage decreased considerably with increased time of ice storage. The solubility decreased markedly in fresh and 3-day ice-stored samples

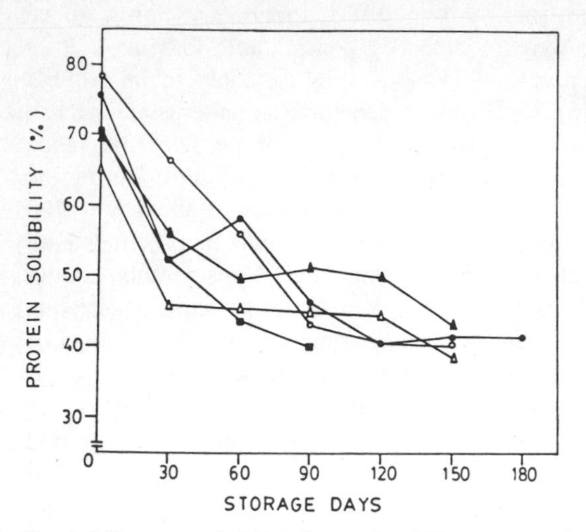

Fig. 1.4. Preprocess holding time on ice: Effects on protein solubility in frozen stored pink perch mince: ■-■ zero day; O-O 3 days; ◆-◆ 5 days; Δ Δ 11 days; ■ - ■ 14 days in ice. From Ref. 54

during the first 90 and 60 days of storage. Damage of proteins during frozen storage was less significant when fish was frozen before the onset of rigor mortis than fish frozen post-rigor and stored under the same conditions.

Cryoprotectants, sodium tripolyphosphate, or sodium hexametaphosphate reduced deteriorative changes in fish proteins during frozen storage. The incorporation of sucrose, sorbitol, polyphosphates and/or sodium glutamate into surimi preserved protein functional properties during long periods of frozen storage [55]. Some amino acids revealed protective effects on the stability of fish muscle proteins.

1.2.7 The Effect of Protein Modification and Irradiation Treatment

Functionality of muscular proteins can be improved by chemical and enzymatic modification. Functional properties of mechanically deboned fowl myofibrillar proteins was improved by partial proteolysis with acid protease [56]. Proteolysis increased myofibrillar protein solubility in 0.6 M NaCl probably by disrupting the aggregated, denatured structure of proteins from mechanically deboned meat. Solubility of proteins was affected by the time of enzyme treatment. Proteolysis for 1 h increased the solubility of myofibrillar proteins in mechanically deboned meat to 46% and of actomyosin to 57% at pH 7.0 in 0.1 M NaCl. Solubility of the modified myofibrils was higher than control myofibrils at pH 3–7. Comparatively low protein solubility of nonmodified protein was related to denaturation during deboning and prolonged frozen storage.

The successful exploitation of new fish protein sources will be not only dependent on nutritional quality, but also to a great extent on the functional properties. Fish protein concentrate is highly denatured, non-soluble and has low swelling properties. Increase in fish protein solubility can be obtained by enzymatic or alkali treatment.

The effect of potassium phosphate and sodium acetate washing on the solubility of the proteins in turkey dark meat was tested [57]. Washing of dark meat resulted in bleaching and loss of specific flavor providing a modified muscle that can be added to products from turkey white meat. Extraction of myoglobin and hemoglobin markedly decreased the protein solubility values of turkey meat at 2%, 3%, and 5% NaCl. Soluble proteins have been extracted mainly during the washing procedure. The protein solubility index increased substantially with the increase in NaCl concentration from 0% to 5%.

Functional properties of alkali extracted and acid precipitated chicken protein from bone residues of mechanically separated poultry were reported [58]. Alkali extraction of proteins may cause damage of some amino acids and influence protein functionality. Extracted chicken proteins exhibited poor functionality when compared with the unextracted control. The values of NSI for extracted protein were lower than for the control – fresh, and frozen chicken protein paste. The NSI was not affected by NaCl content in both fresh and frozen extracted protein fractions.

The Effect of Irradiation

An effect of irradiation on the properties of myofibrillar proteins of bovine longissimus dorsi irradiated with Cs-137 at 0 °C and 4 °C has been reported [59]. It has been demonstrated that there was a significant reduction of heavy chain myosin levels in muscle of beef irradiated to 30 kGy and 50 kGy at 0–4 °C. Myosin was depolymerized with increased fragmentation of myofibrils, and changes of textural properties were found with a decrease of water holding capacity after irradiation treatment above 5 kGy. Irradiation affected textural properties of the meat. Linear decrease in shear force values and tensile strength was observed with an increasing dosage of irradiation. This finding corresponds with sensory evaluation indicating the tenderizing effect of irradiation. This effect was related to the solubilization of collagen.

Formation of crosslinkages between peptides and proteins treated in solution by irradiation was found [60]. When radiation-crosslinked peptides were hydrolyzed, amino acid dimers with unique properties not normally observed in foods were obtained [61]. Schuessler and Schilling [62] reported that a combination of irradiation and high oxygen pressure resulted in extensive protein chain scission.

The main effect of 10 kGy irradiation on myoglobin studied after treatment and after post-irradiation frozen storage was aggregation and protein precipitation [63]. Myoglobin retention increased with a decrease in pH from 7 to 5 and with the presence of propyl gallate (PG). Higher myoglobin retention at lower pH was related to increased antioxidative activity of PG at pH 5 compared to pH 7. Irradiated myoglobins to 10 kGy with glucose, SDS and PG were tested both before and after post-irradiation storage. Myoglobin retention (78–82%) was found to be higher for glucose and SDS than PG. However, both additives in combination with PG produced very high soluble protein retention (78–82%). Markedly lower myoglobin retention was indicated after 3 months storage at –20 °C when the combination of PG and SDS was utilized. Under the same conditions, the glucose and PG system resulted in recovery of approximately 35% of the initial native, untreated protein. The effect of irradiation (10 kGy) on retention of protein solubility with the glucose and PG addition was different for myoglobin, BSA and lactalbumin. The most effective protective effect for myoglobin and BSA was provided by a combination of glucose (0.3%) and PG (0.02%). Almost complete preserving of protein solubility was obtained for lactalbumin with PG alone added, and this PG effect was preserved during 6 months of post-irradiation frozen storage in buffer at –20 °C.

The effect of ascorbic acid on irradiated myoglobin solubility at pH 5.0 was concentration dependent [63]. A clear protective effect was recorded for 0.04% solutions of ascorbic acid. In this case protein loss increased, particularly at high radiation doses and after frozen storage. Use of ascorbic acid as a protective agent in irradiated foods where protein functionality is critical may be questionable since ascorbic acid is oxidized to dehydroascorbic acid by irradiation.

The decrease in the amount of total soluble proteins in irradiated meat was affected by irradiation dose [64]. The most radiosensitive proteins were myofibrillar for which an increase in 1 kGy dose resulted in a decrease of the initial protein solubility by 1.628%. Radiosensitivity of the sarcoplasmic fraction was 3 times lower (0.515%/kGy). Zabielski et al. [64] found a decrease of up to 90.11% of total protein solubility.

Irradiation of dry plasma with 60Co and/or electron beams has potential as a method for reducing the microbial count. Hayashi et al. [65] reported no significant effect of gamma-rays and electron beams on functional and other properties of dehydrated blood plasma. At the same time, heat treatment and fumigation damaged blood plasma protein. Protein solubility of irradiated plasma proteins slightly decreased with an increased dose for both gamma-rays and electron beams, and was in the range 95.0–99.6%. Protein solubility of plasma in a dehydrated state heated at 100 °C for 20 min was significantly lower. However, no change in protein solubility was observed in samples incubated in solution at 63 °C for 30 min. There was no significant difference in protein solubility between gamma-irradiated and electron-irradiated samples. The number of SH groups in dehydrated plasma proteins decreased slightly with an increased radiation dose. Heating decreased SH groups more significantly than irradiation.

1.3 Solubility of Milk Proteins

Proteins from dairy products have traditionally been one of the main protein sources. Large amounts of whey proteins are available due to the increasing production of milk and cheese. This brought about the manufacture of a whole range of undenatured protein concentrates and isolates. The functional properties of milk proteins are influenced by chemical and physicochemical properties, and these in turn are affected by chemical composition, conditions of processing, and storage [66].

Solubility is a primary functional property of milk proteins. High solubility of milk proteins in acidic foods is utilized in fruit juices, carbonated beverages, and others. Compositional and physical properties of milk proteins and processing parameters are modified to manufacture milk proteins with high solubility. Milk protein manufacturers are applying various modification treatments to produce proteins with improved functionality and composition. Application of milk proteins can be dramatically expanded if high solubility is obtained. The casein and whey proteins provide specific functionality in formulated food products. The negative effect on the functionality of milk proteins was developed by complexing with lipids and Ca ions during their isolation and manufacture. The milk protein solubility profile is an excellent index of protein functionality and potential usefulness.

The physicochemical and functional properties of milk proteins have been extensively reviewed [67, 68]. Nonfat dry milk (NFDM) has a very high nitrogen solubility index. Whey proteins are composed of at least five components (a-lactalbumin, β-lactoglobulin, bovine serum albumin (BSA), immunoglobulin, and proteose peptones), which are normally stable to acid but sensitive to heat, especially β-lactoglobulin. The first three components account for 80% to 90% of the protein content. Of these, β-lactoglobulin and α-lactalbumin are present in the highest concentrations and are primarily responsible for the physicochemical properties of whey proteins [69]. Because α-lactalbumin has low solubility but good water absorption characteristics after heating, it is used widely in bakery products, meats, yoghurt, and processed cheese.

The solubility of milk proteins was compared with the solubility of corn germ protein flour (CGPF) [70]. In general, protein solubility of milk proteins increased with increasing pH (Fig. 1.5). This increase was more pronounced for sodium caseinate (SC) than for NFDM and whey protein concentrate (WPC), probably as a result of the protein treatments and properties. NFDM, WPC, and SC contained negligible amounts of insoluble proteins or insoluble carbohydrate materials between pH 6.5 and 8.0. The lowest protein solubility reflects the response to pH and as expected, was at a pH near the isoelectric point. Although SC had excellent protein solubility it was insoluble around the isoelectric pH (3 to 5), which could restrict its application in acid foods [67]. Pearce [71] reported that solubility of whey proteins was pH dependent, with solubility decreasing as pH decreased from about pH 7 to 5 or 4. WPC showed lower protein solubility in the isoelectric range, but protein solubility exceeded 68% at all times (Fig. 1.5C) [70]. One potential area for use of WPC would be acid foods, i.e., carbonated and protein-fortified beverages.

In the low pH range (pH 5), protein solubility of CGPF was markedly lower than milk proteins and increased with increased incubation temperature (Fig. 1.5A)

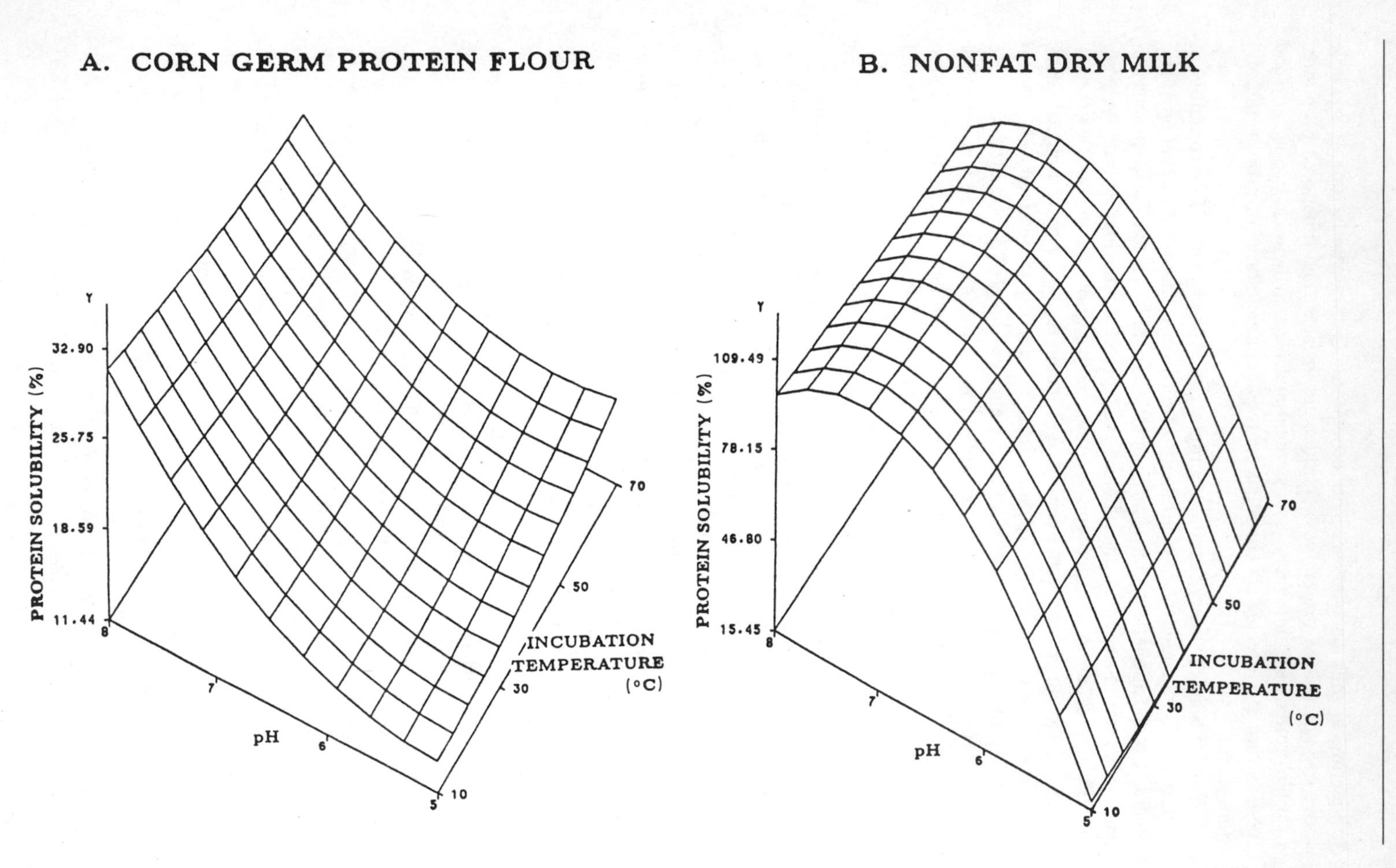

A. CORN GERM PROTEIN FLOUR
PROTEIN SOLUBILITY (%)
Y
32.90
25.75
18.59
11.44
8
7
6
5
pH
10
30
50
70
INCUBATION TEMPERATURE (°C)
B. NONFAT DRY MILK
PROTEIN SOLUBILITY (%)
Y
109.49
78.15
46.80
15.45
8
7
6
5
pH
10
30
50
70
INCUBATION TEMPERATURE (°C)

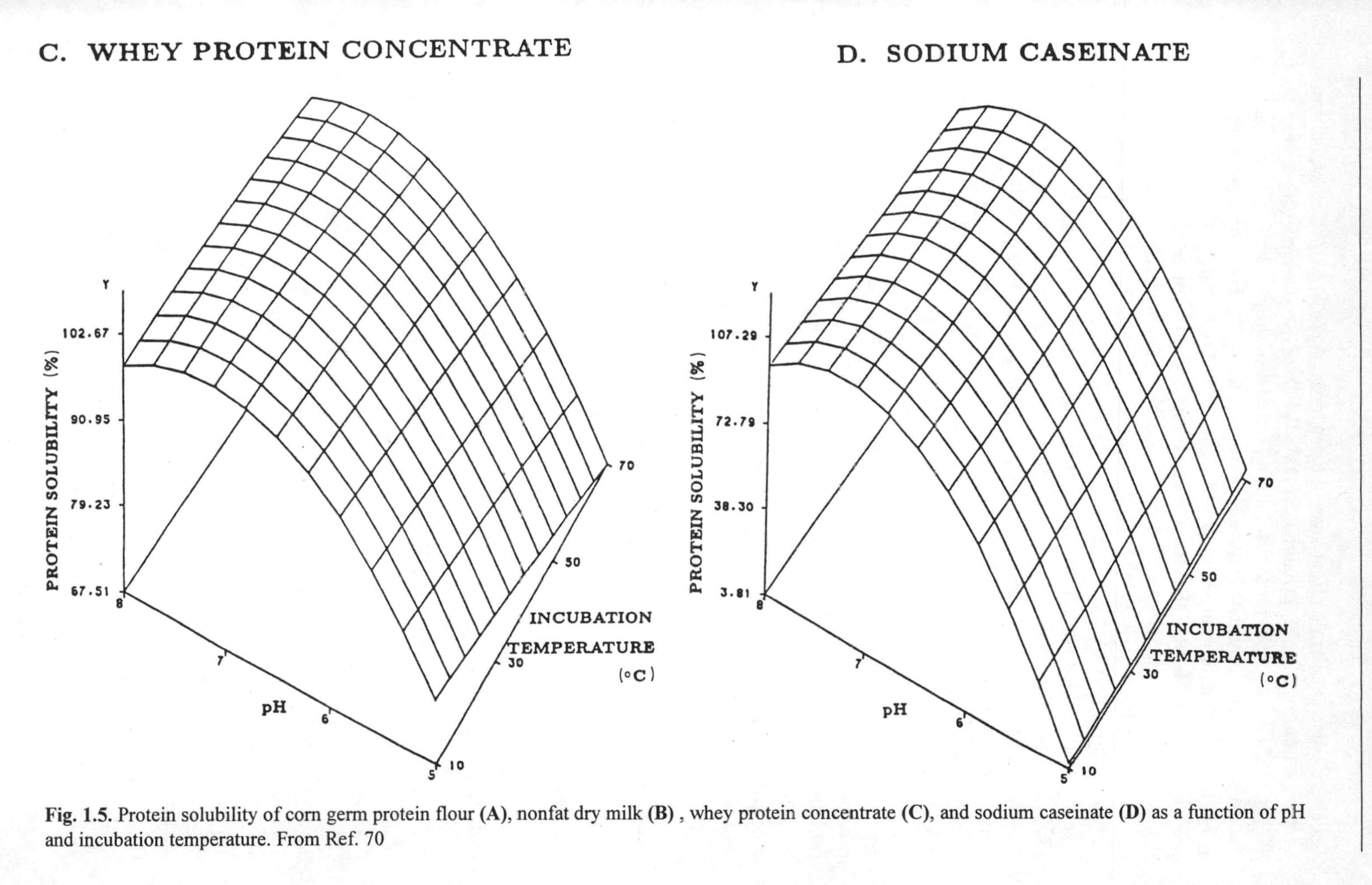

Fig. 1.5. Protein solubility of corn germ protein flour **(A)**, nonfat dry milk **(B)** , whey protein concentrate **(C)**, and sodium caseinate **(D)** as a function of pH and incubation temperature. From Ref. 70

but there was no overall effect over the response surface [70]. Compared with protein solubility of milk proteins (NFDM, WPC, or SC), which was up to 95–100%, CGPF was low in protein solubility, only up to around 30%. This might be due to the highly hydrophobic nature of corn protein [72]. Barbieri and Casiraghi [73] reported that protein solubility of corn germ flour sharply decreased from 28.7% to 15.9% (% total N), after the conditioning step (moistening, heating, and drying) for the production of a food grade flour from defatted corn germ meal.

Protein solubility of WPC was significantly affected by higher temperature during heating at 60 °C, 73 °C and 95 °C [74]. In our study of WPC solubility, only a few effects of incubation temperature were found at low pH (near 5, Fig. 1.5C) [70]. Increase of the NFDM utilization can be obtained by recovering the milk proteins as co-precipitates and using them as food components. Solubility of the milk protein co-precipitates sharply increased at pH's above 9.0; however, co-precipitates prepared from NFDM that were adjusted prior to precipitation to pH 5.5, showed a drastic increase in solubility at pH values above 5 [75].

Functional properties including nitrogen solubility of milk and blood plasma co-precipitates were reported by de Haast et al. [76]. Commercial protein co-precipitates with an increased range of functionality were prepared with high protein content, from 75% for casein to 90% for co-precipitates. During co-precipitate production vegetable and other animal proteins could be incorporated into co-precipitate with casein. The co-precipitate has been prepared from skim milk and blood plasma. Protein solubility of total milk protein (TMP) and milk-blood plasma (MP) co-precipitate with 70/30 MP compared favorably with that of Na caseinate over the pH range 2–10 [76]. Poor solubility profiles of the 30/70 MP co-precipitate and the plasma precipitate showed that complex formation of heat-sensitive whey and plasma proteins with casein had a protective effect against denaturation. The excellent solubility of the 70/30 MP co-precipitate indicated that a certain casein content was needed for formation of a complex between casein and other proteins.

Casein is a phosphoprotein with an ester-bound phosphate and a relatively high content of phosphorus (0.85%) and proline. In normal milk caseins, they exist in the form of large, spherical, colloidally-dispersed micelles with a size in the range of 50–300 nm. These casein micelles together with fat globules, impart the white color to milk. Casein and caseinate derivatives have functional, physicochemical, and nutritive properties which cause their extended utilization in food processing. Commercial caseins after processing are insoluble in water and are converted into the soluble form before utilization by addition of alkali. Sodium and potassium caseinates are generally more soluble than calcium caseinates. The solubility of sodium caseinate was influenced by pH with minimum at pH 4.0 to 4.5. Increase in sodium caseinate solubility was observed at pH $\geq$ 4.6. Sodium caseinate exhibited higher solubility than other studied caseinates [77]. Addition of salts (NaCl, $CaCl_2$, NaH_2PO_4), increase in temperature and pH resulted in increased solubility of caseins and caseinates. Rennet casein remained insoluble under most conditions and solubility markedly increased with the addition of simple phosphate (NaH_2PO_4) at pH 6.2. At 5.6 $\leq$ 6.2, all caseins and caseinates exhibited their greatest solubility when sodium phosphate was added. The dephosphorylated caseins were completely soluble at pH ≥ 5.5 but almost insoluble (over 90% of the protein precipitated) below its isoelectric point [78]. The isoelectric point of modified casein shifted to a higher pH (up to 0.5 pH units).

Solubility of WPI in the range 73.8–93.5% at pH 7 and temperature 25–65 °C was reported by Lee et al. [79]. Sodium caseinate exhibited minimum solubility values at pH 6 and 65 °C. The solubility value 49.5% of WPI at pH 8 and 65 °C was related to protein-protein interactions via disulfide bonds. Formation of new intermolecular disulfide bonds and protein aggregation is causing insolubilization. Murphy and Fox [80] reported the pH-solubility profile of sodium caseinate, α_sK-casein-enriched caseinate and b-casein-enriched caseinate and determined the effect of ultrafiltration on the derived caseinates (Fig. 1.6). The minimum solubility was at pH 3.5–4.5 and the solubility was greater than 90% at pH 5.5. The α_s1-/K-casein-enriched caseinate was more soluble at pH < 3.5 than sodium caseinate and potentially more useful as an ingredient in acid foods. The solubility of sodium caseinate is little affected by salt, however, NaCl influences viscosity of sodium caseinate. Lonergan [81] developed a new method of isolating casein from milk (cryocasein) by ultrafiltration and cryodestabilization that was easily suspended in water. Cryocasein may have unique functionality as compared with sodium and calcium caseinates. Cryocasein resuspended in water retained the micellar structure similar to the native casein micelles in milk. Cryocasein was dispersed rapidly during the first 5 min of agitation [82]. Solubility of cryocasein was affected by temperature and agitation. The pH-solubility profiles of 1% cryocasein, sodium and calcium caseinates are similar and minimum solubility was between pH 4 and 5. In the alkaline pH range of 7 to 9, protein solubility of cryocasein and sodium caseinate was 100% and solubility of calcium caseinate was slightly lower. The solubility of cryocasein at pH 4.6 increased with ionic strength from 6% at 0.0 M NaCl to 96% at 0.6 M NaCl indicating a considerable dissociation of the aggregates.

The required functional property of whey proteins in various applications is rapid dispersibility and dissolution. WPCs are utilized for the formulations of nutritious soft drinks and high-protein milk-based beverages, because of their functional qualities. The primary requirement to WPC in drinks is high solubility, to impart viscosity and colloidal stability. Solubility of WPC is adversely affected by calcium ions; they have been replaced with sodium prior to ultrafiltration and drying. The replacement of calcium by sodium markedly increased solubility of whey proteins at all pH values and especially in the pH range 4.5 to 5.5. Whey protein solubility is influenced by pH, ionic strength, temperatrue, and ion valency. Several researchers found that WPCs prepared by different methods are soluble in the pH range 3 to 8. Beck [83] found a sudden decrease in solubility near pH 4.65. The process of WPC manufacturing causes reduction in WPC solubility as a result of the processing treatment, excessive heat, and pH extremes. WPC with increased protein concentration (up to 90%), minimal denaturation, and high solubility is prepared by ultrafiltration and reverse osmosis [84]. Undenatured proteins with

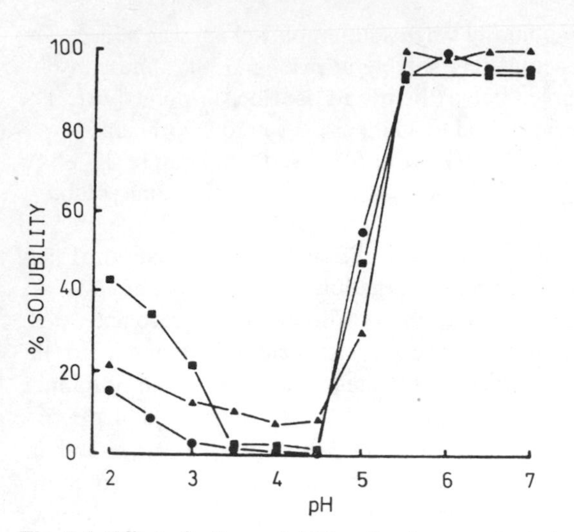

Fig. 1.6. Effect of pH on solubility of sodium caseinate (● - ●), α_s-/k-casein-enriched caseinate (■ - ■) and β-casein-enriched caseinate (▲ - ▲). From Ref. 80

good functionality were prepared by combined reverse osmosis-ultrafiltration feasible method.

The solubility of WPC prepared by gel or membrane filtration techniques is relatively high. Methods of the protein extraction and preparation are limiting factors of protein solubility and are influencing the pH-solubility profile of WPC. Solubility of WPC prepared by ultrafiltration and electrodialysis was not altered by the pH of the bulk solution. The WPCs after vacuum packaging have been stored at -40 °C, 5 °C, 20 °C, and 40 °C for 6 months with determination of protein solubility [85]. Storage of WPC at 40 °C for longer than 3 months causes solubility to decrease by about 7%. Retention of protein solubility during more than 3 months requires temperatures well below 40 °C. Li-Chan [74] stored WPC containing 35% protein at 37 °C and found a 14% decrease in solubility after 42 days storage due to changes in protein structure followed by aggregation. Detrimental effects on whey protein functional preformance during prolonged storage can result from the Maillard reaction, polymerization and loss of solubility.

The protein solubility of ultrafiltration WPCs exceeded 90% with a small drop around pI when protein solubility of egg white was 98% [86]. The protein solubility of whey protein isolate (WPI) with 95% protein exceeded 98% at pH 3.0 and 8.0 and fell to 35% at pH 4.5. Low protein solubility of WPI at pH 4.5 suggests alterations of the native structure of the whey protein including partial unfolding at pH 9 during alkaline recovery process and heating at 50 °C. The WPI was acceptable for usage in carbonated beverages with pH 3.5, because it exhibited almost 100% solubility at pH 3.0.

Functionality of 11 commercial WPCs and WPIs has been tested and protein solubility of 98–100% at pH 3 and 7 and 85–95% at pH 4.5 was established [87]. WPCs had lower solubility ranging from 64% to 84% at pH 3, and other group of

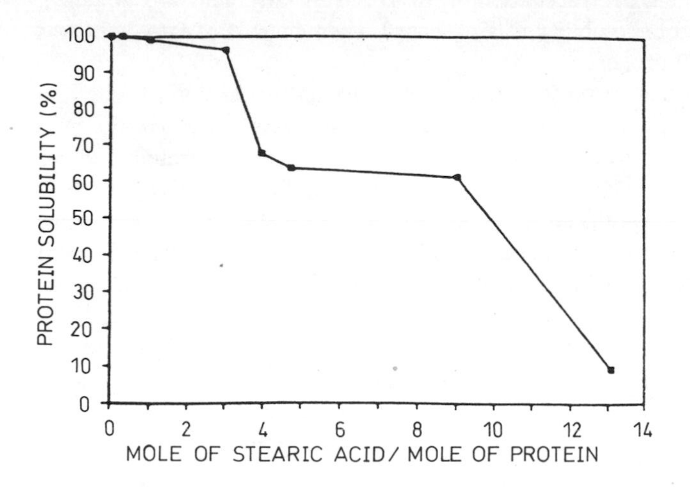

Fig. 1.7. Effect of incorporation of fatty acids on the solubility of β-lactoglobulin. From Ref. 89

WPCs exhibited solubility 49–79% at pH 4.5. However, some WPCs had solubilities of 82–86% at pH 7. The addition of salt (1.0–3.0 M NaCl) at pH 4.65 and 7.0 did not have a major effect on the nitrogen solubility of WPCs [88]. Increased salt concentration in the β-lactoglobulin enriched samples increased their nitrogen solubilities at pH 4.65, especially WPCs prepared from sweet whey. However, at pH 7.0 the nitrogen solubility of β-lactoglobulin-enriched fractions prepared from acid whey decreased. Complete solubilizatin of α-lactalbumin enriched fractions was observed and this was not significantly affected by changes in pH and salt concentration. α-lactalbumin was 100% soluble over the pH range of 3 to 9. This α-lactalbumin was recommended as a nutritional supplement in fruit juices and soft drinks and in coffee whiteners because of its high solubility in acidic conditions.

Protein solubility, frequently used as an index of denaturation, was found to decrease as a result of lipophilization, subsequent denaturation, and increased hydrophobic interactions, which caused precipitation of the proteins [89]. Functionality, i.e. foaming and emulsifying properties, of β-lactoglobulin has been improved by an increase of its hydrophobicity by lipophilization. Lipophilization of proteins includes an increase in their hydrophobicity by modification with fatty acids attached to proteins. The attachment of the fatty acid is between the carboxyl group of the fatty acid and the amino groups of the protein, thus fatty acid attachment should result in a decrease in available amino groups.

Incorporation of fatty acid to relatively hydrophobic β-lactoglobulin influenced solubility of β-lactoglobulin (Fig. 1.7) [89]. The decrease in solubility was found when 0.3 mol stearic acid was attached to 1.0 mol of β-lactoglobulin, and small loss in solubility was observed when 1–3 mol were incorporated. The first sharp drop in solubility was observed between 3–4 mol of incorporated fatty acid with another gradual decrease when up to 9 mol of fatty acid was attached. The second sharp drop in solubility was observed when 13 mol of fatty acid was attached to one mole of protein.

Heat treatment influences protein functionality during production of milk and other protein preparations and in processing of food products. Pasteurization and sterilization may change structure and properties of milk whey proteins, either reversibly or irreversibly. Milk proteins possess diversified sensitivity to heat treatment, e.g., β-lactoglobulins are very sensitive to heat denaturtion and caseins are highly stable. De Wit and Klarenbeek [90] reported the effect of whey protein denaturation on their functional performance in different foods. The thermal denaturation of whey proteins is pH sensitive and the pI ~ 4.6 is used to recover heat-denatured whey proteins. The functionality and utilization of whey proteins as functional ingredients is limited by the extent of protein denaturation. The rate of heat protein denaturation decreases with protein concentration, especially for β-lactoglobulin. Depending on the heating process denaturation of whey proteins ranges from 12% to >90%. Denaturation of commercial milk proteins was in the range from 10% to 50% and should be minimized to maintain desirable functional properties [91]. Heating of milk proteins at above 60 °C causes irreversible changes in protein structure that is also influenced by pH, ionic strength and protein concentration. These changes result in protein denaturation and reduced solubility.

Heat treatment of whey proteins, e.g., β-lactoglobulin at up to 60 °C may affect its structure and protein association or dissociation [67]. The solubility of β-lactoglobulin and α-lactalbumin was improved by heating at pH 6 and 60 °C [92]. Heating can be used to modify and increase viscosity of condensed skim milk, imparting a smoother texture to dairy desserts [93]. Viscosity of condensed skim milk increased from 24 cP to 398 cP after heating at 100 °C as a result of denaturation.

The considerable reduction in the solubility of casein from ultra-high temperature (UHT) pasteurized compared with caseins from raw and high temperature short time (HTST) pasteurized milks [94]. Formation of the β-lactoglobulin-K-casein complex caused a decrease in solubility of UHT-treated milk. The evidence of interactions between whey proteins and casein during UHT-pasteurization was obtained from amino acid composition of raw, HTST- and UHT-heated samples. In raw samples, higher molar ratios of amino acids, such as alanine, aspartic acid, and cystine have been observed. The UHT-pasteurized casein contained a lower ratio of proline and histidine.

Spray-dried and freeze-dried acid WPCs exhibited a decrease in nitrogen solubility as their concentrations increased [88]. At concentrations below 3% the nitrogen solubility of acid WPC prepared by spray-drying was greater than that prepared by freeze-drying. However, at 5% concentration, the nitrogen solubility of freeze-dried WPC was higher. The sweet WPCs prepared by freeze-drying were completely soluble at all tested concentrations, while those prepared by spray-drying showed nitrogen solubility in the range 91% to 93%. The nitrogen solubility of the β-lactoglobulin from electrodialyzed spray-dried WPC increased with concentration, when nitrogen solubility of freeze-dried β-lactoglobulin decreased with concentration. There was a low concentration effect on nitrogen solubility of the sample fractionated from diafiltered freeze-dried WPC.

Functional properties of milk proteins have been modified by physical and/or chemical treatments. For increased utilization of β-lactoglobulin, improvement of its functionality and reduction alergenicity is necessary. β-lactoglobulin was modified, glycosylated with gluconic acid or with melibionic acid, and the effect of glycosylation on the solubility was reported [95]. Native β-lactoglobulin cannot be dissolved in neutral solutions of low ionic strength. Native β-lactoglobulin is soluble at concentrations of 1.0 mg/ml of 0.2 M NaCl. Highly glycosylated β-lactoglobulin remains soluble at much lower ionic strengths, down to a concentration of NaCl equal to 20 mM. The solution of native β-lactoglobulin becomes turbid at pH's close to 5 but remains clear above pH 6 or below pH 4. A slight decrease in transmittance taking place around pH 4.3 suggested that the isoelectric point of β-lactoglobulin was changed by glycosylation. Glycosylation with melibionate was more effective than gluconate for enhancing protein solubility.

1.4 Solubility of Egg Proteins

The functional properties of egg proteins are markedly affected by heat treatment. The functionality of whole eggs is determined by the ability of egg proteins to coagulate which is important in commercial food processing. The solubility of whole egg proteins at 25 °C increased with increasing pH in the range pH 2 to pH 8 and NaCl concentration in the range 1% to 3%. The lower soluble protein contents were found at acidic pH by the addition of NaCl [96]. A significant change in solubility at about 70 °C was observed for the samples at pH 2, 3, 4, and 5. The remaining levels of soluble protein in these samples were less than 17% in comparison with the highest soluble protein content. During heating, phospholipids in yolk might interact with albumin proteins.

The effect of spray- and freeze-drying on the solubility of a 1% dispersion of egg white protein was investigated [97]. Protein solubility increased at higher pH and salt concentrations over the ranges studied. The solubility of freeze- and spray-dried egg white was in the range from 96.53% to 99.33% and a minimum was found at pH 5.0 and increased on both sides of this pH. The spray-dried protein powder was more soluble than freeze-dried protein powder at pH 9.0. An increase in solubility was observed in NaCl solutions with high pH values (pH 5.0 to pH 7.0). A decrease in egg protein solubility at pH 3.0 was related to the reduction in net positive charge.

The protein solubility index of products fractionated from spray-dried and nonspray-dried egg yolk showed distinct similarities [98]. The protein solubility index of nonspray-dried, spray-dried, freeze-dried, and reconstituted spray-dried yolk products increased steadily with increasing salt concentration. Storage of spray-dried whole egg at 3–5 °C for about one year caused a marked reduction in the proportion of insoluble proteins and an increase in the inert lipoprotein fraction of the soluble proteins. This suggests that there is a transformation of some of the high lipid lipoprotein in the yolk low density fraction from an insoluble to a soluble form.

Ultrapasteurization with aseptic packaging of raw, liquid whole egg extended the refrigerated shelf-life and maintained functional properties [99]. There was no significant change observed in protein solubility and functional performance in cakes and custards of the ultrapasteurized whole eggs. The amount of soluble protein of the processed, relative to the control egg was 92.7–104.5%. The representative indicator of protein change during heating is a change in viscosity. There was no difference in viscosity between homogenized ultrapasteurized eggs and raw egg controls. The percentage of soluble protein indicated that ultrapasteurization did not result in extensive damage to the egg proteins although viscosity data showed that some denaturation °Curred. The insoluble apoprotein of egg yolk low density lipoprotein could be solubilized by ultrasonic treatment in 80% ethylene glycol [100]. Apo-low-density lipoprotein complex formed with lecithin during sonication exhibited excellent solubility in the wide range of pH's.

1.5 Solubility of Plant Proteins

1.5.1 Soybean Proteins

The importance of soy proteins, such as soy flour, concentrate, and isolate, is well known. The increasing utilization of soy protein products is due to several factors including abundance, desirable texture, and good nutritional properties. Soy proteins have received much attention because of their high content in soybeans, good functionality and low cost. In the 1980s, cereals provided over two-thirds and oilseeds over one-fifth of the plant proteins. Energy costs of protein production (in MJ/kg protein) are 30 for soybeans, 65 for corn, 170 for single-cell protein, 300 for hen eggs, 530 for fish, 585 for milk, 590 for pork, and 1300 for beef [101]. Plant proteins are cheaper than those of animal origin.

The successful use of soy proteins as a component in a food product depends on their functional properties, e.g. solubility, water retention, swelling, emulsifying capacity and emulsion stability and foaming properties. Solubility of the proteins is required because of its influence on other functional properties. Protein-containing additives with high solubility are easier to incorporate into foods than proteins of low solubility. For these reasons almost all soy concentrates and isolates are neutralized and supplied as the proteinates.

Generally, heat treatment insolubilizes the proteins, and functional properties are impaired. Commercial soy isolates and concentrates exibited significant variations in solubility. Soy flours with various solubilities are produced by food manufacturers and utilized depending on the protein dispersibility index (PDI). In some foods such as high-protein drinks, soft drinks, and soy milk, protein solubility is the most important factor of product quality. Utilization of proteins with low solubility causes protein settling out and problem of redispersion as ingredients in food products. Interaction of soy protein with water influences such functional properties as solubility, hydration, swelling, gelling, and viscosity. Water retention and protein solubility are critical factors in protein functionality because they affect texture, color, and the sensory properties of products. A high protein solubility is required to obtain optimum functionality in gelation, emulsifying, and foaming.

Classification of soy proteins based on ultracentrifuge sedimentation rates has identified four soy globulin fractions of which the 11S globulin (glycinin) represents the major protein in the soybean. The 7S (conglycinin) fraction makes up more than one third of the total protein. Effective processing of oilseed proteins including soy protein has been developed, but products of high quality have been produced only by expensive processing, leading to high cost of the finished product. Soy flours are manufactured from defatted soy flakes. Soy concentrates and isolates are then prepared from „low heat“ undenatured defatted soy flour by removal of soluble sugars and other low molecular-weight components. The protein content is approximately 50% in soy flour, 70% in soy concentrate and 90% in soy isolate. The unique property of soy protein is that by controlling processing conditions, soy protein products with different functional properties can be processed. The unique functional properties of soy proteins have been made use of in a variety of foods. In soy protein preparations, especially soy isolate and concentrate, the principal functional components are proteins, although the carbohydrates may play a role in water retention, hydration, and textural properties. Soy protein isolates are similar in composition. However, functional properties of soy isolates and sensitivity to various treatments differ because of differences in processing conditions, which affect the level of protein denaturation.

Modified soy protein concentrate (SPC) with improved functionality was developed by Howard et al. [102]. SPCs were subjected to successive pressure and cavitation, such as centrifugal homogenization at elevated temperatures and alkaline pH. Soy protein concentrates with a high nitrogen solubility index (NSI) can be obtained with functional properties similar to those of milk proteins. The highest content of protein in SPC is made by an aqueous alcohol process, however, the protein is mostly insoluble. The functionality of SPCs is influenced by the manufacturing process, by their high protein and fiber content, and the quality of the soy flour. The following factors influence soy protein solubility: soy protein processing conditions, size of particles, pH and temperature, concentration of protein, blending and centrifugation conditions, and ionic strength at which measurements were made [103, 104].

Exposed aromatic amino acids may play an important role in the protein solubility which is mostly related to high hydrophobicity of aromatic amino acids calculated from the free energy of transfer [6]. Protein solubility is significantly affected by the net charge on the protein molecules and is usually presented as a pH effect. However, it is more effective to use zeta potential instead of pH as a variable. Protein insolubility was influenced by zeta potential and proteins showed

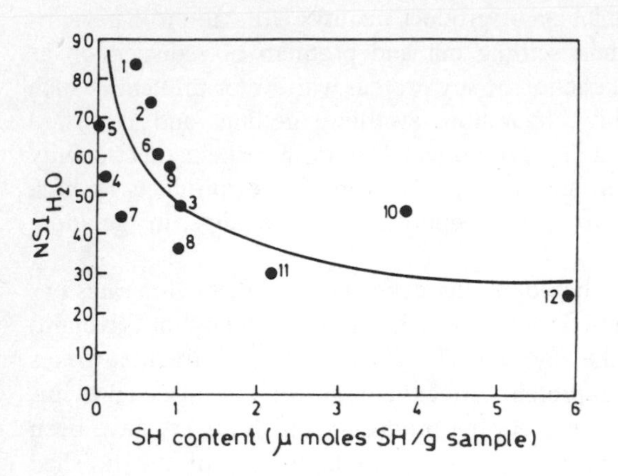

Fig. 1.8. Solubility in water (NSI_{H2O}), as a function of sulfhydryl content of soy isolates. The *numbers* on the figure represent the soy isolates analyzed. From Ref. 106

the maximum insolubility at a zeta potential near zero. Protein insolubility increases with an increase of hydrophobicity and with a decrease in zeta potential [105]. It was found that the insolubility properties of soy proteins are slightly different from that of milk proteins. Proteins tend to decrease in solubility as their hydrophobicities increase. However, hydrophobicity is not the only structural factor that determines the functional behavior of a protein. The content of sulfhydryl groups has been related to the insolubilization of soybean proteins. Correlation between solubility of soy isolate and the content of sulfhydryl groups is shown in Fig. 1.8 [106]. The tendency for soy isolate solubility was found to decrease with an increase in the number of free SH groups.

The Effect of Processing

During soy protein processing, heat treatment (moist and dry) is applied to inactivate lipoxygenase and trypsin inhibitors. This treatment is associated with protein denaturation and insolubilization. Additional deterioration of protein solubility is due to the solvent extraction and subsequent desolventization. The principle of the soy isolate production is the extraction at pH 7–9 and then recovery of protein by acidifying the extracts to pH 4–5. High solubility of extracted soy proteins is a critical characteristic for protein functionality. Significant soy protein insolubilization was observed during isolation by acid precipitation. This insolubilization is caused by polymerization of the protein by formation of disulfide bonds and by the irreversible denaturation of the protein. Solubility of soy proteins was increased by dialysis of soy isolates at pH 9–10 before acid precipitation. Solubility of soy isolates was affected by the initial protein concentration in the range 0.5% to 10%.

The Effect of pH and Ionic Strength

Protein solubility and the pH of the medium influence functionality of soy products. The soybean and peanut protein co-isolates at various pH's exhibited significantly higher water solubility than cottonseed-soy-peanut co-isolates [107]. Among the oilseed flours and protein isolates suspended in water at pH's ranging from 2.5–9.0, soy and peanut proteins showed higher solubility than cottonseed protein. The comparative studies of soy flour (SF), soy concentrate (SC), soy isolate (SI), and corn germ protein flour (CGPF) solubility were carried out [104]. The results showed that the effects of pH and incubation temperature on protein solubility were significant in all samples except SF. Comparison of results from response surface methodology showed that all samples (SF, SC, SI, and CGPF) had the same tendency: protein solubility increased with increase of pH from 6 to 8 and temperature from 10 °C to 70 °C (Fig. 1.9) [104]. A significant increase in protein solubility was found for CGPF from 10 °C to 70 °C (Fig. 1.9, CGPF); for SF from 20 °C to 70 °C (Fig. 1.9, SF); for SC from 30 °C to 70 °C (Fig. 1.9, SC); and for SI from 30 °C to 70 °C (Fig. 1.9, SI). When temperature increases, protein structure may be unfolded to a straight chain, allowing more protein-water interactions which would increase protein solubility.

Exposing soy protein to high temperature (120 °C) in an alkaline medium (pH 11) could greatly increase protein solubility, but the functional properties might be impaired because of disaggregation. Experimental data (Fig. 1.10) [104] showed that protein solubility increased with increasing pH from 6 to 8, whereas incubation time did not affect protein solubility. As pH increased from 6 to 8, protein solubility of SI and CGPF increased sharply. However, a gradual increase was observed from pH 7.3 to 8.0 in SF (Fig. 1.10, SF) and SC (Fig. 1.10, SC). In the tested range of pH, time, and temperature of incubation, protein solubility was affected by increasing the temperature of incubation and the pH of the medium. Shad a higher level of protein solubility than SF and SC. The variation of protein solubility among soy products may have been due to processing methods and conditions and/or to a high degree of protein denaturation. The balance of hydrophobic and electrostatic interactions can be controlled not only by ionic strength and pH, but also by the nature of the salts used. The effects of ions on electrostatic interactions in proteins depend only on the ionic strength, and the type of ion that influences hydrophobic interactions.

The effects of maturation and 6 months storage on protein solubility of isolated soy proteins have been reported [108]. The U-shape pH-solubility profiles were similar irrespective of three maturation stages and six-month storage with minimum solubility at pH 4.5–4.8. The solubility increased when the pH was either increased or decreased. Different solubility profiles of similar shape were obtained for 7S and 11S proteins, but 11S protein was more soluble in an acidic pH range than the 7S protein. The addition of 2% NaCl to aqueous soy protein solutions increased the number of charges on protein molecules and caused salting-in effects. Storage of soy proteins for 10 months at 35 °C and RH 85% mar-

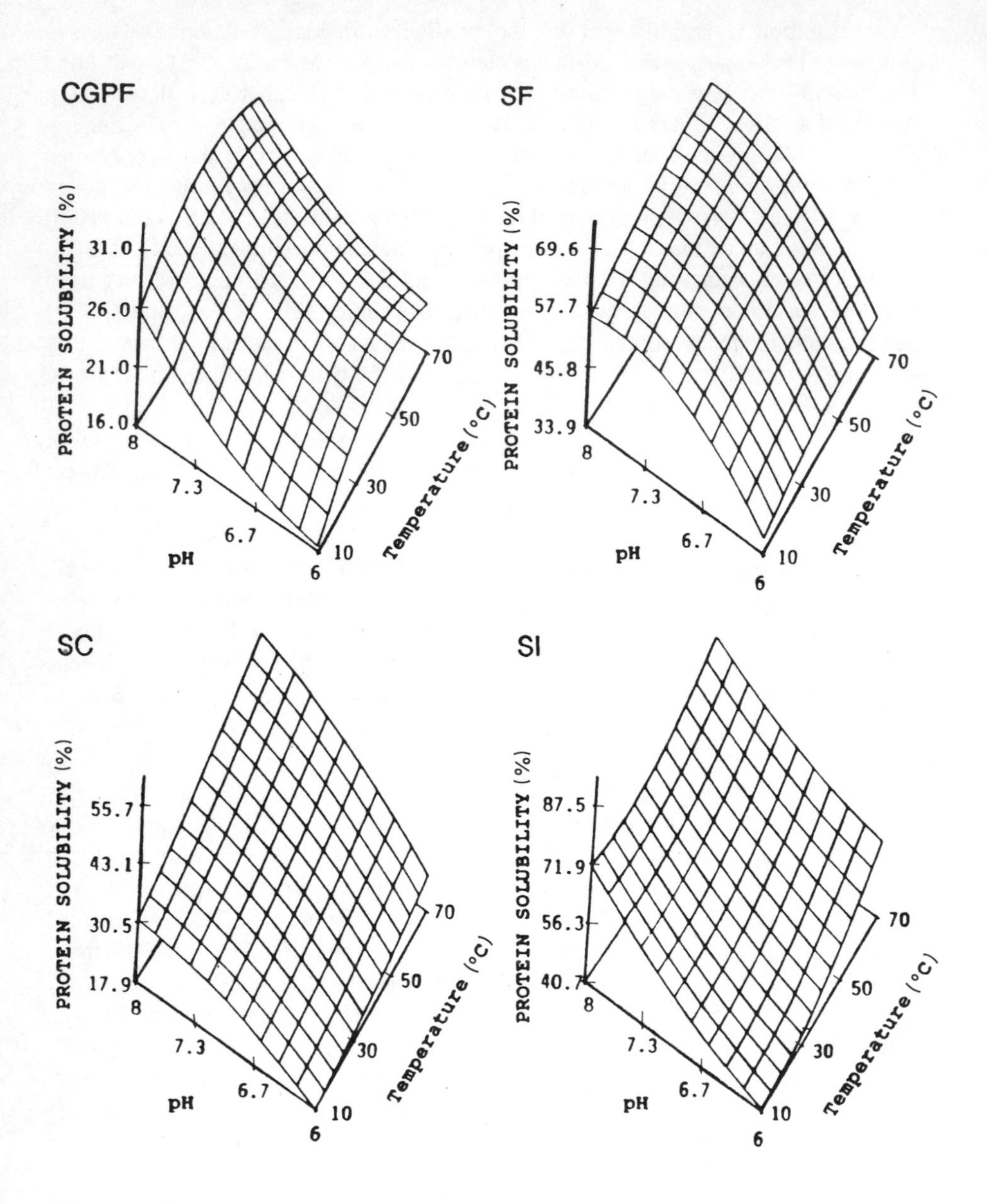

Fig. 1.9. Protein solubility of CGPF, SF, SC, and SI as function of pH and temperature of incubation. (*CGPF*: corn germ protein flour; *SF*: soy flour; *SC*: soy concentrate; *SI*: soy isolate). From Ref. 104

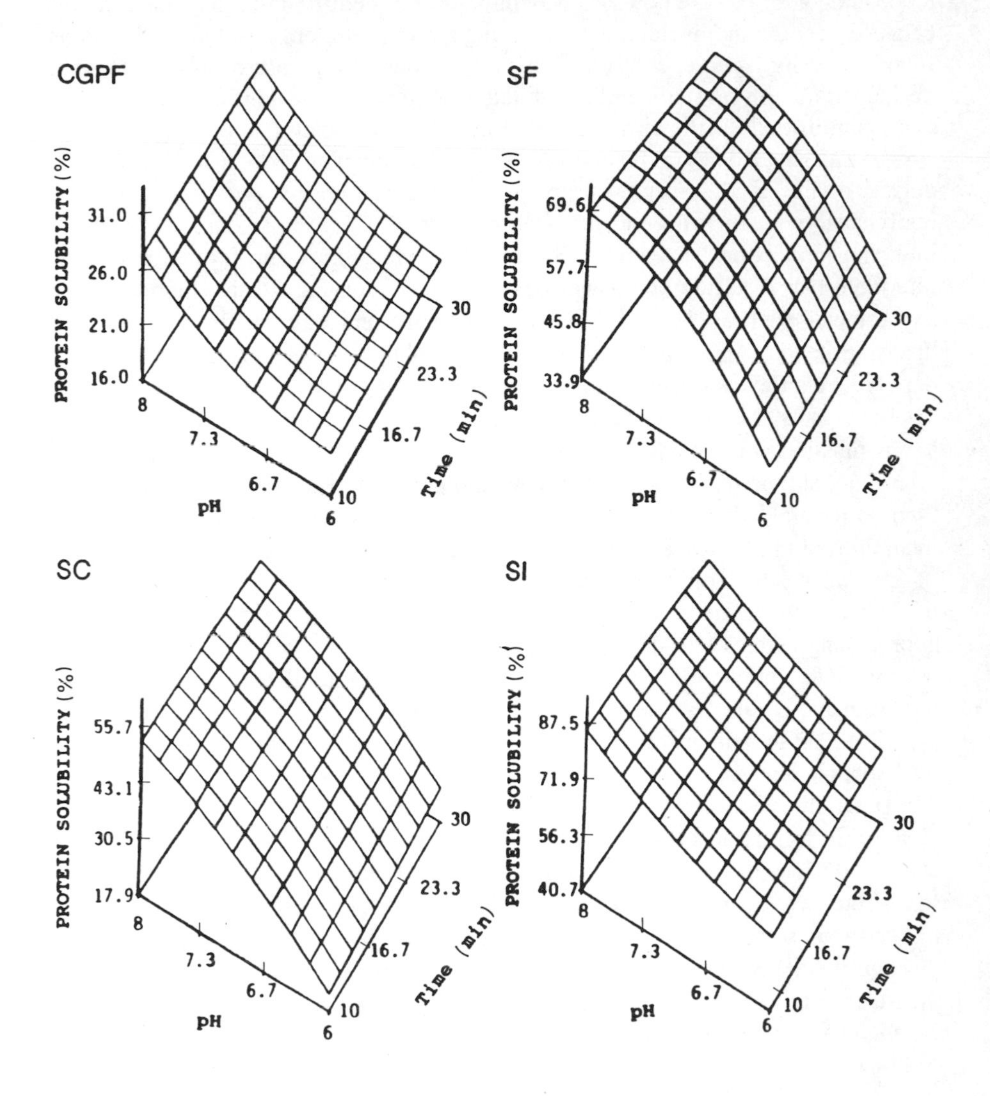

Fig. 1.10. Protein solubility of CGPF, SF, SC, and SI as function of pH and time of incubation. (*CGPF*: corn germ protein flour; *SF*: soy flour; *SC*: soy concentrate; *SI*: soy isolate). From Ref. 104

kedly decreased protein solubility [109]. The proteins most likely to lose solubility during storage were the 7S and 11S globulins.

The Effect of Drying

Chemical and physical changes of proteins during spray drying influence the reconstitution of protein powders. Soy milk powder is difficult to disperse in water because of protein insolubilization during drying. Protein insolubilization was found even during freeze-drying or drying at room temperature. Insolubilization during drying of raw soy milk without heating was nonsignificant [110]. The maximum insolubilization was recorded after 10 min heating and then decreased slowly when heated at 100 °C and 120 °C for 120 min. The protein insolubilization during drying is the result of protein polymerization by the disulfide bond interchange reaction with the increased number of SH groups. As a result of the interchange reaction between the SH and disulfide groups, the formation of new intermolecular disulfide bonds was observed during drying. Basic components of soy proteins such as 7S and 11S can form polymers through the formation of disulfide bonds. An increase in soy protein solubility was obtained by protein depolymerization using reducing agents such as sodium bisulfite and cysteine. These reagents are specific for disulfide bonds and they increased PDI when added to soy milk base before spray drying.

Major soy proteins 7S and 11S have more than two active free SH groups on their native molecular surface with small distances between molecules. Soy protein insolubilization through S-S polymerization increased sharply to a maximum value of 35% after 5 to 10 min of heating at 100° and 120 °C. During further heating, insoluble proteins decreased gradually at 100 °C and more rapidly at 120 °C approaching zero. Thirty-five percent of the total proteins were insolubilized by hydrophobic interactions and 30% were not insolubilized during drying. Hydrophobic bonds that are exposed during unfolding of protein molecules also contribute to protein insolubilization of soy milk as a result of intermolecular hydrophobic interactions [110].

Deteriorative changes of soy milk proteins and cow milk proteins are related to the film formation on the surface of milk, which occurs when milk is heated above 60 °C in the open air. The mechanism of protein insolubilization is the same, however, when molecular concentration of proteins at the surface film markedly increase, the reactive groups interact by hydrophobic interactions and through the SH/SS interchange reaction. As a result of polymerization, a film on the surface is formed, and this phenomenon is utilized in „yuba“ production in Japan. The film is skimmed from the surface and up to 80% of the soy milk solids can be recovered [110].

The Effect of Freezing

Freezing is known to alter the properties of proteins in foods. Freezing of soy milk and subsequent thawing causes reduction in the protein solubility. This reduction is related to the formation of ice crystals and high solutes concentration in the unfrozen phase, which results in denaturation and aggregation of proteins. The deterioration of the physical properties of proteins is mainly caused by irreversible insolubilization that occurs when unfolded protein molecules are brought close enough together during freezing and/or drying. Irreversible insolubilization may occur when free -SH and disulfide bonds are exposed at the surface of proteins for a sulfhydryl/disulfide interchange reaction. In the native protein molecules, SH groups are not exposed significantly to interaction. However, native unfolded proteins can be irreversibly insolubilized when the protein solution is highly concentrated and the protein molecules are brought together, as during drying or freezing. Fukushima [110] suggested that the mechanism of soy milk protein insolubilization during freezing and drying is similar. In both processes, sulfhydryl/disulfide interchange reactions and hydrophobic interactions are responsible for protein insolubilization. During freezing, the molecules are brought close enough to react by removal of water as a result of the formation of ice crystals. Heated soy protein was insolubilized rapidly during frozen storage. The contribution of the S-S bonds to protein insolubilization during frozen storage was more significant in freezing native proteins than in heated ones. In denatured proteins, hydrophobic groups are exposed to interaction with proteins. The protein concentration effect on protein insolubilization during frozen storage is utilized in the production of „kori-tofu“ originally developed in Japan in the regions with cold winters.

The Effect of Modification

Functionality of soy protein concentrate prepared by the water leaching process was improved by chemical modification. Solid's dispersibility of soy protein was improved up to three times, and pH in the range 4.5–8 had a slight effect on dispersibility (Fig. 1.11) [111]. Dispersibility was affected by sodium dodecyl sulfate (SDS) and sodium sulfite concentrations. The treatment effect increased with temperature. When treating soy protein at pH 6 and 0.5 SDS, dispersibility increased 6.5 and 8.4 times at 40 °C and 60 °C, respectively. An increase of dispersibility is caused by negative charges due to the sulfate bound to proteins [112]. Sodium sulfate was more effective at pH 6 and 8 than SDS in increasing the dispersibility of soy protein. Combined effect of SDS and sodium sulfate at 60 °C increased dispersibility up to 10 times, which resulted in soy protein with dispersibility 65% as compared to 6.2% in untreated soy protein.

Enzymatic deamidation with alcalase affected the solubility of soy proteins [113]. The extent of deamidation was 6.0% and 8.2% for two various conditions of hydrolysis. The pH-solubility profile of deamidated soy protein hydrolysates at a 2% degree of hydrolysis was similar to the hydrolyzed controls, but at a 4% degree of hydrolysis was considerably greater than their respective controls. Structural changes of soy proteins during deamidation caused changes in solubility.

The introduction of SH groups (thiolation) into food proteins was said to modify and improve their functionality [114]. Papain-catalyzed acylation was utilized to introduce new SH groups into soy protein. The incubation with papain in the presence of N-acetyl-homocysteine thiolactone (AHTL) under alkaline conditions was utilized for N-acetyl-L-homocystein incorporation and SH group introduction through a peptide bond in soy protein isolates. The papain-catalyzed

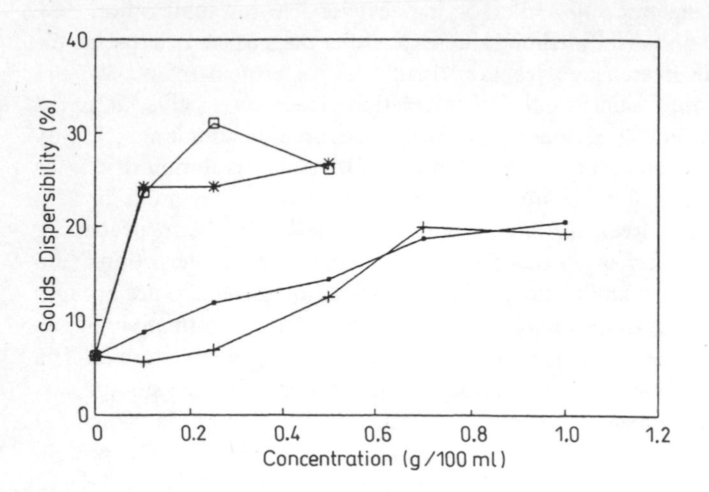

Fig. 1.11. Effect of sodium dodecyl sulfate (*SDS*) and sodium sulfite concentration on solids dispersibility (*SD*) of soy protein concentrate (+SDS, pH 4.5; ■ SDS, pH 6; * sodium sulfite, pH 5; □ sodium sulfite, pH 8). From Ref. 111

acylation increased the solubility of soy protein isolates at all pH levels examined except below pH 3 and at pH 6–10 in the 24-hour acylated sample.

Phosphorylation increased soy protein solubility under neutral and slightly acidic conditions [115]. Phosphorylation was performed with sodium trimetaphosphate by incubation with soy protein isolate in aqueous solution at pH 11.5 and 35 °C for 3 hours. At pH values below pI the solubility decreased. Phosphorylation shifted the pI of soy protein isolates approximately 0.8 pH unit, i.e. from 4.5 to 3.7.

The Effects of Various Processes

Protein solubility is influenced by homogenization, ultrafiltration, irradiation, and ultrasonic treatment. The protein dispersibility index (PDI) was lowest in the unhomogenized soy milk. Homogenization resulted in smaller particles of soy milk, and smaller droplets were produced upon atomization. Smaller droplets were dried faster and were less exposed to heat. Membrane isolated ultrafiltration oilseed proteins can be competitive functionally with the conventionally produced commercial soy isolate. Soy protein isolate's solubility was greatly enhanced by the membrane isolation process due to the inclusion of the highly soluble whey proteins in the isolates [116]. In aqueous extraction soybean protein is separated by precipitation from the aqueous phase at its isoelectric point. Nitrogen solubility profiles of spray-dried protein powders at nine pH's were determined for soybean proteins extracted by combined extraction procedure. Nitrogen solubilities were generally high below pH 3.5 and above pH 7.

The solubility of raw untreated soy protein in deionized water and 0.6 M NaCl were similar (80.3%) and higher than the solubility in $CaCl_2$ (68.1%). Solubility of soy protein irradiated using a 60Co source at the levels from 1.0 to 10.0 kGy was not significantly different [117]. Soy protein solubility decreased at higher doses of irradiation; at 100 kGy it was 67.2%. The decreases in protein solubility in NaCl and $CaCl_2$ solutions were higher than in water (57.8% and 48.8%, respectively).

A negative correlation was found between radiation dose and soy protein solubility in an aqueous medium, 0.6 M NaCl and 0.2 M CaCl solutions and at three levels of moisture content in soy protein blends, 15.33%, 22.48%, and 30.47% [117]. At lower doses of irradiation, an increase in the moisture content of soybeans was accompanied with a decrease in protein solubility. The amount of highly reactive free radicals increased with moisture content, and those free radicals can cause damage to proteins and decrease solubility, leading to the formation of protein aggregates. At high doses of irradiation (100 kGy) and low moisture content (7.4%) trypsin inhibitor activity decreased by 25%. A similar effect was obtained at high moisture content and low levels of irradiation.

The basic effect of sonication in a water system is due to cavitation, a phenomenon that includes formation and collapse of gas bubbles. During the cavitation phenomenon, the acoustic energy is transferred into the liquid to create many effects, such as intense mechanical vibration and agitation, heat generation and chemical changes. Extractability of soybean proteins was markedly affected by sonication, particularly when this treatment was applied after stirring [118]. Acoustic extraction can solubilize 80% of the proteins. The sonication was performed at a frequency of 20 kHz with an output power of 125 W. Sonication was an effective method of soy protein extraction. The specific heat-labile protein fraction of 7S proteins that failed to extract in autoclaved flakes was recovered by sonication. The conventional stirring extraction reached 30% of the total proteins. This effect of sonication is probably related to conformational changes of proteins that involve the b-subunit of 7S proteins and significantly increased water solubility. The ultrasonic method of soy protein isolate modification and improved solubility was probably due to the ultrasonic ability to increase the hydrophilic nature of proteins [119].

1.5.2 Peanut Proteins

In peanut flours protein content ranges from ~28% for full fat to 42% for partially defatted and 57% protein for the fully defatted flour. Peanut proteins have been classified into the globular proteins arachin, conarachin, and nonarachin. Recently, Basha [120] separated peanut proteins into 10 components based on molecular weights. Arachin and conarachin have similar amino acid profiles, except that conarachin contains substantially higher amounts of methionine and cystine. Limiting amino acids for peanut proteins are methionine, lysine, and threonine.

The protein solubility profiles of peanut protein fractions in water and 0.2 M NaCl exhibited a U-shaped pattern and were similar to profiles reported for other proteins (Fig. 1.12) [121].Solubility of globular proteins varies markedly as a function of pH. Minimum solubility of peanut proteins was at the isoelectric point (pH 4–5) and more than 95% of the protein was solubilized at pH below 2.5 and above 7.0.These results showed that solubility properties of isolated peanut protein fractions vary from those of total proteins. The solubility of peanut protein fractions in the presence of 0.2 M NaCl decreased at pH 4 and after reaching a maximum, decreased again (Fig. 1.12B). The presence of NaCl de-creased the solubility of peanut protein fractions particularly at acidic pH due to NaCl competition with protein for water. Solubility of globular proteins varies significantly with change of pH. The solubility of peanut protein isolates at pH 7 ranged from 15.1% to 55.6% [122].

Functional properties and flavor characteristics of peanut concentrates and isolates may be improved during processing. They can be incorporated into a variety of foods because they have no specific color and odor. Peanut protein concentrate may be prepared from flour by extraction most of the oil and the water-soluble nonprotein components. Protein solubility of peanut flour is decreased because of denaturation during extraction. Raw protein flour should be denatured at the minimum level to retain solubility. Peanut protein isolate is used for the production of a blend of equal amounts of whole milk and peanut proteins [123]. Peanut proteins do not have some of the required functional properties such as good emulsifying capacity and dispersibility. Acid extracted peanut proteins showed poor solubility (50%) compared to alkali extracted protein (90%). Highly acidic conditions ($pH < 2.0$) during the isolation of proteins may cause irreversible changes in protein conformation.

Moist heat of various peanut protein extracts causes insolubility as a result of denaturation. Peanut proteins during moist heating form new structural components and become insoluble by interaction with other components of the seeds. The amount of soluble protein in extracts of peanuts heated at 100 °C and kept at this temperature for up to 16 h continually declined from 27.8 to 4.5 mg/ml between 15 and 105 min heating and remained at this level through 180 min [124]. During the same period, the level of insoluble protein increased. Electrophoretic studies [124] showed that wet-heated peanut proteins had solubilities comparable to native proteins at all levels of heat. Samples heated at 110 °C exhibited an intermediate level of solubility. Moist heating of peanut seeds at high temperatures sequentially alters them to subunit polypeptide forms or fragments, then to aggregates, and finally to insoluble components.

The effect of acylating agents on acid and alkali extracted peanut proteins was different [123]. Succinylation slightly increased the solubility of alkali extracted peanut proteins, and acetylation decreased it. The acetylation and succinylation greatly improved the solubility of acid extracted peanut proteins, and the solubility of succinylated peanut proteins was significantly greater than acetylated. The net effect of succinylation is not only the introduction of acidic groups, but also the reduction of positive charge of proteins.

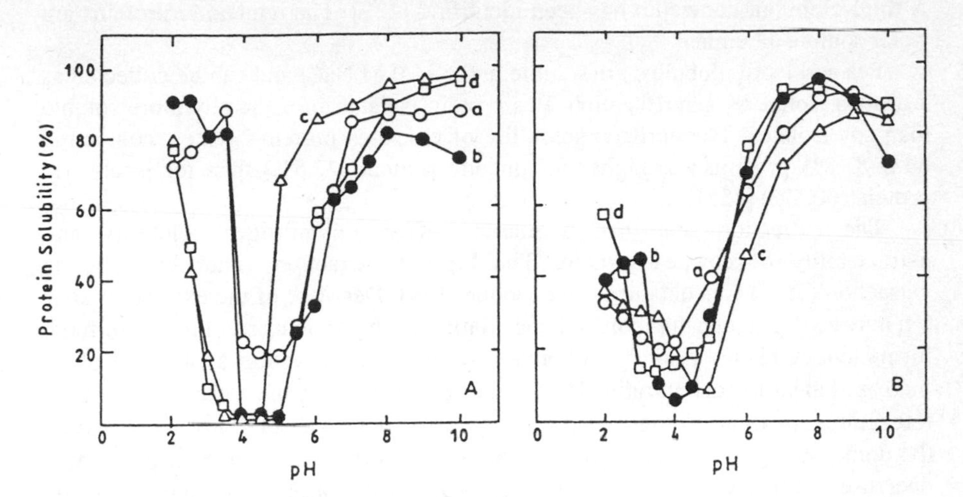

Fig. 1.12. *(A)* Protein solubility curves of peanut protein fractions in water: *(a)* total protein; *(b)* arachin; *(c)* conarachin II; *(d)* conarachin I. *(B)* Protein solubility curves of peanut protein fractions in 0.2 M NaCl: *(a)* total protein; *(b)* arachin; *(c)* conarachin II; *(d)* conarachin I. From Ref. 121

Protein solubility of peanut flour milk hydrolyzed with immobilized papain was higher than the control protein in peanut flour milk [125]. Protein solubility increased in particular at pH 4.5–7.0, which encompasses the isoelectric points of most peanut proteins. Heat treatment at 121 °C for 15 min caused a greater decrease in protein solubility of non-hydrolyzed peanut protein at pH from 5 to 9. An increase in protein solubility in hydrolyzed peanut protein heated at an elevated temperature may be advantageous in its food application. Nonmodified peanut protein flour and modified by controlled fungal fermentation exhibited higher protein solubility at an alkaline pH than at an acidic one, with the lowest solubility at pH 4, within the isoelectric point range of peanut proteins [126]. Fungal fermentation slightly increased protein solubility at pH 4–6, but markedly decreased solubility at pH 8–10.

1.5.3 Pea and Bean Proteins

Legumes have been used as a source of protein, especially when animal proteins were expensive and the supply was limited. Peas, beans, and lentils can be used as protein sources for protein flours, concentrates, and isolates. Air-classified pea protein concentrates differ from soy concentrate in residual starch that can be useful in functional performance. Dry peas contain 20–30% lysine-rich protein. Legume proteins are mainly storage proteins comprised of two globulins, legumin, and vicilin. Legumin/vicilin ratio range from 1.0 to 4.2 and is affected by cultivar.

A third globulin, convicilin has been identified [127]. The remaining proteins are water soluble albumins.

Pea and lentil globulins are soluble in 0.5–1.0 M NaCl and can be collected as a protein isolate by centrifugation. Pea protein isolates are generally more soluble than soy isolates. The nitrogen solubility of chickpea protein isolates, containing 84.8–87.8% protein, was higher for micelle protein (72.5%) than for isoelectric protein (60.4%) [128].

The extraction time had a marked effect on nitrogen solubility and extractability of cowpea protein. The highest extraction yield for a single extraction (76%) was obtained after 1 hour [129]. Decrease of the extraction after 1 h may be due to denaturation and coagulation of the proteins resulting from foam formation. A higher yield was obtained by multiple extraction as a result of an increased meal-to-solvent ratio. The level of extractable nitrogen was 93.4% of the total nitrogen. Ionic strength affected nitrogen solubility; solubility increased as the ionic strength increased from 0.05 to 1.0, after which there is a gradual decrease [129]. Extractability of cowpea proteins was affected by the pH of the extracting solvent. At the isoelectric point (pH 4.4) about 40% of the protein was still extractable. Unlike cowpeas, most of the legume proteins show a protein extractability of about 10% or less at their isoelectric point. Some albumins and globulins of cowpea may not precipitate at the isoelectric pH. Globulin and legumin of pea protein in mixtures were more soluble at pH 3 than at pH 7 [130]. Both legumin and vicilin were 80% soluble at pH 3, but the vicilin was more soluble than legumin at pH 7 (96% vs. 73%). Possibly, this is related to a dissociation into lower molecular weight subunits at acidic pH.

Cowpea proteins showed a higher solubility in water than in 0.5 M NaCl solution [129]. The high extractability in water may be due to the presence of salts in the seed. Protein solubility decreased at low ionic strength of cowpea proteins, and this was related to the formation of ionic bonds within the protein molecule and between adjacent protein molecules that caused formation of protein aggregates. Naczk et al. [131] showed that protein solubility of pea protein (85–86% protein) was moderately higher (30.3–41.9%) than soy concentrate (31.5%) and much higher than for soy isolate (22.2%) and gluten (16.3%).

Faba beans and field pea proteins are rich sources of lysine. High protein solubility (80–86%) at pH 6.6 was reported by Sosulski and McCurdy [132]. Air classification process that doubled the protein content showed a lower NSI, in the range 65%. Protein solubility probably decreased as a result of heat denaturation in grinding equipment. The faba pea and faba bean isolates exhibited low NSI values (38–40%) at pH 6.6. The pH-solubility profiles of faba bean showed 10–15% soluble protein and nonprotein compounds over the isoelectric pH range [132]. High nitrogen solubility values were observed on both sides of pI. These proteins were highly soluble at acid pH's which is important in certain food utilizations.

Tepary beans are indigenous legumes of the Southwestern U.S. and Mexico and contain 19.7% to 32.3% protein. Tepary proteins are soluble in water (84% of recovered protein) and salt solution (14%) [133]. Protein content in tepary bean flour is high and protein concentrate or isolated albumin and globulin fractions can be prepared easily. Tepary albumin had a good solubility profile (up to 80%) over all pH ranges. Minimum solubility (22.44%) was at pH 4.0. Tepary globulin had an excellent protein solubility through all pH's except at its isoelectric point.

The black gram is widely used as a good source of protein in the tropics [134]. Black gram protein concentrate exhibited U-shaped protein solubility curve with minimum solubility at pH 4.0. The solubility of black gram protein is similar to other bean proteins.

Moth bean and horse gram proteins were extracted 89% in 10% NaCl solution and 80% in 0.5% Na2CO3 solution [135].The minimum solubility (10–20%) of horse gram proteins was at pH 4.0 and moth bean at pH 4.5. The solubility increased sharply beyond this region for most bean proteins. The pH-solubility profile of navy beans, pinto beans and chickpea proteins showed that NSI was lowest at pH 4.0 (isoelectric pH region) and highest at pH 10 [136]. The pinto bean and chickpea protein samples gave high NSI values on both sides of the isoelectric pI of 4–5.

The Effect of Processing

Heat treatment of bean protein flour and bean protein isolate at temperatures from 50 °C to 90 °C caused a significant decrease in protein solubility [137]. Complete (100%) loss of nitrogen solubility in bean protein flour and isolate was observed for moist heating (0.111–0.141 g H2O/g dry matter) at 90 °C. When dry heating at 90 °C was applied, loss of nitrogen solubility was considerably lower, 13.4% for bean protein flour and 63.5% for bean protein isolate. Bean protein isolate was more sensitive to heating than bean protein flour.

There have been limited studies to determine the effects of high temperature extrusion on protein solubility and molecular properties of proteins. The effect of different extrusion temperatures on bean protein solubility and electrophoretic behavior has been studied [138]. A high degree of protein insolubilization was found after extrusion, which resulted in a decrease in solubility of albumin and globulin fraction. Extrusion caused a decrease in water and salt-extractable proteins and an increase in insoluble nitrogen compounds for both samples. The water-extractable protein fraction decreased markedly when extrusion temperature increased from 110 °C to 135 °C and 150 °C for navy bean and from 110 °C to 121 °C and 135 °C for pinto bean samples [138]. However, no changes were found in salt-soluble proteins with increasing extrusion temperature. An extrusion temperature of 110 °C had a greater effect on solubility of albumin and globulin fractions for pinto than for navy bean proteins.

The effect of different drying methods on the solubility of pea protein has been reported [139]. Sumner et al. [139] reported a comparative study of spray-, freeze-, and drum-drying on functional properties of pea protein isolates, containing 83–90% of crude protein. The nitrogen solubility of pea protein isolate varied from 0% to 100% over the pH range 3–10. The nitrogen solubility was least at the pI of pH 4.5. Nitrogen solubility values were markedly higher for spray- and freeze-dried protein isolates than for drum-dried. Significantly lower nitrogen solubility values for drum-dried sodium proteinate were probably due to protein heat denaturation.

The nitrogen solubility of acetylated and succinylated winged bean flour in water showed gradual increase up to 80% acetylation [140]. Increase in nitrogen solubility was from an initial value of 80% to nearly 100% at 80% acetylation. Nitrogen solubility decreased sharply at acetylation higher than 80%, and 52% nitrogen was soluble at 90% acetylation. The reduction in soluble nitrogen for succinylated protein was not as steep; 87% succinylated protein flour had a solubility of 85%. Protein solubility of the raw beans was increased by lysine acetylation and succinylation [141].The protein solubility of the raw beans in water increased, up to 90–100%, on both sides of pH 4.0, while in 0.5 M NaCl it was higher in the pH range 2 to 7 [137].

The pH-nitrogen solubility profiles of freeze-dried protein powders of cowpea, peanut and soya prepared from unfermented and fermented extracts are presented in Fig. 1.13 [142]. The pH-nitrogen solubility profiles of the unfermented proteins are typical and similar to pH-solubility profiles of other plant proteins. Disc-gel electrophoresis revealed that fermentation reduced the amount of soluble protein in the extracts. Data presented in Fig. 1.13 showed that fermentation had the effect of flattening the nitrogen solubility curves over a wide pH range. Solubility of fermented proteins was less affected by changes in pH, however, the relative amounts of soluble nitrogen were different at various pH's. In other data [143] fermentation increased protein solubility at various pH values. The lactic acid bacteria used in this study were weakly proteolytic, and large amounts of acid were produced. The acid may irreversibly coagulate some proteins and influence protein solubility.

Guar meal is rich in protein (~55%), however, it contains some toxic and antigrowth substances that can be inactivated by heating or selective extraction with aqueous alcohols or dilute HCl. Effects of different detoxification methods of guar protein on their functional properties have been reported [144]. Extraction with aqueous alcohols gave high yields (73–82%) of the detoxified meal with 77% protein. The level of crude protein of the processed meals obtained by different extractions varied from 67.8% to 84.3%, while for the raw meal it was 60%. The typical U-shaped nitrogen solubility-pH profile for pH range 1–11 was obtained with minimum nitrogen solubility around pH 5.0 [144]. The solubility profile in water and NaCl was similar to those of soy and groundnut proteins. The autoclaved guar protein gave a solubility value of 24% at pH 7 which is lower than that of defatted guar meal. In contrast to other cases the acid extracted guar proteins gave a peak in the pH range 2.0–2.5 and nitrogen solubility was lower compared to other samples. Guar protein showed lower nitrogen solubility in NaCl 1.0 M solutions below the isoelectric pH's, whereas values of nitrogen solubility were much higher in the neutral and alkaline range of pH's.

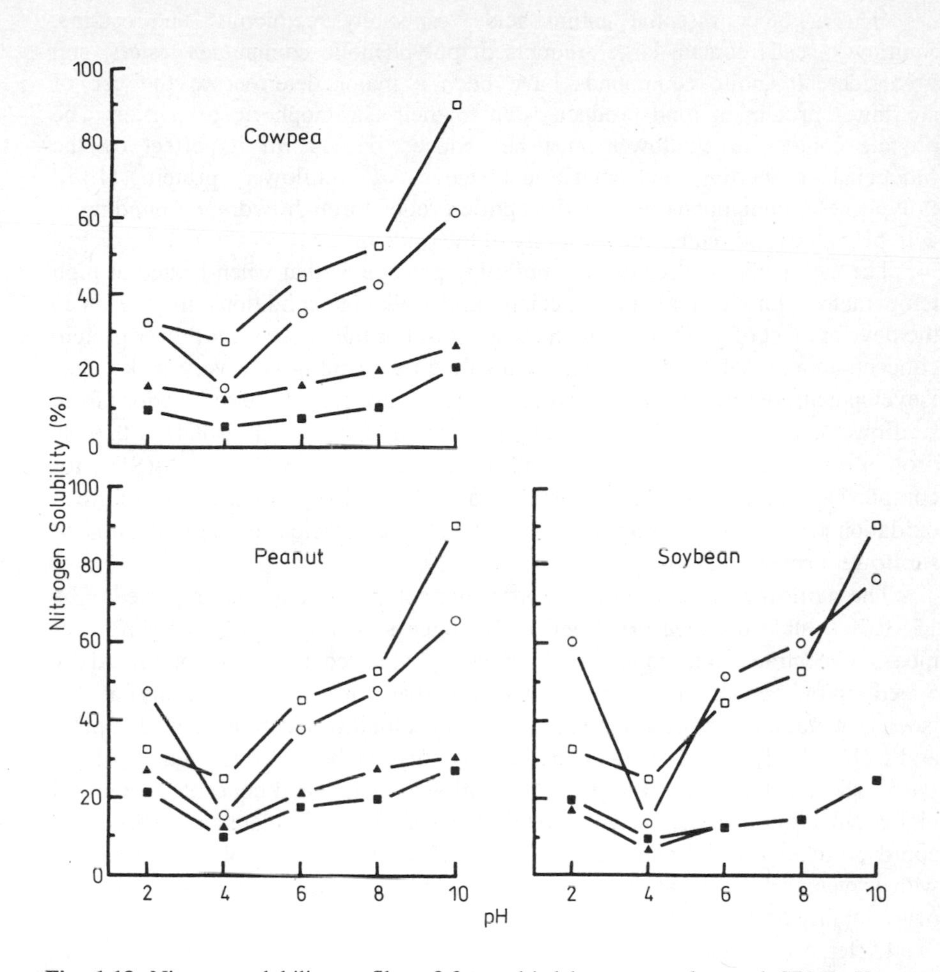

Fig. 1.13. Nitrogen solubility profiles of freeze-dried legume powders and CBMP. Key: ❑, cultured buttermilk powder; ○ , unfermented extract; ◆, extract fermented with L. bulgaricus; and ■, extract fermented with S. thermophilus. From Ref. 142

1.5.4 Sunflower Proteins

Sunflower meals are a rich source of highly soluble protein. Yields of sunflower proteins were 52–54 g/100 g meal, substantially higher than for soybean and other plant proteins. After oil extraction a cake of sunflower rich in protein (about 35%) is obtained. Protein contents of sunflower seed proteins on a dry weight basis are: flour, 60%, concentrate, 70%, and isolate, 90%. An important advantage of sunflower protein products is that, to date, no toxic component has been reported. Sunflower protein products are deficientin lysine and isoleucine. However, they are rich in other essential amino acids, especially methionine and cystine. Sunflower seeds contain large amounts of polyphenolic compounds, esters, and glycosides. Phenolic compounds have been a major deterrent to the use of sunflower protein in food products, due to their chromophoric properties. The phytate content in sunflower must be reduced because of its effect on the functional properties and nutritional value of sunflower protein [145]. Polyphenolic compounds react with peptide groups through hydrogen bonding at acid pH values and reduce the solubility of the protein.

The major obstacle for use of sunflower proteins is that when heated at high temperature with moisture and, especially under alkaline conditions, they result in the development of green and brown complexes. For this reason, sunflower protein concentrate and isolate, which employ alkaline treatment, give a very dark color. Development of an effective method of phenolic compound extraction from sunflower kernels or meal crucially important. Sunflower protein isolate with 95% protein was prepared by extraction with alkali and use of $AlCl_3$ and $A_2(SO)_3$ for complexing and removal of polyphenolic compounds [146]. The presence of green oxidation products of chlorogenic acid is a potential problem in the utilization of sunflower protein in foods.

The sunflower proteins possess poor water solubility; they were reported to be 15–30% soluble between pH 3 and 6. Nitrogen solubility profiles of sunflower meal, concentrate, and isolate with reduced phytate content were presented by Saeed and Cheryan [147]. There was no sharp minimum solubility at the isoelectric point and a broad range of minimum solubility was found between pH 3 and 6 (Fig. 1.14). Good protein extractability (over 90%) was observed at pH 10. Below the pI, and above pI at pH 6–9, sunflower isolate and concentrate showed higher solubility than sunflower meal. Higher solubility of isolate and concentrate could be related to the removal of phytate which can form insoluble complexes with proteins. Phytate has the property of interacting with proteins with shifting pH-solubility profiles.

Differences in protein solubility between sunflower albumin, globulin, and meal were reported, and a contribution of the individual fractions to the variations of the functional properties was found [148]. The albumin fraction was almost completely soluble above pH 7 and exhibited high solubility at acid pH's with minimum solubility of 50% at pH 5.0. The protein solubility of the albumin fraction was about 70% higher at the acidic end of the curve and in the neutral pH regions compared to sunflower meal. The globulin fraction was completely soluble in the neutral pH range, however, in the acid pH region, solubility of globulin fraction was markedly lower than albumin.

Nitrogen solubility profiles of sunflower protein in water and salt solutions were similar. Sunflower meals exhibited a U-shaped pH-solubility profile with minimum solubility at pH 5 [149]. The sunflower protein solubility considerably increased in 1.0 M NaCl and 0.5 M $CaCl_2$ in the pH range 4 to 7 with an increase in NSI to 76% in 0.5 M $CaCl_2$ and 55% in 1.0 M NaCl. These pH conditions are present in several food products. The increase in NSI was observed up to 0.75 M

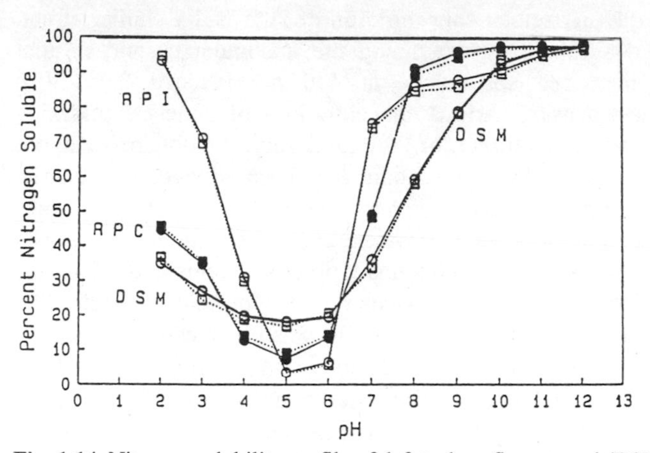

Fig. 1.14. Nitrogen solubility profile of defatted sunflower meal (DSM), reduced-phytate protein concentrate (RPC) and reduced-phytate protein isolate (RPI). O-O and □ - □ represent hybrid cultivars 7000 and 7111, respectively. From Ref. 147

NaCl, and there was no NSI increase at higher concentrations. High solubility in salt solutions is important for the practical use of these proteins.

Protein solubility of sunflower protein was significantly increased by modification, namely treatment with lactic acid [143]. Functional properties of sunflower meal have been improved by fermentation with lactic acid by incubation at pH 5.5 and 38 °C for 3 days. The pH-nitrogen solubility profile of nontreated protein showed minimum nitrogen solubility of about 11% at pH 4 and values of 42% and 78% at pH 7 and 9, respectively.

1.5.5 Corn Proteins

Corn proteins are classified into four classes: albumins (water soluble); globulins (saline solution soluble); prolamine (70% ethanol soluble); and glutenin (soluble in 0.1 M NaOH). Protein content in corn kernel is about 10% and protein content in corn germ is about 20%, but the germ is only 10% of the kernel. Corn gluten containing 68% protein (dry basis) is a by-product of the wet milling industry. Corn gluten is utilized for zein obtaining by extraction with isopropyl alcohol. Zein functionality is determined by its ability to form clear, strong films and is used as a food and pharmaceutical coating material. Corn kernel proteins (zein) are highly hydrophobic and possess limited solubility in an aqueous medium. Therefore, studies of their solubility have been limited to extraction with an alcohol/water solvent. The kind and composition of the solvent had a significant effect on zein solubility [150]. Basic studies of corn proteins have been carried out for improved solubility and future utilization. Highest zein solubility in an ethanol/water mixture was obtained at an ethanol concentration of 70%. An acetone/water mixture with an acetone concentration of 70% had a similar effect. Solubility of zein was determined by examining the maximum protein content obtainable before the measured absorbance at 420 nm reached 2.0 [140]. Solubility of zein was tested using various concentrations of dimethyl sulfoxide and in 100% pyridine. Zein concentration of 90 mg/ml could be obtained without precipitation. Greatest zein solubility was obtained in 100% dimethyl sulfoxide. The solubility of zein in pyridine/water (1:1, v/v) proved to be over three times that of any of the other solvent mixtures with water. Corn germ protein possesses a quite different composition than zein. It contains mainly globulins and albumins, which possess lower viscoelastic properties than wheat proteins (gliadin and glutenin). A defatted corn germ protein flour (CGPF) was obtained by applying a milling and screening process to a commercial defatted corn germ meal. The p the defatting process (in a commercial oil extraction installation) sharply decreased the protein solubility of corn germ flour [151]. At present, conventional oil extraction (by hexane) leaves a certain amount of residual lipids in the flour, which reduces its quality. A new supercritical CO_2 ($SC\text{-}CO_2$) method has been used for defatting corn germ flour [152]. The defatted corn germ protein product had a food-grade quality and considerable potential for use as a supplement in a variety of foods to provide a new protein source. Compositional data and nutritional studies have shown that a good corn germ protein source can be obtained by drying corn germ meal at low temperatures [153]. Solubility and dispersibility of corn germ protein is a physicochemical property that is related to other functional properties.

The relationships between protein solubility and pH of aqueous medium, solubility and time of incubation, and solubility and temperature of incubation have product had good sensory and nutritional qualities. However, the effect of heat in been reported [154]. The percentage protein solubility for two hexane-defatted CGPF are shown in the form of response surfaces in order to emphasize relationships over a continuous range rather than values for specific treatment combinations. Protein solubility response surfaces are shown in Fig. 1.15 and Fig. 1.16 of CGPF defatted by the conventional, hexane extraction method [154]. In Fig. 1.15 protein solubility response surfaces are shown as a function of pH and temperature of incubation and in Fig. 1.16 as a function of time and temperature of incubation. Samples of conventional CGPF had lower protein solubility, which may have been due to the level of protein denaturation (Fig. 1.15 and Fig. 1.16). Higher temperatures used in the conventional hexane-defatting processes resulted in a significant browning of the color and development of a toasted flavor. These data agree with the results of Barbieri and Casiraghi [151], showing a higher CGPF denaturation during defatting with hexane, especially during desolven-tization. The difference in solubility between two hexane-defatted CGPFs, processed by conventional and modified methods of fat extraction, reflected the specificity and technological parameters of the processing methods. Although synergistic effects existed between conventional CGPF and incubation temperature and pH, the highest protein solubility (18.77%) was obtained without significant effect of time of incubation up to 30 min (Fig. 1.15).

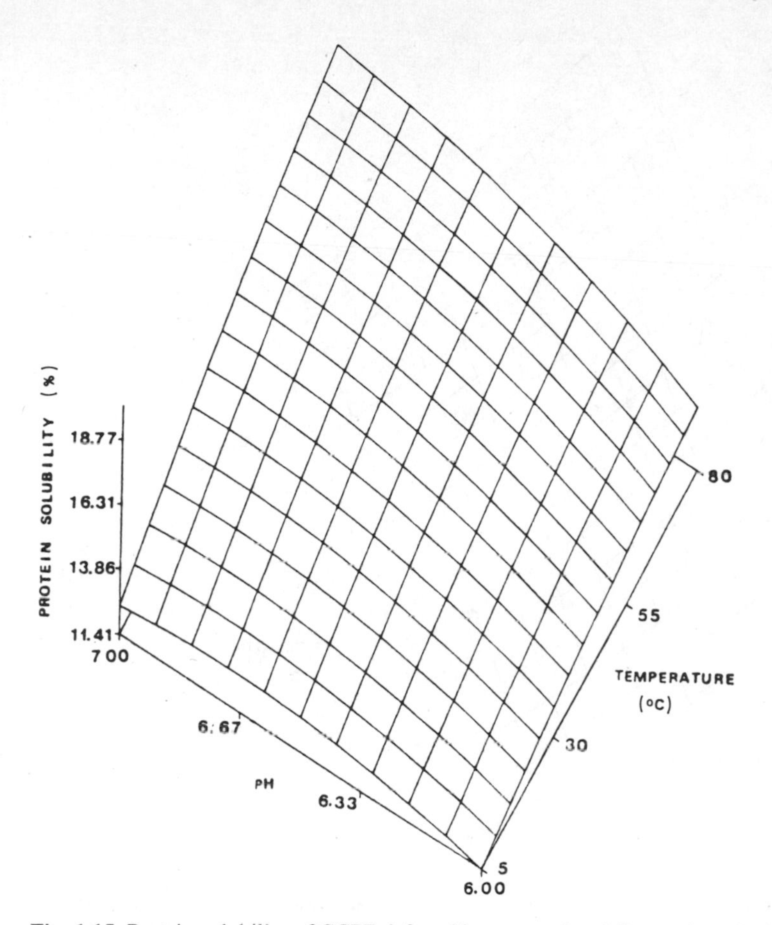

Fig. 1.15. Protein solubility of CGPF defatted by conventional (hexane) method as a function of pH and temperature of incubation. From Ref. 154

Protein solubility response surfaces of CGPF defatted by a modified hexane extraction method was shown as a function of pH and temperature of incubation [154]. Increase in the pH of the aqueous medium in the range 6.0–7.0 slightly increased protein solubility. Increase in the time of incubation also had a positive effect on the protein solubility. The highest protein solubility of 24.58% was obtained when the temperature, time, and pH were at the highest end of the tested ranges. Protein solubility of hexane-defatted CGPF obtained by the modified extraction process was significantly higher than that of CGPF obtained by conventional extraction. Messinger et al. [155] reported that corn germ protein isolate had a very high nitrogen solubility index (95.6%) at pH 8.

The difference in protein solubility between two hexane-defatted CGPF samples undoubtedly reflects variation in processing methods. The samples obtained by the conventional methods were denatured to a greater extent during processing and were less soluble than the samples defatted by the modified method.

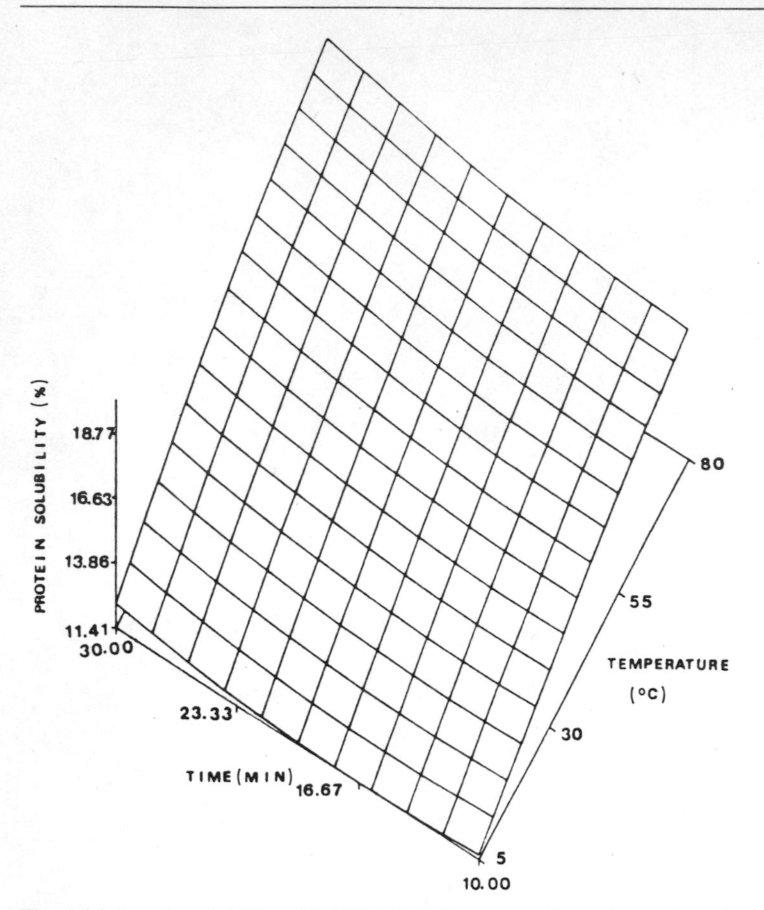

Fig. 1.16. Protein solubility of CGPF defatted by conventional (hexane) method as a function of time and temperature of incubation. From Ref. 154

Other possible contributors to the different solubilities were the nonprotein components of the CGPF preparations. Carbohydrates could compete with other components, particularly proteins, for the available water.

Supercritical-carbon dioxide (SC-CO_2) CGPF had rather poor solubility under experimental conditions [152]. However, low solubility is not always a disadvantage and many proteins of limited solubility find uses in the production of foods. Solubility of SC-CO_2 CGPF as affected by pH, time and temperature of incubation is presented in Fig. 1.17 and Fig. 1.18 [156]. The amount of protein solubilized in solution was lowest at low pH (pH 6.0) and low temperature (10 °C) and highest at higher pH (pH 7.0) and higher temperature (70 °C) for SC-CO_2 CGPF (Fig. 1.17). Protein solubility in general increased with increased pH in the range of pH 6.0 to 7.0. The influence of pH was temperature dependent; both temperature and pH factors affected the protein solubility of the two CGPFs studied. However, the samples of hexane-defatted CGPF had a different response in this experiment. Protein solubility of hexane CGPF decreased slightly at higher

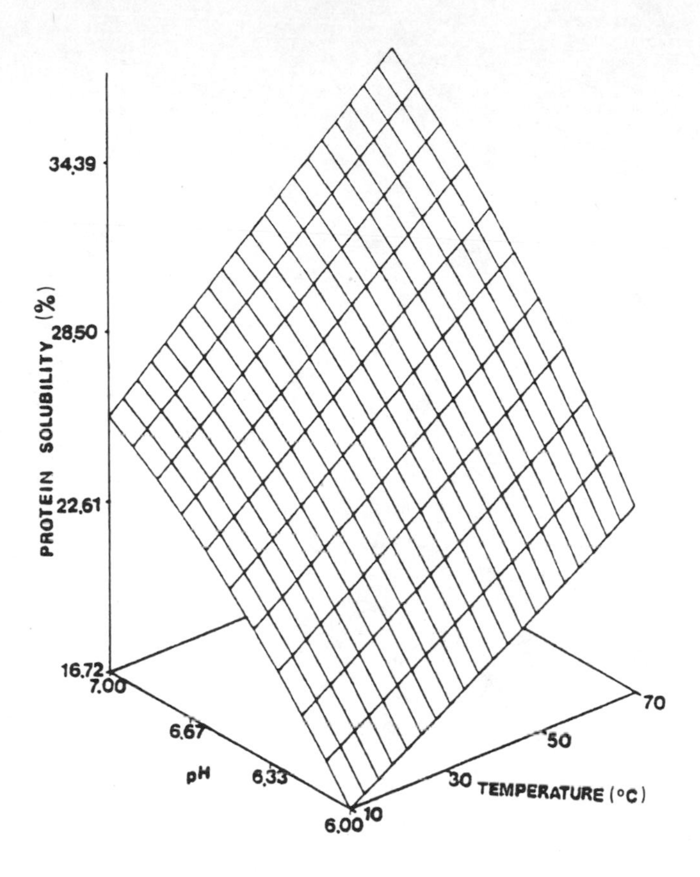

Fig. 1.17. Protein solubility of SC-CO_2-defatted corn germ protein as a function of pH and temperature of incubation. From Ref. 156

temperatures of incubation (50–70 °C). When the pH reached 7.0, the hexane-defatted CGPF still possessed the highest protein solubility at 70 °C. However, protein solubility decreased when the temperature of incubation was at 50 °C and pH was low. The SC-CO_2 CGPF had decreased protein solubility with increasing time of incubation (Fig. 1.18) [156]. A similar level of protein solubility was obtained for SC-CO_2 CGPF and for hexane CGPF. Data obtained in these experiments agreed with results presented in Fig. 1.17, which indicated that functionality of proteins was significantly dependent on their native state. These data agreed with the results of Barbieri and Casiraghi [151] that a higher degree of corn germ protein denaturation occurs during defatting with hexane. The difference in solubility between SC-CO_2 CGPF and hexane CGPF samples reflected the specificity and technological parameters of the processing methods.

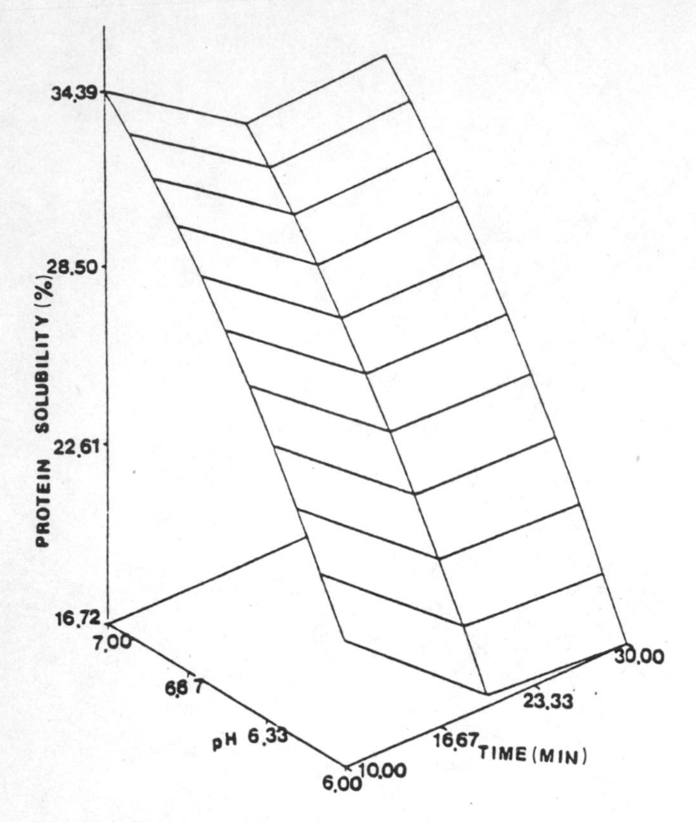

Fig. 1.18. Protein solubility of SC-Co₂-defatted corn germ protein as a function of pH and time of incubation. From Ref. 156

1.5.6 Miscellaneous Proteins

Rice Proteins

Protein is the second constituent of rice grain, and its uniqueness among cereal proteins is caused by an 80% glutelin level. The content of protein in the rice of high-protein variety, BPI-76–1, may range from 8–14% and the low protein variety, Intan, from 5–11%. Rice bran can provide protein of excellent quality and good functionality. The nitrogen solubility profiles of the full-fat rice protein concentrate, as a function of pH, are shown in Fig. 1.19 [157]. Both samples showed minimum solubility at pH 4.5 and maximum solubility in the pH range 9.0–10.5. The full-fat sample had maximum solubility (more than 75%) at pH 9.0–10.5 and minimum (13%) at pH 4.5–5.5 (pI). On the acid side of the isoelectric pH, there was a nonsignificant increase in solubility, whereas on the alkaline side,

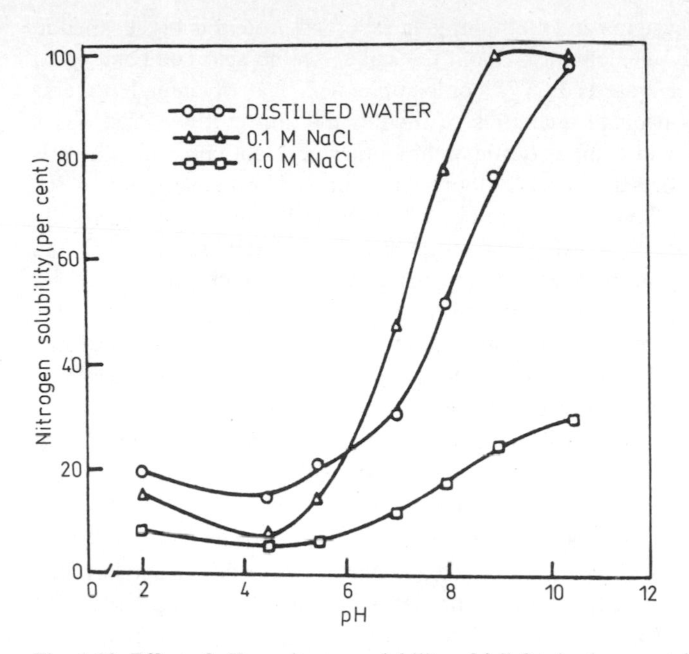

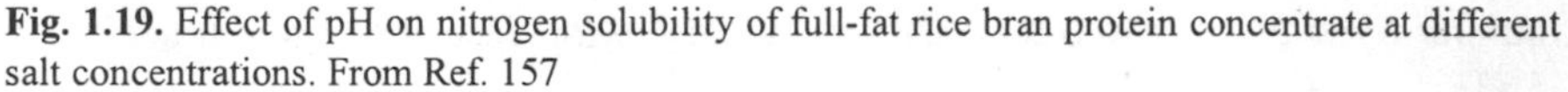

Fig. 1.19. Effect of pH on nitrogen solubility of full-fat rice bran protein concentrate at different salt concentrations. From Ref. 157

a steep increase in solubility was observed. Low solubility of full-fat and defatted rice protein concentrate at low pH's is presumably related to the formation of complexes of cationic proteins with water-soluble compounds, phytates that cause insolubilization of proteins. The nitrogen solubility curve of rice bran protein concentrate resembles the nitrogen solubility curve of other plant proteins, i.e. sunflower [158].

Rice protein solubility is influenced by ionic strenth.At low NaCl (0.1 M) concentration solubility slightly increased when studied at pH 2.0, 4.5, and 9.0, while at the higher (1.0 M NaCl) concentration solubility decreased markedly. Full-fat protein concentrate had a maximum solubility of more than 28% at pH 10.5 in 1.0 M NaCl concentration. Similar data on rice protein solubility were reported by Abdel-Aal et al. [159]. The solubility profile of rice protein flour shows minimum solubility at pH 4.5 and maximum solubility at pH 12. Treat-ment with pepsin and pancreatin increased rice protein solubility by about 15%.

Rapeseed Proteins

Rapeseed meal is the most concentrated vegetable protein produced in the cool temperature zones. The essential amino acid pattern of rapeseed protein compares favorably with that of the FAO/WHO reference protein. The amino acid composition is influenced mostly by the process of rapeseed meal production.High processing temperatures and pressures result in the destruction of basic amino acids. The content of cysteine and methionine in rapeseed protein is higher than in soy protein, whereas lysine content is about the same. Amino acid composition of rapeseed protein concentrate (RPC) is comparable with that of animal proteins [160]. RPC contains adequate quantities of methionine and cystine. RPC has a more balanced amino acid composition than any other vegetable protein. There is no specific off-flavor present in RPC flour. The rapeseed protein isolate (RPI), containing the minimal level of nonprotein components is not yet commercially manufactured. The process of RPI production includes dissolving the RPC and recovering it from solution by precipitation. Samples of RPI with protein content more than 90% have been obtained [160].

Processing conditions and associated nonprotein components influence the functionality of rapeseed proteins. RPIs and meals with different levels of phytic acid (0.9–4.6%) were compared with soybean meal and soy isolate [161]. Nitrogen solubility for rapeseed meal was 40.7%, and for rapeseed protein isolate 50.2%, comparing to soy meal 76.8% and soy isolate 56.0%. Rapeseed protein concentrate showed good nitrogen solubility (56%) at pH 6–7, indicating usefulness in a variety of foods, particularly processed meats and milk-type beverages [162]. Rapeseed protein extracted from commercial meals has low solubility, and its heating with potassium linoleate increased solubility from 60% to 90% while protein isolate maintained high solubility for up to 6 months. A number of studies showed that partial proteolysis increased the protein solubility [163].

Nitrogen solubility of rapeseed protein increased up to 95% by treatment with anionic surfactants, i.e. 10% level of sodium and potassium salts of myristic, oleic, and linoleic acids and sodium dodecylsulfate (SDS) [112]. An increase of solubility was obtained with an increase in temperature from 45 °C to 85 °C and at a higher pH (pH 10 instead of 8) and a higher surfactant to protein ratio (from 10% to 20%). At a concentration of 10%, linoleate RPI was almost completely solubilized at reaction conditions of pH 10, at 65 °C. Solubility of rapeseed protein was increased by enzymatic treatment with protease, pepsin, and trypsin [112]. Enzyme treated RPIs after drying retained high solubility at storage temperatures of 4 °C, 20 °C, and 37 °C, whereas the control as well as the SDS, oleate, and linoleate treated protein isolates decreased in solubility during storage. The mechanism of protein solubilization with fatty acid soaps and SDS includes an interaction of the hydrophobic groups of the anionic surfactants with hydrophobic sites of the protein. This interaction caused an increase in the negative charges on the protein surface, thus intensifying the repulsive forces between protein molecules.

The dispersibilities of rapeseed, soybean, and sunflower proteins determined before and after drying were very high and unchanged with values of 92–100% [112]. Dispersibilities of freeze-and spray-dried protein products were not different because of relatively mild drying conditions. Before oil is extracted from rapeseed meal, heating is applied to destroy myrosinase. Acylation caused improvement of protein functionality as a result of an increase in the net negative charges of the protein and change of the protein conformation. Acylation of rapeseed flour increased the nitrogen solubility of protein isolates [164]. Succinylated and acetylated RPI showed higher nitrogen solubility. Nitrogen solubility of untreated rapeseed isolate was 9.9%, mostly due to extensive aggregation of protein during isoelectric precipitation. Phytic acid also has the ability to decrease the protein solubility. All the acylated protein isolates exhibited better nitrogen solubility than rapeseed protein flour or Promine D. However, more data are needed on the safety and toxicity of acylated proteins.

Cottonseed Proteins

Cottonseed protein can be used in the fortification of products where solubility in water and related functional properties are not important. Cottonseed protein exhibited low aqueous solubility, 25.4–25.9% [165]. However, protein solubility in 5% NaCl solutions was significantly higher, 66.5–75.9%. Utilization of protein isolates prepared from glandless cottonseed flour by the conventional method has been limited because of unfavorable functional properties, i.e. the water solubility and emulsifying capacity required for food formulation. Cottonseed protein solubility in the aqueous medium has been expressed by the precipitability profiles [166]. Maximum precipitation was observed at pH 4.0 for the nonmodified sample while a pH lower than 4.0 was required for modified samples. Maximum precipitation was obtained for acetylated samples at pH's lower than 3.0. As a result of modification more than 90% of the extracted protein was in soluble form at pH ranges of 6–7, while approximately 60% was soluble for control samples.

High water solubility and low heat coagulability of freeze-dried succinylated, maleylated, and dimethylglutarylated protein isolates was obtained. More than 85% of the total protein was solubilized when 1% suspensions were used. Low heat coagulability was demonstrated by less than 5% of the protein in the suspension precipitated by heating at 95–100 °C for 30 min. The protein isolate of the control sample was moderately soluble (40%) in water and moderately heat-coagulable (50%). Solubility of protein isolate was not affected by acetylation. Highly sensitive protein isolate to coagulation was obtained by treatment with sodium sulfite. Choi et al. [166] have suggested a correlation of protein solubility with the amounts of negative charges, but not neutral charges of protein molecules. Increases in protein extractability may be related to rupturing protein bodies. Conversion of amino groups from positive to negative may also be res-ponsible for the increased extractability of proteins [167]. An increase in water solubility of cottonseed protein isolate was obtained depending on the concentra-tion of added succinic anhydride [168]. The protein solubility in unmodified pro-tein isolate was about 20% at room temperature when 2% protein suspension was prepared. However, the 100% sample was completely solubilized at the same pro-tein concentration. Protein solubility exponentially increases with the degree of succinylation of the protein with the sharpest increase at 95% succinylation or higher.

Oat Proteins

Oat proteins have a good nutritional quality; PER is 2.2 compared to casein that has a PER of 2.5. Wild oats contain more than 20% protein, 9% lipid, and 55% starch. Wild oats have provided meal, bran, and flour with 22%, 28%, and 18% protein, respectively. There is limited information on the chemical and functional properties of oat proteins. The four individual oat protein fractions exhibited different solubilities [169]. The highest solubility showed albumins over a pH range 1.5 to 10.5. The pH-solubility curve for the globulin fraction had a bell-shape with minimum at pH 6–7 and high solubility (over 90%) at the alkaline and acidic ends of the curve. The minimum solubility for prolamins was at pH 5–6, and it was higher at alkaline pH than at acidic pH. All oat protein fractions demonstrated higher solubility at alkaline pH; this property was utilized for protein extraction in the oat protein concentrate production.

The domestic and wild groat protein products exhibited low nitrogen solubility within the pH range 4–8 [170]. Steam treatment of the groats affected solubility. The unheated wild groat samples showed higher nitrogen solubilities than the heat treated samples. Oat proteins exhibited a broad range of insolubility over the pH range 3–6. The solubility of wild oat proteins increased from 11% at 0.01 M NaCl to 60% in 0.25 M NaCl solutions. Proteins in commercial meals and flours stabilized by heating to inactivate enzymes would be insoluble over the range of pH commonly used in foods.

Wheat Proteins

Cereal proteins are mostly used in their native form, and a small proportion is used in the form of protein concentrates and isolates. The protein in wheat is concentrated in the aleurone layer (7% of the weight of the kernel) and in the germ (2–3% of the weight of the kernel). However, in commercial conditions the level of protein in dry milled wheat protein concentrate (WPC) products is only in the range of 20–22% and utilization of WPC is limited by poor functionality. Different varieties of wheat have a various protein content, and this affects the functional properties of flours. Wheat proteins in isolated forms are commercially available. Wheat gluten has been produced and utilized because of its functional properties. Wheat gluten exhibits low solubility in aqueous solution. Wheat gluten is the visco-elastic mass composed mostly of proteins (about 70%) classified into two groups: 1) gliadins, soluble in aqueous alcohols and separated according to electrophoretic properties into α-, β-, γ- (sulfur-rich), and w-gliadins (sulfur-poor); 2) glutenins aggregated into complexes of high molecular weight. Gluten proteins are responsible for the visco-elastic properties of wheat dough. Gluten functionality in breadmaking can be understood if the amino acid sequence of gluten proteins, their conformation, and interaction are known. Tatham and Shewry [171] demonstrated that the sulfur-rich α-, β-, and γ-gliadins and sulfur-poor ω- gliadins have different secondary structures with different stabilization by disulfide bonds, hydrogen bonds, and hydrophobic interactions.

Most of the research efforts have focused on the gliadin proteins and the easily extractable glutenin. Proteins of wheat endosperm, gliadin and glutenin, have received most attention from the point of view of functionality in the processing of wheat into bread. The unique feature of the amino acid composition of glutenin is its high content of glycine, proline, glutamine, and leucine. The high content of glutamine is necessary for hydrogen bonding, and leucine promotes hydrophobic interactions. These amino acids are important in determining physical structure and functionality of glutenin. Albumins comprise up to 50% of the total protein of wheat flour solubles. Wheat flour solubles exhibited a typical U-shaped pH-solubility curve with minimum solubility at pH 6.0 [172]. A rapid decrease in nitrogen solubility was observed from pH 4 to 6 and a linear increase from pH 6 to 10.

Protein solubility of wheat germ protein flour (WGPF) at pH 4–8 and temperature of incubation 5–70 °C is presented in Fig. 1.20 [173]. A rapid increase in protein solubility was observed from pH 4 to 8 at all tested temperatures (7, 15, 30, and 70 °C). Overall, protein solubility was maximum at pH 8 and 70 °C and minimum at pH 4 and 5 °C. Electrostatic attractions, hydrophobic and hydrogen bonding in wheat germ protein may have been broken by the changing pH and temperature of the solution. The changes of protein hydrophilic properties and net charge were affected by pH changes. At pH levels 4–5 in the isoelectric range, the wheat germ proteins showed a gradual decrease in solubility due to a decrease in attractive forces between protein and water. Solubility of WGPF, corn germ protein flour (CGPF), soy flour (SF), non-fat dry milk (NFDM) and egg white protein (EWP) was influenced by protein concentration (Fig. 1.21) [173]. For all tested proteins, protein solubility increased with increase in protein concentration from 1 to 8%. Highest protein solubility was found for NFDM and EWP (1–8% level). The lowest protein solubility at levels 1–8% was found for CGPF and WGPF. Intermediate solubility was exhibited by SF. Sodium salts of long chain fatty acids were reported to be effective in solubilizing freeze-dried glutenin. Solubility of glutenin in water is markedly affected by sodium salts of long-chain fatty acids [174]. Glutenin was dissolved in water in the presence of relatively high concentrations of sodium palmitate or so-dium stearate. Salts, e.g. NaCl, inhibit the capacity of the sodium salts of long-chain fatty acids to dissolve glutenin. It is postulated that glutenin insolubility is mostly due to hydrophobic interactions that play an important role in the functio-nal properties of glutenin in breadmaking. The dissolving capacity of fatty acids increased with the length of the carbon chain. The most drastic change in solubi-lizing effect occurs in the C_{10}-$C1_2$ region. Glutenin was dissolved significantly with sodium stearate at the very low concentration of the soap used (1 mg/10 mg glutenin in 2 ml water). The effect of soaps on glutenin solubility can be related to the interaction between their hydrophobic chains and the hydrophobic regions of glutenin. Lipids contained in glutenin as contaminants are very difficult to remove and may participate in the solubilizing effect on glutenin.

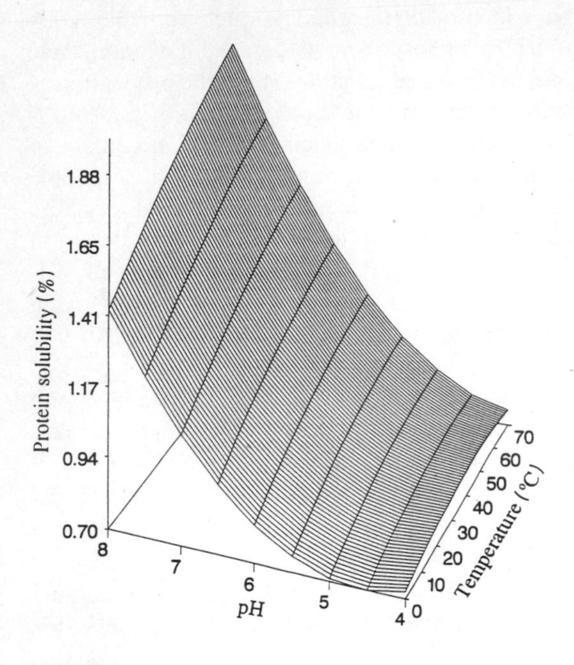

Fig. 1.20. Effect of pH and temperature on protein solubility of wheat germ protein flour. From Ref. 173

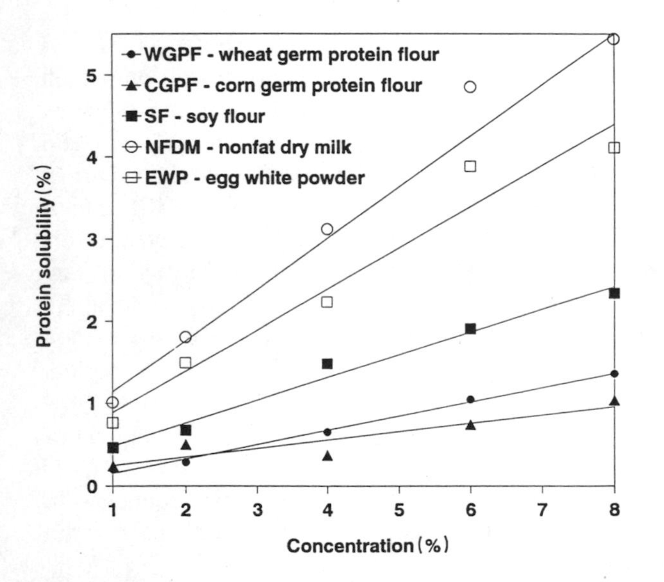

Fig. 1.21. Effect of concentration on protein solubility. From Ref. 173

Solubility of salt-soluble gluten decreased as heating time in boiling water increased (1.5–5.0 min) [175]. However, the ethanol-soluble and mercapto-ethanol-soluble fractions showed different sensitivities to heat treatment. With increased heating time, the amount of ethanol-soluble fraction decreased markedly and the amount of the mercaptoethanol-soluble fraction increased. The ethanol-soluble proteins are insolubilized through the formation of new bonds between polypeptide chains.

Sesame Proteins

Sesame protein hasnot been used as a food component and studies of its properties and functionality are limited. Sesame is cultivated for both oil and protein production. Defatted sesame protein flour contains nearly 50% protein. Sesame proteins have a unique amino acid composition, i.e. a high content of sulfur-containing amino acids. Sesame protein contains a high level of methionine (3.9–4.7 g/16 g nitrogen) [171]. Sesame proteins may become an important source of high-quality food protein. Maximum extraction of sesame protein was obtained at pH above 9.0 with water/meal ratios of 15:1. Sesame protein extractability was increased at pH 6.0 by addition of NaCl to obtain an ionic strength of up to 1.0 M. Solubility was much lower for expeller cake than for whole seed meal, because of heat denaturation of the protein during processing.

Safflower Proteins

The content of safflower protein in safflower seeds is about 15% and it represents a potential for food usage. Lysine is the first limiting amino acid in safflower protein. A high level of protein extraction and yield have been obtained at pH 8.5 or over. The minimum solubility of various safflower protein isolates was between pH 5 to 6 [177]. Protein isolates produced by the micellization technique had a higher solubility than the isolates obtained by the isoelectric precipitation procedure. However, the later protein showed a wider range of minimum solubility. The solubility profile was mostly determined by the isolation technique. The supernatants were diluted with deionized water and after 30 min incThe micellization procedure includes extraction in a 10% NaCl solution, at pH 5.8 and 7.0 for 30 min at 35 °C and centrifugation. ubation, protein was recovered by centrifugation. The pH-extractability profile of safflower protein showed a maximum extractability of 95% at pH 9.3 and a minimum of ~20% between pH 3 and 6 [178]. However, there was no sharp solubility minimum as in other seed proteins.

Linseed Proteins

The solubility of the following samples of linseed were reported [179]: low-mucilage flour (LMF); low-mucilage protein concentrate (LMPC); high-mucilage protein concentrate from seed (HMPC-S) and high-mucilage protein concentrate from expeller cake (HMPC-EC). pH-nitrogen solubility profiles for LMF and LMPC were similar (Fig. 1.22), especially in the acidic region [179]. LMPC showed lower nitrogen solubility than LMF in the alkaline region. Different nitrogen solubility was found for HMPC-S and HMPC-EC samples; the later had a lower nitrogen solubility than HMPC-S at all pH values. The nitrogen solubility profile vs. pH of linseed protein flour has been reported [180]. Linseed protein showed minimum solubility between pH 2 and pH 6 where solubility was about 20%. Solubility increased at pH above 6, and at pH 10 it was 80%. Heating at boiling temperature for 15 min considerably decreased protein solubility at all pH values. Addition of 0.5 M and 1.0 M NaCl caused shifting of the solubility minimum to lower values. However, nitrogen solubility of heated protein in 0.5–1.0 M NaCl or sodium hexametaphosphate was considerably lower (~5%) than in non-heated proteins (~22–24%). The decrease of protein solubility may be related to protein heat denaturation and to the leaching of some of the nonprotein nitrogen.

Alfalfa Proteins

The solubility data of alfalfa protein may be used as an indicator of denaturation during processing. Approval of protein concentrates derived from alfalfa has been impeded by objections related to very low solubility, green color, grassy flavor, and bitter taste. Leaf protein isolates and concentrates to be utilized as ingre-dients in conventional formulations and new food analogs should possess the fol-lowing critical functional properties: solubility, emulsifying and foaming proper-ties, and gelation. A white fraction of leaf protein concentrate was developed with good balance of amino acids and PERs not different from those of casein [181]. Nitrogen solubility profiles of leaf protein concentrate might be improved by spray-drying at lower temperatures. Solubility profiles of leaf protein concentrate were similar to freeze-dried alfalfa leaf protein if an outlet temperature of 85 °C was maintained in the spray-drier. A significant decrease in nitrogen solubility was observed with an outlet spray-drying temperature of 140 °C.The heating effect on the solubility of alfalfa protein is determined mostly by the pH of the medium [182]. Protein samples in solution at pH 2.3 and 8.6 were stable after heating to 100 °C for 15 min; whereas, the sample of the protein in solution at pH 6.9 became cloudy at 55 °C. The solubility of the protein isolate was affected by calcium ions at pH 2.3 and 8.9. Controlled addition of Ca ions can be utilized for limited aggregation of the alfalfa protein isolate.

Lupine Proteins

The differences in protein solubility between the species of lupine proteins have been found [183]. The amount of protein extracted with salt solution varied from 53% to 85%. Defatting of the lupine flour altered protein solubility by increasing the amount of globulins and decreasing the amount of albumins and prolamins.

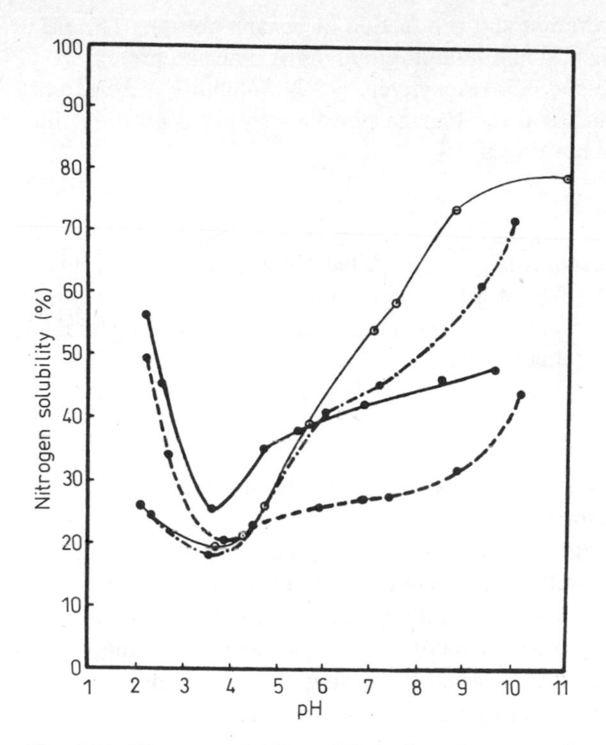

Fig. 1.22. Nitrogen solubility of linseed products as a function of pH: ○–○ LMF; ● – – ● LMPC; ● – ● HMPC-S; ● – – – ● HMPC-EC. From Ref. 179

Guar Protein

Protein isolate prepared from guar meal exhibited U-shaped nitrogen solubility-pH profiles with minimum of ~2% around the pI (pH 5) and very high solubilities of 80–90% at extreme acidic and alkaline regions [184].

Potato Proteins

The potato protein isolate contains 80–85% high quality protein [185]. Potato protein added to whey by wet-blending reduced solubility of composites at pH 4 and 5; that is related to a large proportion of the precipitating fraction in potato protein. Protein solubility of potato protein increased from 79.4% to 99.7% in the pH range 4 to 8. When proportions of potato protein to whey protein increased, the protein solubility of composites was reduced. However, no differences in protein solubility were observed at pH 6 and above [185].

Banana Protein

Content of protein in banana is in the range 4.85–5.00% on a dry weight basis which is one of the highest among fruits. Protein extractability and dispersibility are important in banana dehydration and production of banana powder. The pH-solubility curve indicated that albumin and globulin of banana protein had minimum solubility at pH 5.0 and 4.5, respectively [186]. Solubility increased at pH 4.0 and 7.0 which indicates that the banana powder may have potential for production of acidic or neutral beverages.

Conophor Protein

Conophor protein flour consumed in Nigeria exhibited 50% nitrogen solubility at pH 2 and much higher solubility at pH 9. Minimum nitrogen solubility of conophor protein was at pH 5.5 and 8.0. Two minimum solubilities suggest that there are two major proteins in conophor seeds.

Sorghum Protein

Insolubility of sorghum glutelins can be partly due to the interactions between proteins and nonprotein components. Most of the sorghum proteins are tightly aggregated mainly through hydrophobic bonds. Sorghum proteins could be solubilized up to 95% in distilled water in the presence of sodium salts of fatty acids. The percentage of solubilized proteins increased with the flour protein content. Solubility of sorghum proteins increased by increasing the amount of soaps. The efficiency of the soaps varied according to the length of their hydrophobic chain; those with shorter hydrophobic chains were more efficient to solubilize sorghum proteins [188]. The optimal concentration of soap was defined as the lowest amount of soap necessary to solubilize the highest amount of proteins. With sodium octodecanoate only 25% of sorghum proteins were solubilized, while with hexa-, tetra-, and dodecanoate, 31%, 70%, and 73%, respectively, of the total proteins were solubilized. Formation of the water soluble complexes between soaps of fatty acids and proteins is responsible for the solubilization of proteins. Protein solubility depended also on the stirring time; highest solubility was obtained after 15 hours of stirring. Protein dissolving capacity of a soap was not related to the pH value of the solution. The authors suggested that 95% of the proteins solubilized with soaps are bound noncovalently by hydrophobic interactions [188]. Extractability of sorghum proteins increased with an increase of the temperature at higher levels of soap concentrations.

Yeast Protein

The pH-solubility profile of phosphorylated yeast protein showed an improvement in solubility between pH 5.5 and 7 [189]. An increase in solubility of phosphorylated yeast protein compared to the yeast nucleoprotein was related to the increased hydration of the charged phosphoryl groups and the loosening of the protein structure. Phosphorylated yeast protein may be utilized as a thickening agent in various foods.

References

1. Kinsella, J. E. (1976). Functional properties of proteins in food: a survey, Crit. Rev. Food Sci. Nutr., 7: 219.
2. McGowen, J. C. and Mellors, A. (1979). Relationships between the solubility of amino acids, J. Appl. Biochem., 1: 423.
3. Hayakawa, S. and Nakai, S. (1985). Relationships of hydrophobicity and net charge to the solubility of milk and soy proteins, J. Food Sci., 50: 486.
4. Kinsella, J. E. (1979). Functional properties of soy proteins, J. Am. Oil Chem. Soc., 56: 242.
5. Aoki, H., Taneyama, O., and Inami, M. (1980). Emulsifying properties of soy proteins: characteristics of 7S and 11S proteins, J. Food Sci., 45: 534.
6. Li-Chan, E., Nakai, S., and Wood, D. F. (1984). Hydrophobicity and solubility of meat proteins and their relationship to emulsifying properties, J. Food Sci., 49: 345.
7. Betschart, A. A. (1979). Development of sunflower protein, J. Am. Oil Chem. Soc., 56: 450.
8. International Dairy Federation (1986). Circular 9.
9. Morr, C. V., German, B., Kinsella, J. E., Regenstein, J. M., Van Buren, J. P., Kilara, A., Lewis, B. A., and Mangino, M. E. (1985). A collaborative study to develop a standardized food protein solublitiy procedure, J. Food Sci., 50: 1715.
10. Goll, D. E. (1977). Muscle proteins, In Food Proteins (J. R. Whitaker and S. R. Tannenbaum, eds.), AVI Publishing, Inc., Westport, CT, p. 121.
11. Honikel, K. O., Fischer, C., Hamid, A., and Hamm, R. (1981). Influence of post-mortem changes in bovine muscle on the WHC of beef. Postmortem storage of muscle at 20 °C, J. Food Sci., 46: 1.
12. Buttkus, H. (1974). On the nature of the chemical and physical bonds which contribute to some structural properties of protein foods: a hypothesis, J. Food Sci., 39: 484.
13. Chen, M. T., Ockerman, H. W., Cahill, V. R., Plimpton, R. F. Jr., and Parrett, N. A. (1981). Solubility of muscle proteins as a result of autolysis and microbiological growth, J. Food Sci., 46: 1139.
14. Aberle, G. E. and Merkel, R. A. (1966). Solubility and electrophoretic behavior of some proteins of post-mortem aged bovine muscle, J. Food Sci., 31: 151.
15. Miller, A. T., Karmas, E., and Fu Lu, M. (1983). Age-related changes in the collagen of bovine corium: studies on extractability, solubility and molecular size distribution, J. Food Sci., 48: 681.
16. Asghar, A. and Yeates, N. T. M. (1979). Muscle characteristics and meat quality of lambs grown on different nutritional planes, 2. Chemical and biochemical effects, Agri. Biol. Chem., 43: 437.
17. Wu, J. J., Dutson, T. R., and Carpenter, Z. L. (1981). Effect of post-mortem time and temperature on the release of lysosomal enzymes and their possible effect on bovine connective tissue components of muscle, J. Food Sci., 46: 1132.
18. Cronlund, A. L. and Woychik, J. H. (1987). Solubilization of collagen in restructured beef with collagenases and a-amylase, J. Food Sci., 52: 857.
19. Bonnet, M. and Kopp, J. (1984). Essai d'attendrissage de la viande: de l'njection d'une collagenase bacterienne non pathogene sur la tendrate de muscles riches en collagene, Sci. Aliments, 4: 213.
20. Siegel, D. G. and Schmidt, G. R. (1979). Crude myosin fractions as meat binders, J. Food Sci., 44: 1129.
21. Theno, D. M., Siegel, D. G., Schmidt, G. R., and Norton, H. W. (1978). Meat massaging: The effects of salt, phosphate and massaging on cooking loss, binding strength and exudate composition in sectioned and formed ham, J. Food Sci., 43: 331.

22. Prusa, K. J. and Bowers, J. A. (1984). Protein extraction from frozen, thawed turkey muscle with sodium nitrite, sodium chloride, and selected sodium phosphate salts, J. Food Sci., 49: 709.
23. Knipe, C. L., Olson, D. G., and Rust, R. E. (1985). Effects of selected inorganic phosphates, phosphate levels and reduced sodium chloride levels on protein solubility, stability and pH of meat emulsions, J. Food Sci., 50: 1010.
24. Offer, G. and Knight, P. (1988). The structural basis of water binding in meat. Part 1: General principles and water uptake in meat processing, In Developments in Meat Science, 4th ed., Lawrie R. Elsevier, Applied Science Publishers, London, p. 63.
25. Paterson, B. C., Parrish, F. C., and Stromer, M. H. (1988). Effects of salt and pyrophosphate on the physical and chemical properties of beef muscle, J. Food Sci., 53: 1258.
26. Asghar, A. and Henrickson, R. L. (1982). Functional properties of food-grade bovine hide collagen in coarse bologna. 2. Effect on different protein fraction, J. Food Quality, 5: 271.
27. Etheridge, P. A., Hickson, D. W., Young, C. R., Landmann, W. A., and Dill, C. W. (1981). Functional and chemical characteristics of bovine plasma proteins isolated as a metaphosphate complex, J. Food Sci, 46: 1782.
28. Shahidi, F., Naczk, M., Rubin, L. J., and Diosady, L.L. (1984), Functional properties of blood globin, J. Food Sci, 49: 370.
29. Delaney, R. A. M. (1977). Protein concentrates from slaughter animal blood. II. Composition and properties of spray dried red blood cell concentrates, J. Food Technol., 12: 355.
30. Cheng, C. S. and Parrish, F. C. Jr. (1979). Heat-induced changes in myofibrillar proteins of bovine longissimus muscle, J. Food Sci., 44: 22.
31. Li-Chan, E., Nakai, S., and Wood, D. F. (1985). Relationship between functional (fat binding, emulsifying) and physicochemical properties of muscle proteins. Effects of heating, freezing, pH and species, J. Food Sci., 50: 1034.
32. Kenney, P. B., Henrickson, R. L., Claypool, P. L., and Rao, B. R. (1986). Influence of temperature, time and solvent on the solubility of corium collagen and meat proteins, J. Food Sci., 51: 277.
33. Bouton, P. E., Harris, P. V., and Ratcliff, D. (1981). Effect of cooking temperature and time on the shear properties of meat, J. Food Sci., 46: 1082.
34. Zayas, J. F. and Naewbanij, J. O. (1986). The influence of microwave heating on the textural properties of meat and collagen solubilization, J. Food Proc. and Pres., 10: 203.
35. Peng, I. C. and Nielsen, S. S. (1986). Protein-protein interactions between soybean b-conglycinin (B1-B6) and myosin, J. Food Sci., 51: 588.
36. Hayakawa, S., Ogawa, T., and Sato, Y. (1982). Some functional properties under heating of the globin prepared by carboxymethyl cellulose procedure. J. Food Sci., 47: 1415.
37. Kim, H. J., Loveridge, V. A., and Taub, I.A., (1984). Myosin cross-linking in freeze-dried meat, J. Food Sci., 49: 699.
38. Wagner, J. R. and Anon, M. C. (1986). Effect of frozen storage on protein denaturation in bovine muscle. II. Influence on solubility, viscosity and electrophoretic behaviour of myofibrillar proteins, J. Food Technol., 2: 547.
39. Nusbaum, R. P. Sebranek, J. G., Topel, D. G., and Rust, R. E. (1983). Structural and palatability relationships in frozen ground beef patties as a function of freezing treatments and product formulation, Meat Sci., 8: 135.
40. Park, J. W., Lanier, T. C., Keeton, J. T., and Hamann, D. D. (1987). Use of cryoprotectants to stabilize functional properties of prerigor salted beef during frozen storage, J. Food Sci., 52: 537.
41. Xiong, Y.L. and Blanchard, S.P. (1993). Functional properties of myofibrillar proteins from cold-shortened and thaw-rigor muscles, J. Food Sci., 58: 720.
42. Borderias, A. J., Colmenero, J. F., and Tejada, M. (1985). Viscosity and emulsifying ability of fish and chicken muscle protein, J. Food Technol., 20: 31.

43. Jiang, S. T., Hwang, D. C., and Chen, C. S. (1988). Effect of storage temperatures on the formatin of disulfides and denaturation of milkfish actomyosin (Chanos chanos), J. Food Sci., 53: 1333.
44. Jiang, S. T., Ho, M. L., and Lee, T. C. (1985), Optimization of the freezing conditions on mackerel and amberfish for manufacturing minced fish, J. Food Sci., 50: 727.
45. Owusu-Ansah, Y. J. and Hultin, H. O. (1992). Differential insolubilization of red hake muscle proteins during frozen storage, J. Food Sci., 57: 265.
46. Jiang, S. T. and Lee, T. C. (1985). Changes in free amino acids and protein denaturation of fish muscle during frozen storage, J. Agric. Food Chem., 33: 839.
47. Krivchenia, M. and Fennema, O. (1988). Effect of cryoprotectants on frozen whitefish fillets, J. Food Sci., 53: 999.
48. Jiang, S. T., Hwang, D. C., and Chen, C. S. (1988). Denaturation and change in SH group of actomyosin from milkfish during frozen storage at –20 °C, J. Agric. Food Chem., 36: 433.
49. Matsumoto, J. J. (1980). Chemical deterioratin of muscle proteins during frozen storage, ACS Symp. Series, 123: 95–124.
50. Stefanson, G. and Hultin, H. O. (1992). Fish muscle myofibrillar proteins are soluble in water, Book of Abstracts, IFT Annual Meeting, New Orleans, p. 43.
51. Pai, D. J. K. and Parkin, K. L. (1992). Reactivity of 14C-formaldehyde with proteins of minced frozen red hake muscle, Book of Abstracts, IFT Annual Meeting, New Orleans, p. 43.
52. Ang, J. F. and Hultin, H. O. (1989). Denaturation of cod myosin during freezing after modification with formaldehyde, J. Food Sci., 54: 814.
53. Owusu-Ansah, Y. J. and Hultin, H. O. (1986). Chemical and physical changes in red hake fillets during frozen storage, J. Food Sci., 51: 1402.
54. Vidya Sagar Reddy, G., and Srikar, L. N. (1991). Preprocessing ice storage effects on functional properties of fish mince protein, J. Food Sci., 56: 965.
55. Suzuki, T. (1981). Fish and Krill Protein: Processing and Technology, Applied Science Publishers LTD, London.
56. Smith, D. M. and Brekke, C. J. (1985). Enzymatic modification of the structure and functional properties of mechanically deboned fowl proteins, J. Agric. Food Chem., 33: 631.
57. Elkhalifa, E. A., Graham, P. P., Marriott, N. G., and Phelps, S. K. (1988). Color characteristics and functional properties of flaked turkey dark meat as influenced by washing treatments, J. Food Sci., 53: 1068.
58. Ozimek, G., Jelen, P., Ozimek, L., Sauer, W., and McCurdy, S. M. (1986). A comparison of mechanically separated and alkali extracted chicken protein fat functional and nutritional properties, J. Food Sci., 51: 748.
59. Lakritz, L., Carroll, R. J., Jenkins, R. K., and Maerker, G. (1987). Immediate effect of ionizing radiation on the structure of unfrozen bovine muscle tissue, Meat Sci., 20: 107.
60. Hajos, G. and Delincee, H. (1983). Structural investigation of radiation-induced aggregates of ribonuclease, Int. J. Radiat. Biol., 44: 333.
61. Simic, M. G. (1983). Isolation and characterization of radiation-induced aliphatic peptide dimers, Int. J. Radiat. Biol., 44: 231.
62. Schuessler, H. and Schilling, K. (1984). Oxygen effect in the radiolysis of protein. Part II: Bovine serum albumin, Int. J. Radiat. Biol., 37: 71.
63. Krumhar, K. C., and Berry, J. W. (1990). Effect of antioxidant and conditions on solubility of irradiated food proteins in aqueous solution, J. Food Sci., 55: 1127.
64. Zabielski, J., Kijowski, J., Fiszer, W., and Niewiarowicz, A. (1984). The effect of irradiation on technological properties and protein solubility of broiler chicken meat, J. Sci. Food Agric., 35: 662.
65. Hayashi, T., Biagio, R., Saito, M., Todoriki, S., and Tajima, M. (1991). Effect of ionizing radiation on sterility and functional qualities of dehydrated blood plasma, J. Food Sci., 56: 168.

66. Schmidt, R. H., Packard, V. S., and Morris, H. A. (1984). Effect of processing on whey protein functionality, J. Dairy Sci., 67: 2723.
67. Kinsella, J. E. (1984). Milk proteins: physicochemical and functional properties, CRC Crit. Rev. Food Sci. Nutr., 21: 197.
68. Morr, C. V. (1984). Production and use of milk proteins in food, Food Technol., 38: 39.
69. Schmidt, R. H. and Morris, H. A. (1984). Gelation properties of milk proteins, soy proteins and blended protein systems, Food Technol., 38: 85.
70. Hung, S. C. and Zayas, J. F. (1992). Protein solubility, water retention and fat binding of CGPF compared to milk proteins, J. Food Sci., 57: 372.
71. Pearce, R. J. (1987). Fractionation of whey proteins, Int. Dairy Fed. Bul., 212: 150.
72. Augustine, M. E. and Baianu, I. C. (1987). Basic studies of corn proteins for improved solubility and future utilization: a physicochemical approach, J. Food Sci., 52: 649.
73. Barbieri, R. and Casiraghi, E. M. (1983). Production of a good grade flour from defatted corn germ meal, J. Food Technol., 18: 35.
74. Li-Chan, E. (1983). Heat-induced changes in the proteins of whey protein concentrate, J. Food Sci., 48: 47.
75. Kosaric, N. and Ng, D. C. M. (1983). Some functional properties af milk protein calcium co-precipitates, Can. Inst. Food Sci. Technol. J., 16 (2): 141
76. de Haast, J., Morressey, P. A., and Fox, P. F. (1987). Protein co-precipitates from milk and blood plasma: preparation and investigation of some functional properties, J. Sci. Food Agric., 39: 253.
77. Konstance, R. P. and Strange, E. D. (1991). Solubility and viscous properties of casein and caseinates, J. Food Sci., 56: 556.
78. Van Hekken, D.L. and Strange, E.D. (1993). Functional properties of dephosphorylated bovine whole casein, J. Dairy Sci., 76:3384.
79. Lee, S. Y., Morr, C. V., and Ha, E. Y. W. (1992). Structural and functional properties of caseinate and whey protein isolate as affected by temperature and pH, J. Food Sci., 57: 1210.
80. Murphy, J. M. and Fox, P. F. (1990). Functional properties of as1-K- or b-rich casein fractions, Food Chem., 39: 211.
81. Lonergan, D. A. (1983). Isolation of casein by ultrafiltration and cryodestabilization, J. Food Sci., 48: 1817.
82. Moon, T. W., Peng, I. C., and Lonergan, D. A. (1989). Functional properties of cryocasein, J. Dairy Sci., 72:815.
83. Beck, A. M. (1981). The physical and chemical properties of whey proteins, Dairy Industries International, 46 (11): 25.
84. Pepper, D. and Pain, L. H. (1987). Concentration of whey by reverse osmosis-ultrafiltration, Bull. I.D.F., 212: 25.
85. Hsu, K. H. and Fennema, O. (1989). Changes in the protein functionality of dry whey protein concentrate during storage, J. Dairy Sci., 72: 829.
86. Burgess, K. J. and Kelly, J. (1979). Technical note: selected functional properties of a whey protein isolate, J. Food Technol., 14: 325.
87. Morr, C. V. and Foegeding, E. A. (1990). Composition and functionality of commercial whey and milk protein concentrates and isolates: A status report, Food Technology, 4: 100.
88. Slack, A. W., Amundson, C. H., and Hill, Jr., C. G. (1986). Nitrogen solubilities of β-lactoglobulin and α-lactalbumin enriched fractions derived from ultrafiltered cheese whey retentates, J. Food Proc. and Preserv., 10: 31.
89. Akita, E. M. and Nakai, S. (1990). Lipophilization of β-lactoglobulin: Effect on hydrophobicity, conformation and surface functional properties, J. Food Sci., 55: 711.
90. De Wit, J. N. and Klarenbeek, G. (1984). Effects of heat treatments on structure and solubility of whey proteins, J. Dairy Sci., 67: 2701.

91. Melachouris, N. (1984). Critical aspects in development of whey protein concentrate, J. Dairy Sci., 67: 2693.
92. Nakai, S. and Li-Chan, E. (1985). Structure modification and functionality of whey proteins: Quantitative structure – activity relationship approach, J. Dairy Sci., 68: 2763.
93. Modler, H. W. and Jones, J. D. (1987). Selected processes to improve the functionaltiy of dairy ingredients, Food Technol., 10: 114.
94. Douglas, Jr., F. W., Greenberg, R., Farrell, H. M., and Edmondson, L. F. (1981). Effects of ultra-high temperature pasteurization on milk proteins, J. Agric. Food Chem., 29: 11.
95. Kitabatake, N., Cuq, J. L. and Cheftel, J. C. (1985). Covalent binding of glycosyl residues to β-lactoglobulin: Effects on solubility and heat stability, J. Agric. Food Chem., 33: 125.
96. Watanabe, K., Hayakawa, S., Matsuda, T. and Nakamura, R. (1986). Combined effect of pH and sodium chloride on the heat-induced aggregation of whole egg proteins, J. Food Sci., 51: 1112.
97. Kakalis, L. T. and Regenstein, J. M. (1986). Effect of pH and salts on the solubility of egg white protein, J. Food Sci., 51: 1445.
98. Dyer-Hurdon, J.N. and Nnanna, I.A. (1993). Cholesterol content and functionality of plasma and granules fractionated from egg yolk. J. Food Sci., 58:1277.
99. Ball, Jr., H. R., Hamid-Samimi, M., Foegeding, P. M., and Swartzel, K. R. (1987). Functionality and microbial stability of ultrapasteurized, aseptically packaged refrigerated whole egg, J. Food Sci., 52: 1212.
100. Mizutani, R. and Nakamura, R. (1987). Emulsifying properties of a complex between apoprotein from hen's egg yolk low density lipoprotein and egg yolk lecithin, Agric. Biol. Chem., 51: 1115.
101. Gassman, B. (1983). Preparation and application of vegetable proteins from sunflower seed for human consumption. An approach, Nahrung, 27: 351.
102. Howard, P. A., Campbell, M. F., and Zollinger, D. T. (1980). U.S. Patent 4,234,620.
103. Elgedaily, A., Campbell, A. M., and Penfield, M. P. (1982). Solubility and water absorption of systems containing soy protein isolates, salt and sugar, J. Food Sci., 47: 806.
104. Wang, C. R. and Zayas, J. F. (1991). Water retention and solubility of soy proteins and corn germ proteins in a model system, J. Food Sci., 56: 455.
105. Hayakawa, S. and Nakai, S. (1985). Relationship of hydrophobicity and net charge to the solubility of milk and soy proteins, J. Food Sci., 50: 486.
106. Wagner, J. R. and Anon, M. C. (1990). Influence of denaturation, hydrophobicity and sulfhydryl content on solubility and water absorbing capacity of soy protein isolates, J. Food Sci., 55: 765.
107. Berardi, L. C. and Cherry, J. P. (1981). Functional properties of co-precipitated protein isolates from cottonseed, soybean and peanut flours, Can. Inst. Food Sci. Technol. J., 14(4): 283.
108. Yao, J. J., Tanteerataram, K., and Wei, L. S. (1990). Effects of maturation and storage on solubility, emulsion stability and gelation properties of isolated soy proteins, J.A.O.C.S., 67: 974.
109. Saio, K., Kobayakawa, K., and Kito, M. (1982). Protein denaturation during model storage studies of soybeans and meals, Cereal Chem., 59: 408.
110. Fukushima, D. (1980). Deteriorative changes of proteins during soybean food processing and their use in foods, In Chemical Deterioration of Proteins (J. R. Whitaker, ed.), American Chemistry Society, p.211.
111. Arce, C. B., Pilosof, A. M. R., and Bartholomai, G. B. (1991). Sodium dodecyl sulfate and sulfite improve some functional properties of soy protein concentrates, J. Food Sci., 56: 113.
112. Nakai, S., Ho, L., Tung, M. A., and Quinn, J. R. (1980). Solubilization of rapeseed, soy and sunflower protein isolates by surfactant and proteinase treatments, Can. Inst. Food Sci. Technol. J., 13: 14.

113. Hamada, J. S. and Marshall, W. E. (1989). Preparation and functional properties of enzymatically deamidated soy proteins, J. Food Sci., 54: 598.
114. Sung, H. Y., Chen, H. J., Liu, T. Y., and Su, J. C. (1983). Improvement of the functionality of soy protein by introduction of new thiol groups through a papain-catalyzed acylation, J. Food Sci., 48: 708.
115. Sung, H. Y., Chen, H. J., Liu, T. Y., and Su, J. C. (1983). Improvement of the functionalities of soy protein isolate through chemical phosphorylation, J. Food Sci., 48: 716.
116. Lawhon, J. T., Manak, L. J., Rhee, K. C., Rhee, K. S., and Lusas, W. L. (1981). Combining aqueous extraction and membrane isolation techniques to recover protein and oil from soybeans, J. Food Sci., 46: 912.
117. Hafez, Y. S., Mohamed, A. I., Singh, G., and Hewedy, F. M. (1985). Effects of gamma irradiation on proteins and fatty acids of soybean, J. Food Sci., 50: 1271.
118. Wang, L. C. (1984). Ultrasonic extraction of a heat-labile 7S protein fraction from autoclaved, defatted soybean flakes, J. Food Sci., 49: 551.
119. Furukawa, T. and Ohta, S. (1983). Solubility of isolated soy protein in ionic environments and an approach to improve its profile, Agric. Biol. Chem., 47(4): 751.
120. Basha, S. M. (1988). Resolution of peanut seed proteins by high-performance liquid chromatography, J. Agric. Food Chem., 36: 778.
121. Monteiro, P.V. and Prakash, V. (1994). Functional properties of homogeneous protein fractions from peanut (Arachis hypogala L.), J. Agric. Food Chem., 42: 274.
122. Kim, N. M., Kim, Y. J., and Nam, Y. J. (1992). Characteristics of functional properties of protein isolates from various peanut (Arachis hypogaea L.) cultivars, J. Food Sci., 57: 407.
123. Sundar, R. S. and Rao, D. R. (1978). Functional properties of native and acylated peanut proteins prepared by different methods, Lebensm.-Wiss. u.-Technol., 11: 188.
124. Cherry, J. P., McWatters, K. H., and Holmes, M. R. (1975). Effect of moist heat on solubility and structural components of peanut proteins, J. Food Sci., 40: 1199.
125. Chiou, R. Y. Y., Beuchat, L. R., and Phillips, R. D. (1985). Functional and physical property characterization of peanut milk proteins partially hydrolyzed by immobilized papain in a continuous reactor, J. Agric. Food Chem., 33: 1109.
126. Prinyawaiwatkul, W., Beuchat, L.R., and McWatters, K.H. (1993). Functional property changes in partially defatted peanut flour caused by fungal fermentation and heat treatment, J. Food Sci., 6: 1318.
127. Swanson, B. G. (1990). Pea and lentil protein extraction and functionality, J.A.O.C.S., 67: 276.
128. Paredes-Lopez, O., Ordorica-Falomir, C., and Olivares-Vazquez, M. R. (1991). Chickpea protein isolates: Physicochemical, functional and nutritional characterization, J. Food Sci., 56: 726.
129. Sefa-Dedeh, S. and Stanley, D. (1979). Cowpea proteins. 1. Use of response surface methodology in predicting cowpea (Vigna unguiculata) protein extractability, J. Agric. Food Chem., 27(6): 1238.
130. Koyoro, H. and Powers, J. R. (1987). Functional properties of pea globulin fractions, Cereal Chem., 64(2): 97.
131. Naczk, M., Rubin, L. J., and Shahidi, F. (1986). Functional properties and phytate content of pea protein preparations, J. Food Sci., 51: 1245.
132. 132.Sosulski, F. W. and McCurdy, A. R. (1987). Functionality of flours, protein fractions and isolates from field peas and faba bean, J. Food Sci., 52: 1010.
133. Idouraine, A., Jensen, S. B., and Weber, C. W. (1991). Tepary bean flour albumin and globulin fractions functional properties compare with soy protein isolate, J. Food Sci., 56: 1316.
134. Sathe, S. K., Deshpande, S. S., and Salunkhe, D. K. (1983). Functional properties of black gram (Phaseolus Mungo L.) proteins, Lebensm.-Wiss. u.-Technol., 16: 69.

135. Borhade, V. P., Kadam, S. S., and Salunkhe, D. K. (1984). Solubilization and functional properties of moth bean Vigna Aconitifolia (Jacq.) marechal and horse gram Macrotyloma uniflorum (Lam.) Verdc. proteins, J. Food Biochem., 8: 229.
136. Han, J. Y. and Khan, K. (1990). Functional properties of pinmilled and air-classified dry edible bean fractions, Cereal Chem., 67(4): 390.
137. Pilosof, A. M. R., Bartholamai, G. B., Chirife, J., and Boquet, R. (1982). Effect of heat treatment on sorption isotherms and solubility of flour and protein isolates from bean Phaseolus vulgaris, J. Food Sci., 47: 1288.
138. Gujska, E. and Khan, K. (1991). High temperature extrusion effects on protein solubility and distribution in navy and pinto beans, J. Food Sci., 56: 1013.
139. Sumner, A. K., Nielsen, M. A., and Youngs, C. G. (1981). Production and evaluation of pea protein isolate, J. Food Sci., 46: 364.
140. Narayana, K. and Narasinga Rao, M. S. (1984). Effect of acetylation and succinylation on the functional properties of winged bean (Psophocarpus tetragonolobus) flour, J. Food Sci., 49: 547.
141. Carbonaro, M., Vcchini, P., and Carnovale, E., (1993). Protein solubility of raw and cooked beans (phaseolus vulgaris): Role of the basic residues, J. Agric. Food Chem., 41: 1169.
142. Schaffner, D. W. and Beuchat, L. R. (1986). Functional properties of freeze-dried powders of unfermented and fermented aqueous extracts of legume seeds, J. Food Sci., 51: 629.
143. Canella, M., Bernardi, A., Castriotta, G., and Russomanno, G. (1984). Functional properties of fermented sunflower meal, Lebensm.-Wiss. u.-Technol., 17: 146.
144. Tasneem, R., Ramamani, S., and Subramanian, N. (1982). Functional properties of guar seed (Cyamopsis tetragonoloba) meal detoxified by different methods, J. Food Sci., 47: 1323.
145. Cheryan, M. (1980). Phytic acid interactions in food systems, CRC Crit. Rev. Food Sci. Nutr., 13: 297.
146. Nuzullo, C., Vignola, R., and Groggin, A. (1980). U.S. Patent No. 4,212,799.
147. Saeed, M. and Cheryan, M. (1988). Sunflower protein concentrates and isolates low in polyphenols and phytate, J. Food Sci., 53: 1127.
148. Canella, M. Castriotta, G., Bernardi, A., and Boni, R. (1985). Functional properties of individual sunflower albumin and globulin, Lebensm.-Wiss. u.-Technol., 18: 288.
149. Rossi, M. and Germondari, I. (1982). Production of a food-grade protein meal from defatted sunflower. II. Functional properties evaluation, Lebensm.-Wiss. u.-Technol., 15: 313.
150. Augustine, M. E. and Baianu, I. C. (1984). Analysis of amino acid composition of cereal and soy proteins by high-field carbon-13 nuclear magnetic resonance and ion-exchange chromatography, Proc. Fed. Amer. Soc. Exp. Biol., 43(3): 672.
151. Barbieri, R. and Casiraghi, E. M. (1983). Production of a food grade flour from defatted corn germ meal, J. Food Technol., 18: 35.
152. Christianson, D. D., Friedrich, J. P., List, G. R., Warner, K., Bagley, E. B., Stringfellow, A. C., and Inglett, G. E. (1984). Supercritical fluid extraction of dry-milled corn germ with carbon dioxide, J. Food Sci., 49: 229.
153. Lucisano, M., Casiraghi, E. M., and Barbieri, R. (1984). Use of defatted corn germ flour in pasta products, J. Food Sci., 49: 482.
154. Zayas, J. F. and Lin, C. S. (1989). Protein solubility of two hexane-defatted corn germ proteins and soy protein, J. of Food Proc. and Preserv., 13(3): 161.
155. Messinger, J. K., Rupnow, J. H., Zeece, M. G., and Anderson, R. L. (1987). Effect of partial proteolysis and succinylation on functionality of corn germ protein isolate, J. Food Sci., 52: 1620.
156. Lin, C. S. and Zayas, J. F. (1987). Protein solubility, emulsifying stability and capacity of two defatted corn germ proteins, J. Food Sci., 52: 1615.

157. Bera, M. B. and Mukherjee, R. K. (1989). Solubility, emulsifying, and foaming properties of rice bran protein concentrates, J. Food Sci., 54: 143.
158. Marco, C., Gactano, C., Adriano, B., and Renzo, B. (1985). Functional properties of individual sunflower albumin and globulin, Lebensm.-Wiss. u.-Technol., 18: 288.
159. Abdel-Aal, E. S. M., Yousssef, M. M., Adel-Shehata, A., and El-Mahdy, A. R. (1986). Extractability and functionality of rice proteins and their application as meat extenders, Food Chem., 20: 79.
160. Ohlson, R. (1985). Rapeseed, In New Food Proteins, v. 5 (A. M. Altschul and H. L. Wilcke, eds.), Academic Press, Inc., p. 339.
161. Dev., D. K. and Mukherjee, K. D. (1986). Functional properties of rapeseed protein products with varying phytic acid contents, J. Agric. Food Chem., 34: 775.
162. Thompson, L. U., Liu, R. F. K., and Jones, J. D. (1982). Functional properties and food applications of rapeseed protein concentrate, J. Food Sci., 47: 1175.
163. Nakai, S., Ho, L., Tung, M. A., and Quinn, S. R. (1980). Solubilization of rapeseed, soy and sunflower protein isolates by surfactant and proteinase treatments, Can. Inst. Food Sci. Technol. J., 13: 8.
164. Thompson, L. U. and Cho, Y. S. (1984). Chemical composition and functional properties of acylated phytate rapeseed protein isolate, J. Food Sci., 49: 1584.
165. Rahma, E. H. and Narasinga Rao, M. S. (1984). Gossypol removal and functional properties of protein produced by extraction of glanded cottonseed with different solvents, J. Food Sci., 49: 1057.
166. Choi, Y. R., Lusas, E. W., and Rhee, K. C. (1982). Effects of acylation of defatted cottonseed flour with various acid anhydrides on protein extractability and functional properties of resulting protein isolates, J. Food Sci., 47: 1713.
167. Choi, Y. R., Lusas, E. W., and Rhee, K. C. (1981). Effect of succinylation of cottonseed flour during protein extraction on the yield and some of the quality properties of protein isolates, J.A.O.C.S., 58: 1044.
168. Choi, Y. R., Lusas, E. W., and Rhee, K. C. (1983). Molecular structure and functionalities of protein isolates prepared from defatted cottonseed flour succinylated at various levels, J. Food Sci., 48: 1275.
169. Ma, C.-Y. and Harwalkar, V. R. (1984). Chemical characterization and functionality assessment of oat protein fraction, J. Agric. Food Chem., 32: 144.
170. Chang, P. R. and Sosulski, F. W. (1985). Functional properties of dry milled fractions from wild oats (Avenua fatua L.), J. Food Sci., 50: 1143.
171. Tatham, A. C. and Shewry, P. R. (1985). The conformation of wheat gluten proteins. The secondary structures and thermal stabilities of a-, b-, g-, and w-gliadins, J. Cereal Sci., 3: 103.
172. Oomah, B. D. and Mathieu, J. J. (1987). Functional properties of commercially produced wheat flour solubles, Can. Inst. Food Sci. Technol. J., 20(2): 81.
173. Bolnedi, V., Zayas, J.F. Influence of pH and temperature on water retention and protein solubility of wheat germ protein flour. Annual IFT Meeting, Chicago, 1993, p. 160.
174. Kobrehel, K. and Bushuk, W. (1977). Studies of glutenin. X. Effect of fatty acids and their sodium salts on solubility in water, Cereal Chem., 54(4): 833.
175. Jeanjean, M. F. Damidaux, R., and Feillet, P. (1980). Effect of heat treatment on protein solubility and viscoelastic properties of wheat gluten, Cereal Chem., 57(5): 325.
176. Rivas, R., Dench, J. E., and Caygill, J. (1981). Nitrogen extractability of sesame (Sesamum indicum L.) seed and the preparation of two protein isolates, J. Sci. Food Agric., 32: 565.
177. Paredes-Lopez, O. and Ordorica-Falomir, C. (1986). Production of safflower protein isolates: composition, yield, and protein quality, J. Sci. Food Agric., 37: 1097.
178. Latha, T. S. and Prakash, V. (1984). Studies on the proteins from safflower seed (Carthamus tinctorius L.), J. Agric. Food Chem., 32: 1412.
179. Dev, D. K. and Quensel, E. (1988). Preparation and functional properties of linseed protein products containing differing levels of mucilage, J. Food Sci., 53: 1834.

180. Madhusudhan, K. T. and Singh, N. (1985). Effect of heat treatment on the functional properties of linseed meal, J. Agric. Food Chem., 33: 1222.
181. Knuckles, B. E. and Kohler, G. O. (1982). Functional properties of edible protein concentrates from alfalfa, J. Agric. Food Chem., 30: 748.
182. Fiorentini, R. and Galoppini, C. (1981). Pilot plant production of an edible alfalfa protein concentrate, J. Food Sci., 46: 1514.
183. Oomah, B. D. and Bushuk, W. (1983). Characterization of lupine proteins, J. Food Sci., 48: 38.
184. Tasneem, R. and Subramanian, N. (1986). Functional properties of guar (Cyamopsis tetragonoloba) meal protein, J. Agric. Food Chem., 34: 850.
185. Jackman, R. L. and Yada, R. Y. (1988). Functional properties of whey-potato protein composite blends in a model system, J. Food Sci., 53: 1427.
186. Mowlah, G., Takano, K., Kamoi, I., and Obara, T. (1982). Physico-chemical properties and protein behaviour of banana as effected by processing treatments and conditions, Lebensm.-Wiss. u.-Technol., 15: 211.
187. Ige, M. M., Ogunsua, A. O., and Oke, O. L. (1984). Functional properties of the proteins of some Nigerian oilseeds: Conophor seeds and three varieties of melon seeds, J. Agric. Food Chem., 32: 822.
188. Fliedel, G. and Kobrehel, K. (1985). Studies on sorghum proteins. 1. Solubilization of proteins with soaps, J. Agric. Food Chem., 33: 303.
189. Huang, Y. T. and Kinsella, J. E. (1986). Functional properties of phosphorylated yeast protein: Solubility, water holding capacity, and viscosity, J. Agric. Food Chem., 34: 670.

Chapter 2

Water Holding Capacity of Proteins

2.1 Introduction

The interaction of proteins with water has been expressed interchangeably by using the following terms: water hydration and holding, water retention, water binding, water imbibing, water adsorption, and others. There is controversy regarding the exact meaning of these terms. There is no standardized terminology in protein functionality, for example, bound water is the water which is not freezable at –40 °C. However, bound water is also considered not separable by centrifugation.

The water holding capacity (WHC) of foods can be defined as the ability to hold its own and added water during the application of forces, pressing, centrifugation, or heating. Hermansson [1] defined WHC as a physical property and is the ability of a food structure to prevent water from being released from the three-dimensional structure of the protein. The level of protein hydration and the viscosity of liquid systems in food are interrelated. Water retention is the water adsorbed or retained by a wet or dry mixture of components, for instance, protein or starch. It is one of the hydration properties that determine applications of proteins in food systems. WHC of proteins is the ability to retain water against gravity physically and physicochemically. Most important functional properties of proteins are related to their interaction with water. Consequently, protein-water interactions determine functional properties of proteins in foods such as: water binding and retention, swelling, solubility, emulsifying properties, viscosity, gelation, and syneresis. Water binding capacity is a limiting factor in protein food applications. Water retention is the preliminary step in many food manufacturing operations. WHC plays a major role in the formation of food texture, especially in comminuted meat products and baked doughs. Water retention of plant proteins used as additives in various foods influences quality characteristics of the finished food products.

For incorporation into foods, new protein additives should possess certain water binding and swelling capacity, which vary for different protein materials. Swelling is an important protein functional property because most foods are water-swollen systems. Swelling of proteins is the first step in their solvation and can be defined as the spontaneous uptake of a water by a protein matrix. In processed food formulations when proteins are replaced, WHC of various protein ingredients must be determined. Protein ingredients with excessively high WHC may dehydrate other formula components.

Water-protein interactions were considered important in protein functionality, but there is superficial knowledge on the nature of „bound" water and its relation to protein solubility. Water retention is a critical factor in protein functionality because it affects the texture, color, and sensory properties of products. Functional properties, eating quality, and storage stability of protein foods are related to the water content and water binding. The capacity of protein preparations of animal and plant origin to retain moisture may be transferred to processed foods when these proteins were incorporated.

The water retention is important in formulated foods and can affect the order of dry ingredients incorporation into the formula. The rate of water retaining can be used to determine whether a protein should be added as powder or rehydrated before the addition to the mixture. Information related to protein powder-moisture interaction is utilized in determining and developing kinds of packaging materials necessary to maintain the required moisture content in the product. There is still much to be learned about protein-water interactions that are important to protein structure and functionality.

Measurement of WHC and various types of water in food proteins is necessary to study protein-water interactions and the functionality of proteins. WHC is characterized by the amount of water held by a protein powder or solid material in the presence of excess water. Protein-water interactions in foods may be studied with the aid of the so-called absorption isotherm, which shows the amount of the water sorbed by the protein (g H_2O/g protein) as a function of the relative water vapor pressure. A standard test for determining water hydration capacity of proteins has been approved by the AACC [2]. According to this procedure, distilled water is added incrementally to a weighed amount of sample, and a homogeneous paste is produced by vigorous stirring. The paste is then centrifuged at 2000 g for 10 min before the supernatant liquid is discarded and the slurry weighed. This method fulfills most of the criteria for a simple model test, although the stirring procedure is somewhat time-consuming.

2.2 The Mechanism of Protein-Water Interaction

The character of protein-water and protein-protein interactions will determine whether the protein will function in the food as a gel, insoluble precipitate, or a colloidal dispersion. In order to predict the water binding of proteins, researchers have attempted to determine the number of water molecules bound per protein molecule. Water binding depends on the composition and conformation of the protein molecules. The mechanism of protein-water interactions is important in many natural and formulated foods, and different aspects of the physical chemistry of water retention by food proteins have been studied. Water interacts with proteins in a number of ways, and significant amounts of water bounded by proteins is retained by hydrogen bonding. Interactions between molecules of water and hydrophilic groups of the protein side chains occurs via hydrogen bonding. Structural water is held by hydrogen bonding between polypeptide groups of the proteins. Binding of water to proteins is related to the polar hydrophilic groups, such as imino, amino, carboxyl, hydroxyl, carbonyl, and sulfhydryl groups. The capacity of proteins to retain moisture is affected by the type and number of these polar groups in the protein polypeptide chain. The binding of water is due to the dipolar character of water. Proteins that contain numerous charged amino acids will tend to bind large amounts of water. Water binding of proteins can be predicted from their amino acid composition. The absorbed water is tightly bound to the protein molecules.

The specific sites of the protein molecules where water is bound were found by studying polypeptides of individual amino acids using NMR and frozen samples [3]. According to these findings, nonpolar amino acid side chains such as those of alanine and valine bind one water molecule. At the same time, polar side chains bind 2 or 3 water molecules, and ionic side chains (in aspartic and glutamic acids, and lysine) bind 4 to 7 water molecules/molecule of amino acid. Amino acids have been classified according to their ability to bind water into: 1) polar amino acids with the highest water binding; 2) nonionized amino acids, binding intermediate amounts of water; and 3) hydrophobic groups which bind little or no water. The polar amino groups of protein molecules are the primary sites of protein-water interactions that participate in binding different amounts of water at cationic, anionic, and nonionic sites. Polar amino acid side chains have specific water binding affinity and capacity.

Absorbed water surrounds the protein molecules with several layers of water tightly bound to specific sites in the protein molecules; another layer of water more loosely bound, covers the surfaces of protein molecules; and further layers of essentially nonstructured water, surround the adsorbed layer. A significant part of the water phase consists of water molecules that have lost motional freedom relative to free water. Hydration water is not bound in a true chemical sense, but motionally restricted due to interactions with the macromolecules, it is unfreezable water. This water makes up a nonsignificant part of the total water in gels and fresh tissue foods. Proteins have been found to contain 0.3–0.5 grams of unfreezable water per gram of protein [3]. In raw meat ca. 4–5% of the total water was considered as „bound“ by proton NMR measurements. Hydration water is bound to either myosin or actin, depending on the charge of the protein.

Capillary held water influences the sensory properties of foods. Knowledge about the pore size distributions in food systems and influence of the phenomenon on water retention is very superficial. Loosely bound water in bigger pores may be significant for sensory characteristics such as juiciness and flavor release. This water is lost during processing, especially during heating. Significant heating losses will impair the sensory quality of the product. The smaller the size of the pores, the more evenly they are distributed in the structure of the product, and the better will be the WHC of the whole system. The force of water binding is high enough if the size of pores in gels are in the range 1–100 nm. Separation of water in systems with such a size of pores is due to the syneresis process as a result of protein-protein interactions and aging of proteins. Those processes cause a spontaneous shrinkage of the gel structure without the effect of any external force.

Wettability is an important property when protein powders are dispersed to produce aqueous beverages and batters. The quality of batters is influenced by the rate of protein hydration to produce a smooth and uniform dispersion. In dried powder foods, such as protein concentrate or isolate, we should be concerned mainly with water absorption or water uptake. When a dry protein is exposed to water vapor, a monolayer coverage is formed due to absorption of water molecules onto all available surface polar sites. While water absorption is progressing multilayers of water are formed followed by water-water interaction and solubilization of protein particles. The rate of wettability of protein powders is determined by the surface activity of dry particles, size of particles, and the amount of polar, hydrophilic groups. Native proteins possess higher wettability than denatured.

The size, shape and surface of protein particles as related to interaction with water can be examined by scanning electron microscopy and internal microstructure by transmission electron microscopy. At low water activity (<0.3) each polar group in a protein molecule binds one molecule of water up to the monolayer level [4]. This monolayer is referred to as the BET monolayer because the theoretical calculation of its value was first proposed by Brunauer, Emmet and Teller in 1938. As relative humidity and water activity is increased, the polar groups bind more water, forming multilayers of water around polar groups of proteins.

The unique property of bound water is that it does not freeze at normal freezing temperature and is called unfreezable water. This property of bound water is utilized for measuring bound water content in protein foods. The physical properties of bound water are different from those of „free“ water. The bound water has a lower vapor pressure and lower dissolving power than normal water. „Free“ water freezes at the same temperature as normal water, has the same solvent power, and there is no indication that it is tightly bound to the protein molecules.

2.2.1 Factors Influencing Water Binding of Proteins

Water binding of a protein is influenced by protein concentration, pH, ionic strength, temperature, presence of other components of foods such as hydrophilic polysaccharides, lipids and salts, rate and length of heat treatment, and conditions of storage.

Protein Concentration

Water absorption was attributable to the protein content of the product and viscosity increased exponentially as the protein content increased. Water absorbtion of proteins increased as protein content increased. The amount of water retained by proteins depends upon the amino acid composition, especially the number of polar groups exposed for water binding, conformation of proteins, surface hydrophobicity, and processing history [5]. The amount of bound water increased with protein concentration of protein preparations, and the highest amount of bound water is during water interaction with protein isolate. Food processing conditions accelerate the interactions of the proteins with other food components such as lipids, carbohydrates, and water. Animal proteins have better WHC than vegetable proteins, probably because of the high level of amide nitrogen.

Effect of pH

Water retention is strongly influenced by pH, being least in an isoelectric region where the net protein charge on the protein is zero and protein-protein interactions are maximal. Change in pH of protein solution is causing a change of charged groups on the protein molecule. Changes of pH can affect the conformation of proteins resulting in exposure or burial of the water binding sites. Consequently, with an increased polarity of proteins, the amount of bound water increased. Change in pH influenced the ionization of amino acid groups.

Effect of Ionic Strength

Swelling characteristics of soy isolate, Na caseinate, and whey protein concentrate are of an entirely different character. Swelling is strongly dependent on ionic strength of the system. Addition of salts influences water binding by proteins because of their effects on electrostatic interactions. At salt concentrations higher than 2.0 M there is a decrease in bound water as ions compete with protein groups for water. At high NaCl concentrations, protein dehydration will occur due to the competition of solutes and proteins for the available water. As a result of suppressing the electrical double layer around the protein molecules, there is an alteration of protein conformation, reduction of protein hydration and precipitation („salting out“).

Effect of Heating

Hydration of proteins and water retention capacity is also influenced by conformational changes of proteins during food processing, particularly during heating. Protein water retention usually decrease with an increase in temperature. However, if during heat treatment, changes of protein conformation were found, the effect of temperature on water absorption can be overridden. Commercially produced protein additives are subjected to heating and denaturation. Consequently, the influence of heating on protein water retention will be affected by conditions of processing treatment and product origin. Unfolding of the polypeptide chains due to denaturation and transition of the globular conformation to a random coil conformation may result in a reduction of the availability of polar amino acid groups for binding water.

2.3 Water Holding Capacity of Proteins in Meat and Meat Products

2.3.1 Water Binding Capacity of Muscle Proteins

Water holding capacity of meat and meat products is the ability to absorb and retain water during mechanical treatment (chopping, coarse grinding, comminution, stuffing), thermal treatment, and subsequent transportation and storage. WHC influences the quality of meat and meat products: juiciness, tenderness, taste, and color, which are important in meat processing. Sensory testing of juiciness is related to the impression of moisture running out of the meat as pressure by the teeth is applied. This mouth sensation is determined in sensory testing by the free and bound water. However, perception of juiciness is influenced also by such factors as fat content, flow of saliva, and softness.

Because proteins are primarily responsible for WHC of meat, an increase in cooking losses would be expected with an increase in the fat level. The decrease of WHC with an increased fat level was related to the increase of the moisture to protein ratio. However, high intramuscular fat content tended to yield higher WHC than one of low fat content. Such a relationship has been observed in curing as well as in the freezing and thawing of meat (drip loss). In ground meat, fat apparently increases WHC, e.g. in processing sausages of the frankfurter type. In lean muscle tissue water content is about 75%, of which approximately 5% is bound water, and 70% is entrapped bulk phase water [6].

Interactions between meat proteins and water significantly affect the textural properties of meat. WHC is a sensitive indicator of changes in the charge and structure of muscle proteins.Pork meat has a better WHC than beef and retail cuts of beef have higher drip than cuts of pork. Meat obtained from young animals had higher WHC than from older ones. The meats with highest WHC are the skeletal muscles; the medium WHC meats are those such as tongues and hearts.

Mechanism of Water Binding by Muscle Proteins

The state of free or bound water is affected by the molecular arrangement of myofibrillar proteins. Three-dimensional network of filaments in myofibrils provides an open space for water to be immobilized. The decrease of immobilized water is observed as a result of tightening the space between the myofibrils of the network as a result of contraction or protein denaturation [7].

Content of myosin in various meats varies and some meats have a high content of myosin. White muscle myosin from beef exhibited higher WHC and solubility below pH 5.7 than red muscle myosin [8]. These properties were related to the presence of myosin isoforms in different muscle fibers. The tightly bound water in meat is given as 5–15 g water per 100 g tissue [9]. Most of the water in muscular tissue may be termed „immobilized" or „entrapped bulk phase water" which is held in the lattice spaces between the myofibrils [6]. WHC of lean muscular tissue is significantly determined by the spatial arrangement of protein filaments.

The main component of meat structure are myofibrils which occupy about 70% of the volume of meat. Consequently, the majority of water is retained in myofibrils in the spaces between the thick and thin filaments. Offer and Trinick [10] have made microscopic measurements of the size of single myofibrils in salt solutions of varying concentrations. The effect of pH on the swelling of myofibrils was established. When pH was raised from 7 to 9 the myofibrils swelled and when the pH was lowered from 7 to 5 they shrank by a similar amount. Because bound water includes 0.05 g H_2O/1 g protein only a small fraction of total water is bounded by proteins.

The significant contribution to our understanding of the mechanism of water binding by meat proteins was presented in the works of Hamm [6], Goll [9], and Offer [11]. They suggested that water holding in muscle tissue is due to changes in the intensity of swelling of the myofibrils. Swelling is utilized as the principle of a method for the determination of WHC. The increase in WHC is related to swelling of the myofibrils caused by expansion of the filament lattice. As a result of increased swelling the myofibril diameter increases to 2.5 times the original value. Increase in the myofibril volume caused by extensive swelling is mostly responsible for water binding in meat and for water losses during meat processing. An important factor influencing WHC is swelling of the protein matrix when shrinkage of the protein matrix, for example during heat treatment, results in decreasing WHC. Shrinkage of the myofibrils and muscular tissue causes a decrease in water retention in rigor meat, in the pale, soft, and exudative condition, and during heat treatment.

The meat proteins, especially myosin, actin and to some extent tropomyosin, are the main water-binding components in muscular tissue. Myofibrillar proteins comprise more than 55% of the total protein content of the muscular tissue. Myofibrillar proteins are responsible for water retention in meat, and approximately 97% of the WHC of meat is related to the myofibrillar protein fraction. In meat muscles with low content of connective tissue, myofibrillar proteins contribute significantly to meat tenderness and toughness. There is a relationship between the state of myofibrillar proteins and meat tenderness. High content of acidic and basic amino acids impairs a high electrical charge to these proteins and determines a high water binding capacity. The extent of the protein molecule hydration is basically the sum of the hydration of the amino acid side chains. Water-binding capacity of myosin is related to large amounts of polar amino acids with a large content of aspartic and glutamic acid residues.

A major role of myofibrillar proteins in water retention has been reported by Goll [9]. Entrapment of water in the lattice spacings between the protein filaments of the myofibrils is most important. WHC of meat is determined mostly by the size of the lattice spacings between thick and thin filaments in myofibrils. Only 40–80 g out of 290–380 g of water/100 g protein in muscle is directly bound to proteins and the remaining 250–300 g water/100 g protein is found in the thick and thin filament lattice [9]. From the total amount of water bound in muscle proteins (40–80 g H_2O/100 g protein) only about 10–12% is tightly bound as a monomolecular layer around the thin and thick filaments, another 10–12% water is immobilized as a second layer outside the first layer and 50–62% is loosely bound in two to three layers outside the second layer of immobilized water. The loosely bound water freezes at storage temperatures below 0 °C and is affected by denaturation of myofibrillar proteins. Tightly bound water remains unfrozen at the subzero temperatures. Sarcoplasmic proteins have low water-binding capacity, and low viscosity.

The collagen network in a muscle is stretched by the muscle fiber and is under tension; consequently, muscle fiber is compressed by the collagen network. During rigor mortis, as a result of fiber shrinkage, the collagen network will act on the aqueous solution released from the fiber, and at the cuts of meat, water will appear as drip. However, connective tissue proteins change little during postmortem storage and the possibility of improving meat quality is related basically to myofibrillar proteins. According to Sadowska et al. [12] increased levels of connective tissue produced an increase in cooking loss and influenced firmness of the finished products. Factors that influence the connective tissue's effect on the WHC of meat in meat processing are found to include physiological conditions of the animals, composition of connective tissue and quantity in meat, histological and microphysical structure, pH, added salts, extent of grinding temperature (transformation to gelatin), effect of connective tissue on grinding and on rate of heat transfer, which is normally slower in connective tissue than in muscle tissue [1].

2.3.2 Factors Influencing Water Binding of Muscle Proteins

The Effect of Rigor Mortis and Aging; Effect of pH

The capacity of meat and meat products to retain water is influenced by the characteristics of animals (species, sex, age) and conditions of postmortem treatment (rigor mortis and aging), chilling and freezing, conditions of massaging, blending, comminution, salting, heating, drying, etc. Water retention of muscle is dependent on the volume of the myofibril, pH of meat, reaching a minimum at the isoelectric point and rising on either side of it. A minimal level of water binding capacity was found at pH around 5.0–5.1, and was related to isoelectric points of the quantitatively most important proteins of myofibrils, myosin, pH 5.4 and actin, pH 4.7. If pH is above or below isoelectric point, total negative and positive net charge increases. At the pI protein-protein interactions are maximal, with protein aggregation, and minimal hydration and swelling. The influence of pH on hydration during spoilage of pork at refrigeration temperature has been reported [13]. Hydration capacity of meat proteins during spoilage depended on the pH of raw meat and the extent of the change in pH during storage.

Variation of WHC between muscles might be expected because of types of muscle fibers and the degree of fiber contraction. WHC of meat is also influenced by the rate of carcass cooling after slaughter. Deeper layers of muscle tissue may have a lower pH. Comparative studies of WHC of beef muscles with different levels of connective tissue, 2% in Psoas and 8% in Trapezius, have shown approximately equal WHC [14]. The water binding of connective tissue proteins is less sensitive to change of pH, and is constant at pH between 5.6 and 7.0. At lower pH connective tissue possesses notably higher WHC due to a decrease in the intramolecular hydrogen bonds in the molecules of tropocollagen.

WHC of meat is influenced by rigor mortis. It is very high after slaughter and decreases rapidly during the first 24 to 48 hours after slaughter. At this stage, formation of lactic acid and decrease in pH has effects on decreased WHC. When meat is in the rigor mortis state, there is a loss of water in the form of drip, which occurs mainly from the cut ends of the meat. The shrinkage of the myofibrils during rigor mortis and the release of water negatively influences meat quality and produces significant drip.

The rigor mortis process and actomyosin formation is developed as a complex biochemical process leading to the formation of crosslinks between the myosin and actin filaments of the myofibril. Postmortem glycolysis and rigor state is detrimental to water binding of proteins. WHC of muscle proteins is notably affected by rapid fall of pH after death or by excessively low pH values. The pH of beef muscle in the state of rigor mortis declines to about 5.5–5.8, causing a decrease in the solubility of myofibrillar proteins and a decrease in WHC. Denaturation of the muscle proteins at low pH's is influencing their water binding properties. Decrease in WHC of muscles is determined by the level of protein denaturation which has occurred in the immediate postmortem period in the myofibrillar proteins, mainly as a result of the rate of rapid pH fall postmortem. The effect of muscle contraction during rigor mortis is considerable. At the same time, postmortem metabolic processes resulting in decreased pH influence WHC more significantly through alterations in protein structure than does the degree of contraction [15].

The functional property of proteins in fresh meat is affected by the proteolytic changes during aging process. WHC of meat proteins increases during aging as a result of enzymatic cleavage of the peptide bonds and release more polar groups active for water binding. During aging of meat, a moderate increase in pH is observed and as a result, an increase in protein solubility and WHC of muscle proteins. An increase in WHC of muscular tissue during aging is related to an increase in pH and an increase in repulsive charges of protein molecules. As a result the matrix of proteins will swell and WHC will increase.

Watcriness of meat is separation of the water on the fresh cut surfaces due to decrease of WHC of myofibrillar and sarcoplasmic proteins. In polypeptide chains of these proteins a large amount of water is bound by hydrogen bonds to polar groups. Water binding of proteins is influenced by their solubility and state; WHC decreases if polar, hydrophilic groups are blocked.

The Effect of Freezing and Frozen Storage

Decrease in WHC of frozen meat was related to insolubilization of sarcoplasmic proteins and actomyosin. Comparative studies of two freezing methods showed that slow freezing caused a larger loss of drip on thawing with a larger content of nitrogenous constituents and nucleic acid derivatives to the drip and more significant loss of WHC, than fast freezing. The rate at which changes of proteins occur during frozen storage is mainly dependent on the storage temperature. However, other factors such as rate of freezing and post-storage conditions are also important. Miller et al. [16] demonstrated a significant decline in the capacity of meat proteins to retain water during extended frozen storage. In the production of comminuted meats frozen meat is directly incorporated into thc formulation without thawing.

Frozen storage of fish muscular tissue during 180 days at –18 °C caused an increase in thaw drip of all samples which reflects the degree of protein denaturation [17]. The inverse relationship existed between protein solubility and absorbed moisture in fresh, 3-day, and 5-day ice-stored samples. A decrease in WHC was observed during storage at –18 °C for 180 days that caused an increase in drip loss and cook loss. A marked increase in cook loss was found during the first 60 days of frozen storage of minced fish meat. This increase was affected by the duration of ice storage prior to freezing. An increase in cook loss was related to low WHC of proteins due to their denaturation during frozen storage. The effect of cryoprotectants on the functional properties of salted prerigor and postrigor beef determined after 2, 4, and 6 months of frozen storage at –28 °C has been reported [18]. Added cryoprotectants were 5.6% polydextrose or a mixture of sucrose and sorbitol. The addition of cryoprotectant was much more effective on the percentage of cooking loss for prerigor than for postrigor muscle. Cooking loss increased from 4.5% to 8.1% over 6 months storage in the control, and only from 4.1% to 5.0% in the polydextrose containing samples. Gels have been prepared from prerigor and postrigor muscles stored frozen for 6 months at –28°C. A higher percent of free water occurred in gels prepared from postrigor muscle (Fig. 2.1) [18]. The level of free water remained almost constant prior to 4 months frozen storage. In all postrigor treatments, the percent of free water increased during 6 months frozen storage. The same trend was found in the values obtained for levels of adjusted free water. Addition of cryoprotectants was effective in reduction loss of gelling capacity due to protein denaturation and/or aggregation. Addition of NaCl to comminuted muscle and 6 months frozen storage accelerated destabilization of the system in terms of WHC, gelling capacity, and protein solubility.

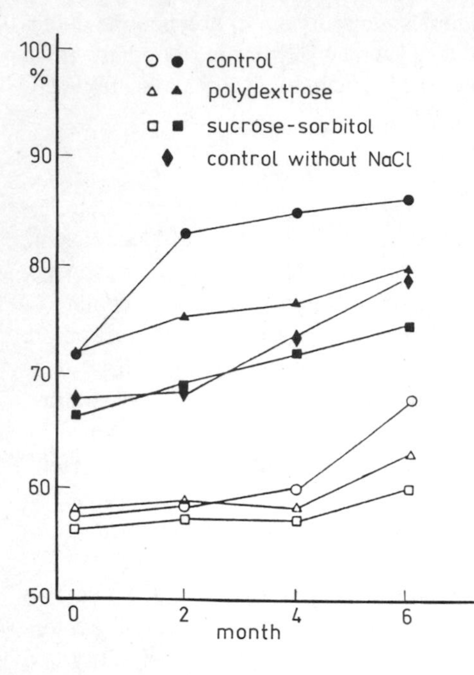

Fig. 2.1. Changes in % free water in cooked gels during frozen storage (○ = prerigor muscle, ● = postrigor muscle). From Ref. 18

The Effect of Heat Treatment

The effect of heat on the water binding within meat influences its quality properties. Juiciness is affected mostly by the amount of bound water in tissues. WHC of meat is closely related to taste, tenderness and color. In the raw meat tissue a certain amount of water is bound to proteins and the remainder of the moisture is free or loose. Heating of meat at temperatures higher than 60°C caused a decrease in tenderness due to the drying, hardening, and coagulation of the myofibrillar proteins. A decrease in tenderness may originate from the loss of water of hydration which is held in the thick and thin filaments.

When muscular tissue is heated, denaturation of myofibrillar proteins is observed followed by their coagulation, shrinkage of the myofibrils, and a tightening of the microstructure. The amount of free water in muscular tissue and the level of cooking loss is affected by temperature and cooking time. The process of collagen shrinkage is probably responsible for squeezing hydrated water from muscular tissue with conversion to the free form. The most tightly bound „hydration water“ is mostly bound to myosin and actin, depending on the shape and charge of the protein. The major percentage of the total water in meat is represented by free water. „Hydration water“ is bound primarily to hydroxyl, carboxyl, and amino groups of meat proteins. „Free water“ is immobilized within the muscle tissues and is influenced by the spatial molecular arrangement of the myofibrillar proteins. The quantity of immobilized water is affected by the space available between the filaments of the three dimensional network. The amount of expressible water increases and immobilized water decreases by the tightening of the spatial network of the myofibrillar proteins during rigor mortis, freezing, and protein denaturation by heating. Collagen gelation is accompanied by a marked increase in water binding, but this effect is not enough to offset the decrease caused by denaturation of muscular tissue proteins. Collagen shrinkage is presumed to be the main factor that squeezes „hydrated water“ from muscle tissue to its „free form“. Intensity of denaturation is influenced by temperature and moisture content. Sarcoplasmic and contractile proteins are denatured maximally at high temperature and the moisture content of the tissue is lower than 20–30% [19].

The Effect of Dehydration

Dehydration can be employed for raw and cooked meat, however better quality of dry product was obtained when meat was dehydrated after cooking. Conventional air drying at relatively high temperatures is more detrimental to muscle proteins than the freeze-drying process. As a result of moisture evaporation, more moisture and salts are drawn to the surface. The resulting increase in salt concentration and change in pH affects protein properties, especially solubility and water binding properties. The change in salt concentration in the freeze-drying process is less significant and the product is at a lower temperature. In freeze-drying there is no significant denaturation of proteins although some meat proteins are denatured. However, WHC of freeze-dried meat is lower than fresh tissue. Freeze-drying of hot meat, after slaughter of animal and before onset of rigor mortis, inhibits the breakdown of ATP and glycogen. Rapid glycolysis of this meat occurs during freeze-drying, and as a result of muscle contraction, a marked decrease of WHC was observed [20]. Rigor mortis can be prevented and high WHC retained if salted ground meat was freeze-dried. Incorporation of sodium chloride resulted in the binding of salt ions by proteins and an increase in electrostatic repulsion between proteins.

Dry texture and the decrease of meat hydration is one of the principal problems in freeze-drying of meat. Decrease of meat hydration (percent bound water) caused by drying occurred only in the pH range from 4.5 to 6.0. The lowest bound water is in the pI of muscle. Decrease in water binding capacity is due to the phenomenon that polypeptide chains are located more closely together after drying than before. Tightening of the network of protein structure that is related to unfolding of polypeptide chains and a formation of new salt and/or hydrogen bonds is observed. This reaction is not identical with the changes occurring during heat denaturation of muscle. Some meat proteins, particularly myosin, coagulate without previous denaturation.

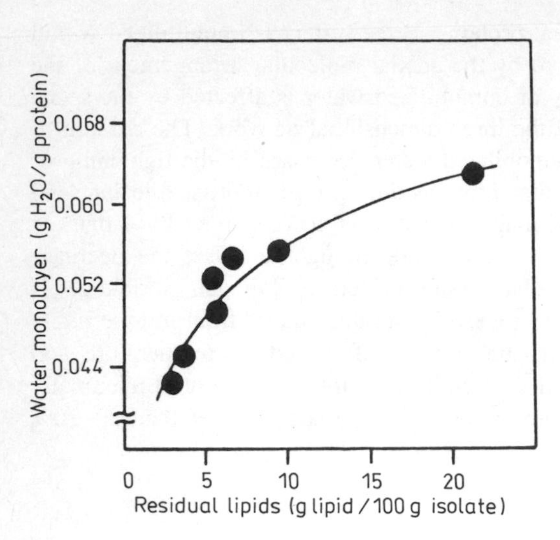

Fig. 2.2. Lung isolate-monolayer values for water adsorption as a function of the amount of residual lipid after solvent defatting. From Ref. 25

Effects of Various Treatments

The influence of ionizing radiation on meat proteins is determined by the dose of radiation and nature of proteins in meat. An important effect of irradiation influencing meat proteins' functionality is an increase in water exudation caused by a decrease in WHC. Tenderizing effect of irradiation at 10°C with either 40 or 60 kGy was reported [21]. At high doses of irradiation an excessive tenderizing effect can cause mushy texture of meat [22]. Effect of irradiation on meat proteins is caused by radiolytic changes in the proteins. However, data related to the irradiation effect on the functional properties of food proteins are limited. Zabielski et al. [23] reported changes in ^{60}Co-irradiated muscular tissue of chicken breast related to hydration and water retaining capacity. In irradiated samples an increase in free water by over 1% content and a decrease in WHC was found in both fresh and heat processed samples. WHC decreased 9.3% for fresh and 18.2% for heat processed samples.

The effect of potassium phosphate and sodium acetate washing on the water holding capacity of the proteins in turkey dark meat was reported [24]. Washing of dark meat resulted in bleaching and loss of specific flavor.Significant differences in WHC between the washed and control tissues were found when tissues were cooked in water. Lower cooking loss for extracted tissues can be explained by the increase in pH produced by phosphate and improvement of WHC. An additional factor of increasing WHC of extracted tissue is loss of lipid component. The presence of NaCl caused an increase in the amount of water lost by the extracted samples and substantially improved the WHC of the control samples.

The effect of various solvents used for fat extraction of lung protein isolates was reported [25]. The water vapor isotherms obtained at 25°C and water monolayer value were calculated (in g H_2O/g protein) for each isotherm. During extraction nonsignificant changes of conformation with exposure of polar sites were observed, i.e. greater effective protein area for water binding. The relationship between water monolayer values of the lung protein isolates and the amount of residual lipids is shown in Fig. 2.2 [25]. With increased lipid residues in protein after extraction, the amount of polar sites exposed increased. During solvent extraction, interaction between proteins and residual lipids occurred in such a way that more polar sites of the protein were exposed enhancing protein hydration capacity.

Alkali extraction and acid precipitation of proteins from bone residues of mechanically separated poultry meat affected water absorption capacity (WAC) of frozen and fresh samples [26]. The WAC values for extracted samples were all negative which showed a decrease in WAC and loss of water. Added NaCl slightly reduced water loss and improved WAC of fresh and frozen extracted proteins.

Blood globin prepared by CMC precipitation exhibited good WHC [27]. Spray dried globin showed significantly higher WHC than lyophilized globin probably due to variations in porosity. The WHC of globin was decreased by an addition of NaCl, especially at concentrations higher than 0.2 M NaCl. The WHC decreased markedly in the pH range from 5 to 7.

2.3.3 Water Binding in Comminuted Meat Products

The batters of comminuted meat products (CMP) are complex colloidal suspensions of meat and fat particles partially extended with solubilized proteins. Myosin is the primary constituent responsible for binding of the water and fat particles. Manufacturing of CMP with the proper textural properties is related to the functionality of muscle proteins in the three-dimensional matrix. Formation of this matrix in sausage batter is due to the interactions between protein-water, protein-protein, and protein-lipid. Proteins are the major structural components in the system; they combine and develop the structure by binding water and fat. Various proteins are added to emulsion-type sausage batter to balance the quality and quantity of protein with processing functionality, nutritional value, and cost. The functional properties of milk and plant proteins in CMP reflect their composition, their nature and reactivity, their native structure, and interactions with other components of these products.

Incorporation of foreign proteins of plant and animal origin is justified not only by their nutritional value but also by their technological and functional properties. The problem of testing protein functionality may be resolved by using them directly in food products. A wide variety of protein additives of plant and animal origin are available for use in CMP to increase water retention, improve emulsifying and gelling properties and fat binding, increase processing yields, and reduce formulation costs. Nonmeat proteins from a variety of plant and animal sources are used extensively as binders, fillers, and extenders in CMP. Stability, palatability, textural properties, nutritive value, and yield of CMP are the major criteria for implementation of nonmeat proteins. These nonmeat proteins assist the meat myosin and actin in stabilizing the emulsion system. Nonmeat proteins act as emulsion stabilizers to reduce the possibility of breaking of the emulsion and decreasing fat and moisture losses during cooking [28, 29].

Different protein sources have unique individual functionalities. These protein sources can improve physical and chemical properties in emulsion-type sausage. The use and functionality in comminuted meats of the following extenders have been well documented: soy protein [30], milk proteins [31], cottonseed flour [32], sunflower protein [33], corn germ protein flour [29], and wheat germ protein [34]. Kinsella [30] reported that different soy products used in CMP produce different functionalities in the final products. Although some soy proteins currently used in the meat industry have a specific unpleasant, bitter off-flavor, they have been used as ingredients of CMP in the past several decades. Soy proteins are used mainly for enhancing water retention, emulsifying capacity and stabilizing emulsions [31]. Simultaneously, soy proteins can increase viscosity and gel forming in the CMP. These functional properties contribute to the formulation of stable sausage batters [31], improved appearance, and yield of final products. Studies of the functionality of nonmeat proteins in CMP have focused on such functional properties as water retention, fat binding, emulsifying capacity, sensory properties, textural properties, storage stability, and nutritional quality [32, 33].

Sodium chloride treatment increased stability of protein to heat denaturation. This is of practical importance because in many protein foods, salt concentrations are ~0.2–0.3 M. The negative effect of increasing thermal stability of NaCl-treated proteins is that at the commercial heating temperatures some important functions of plant (soy) proteins cannot be performed. In processed meats myosin performs various important functions (water binding, emulsifying and gelling capacity), and myosin transition temperature is 55–65 °C. In processed meats containing NCl, extended with soy protein and heated at conventional heating temperatures (70–72 °C), some important functions of soy proteins cannot be performed because the transition temperature of soy protein is higher than 60 °C. There is no complete unfolding of soy proteins even at 90 to 100 °C. The resistance of conglycinin of soy protein to heat coagulation may be due to its electrophilic/hydrophobic ratio.

In the future, the use of plant proteins as additives in foods will increase significantly. According to USDA regulations, 3.5% added proteins are permitted in comminuted meats. Sausage bologna and frankfurters containing higher levels of plant proteins cannot be named as such. Currently, comminuted meats containing levels of nonmeat proteins higher than 3.5% are available under different trade or fancy names. Use of high levels of plant proteins in comminuted meats with protein, fat, and moisture content equivalent to those of an all-meat product may cause problems in emulsion stability, texture, flavor, and color. Stable sausage emulsions, without fat and moisture separation during the cooking process, are of paramount economic importance to the meat industry. It was established that a preemulsified fat preparation utilizing nonplant stabilizers, including sodium caseinate, nonfat dried milk, and gelatin can improve the quality and yield of sausages [35].

Proteins of animal and plant origin are recognized as performing three basic functions in the formation of sausage emulsions. The first function is fat emulsification, the second is water retention, and the third is formation of the structure of meat products. The basic functionality of proteins in meat products is the ability of muscle, plant, or milk proteins to swell and bind water, resulting in a mechanical fixing of fat particles in the protein filament structure. Muscle proteins are better emulsion stabilizers in comminuted meats than most of the plant proteins. Commercial use of soybean proteins as binders in comminuted meats has shown them to be compatible with chopped meat systems [32].

The main effect of blood plasma proteins incorporated in sausage batters is to increase WHC, including water added in the formulation. WHC of blood proteins is even higher than meat proteins. Blood plasma proteins have pH ~7.8 and they increase pH of batter by about 0.2 units when added at the 10% level [36]. Blood plasma proteins form a solid stable gel when heated to 75 °C, and required endpoint interior temperature of frankfurters is 68–70 °C. The WHC of blood globin was 3.27 g H_2O/g that is comparable to the WHC of oilseed protein concentrates [37]. Heating of the blood globin in water in the temperature range from 50 to 100 °C for 30 min decreased WHC from 5 to 3–4 ml/g [27]. The most significant decrease of WHC was observed in the temperature range 50–60 °CWHC of globin was higher for 0.2 M NaCl than for 1.0 M NaCl; this trend was maintained in heated samples at 50–100 °C for 30 min. However, NaCl 0.2–1.0 M stabilized globin to the heating process and no significant change in WHC was observed at 50–100 °C heating temperatures. Heating of globin (5%) water solution at 80–95 °C for 30 min caused an increase in viscosity and gelling capacity in the pH range 5.4–5.8. Globin has potential to be used in semisolid food systems because of high WHC.

The Mechanism of Water Binding in Comminuted Meats

Proteins in comminuted meats must bind water and fat and form a firm, elastic gel. The WHC of CMP is affected by pH, temperature, ionic strength, extent of muscular and connective tissue disruption, and other factors. The amount of water held is affected by the degree of comminution. At higher degrees of comminution and tissue disruption there is a greater amount of protein extraction, more protein-water interaction, and an increased amount of water binding. In sausage batter the cellular structure of myofibrils is destroyed during comminution, and myosin and actin can be characterized as a sol when sarcoplasmic proteins are in solution, and collagen and elastin are in suspension. Comminution process can be considered as effective if the maximum amount of proteins is released from myofibrils. During comminution and physical disruption of muscular tissue at ionic strength above 0.6, intense fiber swelling was observed with myosin depolymerization and solubilization [38]. Efficient comminution of lean muscular tissue must disrupt membranes and sarcolemma to release myofibrils and myofilaments and to accelerate swelling and extraction of myofibrillar proteins [39]. Extraction of myosin and actomyosin accelerated by the presence of sodium chloride and phosphates increase protein-water interaction and water binding.

During comminution a local increase in temperature to 40 °C and higher at the edge of the knife blades can decrease WHC of sausage batter. If comminution is performed in a cutter, not all of the water must be added at the beginning. Comminution of the meat in a cutter is hindered if too much water is added at once. Meat preblending increased the level of protein extraction and improved water- and fat binding properties. Grinding and comminution increases WHC of meat as a result of increasing the number of polar groups available for binding water molecules. A decrease in water binding capacity of sausage batter is possible if the time after comminution and heat treatment is prolonged and the holding temperature is too high. The reason for the WHC decrease is the change in pH. As a result of fast microbial growth of lactobacilli and micrococci that are predominant in sausage batter, water binding capacity could decrease markedly. The pH of sausage batter can drop notably within a few hours as a result of accumulation of acids, especially if sugar was added.

Changes of proteins during prerigor, rigor, and postrigor state significantly influence their functionality in comminuted meat products, especially, WHC, swelling and tenderness. Quality of cooked sausages made with rigor or postrigor beef is much lower than that from prerigor beef. Water binding capacity of sausage batter is significantly affected by pH and is lowest at pH 5.0–5.2 (isoelectric point of actomyosin).

During development of rigor mortis, as the pH drops, the proteins increasingly repel water. This is desirable in summer sausages to create a low water binding capacity, and to form a firm structure with good slicing. Low pH level of meat is unacceptable in CMP production because of very low water binding capacity. After the onset of rigor the swelling and extractability are influenced mainly by temperature and ultimate pH. Wismer-Pedersen [14] demonstrated that proteins in swollen fibers and solubilized monomeric myosin enhance water binding capacity because water binding groups are exposed to solvent instead of participating in protein-protein interactions. Repulsion of the net charges result in increased myofibrillar volume and increased capacity to retain water. The protein-protein (myosin and actomyosin) interactions in sausage batters are affected by temperature, and they reduce the availability of polar groups due to binding between protein molecules [40]. This phenomenon results in a loss of water binding capacity of sausage batter.

In the non-comminuted meat, an increase in volume is restricted by the fibers of connective tissue and membranes surrounding the muscle fibers. As a result of meat comminution this restriction to increase volume of myofibrils is nonsignificant, and water binding capacity is notably higher [41]. In comminuted meats a lower level of water was released by pressure. The rapid drop of pH in pale, soft, exudative (PSE) meat leads to a reduction of WHC; it is recognizable from the wet, watery cut surface of the meat. Consequently, PSE meat has a poor functionality, especially water retention in CMP. PSE meat should be utilized as an additive in the mixture of various meats. It is most effective to select meat carcasses according to pH and to utilize carcasses with high pH values for CMP production.

The usage of minced fish as a raw material for cooked sausages and jellied products depends upon the fish protein's capacity to bind water and fat and the ability to form stable gels after heating in a hydrated form [42]. The level of fat (5% and 30%) in formulation had no significant effect on the water binding capacity [43]. The difference in water binding ability between the high and low fat products was nonsignificant.

The Role of Collagen

A high level of collagen in CMP formulations is undesirable. Under the conditions of sausage batter processing, collagen is not soluble. When heated to 60–65 °C in the aqueous medium, the fibers of collagen shrink to about one-third of their original length. When collagen is heated to temperatures of 65 °C and higher, it is transformed to gelatin. The optimal temperature of collagen conversion to gelatin is related to various factors, such as specific muscle, species, age, and sex of animal. It is recommended for CMP to have no more than 25% of the total protein present as collagen.

The hydration capacity of unheated collagen was minimal in the pH range from 6 to 9 and consisted of 4 g of water by 1 g of freeze-dried collagen [44]. The hydration increased on the acidic side of the isoelectric point. The least hydration near the isoelectric pH values resulted in the unavailability of the charged groups. Hydration increased markedly after heating of the collagen in aqueous media at various pH values. Ten grams of water were immobilized by 1 g of collagen in the pH range 6.0–9.0. Morrissey [45] reported that WHC of freeze-dried bovine collagen increased twofold on heating to 70 °C at pH 5 to 7 compared to unheated collagen. The addition of polyphosphate to 0.5 M did not affect WHC under these conditions. Hydration of collagen was not influenced by NaCl up to 4% in aqueous phase at pH 5–7 as compared to solutions without salt. However, the presence of salt enhanced collagen solubility on heating in this pH range. Utilization of wet fibrous collagen in meat products would increase water binding and serve as an extender-filler in meat products. The extent of hydration of unheated collagen in the pH range 5–8 was less in the presence of different phosphate salts than in their absence.

Effect of Sodium Chloride and Phosphates on Water Binding in Comminuted Meats

The physical and chemical properties of meat proteins are influenced by ionic strength. Sodium chloride affects the WHC of the proteins and decreases water activity. The studies of the influence of various salts showed that protein functionality is dependent on the balance of interactions between protein, water, and salt. The capacity of meat proteins to retain water is significantly affected by the ionic strength of the medium. The amount of water bound by proteins is influenced notably by the concentration of neutral electrolytes. As the concentration of neutral electrolytes is reduced the water holding capacity is increased. The effect of NaCl on the WHC of meat is utilized in the manufacturing of sausage batters and during the curing of meat. Salt not only increases WHC but also liberates the proteins of myofibrils (salt soluble proteins); they can function as emulsifiers.

The effect of NaCl on the water retention is related to the preferential binding of chloride anion (Cl^-) to the positively charged amino groups of the proteins. The increase in WHC on the addition of NaCl is considered to be related to binding of chloride ions to the myofibrillar and sarcoplasmic proteins. More water can be retained by proteins if chloride ions are bound to proteins at pH's above the isoelectric point causing an increase in negative charge [39]. This preferential binding of chloride ions by the protein molecule at pH's above the isoelectric point increases its net negative charge and its resulting repulsive forces. An increase in repulsion between negatively charged groups caused an opening of the proteins and increased interaction with water. As a result additional water imbibition within the protein network is obtained [6]. The effect of NaCl and pH on WHC can be explained by changes in the electrical charges of the myofibrillar and sarcoplasmic proteins.

After ionic strength had reached an optimum (0.8–1.0), the further increase of ionic strength resulted in a decrease of bound water and was related to the „salting out" effect involving strong binding of water by the salt and consequent dehydration of the proteins. In meat systems high concentrations of NaCl caused depolymerization of the thick filaments to myosin molecules and other components. Water-meat protein interactions of isolated myofibrils at pH 5.5 in the presence of various levels of NaCl and combinations of pyrophosphates and NaCl have been reported [10]. The intensity and degree of swelling of myofibrils is responsible for water binding of meat systems. Water binding in the myofibrils is by capillary forces, basically in interfilament spaces within the myofibril, in extracellular spaces, and spaces between the myofibrils.

The ability of meat to hold water is greatest during the few hours after slaughter (hot meat), and it rapidly declines to the minimum level after 24 h [36]. The effect of „hot meat" with prevention of actomyosin formation and high WHC can be preserved by blending with NaCl and nitrite. When salt ions are absorbed they influence electrical charges between the protein molecules and prevent actomyosin formation, in spite of ATP breakdown. Raising the NaCl concentration weakens the binding of actin and myosin and partly prevents actomyosin formation [10]. Effective prevention of actomyosin formation is possible if salt is evenly redistributed in the muscular tissue, for example by grinding and blending. However, storage time of such blends is strongly limited because of high microbial count of these blends.

The high WHC of prerigor and postrigor beef can be retained for several days by the addition of salt and phosphates in meat and an even redistribution. High

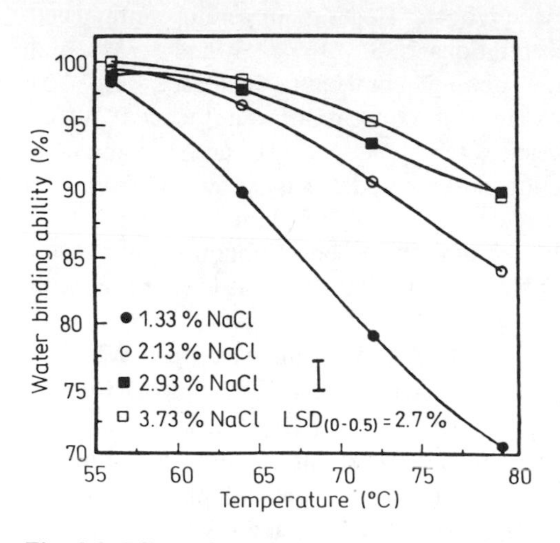

Fig. 2.3. Effect of cooking temperature and effective salt concentration on the water binding ability of finely comminuted meat procucts. From Ref. 43

WHC of prerigor meat can be preserved for several months by rapid freezing of prerigor meat with salt addition and processing without prior defrosting [36]. The freezing temperature in the core of the meat must be reached within 4 h for beef and 1 h for pork after slaughter. The properties of hot meat are lost if meat is thawed before comminution. Increased WHC of meat salted in the prerigor state is related to the strong electrostatic repulsion between the dissociated myofibrillar proteins, myosin and actin, as a result of influence of high pH, high ionic strength and ATP [46]. Honikel et al. [47] showed that at onset of rigor mortis at pH 5.9, WHC of salted meat homogenate decreased strongly. Consequently, sausage batter with high WHC can be produced from the prerigor muscle, but salt must be added and meat comminuted before pH of meat reaches 5.9 or before the ATP level declines sufficiently to allow development of rigor mortis. Honikel [47] suggested that formation of relatively few crosslinks between actin and myosin filaments causes a significant loss of water retaining capacity.

WHC of unsalted homogenates was relatively low (high cooking losses) and increased slowly and linearly with decreasing pH until the final pH of about 5.4 was reached. A slow decrease in WHC (slow increase in cooking losses) occurred at pH 5.9 in salted homogenates. In the pH range between 7.0 and 5.9 the cooking losses of the salted homogenates were much smaller (higher WHC) than unsalted. Increase of the WHC of meat at high pH values can be explained by the electrostatic theory of swelling.

Reduction of the fat level in sausage batters affected WHC [48]. Reduction of the fat level decreased the effective NaCl concentration and hence WHC because salt is not soluble in the nonpolar fat. A decrease of the concentration of NaCl in the muscle proteins influenced the thermal denaturation of muscle proteins as shown by DSC studies. The lowering of NaCl concentration may change the temperature at which the meat proteins aggregate. Heating temperature influenced WHC of sausage batters for salt concentrations 1.33, 2.13, 2.93, and 3.73%, and WHC decreased progressively as the cooking temperature increased (Fig. 2.3) [43]. At the same time an increase in salt concentration influenced WHC. Maximum WHC was obtained at heating temperature 56 °C and minimum at 80 °C. The effective NaCl concentration de-termined the maximum temperature (57 - 68 °C) the product could be heated to before WHC started to decrease. A significant increase of water and fat binding have been obtained in canned frankfurters by adding a high level of NaCl, up to 5%. The can is filled up with brine, and the product flavor is preserved [36].

Alkaline phosphates added to lean meat effectively increased the WHC in proportion to their ionic strength and increased the pH of the meat. Certain phosphates (diphosphates) have synergistic effects. Phosphates cause dissociation of actomyosin, increase solubility of myosin, and as a result, increase extraction of proteins. Pyrophosphate formed by the breakdown of larger polyphosphates and NaCl influence the strength of binding of myosin heads to actin, which leades to the dissociation of actomyosin [49]. The phosphates tetra sodium pyrophosphate and sodium tripolyphosphate increase the pH about 0.2 units, which accelerates the water retention. The effects of the phosphates on proteins include increased pH and ionic strength and interaction with proteins that caused dissociation of actomyosin by pyrophosphate.

2.3.4 Milk Proteins in Comminuted Meats

In the manufacture of comminuted meats, it is important to balance the quality and quantity of protein with processing functionality, nutritional value, and cost. Milk protein ingredients have been used as fillers, binders, and extenders in CMP to improve not only emulsifying properties but also immobilization of water and sensory properties. Batter stability, yield, texture, palatability, and cost of CMP are the major criteria for nonmeat proteins. Other important factors might include differences in particle size, color, quantity, and quality of the protein. A number of nonmeat ingredients can be utilized to improve sensory properties and nutritional value of CMP, as well as providing a greater flexibility of formulation to minimize processing losses [50]. Such ingredients are especially useful for cases of limited salt-soluble myofibrillar proteins in low meat or cheap meat products.

Nonfat dry milk (NFDM) improves yield, color, flavor, and stability of sausage batter upon cooking [51]. NFDM has a very high nitrogen solubility index, high emulsifying capacity, and moderate WHC compared with other nonmeat proteins. Whey protein concentrate (WPC) is highly nutritious with a protein efficiency ratio (PER) of 3.0–3.2 (2.5 for casein) [52], and contains significant amounts of essential amino acids such as lysine, tryptophan, methionine, and cystine. WPC could replace lean beef in the fat stabilization of meat emulsions.

Caseinate and WPC incorporated into meat batters resulted in higher cooking losses than when the exchange was for Promine-D, probably because of higher solubility of caseinate and WPC and different gelling properties. Caseinates contribute to better water binding and consistency of CMP. Replacement of calcium by sodium in casein increased its water retention and viscosity. In CMP, the capacity of sodium caseinate (SC) to retain moisture is increased by the salt addition. Utilization of caseinate saves part of the soluble meat proteins from denaturation in the interface, leaving them available for gel formation. With the incorporation of caseinates, the sausage batter stability becomes less sensitive to chopping temperatures, resulting in a more reliable and flexible process. Increase in heat stability occurs because caseins are rather unique in having very little, if any, secondary structure, because of high levels of proline, resulting in extremely high heat stability [53].

Functional properties of milk proteins and corn germ protein flour (CGPF) as extenders in meat products have been reported [54]. Batter properties of WHC, viscosity, and cooking losses are presented in Table 2.1. WHC of sausage batters with milk proteins and CGPF added were higher than that of the control batter. There were no differences in WHC among samples containing CGPF, NFDM, and SC. The WHC of sausage batters affects important quality characteristics of the finished products. Water absorption by milk proteins and CGPF decreased the amount of free water and increased immobilized water in the sausage batters. WHC of sausage batters affected the cooking losses similarly. The control batter with lower WHC had higher cooking losses of 11.75% than all four treatment batters (from 7.17% to 9.66%). The sausage batters with NFDM added showed the lowest cooking loss. This trend corresponds with the report of Holland [51] that NFDM significantly improves sausage batter stability upon cooking. Viscosity was lower for batters with milk proteins than for batter of the all-meat control (Table 2.1). This low viscosity may have been due to the high protein solubility of milk proteins that would not increase viscosity of the batter and/or to the dilution effect of the additional water (2.0–3.5% more) in the extended batters. As the percentage of added water increased in sausage batters at 10°C, the apparent viscosity decreased.

2.3.5 Soy Proteins in Comminuted Meats

Isolated soy proteins significantly improved emulsification, water and fat binding, thickening, cohesiveness, adhesiveness and gelation of comminuted meat systems. These functional properties demonstrate the desirable uses of soybean proteins in food applications. Soy products provide a variety of protein ingredients: flour, concentrate, isolate and their modified products. Soy proteins are used widely in comminuted meats as meat extenders and replacers of meat proteins. Textured soy flours are used as extenders in chopped meats. Texturization of isolated and purified soy proteins should be considered as one of the most important technological achievements. Soluble soy proteins have been injected into pork, hams, and poultry rolls to reduce losses during heat treatment. Augmentation of soy pro-

Table 2.1. Water holding capacity (WHC), viscosity, and cooking losses of sausage batters containing milk proteins and corn germ protein flour

Treatment[1]	Added water[2] (%)	WHC	Viscosity x 10^5 (cps)	Cooking losses (%)
Control	25.0	0.375^{c}	3.40^{a}	11.75^{a}
CGPF, 3.5%	$28.5^{a,b}$	$0.559^{a,b}$	$3.32^{a,b}$	$8.57^{b,c}$
NFDM, 3.5%	28.5	0.586^{a}	2.26^{c}	7.17^{c}
WPC, 3.5%	28.5	0.507^{b}	2.77^{c}	9.66^{b}
SC, 2.0%	27.0	0.594^{a}	$2.99^{b,c}$	$9,51^{b}$

[a-d] Means in the same column with different superscripts are different ($P<0.05$).

[1] *CGPF, NFDM, WPC,* and *SC* are corn germ protein flour, nonfat dry milk, whey protein concentrate, and sodium cascinate, respectively.

[2] Sausage batters were formulated with 25% added water for control samples; 1.0% water was added with each 1.0% additive used. From Ref. 54

teins in intact muscles, such as beef rounds, turkey breasts, or pork, represents some of the major developments in the meat industry. The granular soy isolate is produced by an extrusion process and is utilized as a meat replacer in ground meat formulations.

Relative functional ranking of eight nonmeat additives based on cooking stability and moisture separation was reported by Comer [55]. Soy isolate, soy concentrate, and a textured soy protein ranked above wheat flour, sodium caseinate, skim milk powder, and potato starch. Using an isoelectric wash process, soy concentrates and isolates with desirable functional properties have been produced. The replacement of animal proteins in food processing by such alternative protein ingredients as soy proteins is caused by their favorable functional properties. In the future, manufactured proteins originating from plants will be used widely to extend and replace meat and dairy proteins and to lower food prices. Various soy products used in CMP produce different functionalities in final products. These functional properties contribute to the formation of stable meat emulsions and improve moisture retention, appearance, and yield of final products.

The most significant changes were in soy protein isolate production, which has increased steadily. This trend is caused by the fact that more refined plant proteins create fewer flavor problems and do not contain oligosaccharides that may cause stomach discomfort. The higher water retention in CMP extended with soy proteins enhances desired sensory properties. However, increased water retention can also induce an off-flavor indirectly by a higher microbial growth. The higher WHC of soy isolates may be related to better swelling capacity, to the capacity to dissociate, and to exposing water binding sites, while in the soy concentrates, carbohydrates and other components may impair the ability to retain water. For vegetable proteins to be fully utilized as a substitute of animal proteins they must possess the same or similar qualities as animal proteins. Soy proteins can be incorporated into foods in various forms and can play a variety of functions. Complete analogs of meat and milk products made exclusively from textured plant proteins probably will not be accepted. The greatest potential for success has mixture of plant and animal proteins in processed foods, where the animal ingredients can provide specific sensory characteristics, and the plant proteins can reduce the cost and content of some of the undesirable components such as fat.

Most acceptable as an extender in CMP is soy isolate because it has myosin-like functional properties. A soy isolate has been developed which simulates the myosin of meat proteins in fat and water binding capacity, even in the presence of salt and heat. The development of such a soy isolate has enabled meat processors to produce CMP in which the meat and soy proteins work together to improve their functional and nutritional properties. Isolated soy protein must be properly hydrated before certain functional properties are obtained in many foods. Soy protein isolate should be added to the formulation of foods as a prehydrated slurry before salt addition. A slurry of soy isolate, water and salts can be injected into the muscle using a stitch pump. In the preparation of a brine with soy protein isolate, the protein must be fully hydrated before other brine ingredients are added.In the comminuted meats processing, soy isolate hydration can be carried out in a bowl chopper before meat ingredients are added by blending one part soy isolate with five parts water [56]. When CMP formulation is prepared in a blender, following comminution in an emulsifier, soy protein isolate and water should be blended several minutes with meat before salt and other ingredients are added. Coarse ground meat products are traditionally extended with soy isolate or concentrate hydrated for 15 min with three parts cold water.

Soy isolate has been used to upgrade mechanically deboned poultry meat, especially to improve the texture of this meat [56]. Incorporation of 20% soy isolate caused textural properties of mechanically deboned meat to be similar to those of a chicken breast roll. The cost of new soy protein ingredients must be considered not only in terms of cost per pound, but in terms of protein levels and the cost per pound of protein. For instance, in the sausage batter, it takes nearly twice as much soy concentrate to obtain the same functionality as with soy isolate.

Functionality of soy protein flour (SF), soy protein concentrate (SC), soy protein isolate (SI), and corn germ protein flour (CGPF) in comminuted meats was reported [57]. Mean values for WHC, viscosity, and cooking losses of meat batters are presented in Table 2.2. Meat batters extended with tested plant proteins had higher WHC than those without protein addition. There were no differences in WHC among samples containing SF, SC, SI, and CGPF. No differences were found in viscosity of meat batters for control and experimental treatments. WHC and viscosity could be affected not only by protein but also by carbohydrate components of the protein additives. The highest cooking losses of meat batters were observed in the control all-meat treatment containing 25% water. No differences in cooking losses were found among samples containing SF, SC, SI, and CGPF. The excess of water added with these additives was absorbed by soy

Table 2.2. Water holding capacity (WHC), viscosity, and cooking losses of sausage batters containing soy and corn germ proteins[1]

Treatment	Addes[2] water (%)	WHC	Viscosity $x10^{-5}$ (cps)	Cooking losses, %
Control, 0%	25.0	0.61^a	3.66^a	10.09^a
SF, 3.5%	28.5	0.66^b	3.41^a	7.68^b
SC, 3.5%	28.5	0.69^b	4.46^a	7.67^b
SI, 2.0%	27.0	0.69^b	3.47^a	7.94^b
CGPF, 3.5%	28.5	0.66^b	3.57^a	8.10^b
LSD		0.044	0.322	0.681

[a,b] Means within the same column with different superscripts are significantly different ($P < 0.05$).

[1] SF, SC, SI, and CGPF are soy flour, soy concentrate, soy isolate and corn germ Protein flour, respectively.

[2] Sausage batters were formulated with 25% added water for control sample; in experimental samples, 1% water was added for each 1% plant protein source. From Ref. 57

and corn proteins. Smith et al. [32] reported that high cooking losses and low cook yield could result from low emulsion stability in meat batters. However, other researchers have reported that SI had a greater emulsion stabilizing effect than SF and SC [31, 32]. Consequently, an increased addition of plant protein can increase WHC and decrease cooking losses, thereby increasing product yield. Utilization of soy and other vegetable proteins as extenders in meat and other products requires development and implementation of the method to determine the content of these proteins in meat (food) products. Using the sodium dodecyl sulfate-polyacrylamide gel electrophoresis method, the presence of sunflower proteins in meat could be detected at levels as low as 5%.

2.3.6 Corn Germ Protein in Comminuted Meats

The corn milling industry produces more than 1 million metric tons of corn germ as a by-product of grit, meal, and flour production [58]. After the process of starch and oil removal, more than 6 million metric tons of wet-milled germ meal is produced every year in the United States [59]. Industry's selection of a protein raw material is based on data measuring its functional properties. The production of protein materials with desirable functional properties is of particular interest to food processors. Defatted corn germ protein flour (CGPF) has considerable potential for use as a supplement in a variety of foods to provide a new protein source. CGPF has been reported to have high water retention, fat binding, emulsifying capacity, and emulsion stability [29, 60, 61].

Corn germ represents 12% of the total corn kernel. It contains 22% protein, whereas the corn seed itself only about 9–10% protein. The proteins in corn germ are mostly albumins (water soluble, coagulated by heat) and globulins (soluble in dilute salt solutions) with a good balance of essential amino acids [62]. Corn germ is a source of other valuable nutrients including vitamins and minerals. The PER value of 2.2 for corn germ protein is comparable to that of casein and soya proteins. Quality of corn germ protein is better retained if oil is removed by solvent rather than expeller processing. The high protein and ash contents indicate that CGPF could be a valuable supplemental source of nutrients in food products.

A study was conducted to establish in commercial conditions the effect of CGPF as an extender and replacer of meat proteins for CMP and to determine the functionality of CGPF in those meat products, particularly its effect on water retention and yield, textural properties of the sausage emulsion and finished product, color, and sensory characteristics of the products. Mean values for proximate composition, WHC and yield of frankfurters containing CGPF are shown in Table 2.3 (Zayas, J. F. and Lin, C. S., unpublished data). Protein contents in experimental samples with 3% CGPF added were no different, except lower protein content in frankfurters with hexane-defatted CGPF. However, experimental samples of frankfurters with 30% water added during formulation had lower protein contents than the control frankfurters or those with 28% water added. Incorporation of the supercritical carbon dioxide- (SC-CO_2-) CGPF and hexane-defatted CGPF in the frankfurters formulations significantly increased the yield of the product (Table 2.3). The yield for samples containing hexane-defatted CGPF was 3.0–3.5% higher than for samples containing SC-CO_2-defatted CGPF. A significantly higher yield of the finished product was obtained as the result of increased water retention in batter formulations containing CGPF. During comminution and heat treatment, the higher amount of water used in the formulation of experimental samples was bound effectively by the meat proteins and added CGPF.

WHC of the control and experimental frankfurters with hexane-defatted CGPF and 28% added water was not different, however during comminution in experimental batters, 5% more water was added (Table 2.3). WHC of the frankfurters containing SC-CO_2-defatted CGPF was lower than those with hexane-defatted CGPF at the same level of added water (28%). Lower WHC of the experimental frankfurters with hexane-defatted CGPF was due to the 30% added water during formulation. The lower WHC of the frankfurters with SC-CO_2-defatted CGPF affected the yield of the finished product. Fat caps were observed more frequently in formulations without CGPF. Taste panel scores showed that juiciness was not different between control and experimental frankfurters containing 28% and 30% water added during formulation. Storage for 30 days did not significantly affect the juiciness of frankfurters.

As established by the experiments carried out under laboratory and commercial conditions, some of the advantages of CGPF in comminuted and coarse ground meat products are: to replace meat proteins to meet standard requirements for protein content; to decrease saturated fat and cholesterol content in meat products; to increase yield as a result of increased water holding capacity of sausage

Table 2.3. Composition and yield of frankfurters containing $SC\text{-}CO_2$ and hexane-defatted corn germ protein

CGP	Fat (%)	Protein (%)	Ash (%)	Moisture (%)	WH	Yield (%)
Control[e]	24.90^a	12.00^a	3.30^a	55.81^a	0.686	111.52^a
$SC\text{-}CO_2$ CGP, 3%[f]	23.62^b	12.24^b	3.16^a	55.36^b	0.730	114.97^b
Hexane CGP, 3%[f]	23.29^b	$12.07^{a,b}$	3.17^a	56.23^c	0.655	118.19^c
Hexane CGP, 3%[g]	24.39^c	11.76^c	3.21^a	55.19^d	0.611	118.75^c

^{a,b,c,d} Means in the same column with different superscripts are different ($P < 0.05$).
[e] Control, without CGP, 23% water added.
[f] Experimental sample, 28% water added.
[g] Experimental sample, 30% water added. Zayas, J.F. and Lin, C.S. Unpublished data

Table 2.4. Content and retention of water in raw, conventionally and microwave heated beef patties with and without added corn germ protein flour (CGPF) slurry

Added CGPF Slurry (%)	Water content (%)				Water Rentention (%)	
	Conventionally heated		Microwave heated		Conven-tionally heated	Microwav e heated
	Raw	Cooked	Raw	Cooked		
0	62.86^a	56.47^a	$62.71^{a,b}$	55.70^a	53.47^a	55.47^a
10	61.49^b	54.86^a	61.94^a	$56.44^{a,b}$	59.95^b	60.56^b
20	62.64^a	55.85^a	$62.17^{a,b}$	56.93^b	61.62^b	63.97^c
30	64.71^c	55.14^a	63.92^b	57.29^b	60.05^b	64.40^c

^{a,b,c} Means in the same column with the same superscript letters are not significantly different ($P < 0.05$). From Refs. 63 and 64

and beef patty batters and of decreased weight loss during heat treatment; and to increase the stability and emulsifying capacity of the meat systems.

The functional properties of hexane-defatted CGPF added to the formulation of beef patties were reported [63, 64]. Beef patties were extended with CGPF to investigate the effect of CGPF on water holding capacity, cooking losses, yield, sensory characteristics, textural properties and color of beef patties.

Beef patties were extended with CGPF at levels of 2.5, 5.0. and 7.5% of the dry weight of the total mix (10, 20, and 30% when rehydrated 1:3 with water) (Table 2.4) [63,64]. Control samples were prepared without CGPF. To prepare the CGPF slurries, dry CGPF was hydrated with distilled water at a ratio of 1:3. Hydrated CGPF was mixed with the meat and ground through a 1.27 cm plate, to incorporate more thoroughly the CGPF throughout the product. Addition of CGPF

Table 2.5. Effect of heating and added corn germ protein flour (CGPF) on pH, Water holding capacity (WHC), and yield of beef patties

Added CGPF	pH		WHC	Yield (%)	
Slurry (%)	Raw	Cooked	Conventionally heated	Conventionally heated	Microwave heated
0	5.80[a]	5.96[a]	.05[a]	59.58[a]	63.68[a]
10	5.89[a]	6.04[a]	.48[b]	67.13[a]	65.98[a]
20	6.03[b]	6.13[c]	.56[b]	68.58[b]	69.79[b]
30	6.11[b]	6.14[c]	.63[b]	70.37[b]	71.06[b]

[a,b,c] Means in the same column with the same superscript are not significantly different (P > 0.05). From Refs. 63 and 64

slurries increased the water content of microwave heated beef patties at the 20% and 30% extension levels (Table 2.4). There was no increase of moisture content at the 10% extension level. Conventional heat treatment caused an equilibrium across extension levels, so there were no significant differences in the moisture content of broiled beef patties. Significant increases in water retention were observed in conventionally heated beef patties containing 10, 20, and 30% hydrated CGPF, when compared to the control, but no differences were found when comparing the three levels of added CGPF slurry. An increase in water retention in microwave heated experimental beef patties was up to 9%. A significant difference was found between beef patties containing 10% and 20–30% CGPF. CGPF has the ability to increase water retention, with a lower value for water retention (60.05%) than for fat retention (82.11%) at 30% hydrated CGPF extension level.

Water holding capacity of meat and meat products is known to be affected by changes in pH and by other factors. The addition of CGPF slurry increased WHC of CGPF-extended beef patties over that found in the control beef patties (Table 2.5). There were no significant differences found among any of the WHC values for experimental patties containing the three levels of CGPF additive.

The pH of raw beef patties increased with the addition of 20 and 30% CGPF slurry. The level of change in pH in raw and cooked beef patties decreased as the level of CGPF extension increased (Table 2.5) [63, 64]. Cooked, CGPF-extended patties had higher pH than the all-meat control. Effect of the CGPF on the pH values probably influenced water retention in raw and cooked beef patties.

Addition of CGPF was found to increase cooking yields in broiled beef patties (Table 2.5). Mean values for percent yield were increased over the control patties at the 10% and 20% CGPF treatment level. Although no significant differences were noted among patties containing the various levels of added CGPF, the trend was for increasing percent yield as the levels of CGPF extension increased. Yield values increased as the values for the WHC of raw beef patties increased. A correlation factor of 0.68 was found for the increased WHC and yield of beef patties and a factor of 0.85 was found for the correlation of water retention and yield of broiled beef patties.

A significant difference in juiciness values for experimental samples was found at the 30% CGPF slurry addition level. Retention values of water discussed earlier would suggest that this difference in the degree of juiciness perceived by panelists was due to an increased retention of fat rather than water.

2.4 Water Holding Capacity of Milk Proteins

The concentration of milk proteins in dairy foods to achieve required functionality is in the range 3.5 to 30%. However, as functional ingredients they are used in the range 0.5 to 10% [5].The most important functionality of milk proteins is affected by interaction with water: hydration, swelling, gelation, and viscous properties.

Milk proteins are widely utilized in the manufacturing of different foods due to their favorable functional properties, and nutritional and sensory quality. The proteins of milk exhibit unique water holding properties in addition to behavior common to most proteins. The capacity of milk proteins to retain moisture may be transferred to processed foods when those proteins are incorporated. Practical concerns pertain to water binding by milk proteins in complex food systems, including sausage batters and whey-soy combinations. Milk proteins added to certain processed foods increase their WHC and improve other functional properties, such as emulsification and foaming. Proteins are the main component affecting the dispersibility and hydration of dried milk powders.

Protein content exerts the major influence on the quality of milk products, including yoghurt, sour cream, and others. In yoghurt processing, the milk proteins may be modified in such a way that the physicochemical properties of the yoghurt will be altered to the benefit of the consumer. The protein functional properties in yoghurt affect the physical, chemical, nutritional and dietetic properties of the product. Proteins in yoghurts determine sensory characteristics as a result of: 1) hydration, solubility, wettability, water absorption, thickening, and gelling; and 2) aggregation and adhesion. In the yoghurt formulation, animal, vegetable, and cereal proteins are incorporated depending on the availability and effect on the overall quality of yoghurt. The protein constituent of milk plays an important role in the formation of the coagulum and the consistency of yoghurt.

WHC of dried milk powders was affected by methods of drying and grinding that influenced size, porosity, and topography of powders [65]. The surfaces of whey powders dried in a fluidized bed dryer were not porous with a low rate of water adsorption. The surfaces of spray- and freeze-dried whey powders were porous with a high rate of water adsorption. The WHC of whey powders ranged from 70 to 147 g water/100 g powder, or 33 to 180 g water/100 g protein. WHC of whey powders was affected by protein content. Adsorption of moisture by dairy protein powders influences their stability, i.e. gradual loss of solubility, browning, chemical deterioration and rate of rehydration [5]. The water absorption capacity of milk proteins affects the drying process, selection of packaging materials and storage conditions.

The knowledge of the syneresis mechanism at the molecular level is related to moisture binding and will optimize moisture release during cheese production. During the manufacturing of Cheddar cheese and other dairy products, the rate of syneresis is influenced by temperature, pH, ionic strength, degree of agitation, and volume of whey surrounding the curd. Syneresis is not a simple physical process in which water is physically squeezed from the curd. Rather, syneresis is considered a result of chemical interactions. The responses of syneresis to temperature, ionic strength, and pH are similar to those that occur during aggregation of denatured proteins. The rate of syneresis accelerates with an increased temperature. Direct effect of fat concentration on syneresis has been observed [66]. Pearse et al. [67] suggested that limited dephosphorylation of β-casein can affect syneresis. Microstructure and physical properties of yoghurt is influenced by the level of protein and ratio of casein to non-casein [68]. At the ratio 1.08/1.0 a softer yoghurt coagulum with a trend to syneresis was produced compared with that prepared from whey protein concentrates (WPCs). It was concluded that yoghurt with good textural properties may be produced at a ratio of 3.2–3.4:1.0. Yoghurts from WPCs had a weaker or softer coagulum with a higher sensitivity to syneresis than casein-containing yoghurts.

Pulsed low-resolution NMR was used to study hydration mechanism and water binding to casein, albumin and γ-globulin [69]. The non-exponentiality of the spin-echo decay curves of water was utilized to determine the water mobility as well as the levels of bound and free water in samples, containing progressive amounts of water to 1 g of dry protein. The relationship between water binding and water activity is important in food quality control and storage. Binding of the molecules of free water reduces vapor pressure and hence water activity.

Structure and properties of commercial sodium, potassium, and calcium caseinate preparations are different than milk casein micelles. Acidification and neutralization processes influence the structure of resulting caseinate particles. Acidification to pH 4.5–5.0 dissipates the colloidal phosphate structure and frees the casein of Ca and other inorganic ions. During casein neutralization there is resolubilization of casein. In this system subunits of casein are arranged with the orientation that provides hydrophilic properties. As a result sodium and potassium caseinates possess improved functional properties, solubility and water binding and surface active properties. Sodium caseinate prepared by solubilizing acid casein with NaOH is the water soluble casein most widely used in foods. The most important functional properties of casein in baked products has been found to be water binding. High water binding capacity is produced by sodium and potassium caseinate and soluble co-precipitates. Sodium caseinate produced from the preheated milk showed higher WHC, probably due to adsorption of whey proteins on casein and formation of a sponge-like surface on the casein [70]. The WHC of sodium caseinates can be located between that of egg white (poor) and soy isolate (excellent).

Functional and physicochemical properties (including water binding) of whey proteins, β-lactoglobulin and α-lactalbumin are determined by their primary structures. Whey proteins possess uniform distribution of acidic/basic and hydrophobic/hydrophilic amino acids along their polypeptide chains. Globular conformation with a substantial helical content explains strong susceptibility of whey proteins to denaturation by heat and other treatments.

In WPC low molecular weight components other than lactose are mainly responsible for water binding. There was no close relationship between the protein content of whey preparations and the amount of bound water. Whey proteins showed great differences in the capacity to bind water, but mostly they exhibit low water-retaining capacity. The whey proteins exhibit the capacity to remain hydrated and soluble at their isoelectric pH. WHC of whey powders and demineralized whey powders are generally low [71]. However, the protein component of these powders has a high WHC. Heating of whey proteins did not notably improve WHC compared with unheated proteins [65]. An increase in WHC was found only after severe heat treatment of whey proteins.

Water retention of milk proteins in model systems has been reported [72]. Investigation of functional properties of proteins in model systems comparing milk proteins and CGPF might have implications for real food systems. Variations in pH and temperature are commonly imposed on these systems. Because of the limited profiles of water retention as functions of pH, and/or incubation time and temperature, and/or protein concentration, this model study was conducted to determine water retention of nonfat dry milk (NFDM), whey protein concentrate (WPC), sodium concentrate (SC), and corn germ protein flour (CGPF). Response surfaces of water retention (WR) are presented in Fig. 2.4 for CGPF, NFDM, WPC, SC, and CGPF [72]. The lowest WR for three milk proteins was obtained at the highest protein solubility (without precipitation) at pH 7.0–7.3, agreeing with the results from the response surfaces of protein solubility for three milk proteins, Ch. I „Solubility of Proteins", Fig. 2.5, for three milk proteins. CGPF had a low WR (Fig. 2.4) at low incubation temperature (10–25 °C) and low pH (pH 5.0–5.5). Water binding increased rapidly in the range of 30–70 °C, possibly because of increased protein solubility. The maximum WR occurred at 70 °C at both ends of the tested pH range (pH 5 or 8) with 30 min of incubation. At low incubation temperatures (10–30 °C), WR of CGPF increased slightly as pH increased from 6 to 8, but no overall pH effect was detected.

The WR response surfaces of NFDM, WPC, SC, and CGPF showed the curvilinear nature of the response to both pH and incubation temperature. However, the milk proteins (NFDM, WPC, and SC) had more similar response surfaces. When pH decreased from 6 to 5, the precipitation of SC increased most notably, whereas that of WPC increased gradually (Fig. 2.4, WPC and SC). The precipitation of NFDM was between that of WPC and SC. This is because SC components – caseins – are susceptible to low pH (close to the isoelectric point, pH 3–5), whereas whey proteins – α-lactalbumin and β-lactoglobulin – are less susceptible to low pH. NFDM contains proteins of caseins and whey proteins and had intermediate precipitation reactions., DeWit [73] also reported that β-lactoglo-

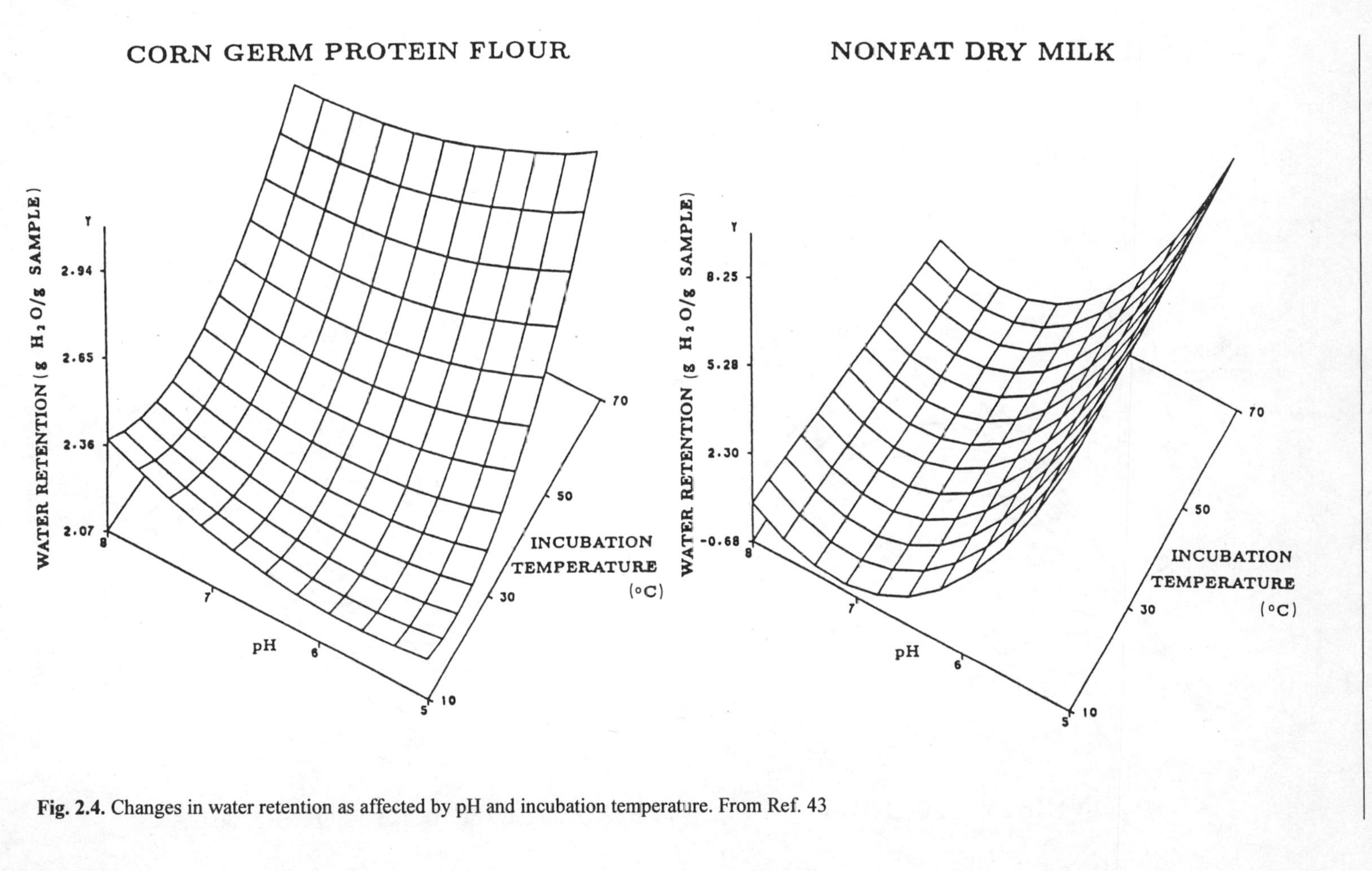

Fig. 2.4. Changes in water retention as affected by pH and incubation temperature. From Ref. 43

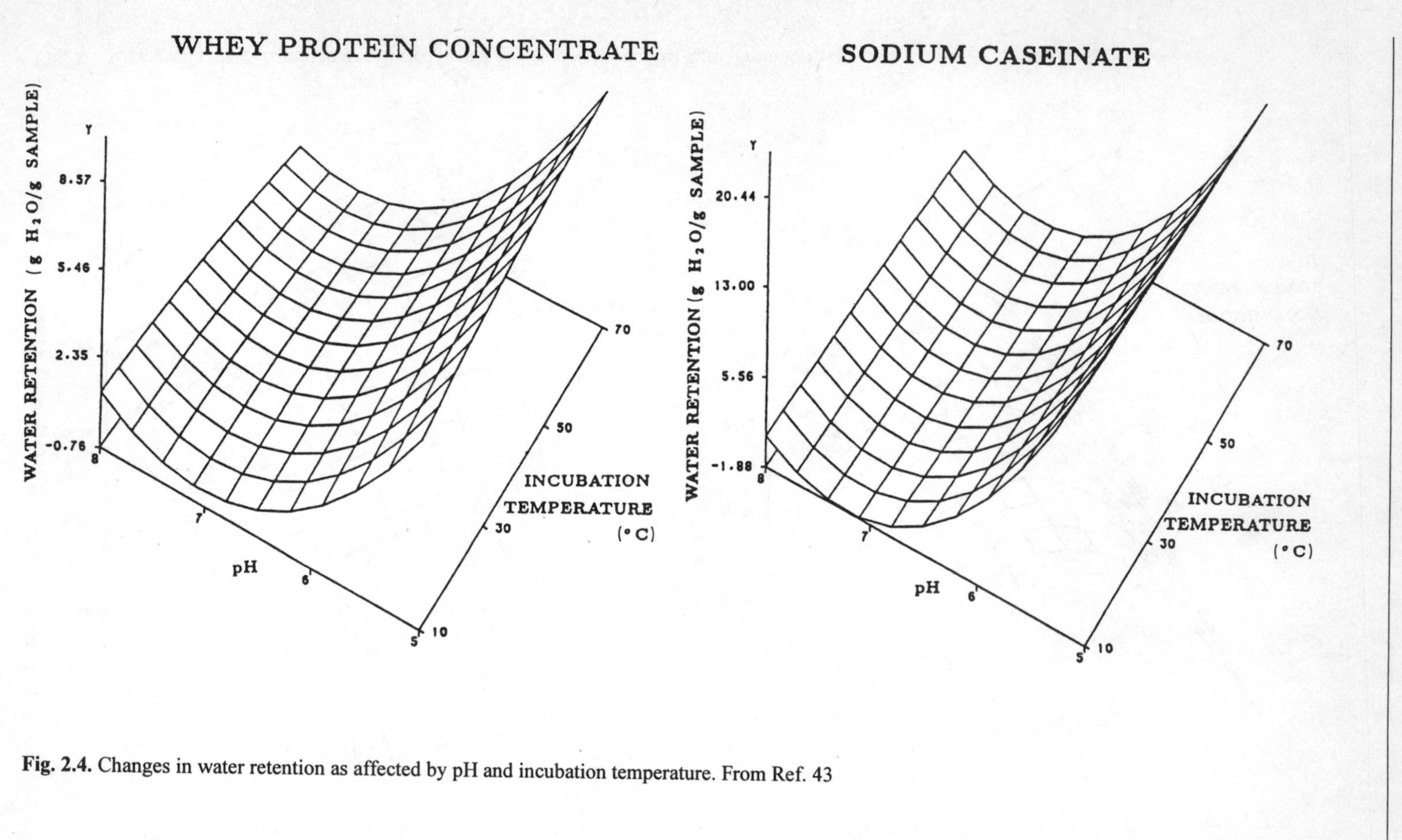

Fig. 2.4. Changes in water retention as affected by pH and incubation temperature. From Ref. 43

bulin is insoluble in water at pH 4.5, whereas Kinsella [5] indicated that the conformation of β-lactoglobulin was pH and temperature sensitive.

Unheated skim milk and milks heated from 60–120 °C were used for the manufacture of the caseinates, and the effect of heating temperature on the WHC was measured [70]. Water retention of protein powders is influenced by their microstructure and surface properties. Preheating of the milk leads to alterations of the functional properties of the milk proteins, mainly caused by denaturation of whey proteins. Sodium caseinate prepared from unheated milk showed poor WHC, and heating of milk at 90 °C for 1 min gave a fourfold increase in WHC. Heating of milk at 120 °C for 1°min improved WHC of Na caseinate. The relationship was established between WHC and whey protein concentration of the caseinates, r = 0.984 [70]. WHC of denatured whey proteins is affected mostly by their porosity and can be explained as a phenomenon of so-called „water affinity".

2.5 Water Holding Capacity of Egg Proteins

Quality of egg products such as omelets and egg burgers that have been precooked, frozen, and thawed prior to serving, has been limited by poor water retention and textural properties. Albumen was found to contribute more to expressible moisture of coagulated whole egg mixture than yolk [74]. Expressible moisture (EM) increased markedly as the percentage of albumen in the mixture increased in the range from 40 to 80%. The increase of % EM was from about 40% for 40% albumen content to 57–58% for 80% albumen content. The percentages of expressible moisture (% EM) for coagulated, frozen, and thawed albumens stored in a frozen state for 1, 3, 5, 8, or 12 days were significantly higher that the % EM for the non-frozen (67.6%) and equaled 74.7, 75.1, 74.3, 73.4, and 76.9%, respectively [74]. The difference in % EM was recorded after at least one day of frozen storage; and there was no significant difference in % EM between one and other tested storage periods. The expressible moisture of the raw, frozen, and thawed albumen was affected by pH in the range pH 4–11. The % EM of both frozen and nonfrozen albumen sharply decreased when the pH of raw albumen increased above 9.0 [74]. The % EM was significantly higher for frozen albumen than nonfrozen at pH's above 9.0. The control nonfrozen albumen at pH 9.0 exhibited approximately 5% difference in % EM comparing to frozen treatments.

Succinylation and pH affected the % EM of the coagulated albumen in the unfrozen and frozen state [74]. The % EM decreased with increasing levels of succinylation up to 0.4% (g anhydride/g albumen). Further levels of succinylation exhibited no decrease in % EM of albumen. Succinylated albumen at a level of 0.2% showed lower % EM than the control albumen for both frozen and nonfrozen treatments. The extent of succinylation for the 0.1, 0.2, 0.3, 0.4, 0.5, 0.6, and 0.7 (w/w) levels was estimated as 1, 11, 21, 32, 41, 56, and 70%, respectively. In succinylated protein at a level of 0.4 g anhydride/g albumen the succinyl groups and their negative charges contributed to greater water retention.

Chemical modification of egg white proteins with oleic acid was performed to improve functionality as a result of changed biophysical characteristics [75]. Oleic acid did not selectively precipitate ovalbumin, conalbumin, or lysozyme, but an increase in negative charge of proteins was observed. Chemically modified egg proteins showed an increase in water retention when compared with untreated material. Cooked, frozen egg white has poor freeze-thaw charac-teristics: it is rubbery, and water separation is observed. Water separation can be prevented by adding 2–4% carbohydrate which binds water [75]. Modification of egg white proteins with oleic acid or sodium dodecyl sulfate (SDS) markedly increased the water retaining index of supernatant from egg white. A similar effect was observed for oleic acid and SDS on the stability of freeze-thawed egg white. No visual evidence of structural damage was observed for modified cooked and frozen albumen. At the concentration 5 and 10 mol, SDS showed a greater effect than oleic acid. At 25 and 50 mol of oleic acid, the water retention index for cooked, frozen, and thawed egg white was 10 times greater than that of the control. The water retention index increased also for cooked unfrozen egg white.

2.6 Water Holding Capacity of Plant Proteins

2.6.1 Soybean Proteins

The increasing utilization of soy protein products is due to several factors including abundance, low cost, good functional properties including water retention, and nutritional quality. Partial denaturation of soy protein during extraction and subsequent drying may influence WHC and other functional properties of soy concentrate (SC) and soy isolate (SI). Soy isolates are utilized in traditional foods and in new food products as nutritional supplements and in formulated foods. Utilization of SC is permitted in the School Lunch Program. Quality characteristics of traditional foods containing soy proteins should be maintained. The soy isolate is produced from soy flour by dissolving the protein at pH 8.0, separating the insoluble components by centrifugation, and precipitating protein at pH 4.5. Precipitate is neutralized and spray dried. Soy isolate can also be produced from soy concentrate. Several forms of textured isolated soy protein are commercially available: heat-structured fibers, spun fibers, and granules.

The hydration rule was implemented that SC is normally hydrated to 75% moisture. At this hydration level the product is 18% protein. Hydration of SC to less than 75% moisture may result in a dry, mealy food. Kinsella [30] summarized the functional properties of soybean proteins, i.e., solubility, water retention, fat absorption, and emulsification. These functional properties demonstrate the desirable uses of soybean proteins in food application. Functional properties of soy proteins are affected by processing conditions and composition of proteins.

Soy protein concentrates and isolates have water absorption common to most proteins. SC contains polysaccharides which also absorb water. Soy protein-water interactions affect functional properties such as swelling, viscosity, gelation, emulsification, and foaming. Soy proteins additionally to water sorption exhibit the capacity to hold water against gravity.

Swelling of protein particles have been used as an indicator of water absorption. Water absorption and holding without protein dissolving is an important function of soy proteins in doughs, comminuted meats, dairy foods, and custards. The addition of 8% soybean flour to tortillas slightly increased water retention [76]. Higher water retention is desirable because sensory properties are retained longer. Soy isolate functionality in baked products includes moisture binding, improvement of moistness of the product, and protein fortification. The loaf of bread made with added soy isolate contained additional water and was softer than a standard bread.

Size of the protein particles and surface topography influences protein-water interactions and water binding. However, samples of commercial soy isolate and soy concentrate showed similar sorption isotherms, and the size of particles of soy isolate had a nonsignificant effect on water retention [5]. Differences in the protein-water interaction in soy proteins are related to sources of protein, degree of protein denaturation, and amino acid composition. Although proteins are the main components of soy preparations participating in water retention during food processing, carbohydrates also play significant roles in swelling and gelation. Protein solubility in soy isolate that affects WHC can be increased by mild heating conditions (25–60 °C) or by increasing the speed of blending.

Water binding data (g of water unfrozen at –50 °C/g of solids) were obtained by NMR studies for four soybean proteins of different preparation/isolation procedures and protein contents: soy isolate, 90% protein; soy concentrate, 70% protein; soy flour, 50% protein; and a carbohydrate-enriched soy fraction, 32% protein [77]. A general trend showed increased water binding with increased protein content, particularly in the protein level range of 60–90%. Increase in water binding with protein content is not linear, suggesting that additional factors to protein content contribute to water binding. The effect of total water content on water binding (unfrozen water content) for soy proteins and ovalbumin was reported [77]. Two different effects of total water on water binding were found. Water binding capacity was invariant with total water content from ~0.3 to ~2.0 g of water/g of solids. Soy isolates markedly increased the water binding capacity between 0.3 and 3 g of water/g of solids. The author related this increase to a swelling phenomenon of the hydrated protein matrix and/or to a change in protein conformation with increasing binding sites. Water binding capacity for soy isolates approaches the value 0.49 g/g calculated from the water binding capacity of the individual amino acid residues, according to the method of Kuntz [3]. Higher water binding capacity of soy isolate compared with soy concentrate may be related to high swelling capacity and possibly dissociation with exposition of additional binding sites. The absence of increased water binding capacity for soy concentrate may be due to the presence of the soybean cellular structure, which inhibits swelling of the hydrated protein. This cell structure is absent in the protein isolates.

Table 2.6. Nitrogen solubility (NSI), spontaneous water absorption capacity (WAC), and oil absorption capacity (OAC) of soy protein isolates

Soy protein isolate	NSI (g sol. protein/-100g protein)	WAC (ml water/g protein)	OAC (ml oil/g protein)
630	16.6	6.2	1.3
610	20.9	6.8	1.2
660	30.9	6.0	1.4
90HE	31.7	8.4	1.4
90HG	41.2	11.3	1.5
90NB	58.0	10.0	1.9
500E	73.5	7.9	1.8
760	82.2	8.8	1.4

Commercial soybean protein isolates were used: Proteinmax 90NB, 90HG and 90HE from Sambra S.A. (Sao Paulo, Brazil) and Purina Protein 500E, 660, 610, 630 and 760 from Raston Purina Co. (St. Louis, MO). From Ref. 78

An increase in water binding capacity was found for sonicated soy concentrate due to disruption of cellular structure. The water binding capacity of sonicated soy concentrate greatly increased upon heating at 100 °C. The effect of sonication on the water binding of soy concentrate could also be due to a change in protein conformation or aggregation, although the ultracentrifuge sedimentation velocity pattern of the soluble protein was not changed by sonication. Water sorption over the water activity range 0.15 to 0.95 was invariant with particle size of soy concentrate. Methods of soy protein isolation and a number of treatments can change their water binding properties.

Nitrogen solubility index (NSI), water absorption capacity (WAC) and oil absorption capacity for various soy protein isolates are shown in Table 2.6 [78]. NSI ranged from very low values as for the 630 and 610 isolates to a very high value as for the 760 isolate. Comparatively high WACs were found for 90HG and 90NB due to different degrees of protein denaturation. Hansen [79] found that the amount of bound water interacting with the soy protein was 0.26 g/g protein. The water binding capacity of the 7S protein was superior to that of the 11S protein over the water activity range 0.10 to 0.97. Bound water associated with soy concentrate was analyzed by a combination of sorption isotherm measurements and pulsed NMR. Bound water was successfully distinguished by NMR and sorption data.

WHC of soy proteins is influenced by ionic strength. The effect of ionic strength is important because sodium chloride is a frequently used component in the formulation of foods. The effect of salts on water retention by proteins is related to the influence on electrostatic interactions. Lopez de Ogara et al. [80] reported that WAC of soy protein isolate was not greatly affected by the addition of solutes: NaCl (2–15%), $CaCl_2$ (2.5–10%), Na_2HPO_4 (2–10%), H_2O glycerol

Table 2.7. Water retention of corn germ protein flour (CGPF), soy flour (SF), soy conentrate (SC), and soy isolate (SI)

	g water/g protein[1]				g water/g sample			
pH	CGPF	SF	SC	SI	CGPF	SF	SC	SI
6	13.43[a]	3.13[c]	5.79[b]	2.00[d]	2.64[b]	1.64[c]	3.67[a]	1.68[c]
7	13.69[a]	3.23[c]	5.32[b]	1.87[d]	2.69[b]	1.70[c]	3.37[a]	1.57[c]
8	14.19[a]	3.27[c]	4.49[b]	0.10[d]	2.79[a]	1.72[b]	2.84[a]	0.08[c]

[a,b,c,d] Means within same row not followed by same letter significantly different at $P < 0.05$.
[1] Water rentention determined at 70 °C and 30 min incubation. From Ref. 82

(6–40%), and sucrose (10 and 20%), but the time required to reach equilibrium was modified. Electrolytes slowed the hydration rate of soy isolate to a greater extent than non-electrolytes. WAC was decreased approximately 10% at low studied concentrations of electrolytes (2–2.5%). At higher concentrations an increase in WAC was observed.

Soy protein isolate water absorbing capacity was affected by protein solubility and the absence of salt [81]. The WAC was high in samples with NSI in the range from 36.3 to 68.1% leading to the formation of viscous dispersions. The presence of 0.2 M NaCl produces a decrease of WAC as a result of increased protein aggregation in all samples. Comparison of solubility values, NSI with WAC (in different solutions) showed that protein solubility and water absorption phenomena involved different mechanisms, and they depended on different factors. Soy isolate samples with a higher WAC corresponded to those having a lower NSI NaCl in a 0.2 M NaCl solution (Fig. 2.5) [81].

Water retention of soy flour (SF), soy concentrate (SC) and soy isolate (SI) was compared with corn germ protein flour (CGPF) [82]. Water retention of soy and other plant proteins is an important factor that influences basic quality characteristics and yield of comminuted meat products. High water retention of samp-

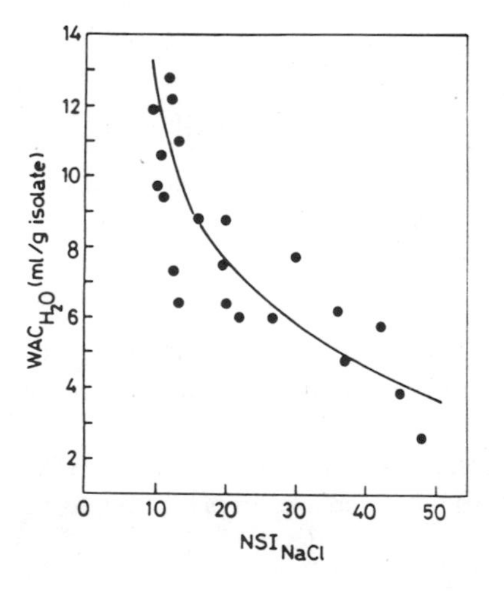

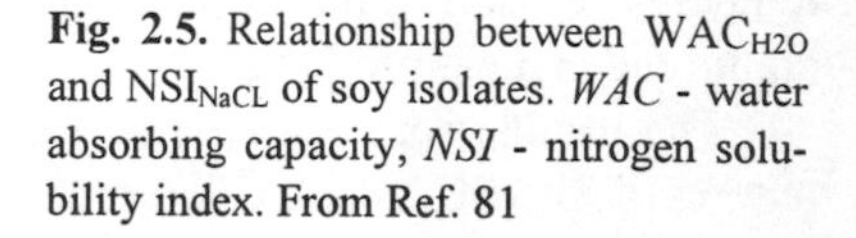
Fig. 2.5. Relationship between WAC_{H2O} and NSI_{NaCL} of soy isolates. *WAC* - water absorbing capacity, *NSI* - nitrogen solubility index. From Ref. 81

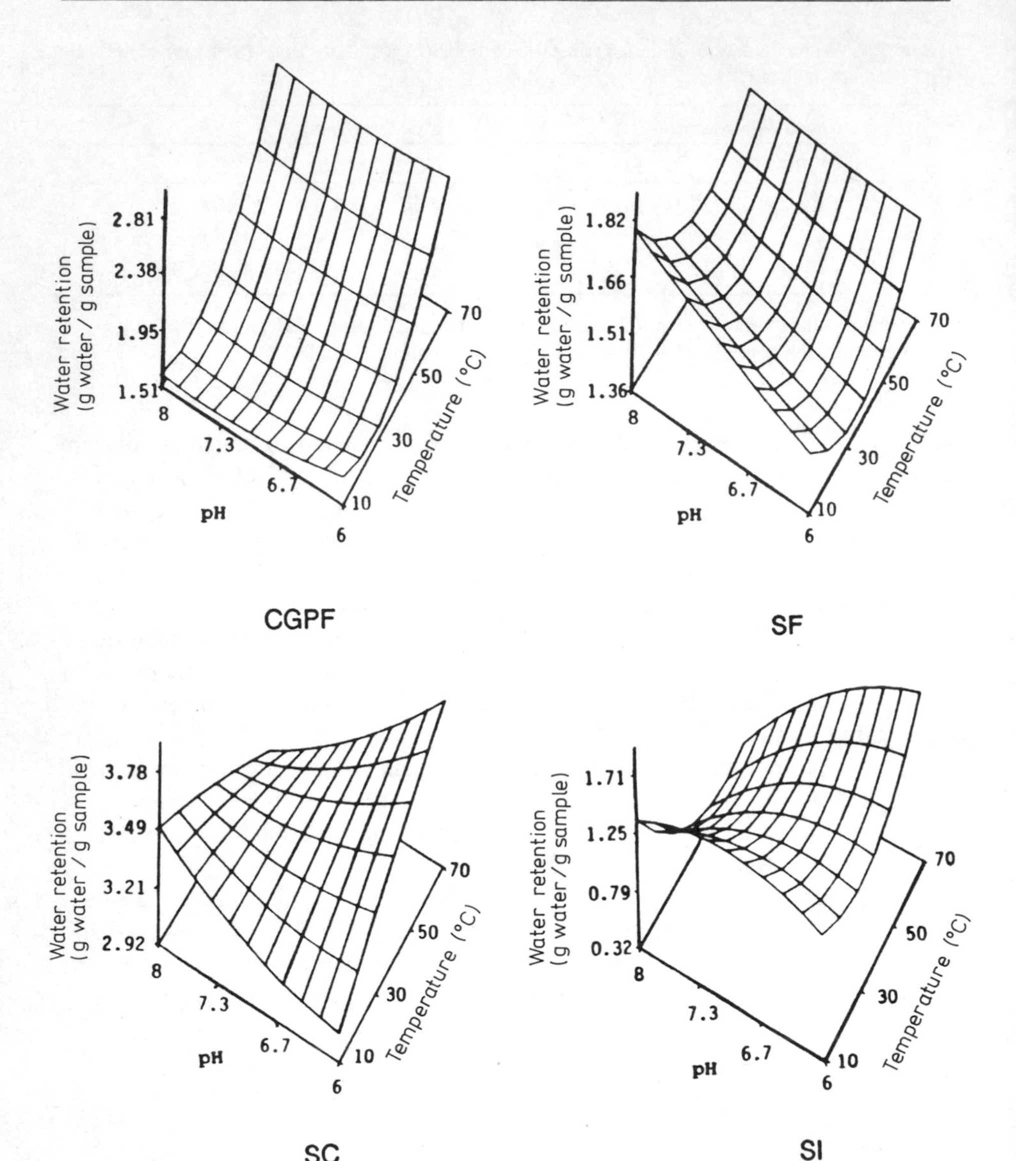

Fig. 2.6. Water retention of CGPF, SF, SC, and SI as function of pH and temperature of incubation. (*CGPF*: corn germ protein flour; *SF*: soy flour; *SC*: soy concentrate; *SI*: soy isolate). From Ref. 82

les. Water retention of SF, SC, SI, and CGPF was determined for ranges of pH, temperature, and time of incubation that could be applied to processing of comminuted meat products.

The results from response surface methodology (RSM) showed that water retention of SF, SC, and CGPF tended to increase with increases in pH from 6 to 8

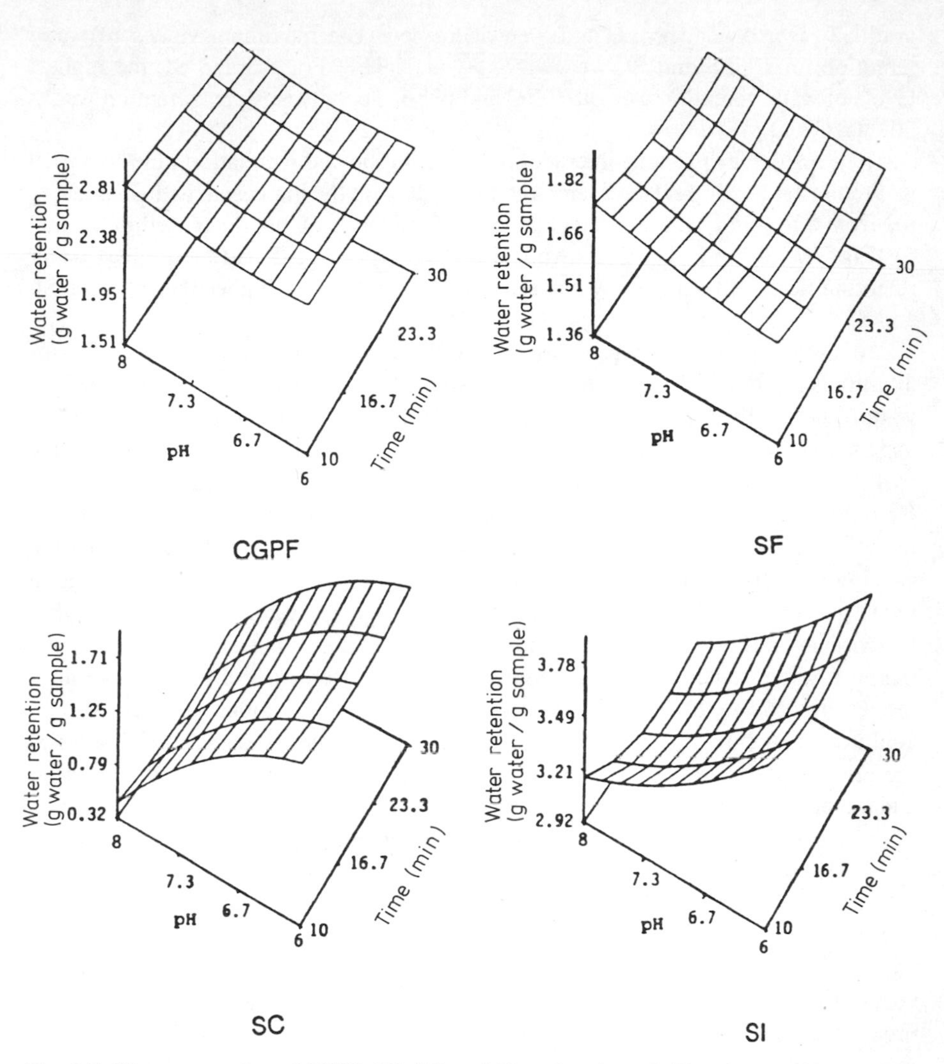

Fig. 2.7. Water rentention of CGPF, SF, SC, and SI as function of pH and time of incubation. (*CGPF*: corn germ protein flour; *SF*: soy flour; *SC*: soy concentrate; *SI*: soy isolate). From Ref. 82

(Fig. 2.6, CGPF, SF, and SC) [82]. With increasing incubation temperature, the pH effect became less significant. This might be related to the increase of starch gelatinization because of the high content of starch and crude fiber in CGPF. Water retention of SF, SC, SI, and CGPF samples notably increased at pH 6 and at incubation temperatures in the range of 40–70 °C (Fig. 2.6). However, only SF showed increased water retention with increasing pH from 6 to 8 and temperatures from 40 to 70 °C (Fig. 2.6, SF). Water retention of SI slightly increased with pH at the low incubation temperature (10 °C). However, at 20 to 70 °C, water retention gradually increased with pH and then decreased. The maximum values of water retention for CGPF and SF were at 70 °C and pH 8. For SC and SI, the highest levels of water retention were at 70 °C and pH 6. The lowest water retention was at 20–30 °C and pH 6.

The results for pH-time interaction (Fig. 2.7) showed incubation times from 10 to 30 min had no effect on water retention. It slightly increased with increasing pH from 6 to 8 in CGPF and SF (Fig. 2.7, CGPF and SF), but decreased in SC and SI (Fig. 2.7, SC and SI). As shown in Figs. 2.6 and 2.7, SC had the highest water retention per gram sample. Water retention of CGPF was higher than that of SF and SI.

Water retention of soy proteins and CGPF preparations incubated at 70°C for 30 min as related to protein content (g water/g protein) and weight of samples (g water/g sample) is presented in Table 2.7 [82]. In relation to protein content,CGPF appeared to have the highest water retention regardless of pH. However, protein was not the only factor that influenced water retention during heating. Carbohydrates played a significant role during incubation. Particularly, starch granules can swell extensively in the presence of excess water during heating causing gelatinization. Therefore, water retention of plant protein presented as g water/g sample was more meaningful than as g water/g protein. Higher temperature and longer time of incubation also could enhance starch swelling and increase water retention. The gelatinization temperature of corn starch granules ranged from 62 to 72 °C [83]. Therefore, the incubation temperature of a system could increase starch gelatinization in high-starch samples, such as CGPF and SF, thereby increasing water retention. Experimental data related to protein solubility and water retention of three soy proteins and CGPF could be used in development of processes for their utilization as protein supplements in meat products.

WHC of soybean protein was increased by modification. The most effective methods were addition of urea and heating in dry form at high temperature in a vacuum [84]. The optimum concentration of urea allowing maximum WHC of soybean protein was around 0.5 M, and the optimum heating temperature in the vacuum was 150–155 °C. Disruption of the structure of native soybean protein by urea treatment increased aggregation during drying and heating resulting in the formation of a large network of protein molecules. Using modified soybean protein, considerably softer gels have been obtained by mixing with water. An increase was found in water absorption capacity (WAC) of SC prepared by the water leaching process and modified by treatment with SDS [85]. This was related to disruption of protein-protein hydrophobic bonds and an increased negative charge on the protein by SDS treatment. The maximum WAC increase (1.3 times) was found at pH 6 and 0.5% SDS or 7 ml H_2O/g soy concentrate.

The mechanism of bromelain action on soy 11S globulin was reported [86]. Acidic subunits surrounding basic subunits in 11S globulin molecules are unfolded but still kept globular. These molecules associate and form the aggregates with M.W. 8|000|000. After bromelin treatment, dissociation occurred with degradation products recombining and with the appearance of smaller fragments of M.W. 15|000. These fragments, associated mainly through hydrophobic interactions and disulfide bonds, form a network structure. WHC of this coagulum increased 2–2.5 times more than that of the native and heated protein.

2.6.2 Pea and Bean Proteins

Pea seeds contain 20–30% protein and the major proteins in pea are legumin and vicilin. Legumin contains a high level of essential sulfur containing amino acids. There are several similarities between pea legumin and vicilin, and soy glycinin and conglycinin. Pea protein isolate containing 85–86% protein exhibited lower water hydration capacity (2.80–3.21 g H_2O/g) than soy concentrate and isolate (5.52 and 5.85 g H_2O/g) respectively [87]. However, WHC of pea protein was markedly higher than in gluten (1.37 g H_2O/g). Paredes-Lopez [88] reported that WAC of chickpea protein isolates with 84.8–87.8% protein was higher for micelle protein (4.9 ml/g protein) than for isoelectric protein (2.4 ml/g protein) but lower than for soy isolate (5.7 ml/g protein). Conformational changes of pea protein molecules during processing may increase the amount of polar sites that interact with water. WHC increased in proportion to protein contents of the flour, protein fraction and isolate of field peas and faba bean [89]. WHC increased substantially by increasing temperatures from 21 to 71 °C. Water absorption values for pea and soy isolate sodium proteinates were much greater (205–282% and 698%, respectively) than their corresponding isoelectric protein isolates (112–191% and 155%, respectively) [90]. The highest water absorption value of 698% was obtained with spray-dried soy sodium proteinate, followed by pea sodium proteinates drum-dried, 283%; spray-dried, 250%; and freeze-dried, 205%.

The chickpea samples exhibited the lowest WHC in all milling fractions, probably due to its lipid content which is higher than in other legumes [91]. The roasted fractions showed higher WHC (1.35 ml/g) than nonroasted (1.19 ml/g), due most likely to differences in the finer particle size distribution of the roasted samples. Increase of WHC in heated bean protein can be explained by the protein dissociation and denaturation with unfolding of the protein polypeptide chains [92].

The WAC of freeze-dried protein powders of cowpea, peanut, and soybean prepared from unfermented and fermented extracts was generally similar [93]. The WAC of peanut protein powder was lower than other tested proteins. The differences in WAC are probably related to different residue lipids in the samples. Cowpea milk contained 0.1% lipid, soybean milk 0.6% and peanut milk 1.6%, respectively.

Water absorption capacity of black gram bean protein flour and concentrate were 1.6 and 5.9 g/g, respectively, and were similar to WAC of other bean proteins [94].

Acylation of winged bean protein increased its WAC, and the effect was influenced by the degree of acylation [95]. Acetylation and succinylation had the same effect on the WAC of winged bean protein. The highest degree of acylation gave a twofold increase in WAC. Chemical and physical changes of protein during acylation resulted in an increase of WAC. Physical entrapment of water is obtained by unfolded protein molecules. In acetylated protein, despite an increase in the number of hydrophobic groups, WAC increased, probably due to the dissociation that exposed a greater number of other hydrophilic groups.

2.6.3 Sunflower Proteins

Defatted sunflower meal has potential in the food industry because of the high level of protein and an absence of toxic substances. Limiting amino acids in sunflower protein are lysine and threonine; sulfur amino acids are not limiting. The PER of sunflower protein concentrate is 2.0, compared with casein, 2.5. The water binding capacity of sunflower proteins tended to increase with the concentration of protein in the product. The sunflower protein isolates were high in protein; they exhibited low viscosity values. Defatted sunflower meals obtained from different varieties exhibited similar water sorption at different relative humidities. WAC of the albumin denatured fraction of sunflower protein (223.8 ml/g) was markedly higher than native globulin (70.0 ml/g), denatured globulin (70.9 ml/g) and sunflower meal (95 ml/g) [96]. WAC of the globulin fraction was not affected by the denaturation process.

Hydration characteristics of sunflower flour and protein isolates obtained from aqueous, alkali, and salt solutions were reported [97], and sunflower flour had the highest rate of absorbed water. Water-soluble sunflower protein isolate adsorbed the highest amount of moisture at all levels of relative humidity. However, at the relative humidity range 23–68%, the alkali- and salt-soluble sunflower protein isolate adsorbed a similar amount of moisture. Studies on the level of bound and free water with DSC showed that the limit of water was highest in sunflower flour (29.2 g/100 g) and water-soluble sunflower protein isolate (28.3 g/100 g). The levels in alkali-soluble and salt-soluble sunflower protein isolates were similar (25.4 and 24.7 g/100 g, respectively). Hydration characteristics of the various sunflower proteins were similar to soy protein isolate (24.8 g/100 g).

The water hydration capacity of sunflower, canola, and soybean proteins was tested in comparative studies [98]. The products had significantly differing water hydration capacities, with sunflower, 2.9; canola, 1.8; and soybean, 3.4 g water/g product. The carbohydrate component in sunflower and soybean products contributed to water hydration. Residues of fiber possess high WHC, and small amounts of crude fiber would considerably increase WHC. WHC of sunflower and canola products was significantly increased by trypsin treatment, however WHC of soybean product decreased. Possibly, a more close packing arrangement of the trypsin-treated soybean particles limited the WHC. Partial hydrolysis of sunflower protein isolate by trypsin and pepsin affected functional properties [99]. Moisture adsorption of sunflower isolate hydrolyzed for 1 hr (5% hydrolysis) with pepsin markedly increased, from 15.8 to 94.2 g H_2O/100 g sample, and longer hydrolysis periods resulted in a small increase in moisture adsorption. The hydrolysis of a few key peptide bonds at the beginning of the hydrolysis resulted in a significant change in moisture adsorption.

2.6.4 Corn Proteins

The functional properties of corn protein preparations reflect the composition of the sample, the nature and reactivity of proteins, their native structure, and interactions with nonprotein components of corn germ protein as carbohydrates and lipids. They are affected also by environmental and processing conditions. The model systems used in our experiments for testing the functionality of defatted corn germ protein flour (CGPF) provided valuable information about use of CGPF. Water retention of defatted CGPF can be used to define how this protein can be added to existing foods and how it can replace more expensive proteins traditionally used. The amount of water present in defatted CGPF will depend on the extent to which the dry CGPF absorbs or desorbs water under various environmental conditions.

Food additives of plant origin with a high content of carbohydrates were reported to neither participate in the water binding property nor improve the emulsifying capacity of meat emulsions [50]. It was found that carbohydrate-rich corn protein had high water retention [100]. The functional properties of protein additives from cereal sources justify the use of these relatively low cost raw materials.

A good CGPF source can be obtained by drying corn germ meal at low temperatures and extracting the oil by solvent extraction [100]. At present, the conventional oil extraction process with hexane as a solvent leaves certain lipids in the flour, which reduce its quality. Oxidation of the residue oil in defatted CGPF will produce an unpleasant off flavor in the product. Sufficient storage stability of defatted CGPF can be obtained as the result of a high level of fat extraction.

Corn germ protein is totally different from that of corn gluten meal. Proteins of the wet milled germ are modified during sulfurous acid steeping. Commercially produced wet corn gluten meal is flash dried at a high temperature to produce dry corn gluten meal (60% protein and 10% moisture). In spite of its high protein and abundance of methionine, corn gluten meal has poor functional properties and nutritional quality. Christianson et al. [101] studied a supercritical CO_2 method to extract oil from corn germ meal and obtained a food-grade quality, defatted corn germ product. However, currently the corn milling industry is using the hexane extraction process.

In the production of defatted CGPF, particularly during fat extraction, removal of hexane residues from the defatted CGPF by the conventional desolventization procedure can affect the quality of the finished product. Quality deterioration is related to changes of flavor and color properties as the result of protein denaturation and a browning reaction. Although studies have been reported on the refining processes and properties of corn gluten and corn germ flour [101] and their use in fortifying some foods [100], very little work has been done on the functional properties of the corn germ proteins. The author has conducted extended studies on the basic factors affecting functionality, including water retention of defatted CGPF in model system and as an additive in comminuted meat products [102].

The water retention of hexane defatted CGPF obtained by modified and conventional processes was determined over a range of temperatures and pH values corresponding to various usage in food processing. Water retention of defatted CGPF obtained by the modified process significantly increased as the temperature of incubation increased from 30 °C to 80 °C (Figs. 2.8 and Fig. 2.9) [102]. Times of incubation ranging from 10 to 30 min had no significant effect on water retention of modified CGPF (Fig. 2.8). However, the highest water retention was reached at 80 °C and 30 min of incubation. A pH of the solution in the range from 6 to 7 did not have effect on water retention of modified CGPF (Fig. 2.9).

Similar surface responses were obtained for CGPF processed by the conventional method. Temperature of incubation also was an important factor in water retention of conventionally processed CGPF. Water retention of the CGPF increased as the temperature of incubation increased from 35 °C to 80 °C (Figs. 2.10 and 2.11) [102]. At incubation temperatures of 40 °C to 80 °C, CGPF processed by the conventional procedure retained 15–20% more water than CGPF processed by modified procedure. However, water retention of CGPF processed by the conventional procedure at low incubation temperatures of 5 °C to 35 °C was significantly less than for CGPF processed by the modified method. Surface responses for water retention at the temperatures from 5 °C to 30 °C are of particular importance because temperatures from 5 °C to 18 °C are applied during sausage batter processing. Water absorption by corn germ preparations, attributed to the protein content, also is affected by starch content and a number of other factors. A pH between 6 and 7 in the solution did not have a significant effect on water retention of CGPF processed by the conventional procedure (Fig. 2.11). In fact, CGPF preparations showed a particularly low response to pH activation. The pH activation process in the range pH 6 to 7 did not improve the water retention properties of CGPF processed by either the modified or the conventional method.

Water retention was higher for supercritical extracted carbon dioxide (SC-CO_2 CGPF) than for hexane CGPF [102]. Both proteins had a similar response surface, which was low at low incubation temperature around 20 °C to 25 °C. The maximum value of water retention was obtained after 10 min incubation at 70 °C. Water binding of SC-CO_2 CGPF increased significantly at the range of temperatures 35–70 °C. This was probably due to increased protein solubility.

The pH of solution affected the water retention of SC-CO_2 CGPF; water retention slightly increased as pH increased from 6.0 to 6.8.

Water retention capacity of defatted CGPF preparations reflect the composition of the sample. The overall pattern showed that in model systems CGPF processed by a modified procedure had better functional properties than CGPF processed by the conventional method. CGPF processed by the conventio-

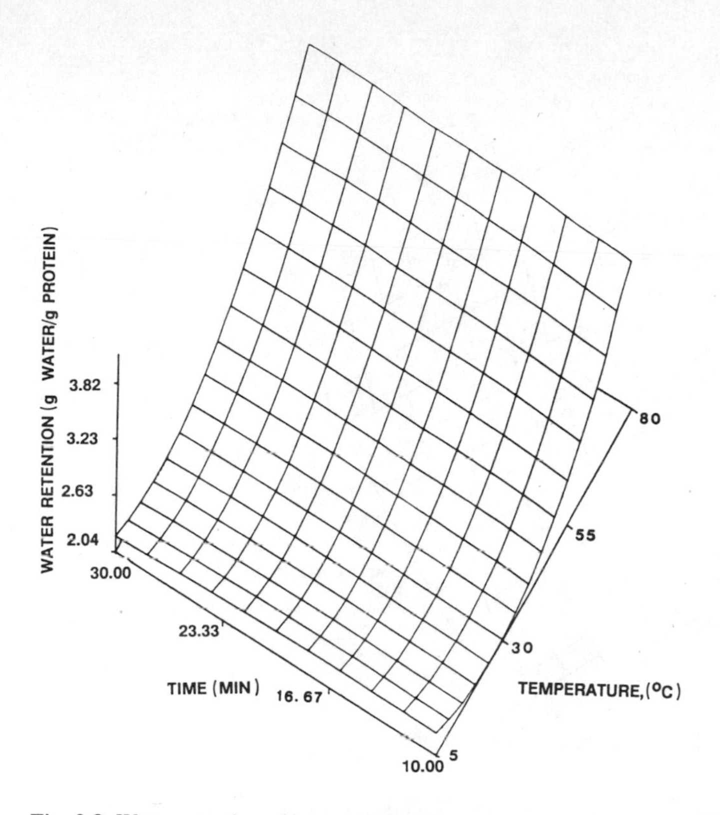

Fig. 2.8. Water retention of hexane-defatted corn germ protein processed by modified procedure as a function of incubation time and temperatur. From Ref. 102

nal procedure had more yellow color, more fat residue, and a higher level of protein denaturation by heat during fat extraction and desolventization. Because of the better functional properties, blend flavor and white color, higher water retention, lower fat and higher protein contents, CGPF processed by the modified procedure was recommended for utilization as an additive in comminuted and other meat products. Defatted CGPF has the potential for use in the food industry because of its high protein content, nutrional quality, functionality, and absence of toxic substances.

2.6.5 Wheat Proteins

Water is one of the major ingredients in a dough. The interactions between flour proteins and water are critical, as dough is hydrated and mixed. Hydration of all flour components, mainly protein and starch, is a prerequisite for proper dough formation. After mixing flour and water during dough preparation, gliadin and glutenin interact with water, with each other, and with other flour components,

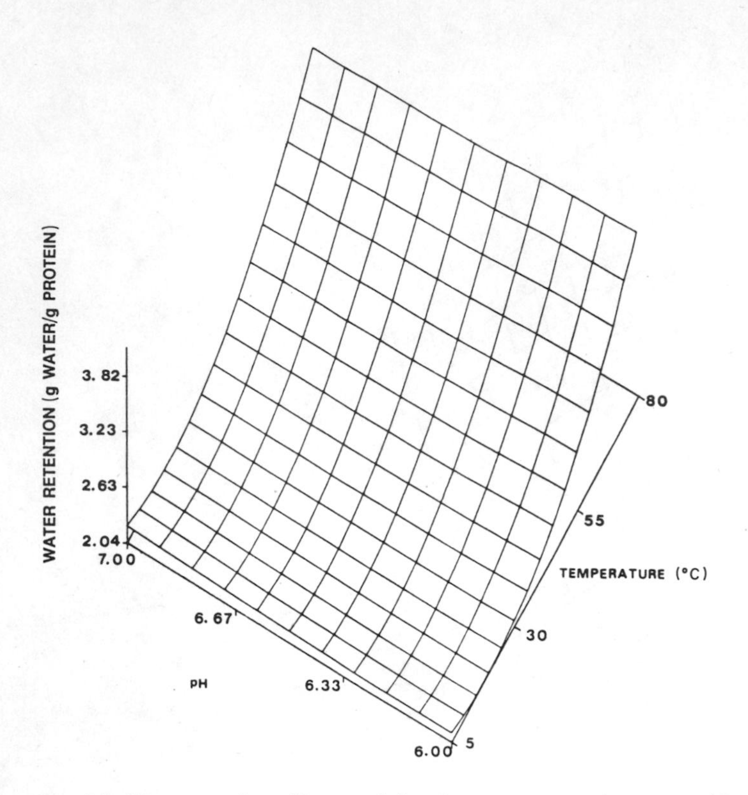

Fig. 2.9. Water retention of hexane-defatted corn germ protein processed by modified procedure as a function of pH and temperature of incubation. From Ref. 102

such as lipids, starch, sugars, pentosans, and soluble proteins. The water hydra-hydration capacity of wheat flour solubles ranges from 2.25 to 3.15 ml/g and is similar to that of soy protein isolate Supro 710 [103]. During dough mixing gluten can retain a significant amount of water in the structure, producing a dough with up to 60% water content. The rheological properties of dough are affected by the ratio of proteins in flour and the formation of bonds during mixing.

Functionality of glutenin in breadmaking is determined by physical properties (molecular size and shape) and chemical properties, i.e. amino acid composition, sequence and tendency to aggregate. Elastic and viscoelastic properties of gluten (and dough) are influenced by glutenin which possesses high molecular weight and hydrophobic interactions by nonpolar amino acid residues. Elastic properties of glutenin are developed by the formation of interpolypeptide disulfide bonds in the glutenin. Glutenin contains a relatively high amount of hydrophobic acids such as leucine, which contributes to the formation of hydrophobic bonds.

During mixing of dough components and water, a three dimensional structure of dough is formed in which particles of gluten are incorporated into thin membranes within which are embedded the granules of starch and other components

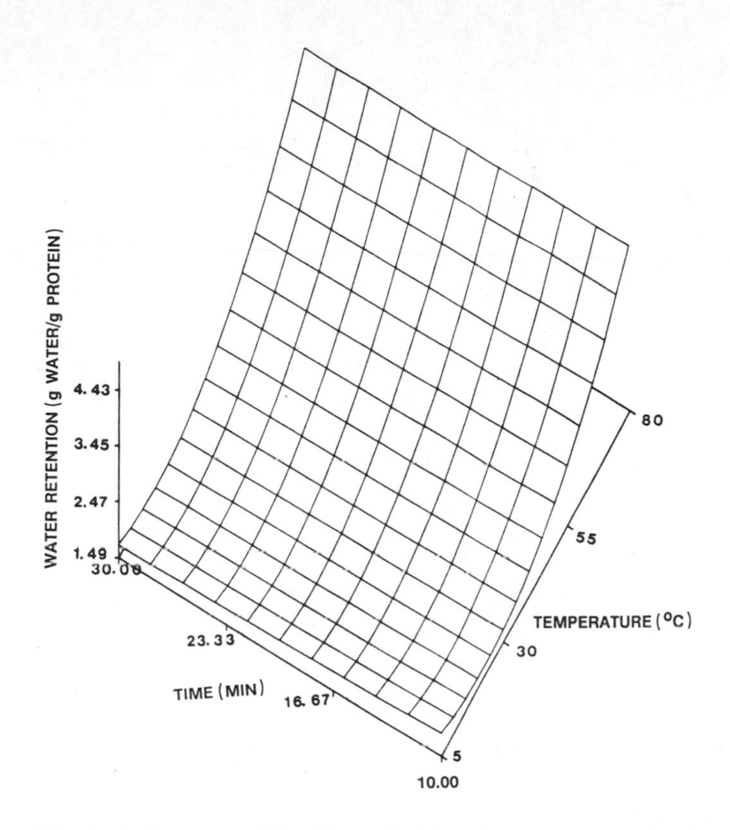

Fig. 2.10. Water retention of hexande-defatted corn germ protein processed by conventional procedure as a function of incubation time and temperature. From Ref. 102

of flour. Dough properties and optimum gluten structure is formed as a result of the presence of many weak secondary bonds and interactions. In the three dimensional structure of dough, carbon dioxide produced by the yeast cells is retained. Membranes in three dimensional structure can expand to a certain degree because gluten proteins possess viscoelastic properties. Wheat gluten in breadmaking is the basic film-forming component. In bread, proteinaceous films with starch granules are formed around gas bubbles. During baking, thermal denaturation of proteins reduces their water binding capacity when the temperature of the crumb reaches about 60 °C to 70 °C. Wheat gluten has been commercially produced. Small amounts of vital wheat gluten are used as extenders in comminuted meats and fish products. Incorporation of vital gluten is related to its high water retention and emulsifying capacity. Extremely high WHC was found for phosphorylated gluten [84]. Groups in gluten molecules which react with phosphoric acid were aliphatic hydroxyl groups.

Water retention of wheat germ protein flour (WGPF) at pH 4–8 and temperature of incubation 5–70 °C is presented in Fig. 2.12 [104]. A rapid increase in

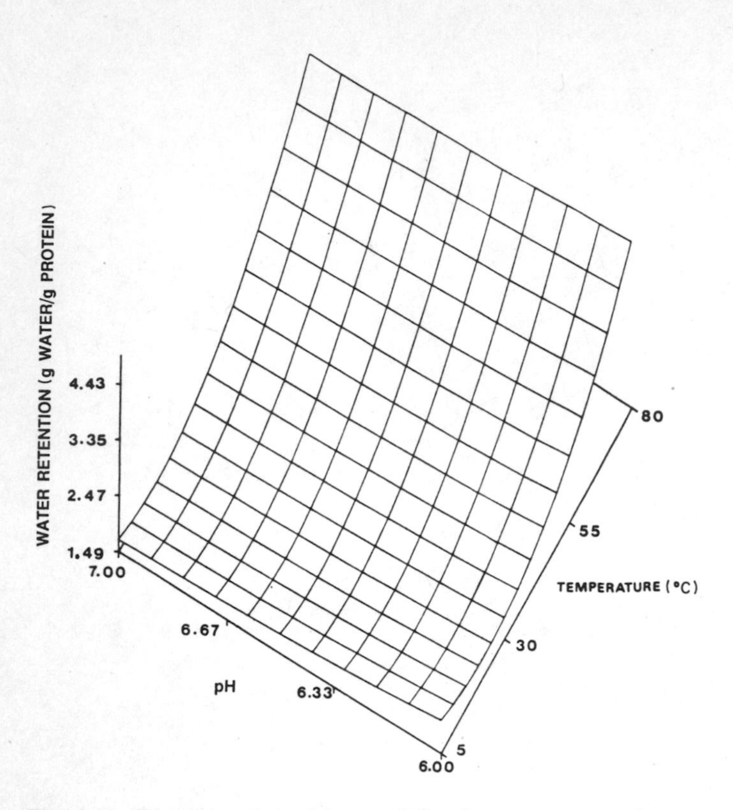

Fig. 2.11. Water retention of hexane-defatted corn germ protein processed by conventional procedure as a function of pH and temperature of incubation. From Ref. 102

water retention was observed in pH range 5–8 at all tested temperatures (5, 15, 30, and 70 °C). Water retention of WGPF was at a maximum at pH 8 and at a minimum at pH 4 (isoelectric pH range). Since the networks of proteins are tightened in the isoelectric pH range, less water could be immobilized. The pH also affects the magnitude of the net charge on the protein molecules. The protein component in WGPF is not the only factor that influences water retention during heating and incubation. Carbohydrates and fiber residues which account up to 60.5% appeared to play an important role in increasing water retention of WGPF. Heating of WGPF up to 70 °C induced a considerable increase in swelling capacity. Water retention of other plant proteins such as sunflower protein concentrate and soy concentrate increased during heating at 97–100 °C.

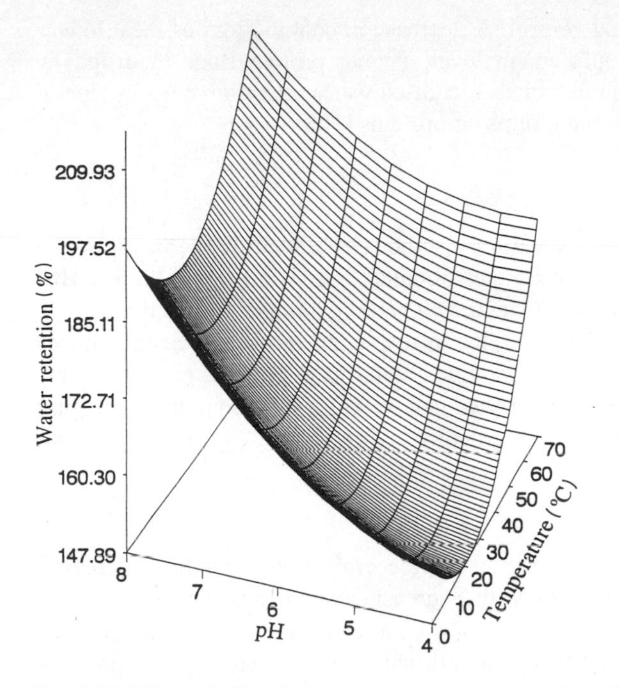

Fig. 2.12. Effect of pH and temperature on water retention of wheat germ protein flour. From Ref. 104

2.6.6 Miscellaneous Proteins

Rapeseed Proteins

Rapeseed protein concentrate (RPC) with light color, using sodium hexametaphosphate has been produced [105]. The water absorption capacity of RPC was half (137%) that of rapeseed flour (266%) and 3–4 times less than that of soy isolate (478%). Insoluble proteins of rapeseed flour and soy isolate retained more water than RPC. The RPC containing beef patties retained more of the initial moisture and gave higher yield than the control after cooking.

Peanut Proteins

Water retention of peanut protein isolate is comparable to those of other oilseed proteins. Peanut flour with marked increase in water retention was obtained. WHC of peanut protein isolates from various cultivars ranged from 1.30 to 1.60 g H_2O/g protein isolate [106]. Mild heating of peanut protein (100 °C for 30 min or 120 °C for 10 min) enhanced WAC, but excessive heating (120 °C for 45 or 150 min) impaired WAC [107]. These heating conditions caused inactivation of trypsin inhibitor. At the 20% replacement in meat loaves, ground peanut protein cooking loss was similar to the all-meat control. A decrease in cooking loss of meat loaves was observed at the 30% replacement level. Peanut protein flour modified by controlled fungal fermentation exhibited increased water retention capacity due to increased number of hydrophobic groups on proteins [108].

Guar proteins

Guar protein isolates prepared by various methods exhibited WAC in the range 74–187 g/100 g sample [109]. The isolates obtained from the 1.0 and 0.25 N HCl extracted meal showed WAC of 187 and 180 g/100 g sample, respectively. The protein isolates from the aqueous alcohol-extracted and autoclaved meals exhibited lower WAC, 74–98 g/100 g sample. Water absorption of guar meal was significantly higher than that of guar protein isolate. Guar meal contains some guar gum, a highly effective hydrophilic ingredient.

Rice Proteins

Water retention capacity of heat-coagulated rice bran protein slightly increased with increasing pH from 5 to 7 and with decreasing temperature damage during dehydration [110]. The effect of dehydration of already heat-coagulated proteins on water retention capacity is due to additional structural changes of proteins during dehydration.

Cottonseed Protein

WAC of cottonseed protein flour with 47.9–51.6% protein was in the range 280–291 g H_2O/100 g sample [111]. Co-precipitation of cottonseed with soybean and peanut proteins increased their WAC over the cottonseed isolate and flour [112]. Co-isolates exhibited higher hydrophilic than lipophilic properties.

Oat Proteins

The water hydration capacity (% water) of the oat meals and flours were similar, 118–130% of the sample weights, and the oat brans showed water hydration capacity of about 200% [113]. The process of defatting did not affect water hydration capacity of oat products. Water hydration capacity of oat flours was higher than wheat, corn, and rice flours. Oat proteins had water hydration in the same range as soy isolate. Water hydration of individual oat protein fractions was tested and it was shown that albumins and glutelins exhibited higher water hydration than globulins and prolamins [114].

Safflower Proteins

Functionality of safflower protein isolates obtained by the micellization procedure (dilution in cold water) and isoelectric precipitation were compared [115].

Safflower protein isolate obtained by micellization procedure showed higher water absorption (1.80 ml/g protein) than those obtained by isoelectric precipitation (1.29 ml/g protein). This was related to more polar and ionic groups exposed, allowing accelerated interaction with water.

Linseed Proteins

The WAC of linseed protein flour and protein boiled for 15 min was 345 and 443 g/100 g of flour, or 717 and 630 g/100 g protein, respectively [116]. Polar amino acids of linseed protein contribute considerably to water absorption properties. Heating facilitates additional binding of water as a result of available additional binding sites, and due to gelation of carbohydrates and swelling of crude fiber. High-mucilage linseed proteins showed a water binding capacity twice as high as their low-mucilage counterparts [117]. Enhanced water absorption was probably related to an increased number of hydrophilic groups in pentosans.

Potato Proteins

Wojnowska et al. [118] reported different swelling capacities for dried potato proteins with different treatments after ultrafiltration concentrations to 1:5, 1:10, and 1:15; the swelling capacity was 0.33, 0.55, and 0.49 ml H_2O/g protein, respectively.

Spirulina and Yeast Proteins

Spirulina protein flour (18% protein), concentrate (27% protein) of *Spirulina* cells, and soybean meal exhibited WACs of 220 g, 109 g, and 230 g/100 g sample, respectively [119]. The WAC was affected by pH and protein denaturation. Methods based on chemical phosphorylation for the separation of proteins from nucleic acids with protein recovery of 75–80% have been developed [120]. WHC of phosphorylated yeast protein was markedly enhanced as a result of phosphorylation of 30% of the lysine of yeast nucleoprotein. WHC increased from 10 to 25 g water/g of protein between pH 6 and 7.5. An increase in WHC was probably related to the increased hydration of the phosphoryl groups added by modification that can bind 5–6 H_2O molecules per group. Increased WHC of phosphorylated yeast protein could be partly caused by the ionization of the phosphoryl groups. Another factor of increased WHC was formation of the gellike network structure with water entrapped in the network.

References

1. Hermansson, A. M. (1986). Water- and fat holding, In *Functional Properties of Food Macromolecules* (J. R. Mitchell and D. A. Ledward, eds.), Elsevier Appl. Sci. Publ., London and New York, p. 273.

2. American Assoc. of Cereal Chemists Technical Committee (1981). *Cereal Foods World*, *26*: 291.
3. Kuntz, Jr., I. D. (1971). Hydration of macromolecules. 3. Hydration of polypeptides, *J. Am. Chem. Soc.*, *93*: 514.
4. Fox, P. F. and Condon, J. J. (1981). *Food Proteins*, Applied Science Publishers, London and New York.
5. Kinsella, J. E. (1984). Milk proteins: Physicochemical and functional properties, *CRC Crit. Rev. Food Sci. Nutr.*, *21*: 197.
6. Hamm, R. (1975). Water holding capacity of meat, In *Meat* (D. J. A. Cole and R. A. Lawrie, eds.), Butterworths, London, p. 321.
7. Hamm, R. Properties of meat proteins. In *Proteins as Human Food* (R. A. Lowrie, ed.), AVI Publishing Co., Inc., Westport, CT, p.167.
8. Fretheim, K., Samejima, K., and Egelandsdal, B. (1986). Myosins from red and white bovine muscles: Gel strength (elasticity) and water holding capacity of heat-induced gels, *Food Chem.*, *22*: 107.
9. Goll, D. E., Robson, R. M., and Stromer, M. H. (1977). In *Food Proteins* (J. R. Whitaker and S. R. Tannenbaum, eds.), AVI Publishing Co, Inc., Westport, CT, p. 121.
10. Offer, G. and Trinick, J. (1983). On the mechanism of water holding in meat: the swelling and shrinking of myofibrils, *Meat Sci.*, *8*: 245.
11. Offer, G. (1984). The nature of the repulsive forces between filaments in muscle and meat, *Meat Sci.*, *10(2)*: 155.
12. Sadowska, M., Sikorski, Z. E., and Dobosz, M. (1980). The influence of collagen on the rheological properties of meat homogenates, *Lebensm.-Wiss. u.-Technol.*, *13(5)*: 232.
13. Murthy, T. R. K. and Bachhil, V. N. (1980). Influence of pH on hydration during spoilage of pork at refrigeration temperature, *J. Food Sci. Technol.*, *17*: 201.
14. Wismer-Pederson, J. (1978). Water, In *The Science of Meat and Meat Products* (J. F. Price and B. S. Schweigert, eds.), Food and Nutr. Press, Westport, CT, p. 177.
15. Honikel, K. O., Hamid, A., Fischer, C., and Hamm, R. (1981). Influence of postmortem changes in bovine muscle on the water-holding capacity of beef. Postmortem storage of muscle at various temperatures between 0° and 30 °C, *J. Food Sci.*, *46*: 23.
16. Miller, A. J., Ackerman, S. A., and Palumbo, S. A. (1980). Effects of frozen storage on functionality of meat for processing, *J. Food Sci.*, *45*: 1466.
17. Vidya Sagar Reddy, G. and Srikar, L. N. (1991). Preprocessing ice storage effects on functional properties of fish mince protein, *J. Food Sci.*, *56*: 965.
18. Park, J. W., Lanier, T. C., Keeton, J. T., and Hamann, D. D. (1987). Use of cryoprotectants to stabilize functional properties of prerigor salted beef during frozen storage, *J. Food Sci.*, *52*: 537.
19. Shenouda, S. Y. K. (1980). Theories of protein denaturation during frozen storage of fish flesh, In *Advances in Food Research*, v. 26 (C. O. Chichester, ed.), Academic Press, New York, p. 275.
20. Hamm, R. (1981). Post-mortem changes in muscle affecting the quality of comminuted meat products, In *Development in Meat Science* (R. Lawrie, ed.), Applied Science Publishers, London, p. 93.
21. Segars, R. A., Cardello, V., and Cohen, J. S. (1981). Objective and subjective texture evaluation of irradiation sterilized meat products, *j. Food Sci.*, *44*: 999.
22. Wierbicki, E. (1980). „Technology of irradiation preserved meats," Proceedings of 26th European Meeting of Meat Research Workers, Colorado Springs, Colorado, *1*: 194.
23. Zabielski, J., Kijowski, J., Fiszer, W., and Niewiarowicz, A. (1984). The effect of irradiation on technological properties and protein solubility of broiler chicken meat, *J. Sci. Food Agric.*, *35*: 662.
24. Elkhalifa, E. A., Graham, P. P., Marriott, N. G., and Phelps, S. K. (1988). Color characteristics and functional properties of flaked turkey dark meat as influenced by washing treatments, *J. Food Sci.*, *53*: 1068.

25. Alcocer, M. J. C. and Areas, J. A. G. (1990). Lipid composition and hydration characteristics of lung protein isolates defatted by several solvents, *J. Food Sci.*, *55*: 19.
26. Ozimek, G., Jelen, P., Ozimek, L., Sauer, W., and McCurdy, S. M. (1986). A comparison of mechanically separated and alkali extracted chicken protein for functional and nutritional properties, *J. Food Sci.*, *51*: 748.
27. Autio, K., Kiesvaara, M., Malkki, Y., and Kanko, S. (1984). Chemical and functional properties of blood globin prepared by a new method, *J. Food Sci.*, *49*: 859.
28. Schmidt, R. H. (1981). Gelation and coagulation, In *Protein Functionality in Foods* (J. P. Cherry, ed.), Amer. Chem. Soc., Washington, D.C., p. 131.
29. Lin, C. S. and Zayas, J. F. (1987). Influence of corn germ protein on yield and quality characteristics of comminuted meat products in a model system, *J. Food Sci.*, *52*: 545.
30. Kinsella, J. E. (1979). Functional properties of soy proteins, *J.A.O.C.S.*, *56*: 242.
31. Parks, L. L. and Carpenter, J. A. (1987). Functionality of six nonmeat proteins in meat emulsion systems, *J. Food Sci.*, *52*: 271.
32. Smith, G. C., Juhn, H., Carpenter, Z. L., Mattil, K. F., and Cater, C. M. (1973). Efficacy of protein additives as emulsion stabilizers in frankfurters, *J. Food Sci.*, *38*: 849.
33. Wills, R. B. H. and Kabirullah, M. (1981). Use of sunflower protein in sausages, *J. Food Sci.*, *46*: 1657.
34. Gnanasambandam, R. and Zayas, J. F. (1992). Functionality of wheat germ protein in comminuted meat products as compared with corn germ and soy proteins, *J. Food Sci.*, *57*: 829.
35. Zayas, J. F. (1985). Structural and water binding properties of meat emulsions prepared with emulsified and unemulsified fat, *J. Food Sci.*, *50*: 689.
36. Wirth, F. (1985). Frankfurter-type sausages. Water binding, fat binding, development of structure, *Fleischwirtschaft*, *65(8)*: 937.
37. Shahidi, F., Naczk, M., Rubin, L. J., and Diosady, L. L. (1984). Functional properties of blood globin, *J. Food Sci.*, *49*:370.
38. Wilding, P. Hedges, N., and Lillford, P. J. (1986). Salt-induced swelling of meat: The effect of storage time, pH, ione-type and concentration, *Meat Sci.*, *18*: 55.
39. Acton, J. C., Ziegler, G. R., and Burge, Jr., D. L. (1982). Functionality of muscle constituents in the processing of comminuted meat products, *CRC Crit. Rev. Food Sci.*, *18–19(2)*: 99.
40. Deng, J. C., Toledo, R. T., and Lillard, D. A. (1981). Protein-protein interaction and fat and water binding in comminuted flesh products, *J. Food Sci.*, *46*: 1117.
41. Wismer-Pedersen, J. (1987). Water, In *The Science of Meat and Meat Products*, Third Edition (J. F. Price and B. S. Schweigert, eds.), Food and Nutrition Press, Inc., Westport, CT, p. 141.
42. Sikorski, Z., Scott, D. N., and Buisson, D. H. (1984). The role of collagen in the quality and processing of fish, *CRC Crit. Rev. Food Sci. Nutr.*, *20(4)*: 301.
43. Trout, G. R. and Schmidt, G. R. (1986). Water binding ability of meat products: effect of fat level, effective salt concentration and cooking temperature, *J. Food Sci.*, *51*: 1061.
44. Ranganayaki, M. D., Asghar, A., and Henrickson, R. L. (1982). Influence of anion and cation on the water-holding capacity of bovine hide collagen at different pH values. Effect of sodium chloride and polyphosphates on hydration, *J. Food Sci.*, *47*: 705.
45. Morrissey, P. A. and Fox, P. R. (1981). Tenderization of meat: A review, *Irish J. Food Sci. Technol.*, *5*: 33.
46. Hamm, R. (1982). Postmortem changes in muscle with regard to processing of hot-boned beef, *Food Technol.*, *35(11)*: 105.
47. Honikel, K. O., Fischer, C., Hamid, A. and Hamm, R. (1981). Influence of postmortem changes in bovine muscle on the WHC of beef. Postmortem storage of muscle at 20 °C, *J. Food Sci.*, *46*: 1.
48. Puolanne, E. and Turki, P. (1984). „The effect of level of hot boned pork fat on water binding capacity and fat retention in cooked sausage," In The Proceedings of the 30th European Meeting of Meat Research Workers, Bristol, U.K., p. 373.

49. Offer, G. and Knight, P. (1988). The structural basis of water-holding in meat. Part 1: General principles and water uptake in meat processing, In *Developments in Meat Science*, 4th edition (R. Lawrie, ed.), Elsevier Appl. Sci. Publishers, London, p. 63.
50. Mittal, G. S. and Usborne, W. R. (1985). Meat emulsion extenders, *Food Technol.*, *39*: 121.
51. Holland, G. C. (1984). „A meat industry perspective on the use of dairy ingredients," Proc. Conf. Canadian Dairy Ingredients in the Food Instry, Ottawa.
52. Kirkpatrick, K. and Walker, N. J. (1985). Casein and Caseinates: manufacture and utilization, In *Milk Proteins '84*, Proc. Int. Congress on Milk Proteins (T. E. Galesloot and B. J. Tinbergen, eds.), Pudoc, Wageningen, the Netherlands, p. 196.
53. Schmidt, R. H. and Morris, H. A. (1984). Gelation properties of milk proteins, soy proteins, and blended protein systems, *Food Technol.*, *38(5)*: 85.
54. Hung, S. C. and Zayas, J. F. (1991). Functionality of milk proteins and corn germ protein flour in comminuted meat products, *J. Food Quality*, *15(2)*: 139.
55. Comer, F. W. and Dempster, S. (1981). Functionality of fillers and meat ingredients in comminuted meat products, *Can. Inst. Food Sci. Technol. J.*, *14(4)*: 295.
56. Lyon, C. E., Lyon, B. G., Davis, C. E., and Townsend, W. E. (1980). Texture profile analysis of patties made from mixed and flake-cut mechanically deboned poultry meat, *J. Poultry Sci.*, *59*: 69.
57. Wang, C. R. and Zayas, J. F. (1992). Comparative study of corn germ and soy proteins utilization in comminuted meat products, *J. Food Quality*, *15(2)*: 153.
58. Alexander, R. J. (1987). Corn dry milling: process, products, and applications, In *Corn: Chemistry and Technology*(S. E. Watson and P. E. Ramstad, eds.), Am. Assoc. Cereal Chemists, St. Paul, MN, p. 351.
59. May, J. B. (1987). Wet milling: Process and products, In *Corn: Chemistry and Technology* (S. E. Watson and P. E. Ramstad, eds.), Am. Assoc. Cereal Chemists, St. Paul, MN, p. 377.
60. Lin, C. S. and Zayas, J. F. (1987). Functionality of defatted corn germ proteins in model system: fat binding capacity and water retention, *J. Food Sci.*, *52*: 1308.
61. Lin, C. S. and Zayas, J. F. (1987). Protein solubility, emulsifying stability and capacity of two defatted corn germ proteins, *J. Food Sci.*, *52*: 1615.
62. Zayas, J. F. and Lin, C. S. (1989). Frankfurters supplemented with corn germ protein: Sensory characteristics, proximate analysis and amino acid composition, *J. Food Quality*, *11(6)*: 461.
63. Brown, L. M. and Zayas, J. F. (1990). Effect of corn germ protein flour on quality characteristics of beef patties heated by microwave, *J. Food Proc. and Preserv.*, *14(2)*: 155.
64. Brown, L. M. and Zayas, J. F. (1990). Corn germ protein flour as an extender in broiled beef patties, *J. Food Sci.*, *55*: 888.
65. Knightbridge, J. P. and Goldman, A. (1975). Water absorptive capacity of dried milk products, *N.Z. J. Dairy Sci. Technol.*, *10*: 152.
66. Marshall, R. J. (1982). An improved method for measurement of the syneresis of curd formed by rennet action on milk, *J. Dairy Res.*, *49*: 329.
67. Pearse, M. J., Linklater, P. M., Hall, R. J., and Mackinlay, A. G. (1986). Effect of casein micelle composition and casein dephosphorylation on coagulation and syneresis, *J. Dairy Res.*, *53*: 381.
68. Modler, H. W., Larmond, M. E., Lin, C. G., Froehlich, D. and Emmons, D. B. (1983). The role of protein in yoghurt, *J. Dairy Sci.*, *66*: 422.
69. Brosio, E., Altobelli, G. and Di Nola, A. (1984). A pulsed low-resolution NMR study of water binding to milk proteins, *J. Food Technol.*, *19*: 103.
70. Kneifel, W. T., Abert, T. and Luf, W. (1990). Influence of preheating skim milk on water-holding capacity of sodium salts of caseinates and coprecipitates, *J. Food Sci.*, *55*: 879.

71. Moor, H. and Huyghebaert, A. (1983). Functional properties of dehydrated protein-rich milk products, In *Physico-chemical aspects of dehydrated protein-rich milk products*, Proc. Int. Dairy Fed. Symp., Danish Res. Inst. Dairy Ind., Helsingor, D.K.
72. Hung, S. C. and Zayas, J. F. (1992). Protein solubility, water retention and fat binding of corn germ protein flour compared with milk proteins, *J. Food Sci.*, *57*: 372.
73. DeWit, J. N. (1981). Structure and functional behavior of whey proteins, *Neth. Milk Dairy J.*, *38*: 47.
74. Gossett, P. W. and Baker, R. C. (1983). Effect of pH and of succinylation on the water retention properties of coagulated, frozen, and thawed egg albumen, *J. Food Sci.*, *48*: 1391.
75. King, A. J., Ball, Jr., H. R., Catignani, G. L., and Swaisgood, H. E. (1984). Modification of egg white proteins with oleic acid, *J. Food Sci.*, *49*: 1240.
76. Bressani, R. (1981). The role of soybeans in food systems, *J.A.O.C.S.*, *3*: 392.
77. Hansen, J. R. (1978). Hydration of soybean protein. 2. Effect of isolation method and various other parameters on hydration, *J. Agric. Food Chem.*, *26(2)*: 301.
78. Fiora, F. A., Pilosof, A. M. R., and Bartholomai, G. B. (1990). Physicochemical properties of soybean proteins related to low, viscoelastic, mechanical and water-holding characteristics of gels, *J. Food Sci.*, *55*: 133.
79. Hansen, J. R. (1976). Hydration of soybean protein, *J. Agric. Food Chem.*, *24*: 1136.
80. Lopez de Ogara, M. C., Pilosof, A. M. P., and Bartholomai, G. B. (1987). Technical note: Effect of solutes on the hydration characteristics of soy protein isolate, *Internat. J. Food Sci. Technol.*, *22*: 153.
81. Wagner, J. R. and Anon, M. C. (1990). Influence of denaturation, hydrophobicity, and sulfhydryl content on solubility and water absorbing capacity of soy protein isolates, *J. Food Sci.*, *55*: 765.
82. Wang, C. R. and Zayas, J. F. (1991). Water retention and solubility of soy proteins and corn germ proteins in a model system, *J. Food Sci.*, *56*: 455.
83. Whistler, R. L. and Daniel, J. R. (1985). Carbohydrates, In *Food Chemistry*, 2nd ed. (O. R. Fennema, ed.), Marcel Dekker, Inc., New York, p. 69.
84. Ochiai-ynagi, S., Miyauchi, H., Saio, K., and Watanabe, T. (1978). Modified soybean protein with high water-holding capacity, *Cereal Chem.*, *55(2)*: 157.
85. Arce, C. B., Pilosof, A. M. R., and Bartholomai, G. B. (1991). Sodium dodecyl sulfate and sulfite improve some functional properties of soy protein concentrates, *J. Food Sci.*, *56*: 113.
86. Mohri, M. and Matsushita, S. (1984). Improvement of water absorption of soybean protein by treatment with bromelain, *J. Agric. Food Chem.*, *32*: 486.
87. Naczk, M., Rubin, L. J., and Shahidi, F. (1986). Functional properties and phytate content of pea protein preparations, *J. Food Sci.*, *51*: 1245.
88. Paredes-Lopes, O., Ordorica-Falomir, C., and Olivares-Vazquez, M. R. (1991). Chickpea protein isolates: Physico-chemical, functional and nutritional characterization, *J. Food Sci.*, *56*: 726.
89. Sosulski, F. W. and McCurdy, A. R. (1987). Functionality of flours, protein fractions and isolates from field peas and faba bean, *J. Food Sci.*, *52*: 1010.
90. Sumner, A. K., Nielsen, M. A., and Youngs, C. G. (1981). Production and evaluation of pea protein isolate, *J. Food Sci.*, *46*: 364.
91. Han, J.-Y. and Khan, K. (1990). Functional properties of pin-milled and air-classified dry edible bean fractions, *Cereal Chem.*, *67(4)*: 390.
92. Abbey, B. S. and Ibeh, G. O. (1987). Functional properties of raw and heat processed brown bean (*Canavalia rosea DC*) flour, *J. Food Sci.*, *52*: 406.
93. Schaffner, D. W. and Beuchat, L. R. (1986). Functional properties of freeze-dried powder of unfermented aqueous extracts of legume seeds, *J. Food Sci.*, *51*: 629.
94. Sathe, S. K., Deshpande, S. S., Salunkhe, D. K. (1983). Functional properties of black gram (*Phaseolus Mungo L.*) proteins, *Lebensm.-Wiss. u.-Technol.*, *16*: 69.

95. Narayana, K. and Narasinga Rao, M. S. (1984). Effect of acetylation and succinylation on the functional properties of winged bean (*Psophocarpus tetragonolobus*) flour, *J. Food Sci.*, *49*: 547.
96. Canella, M., Castriotta, G., Bernardi, A., and Boni, R. (1985). Functional properties of individual sunflower albumin and globulin, *Lebensm.-Wiss. u.-Technol.*, *18*: 288.
97. Kabirullah, M. and Wills, R. B. H. (1982). Hydration characteristics of sunflower protein, *Lebensm.-Wiss. u.-Technol.*, *15*: 267.
98. Jones, L. J. and Tung, M. A. (1983). Functional properties of modified oilseed protein concentrates and isolates, *Can. Inst. Food Sci. Technol. J.*, *16(1)*: 57.
99. Kabirullah, M. and Wills, R. B. H. (1981). Functional properties of sunflower protein following partial hydrolysis with proteases, *Lebensm.-Wiss. u.-Technol.*, *14*: 232.
100. Lucisano, M., Casiraghi, E. M., and Barbieri, R. (1984). Use of defatted corn germ flour in pasta products, *J. Food Sci.*, *49*: 482.
101. Christianson, D. D., Friedrich, J. P., List, G. R., Warner, K., Bagley, E. B., Stringfellow, A. C., and Inglett, G. E. (1984). Supercritical fluid extraction of dry-milled corn germ with carbon dioxide, *J. Food Sci.*, *49*: 229.
102. Zayas, J. F. and Lin, C. S. (1989). Water retention of two types of hexane-defatted corn germ proteins and soy protein flour, *Cereal Chem.*, *66(1)*: 51.
103. Oomah, B. D. and Mathieu, J. J. (1987). Functional properties of commercially produced wheat flour solubles, *Can Inst. Food Sci. Technol. J.*, *20(2)*: 81.
104. Vani, B. and Zayas, J. F. Influence of pH and temperature on water retention and protein solubility of wheat germ protein flour. *Annual IFT Meeting. Book of Abstracts,* 1993, p.160.
105. Thompson, L. U., Liu, R. F. K, and Jones, J. D. (1982). Functional properties and food applications of rapeseed protein concentrate, *J. Food Sci.*, *47*: 1175.
106. Kim, N. S., Kim, Y. J., and Nam, Y. J. (1992). Characteristics and functional properties of protein isolates from various peanut (*Arachis hypogaea L.*) cultivars, *J. Food Sci.*, *57*: 406.
107. Perkins, D. and Toledo, R. T. (1982). Effect of heat treatment for trypsin inhibitor inactivation on physical and functional properties of peanut protein, *J. Food Sci.*, *47*: 917.
108. Prinyawiwatkul, W., Beuchat, L.R., and McWatters, K.H. (1993). Functional property changes in partially defatted peanut flour caused by fungal fermentation and heat treatment, *J. Food Sci.*, *6*: 1318.
109. Tasneem, R. and Subramanian, N. (1986). Functional properties of guar (*Cyamopsis tetragonoloba*) meal protein isolates, *J. Agric. Food Chem.*, *34*: 850.
110. Knorr, D. and Betschart, A. A. (1983). Effect of dehydration methods on the functionality of plant protein concentrates, *Starch*, *35*: 23.
111. Rahma, E. H. and Narasinga Rao, M. S. (1984). Gossypol removal and functional properties of protein produced by extraction of glanded cottonseed with different solvents, *J. Food Sci.*, *49*: 1057.
112. Berardi, L. C. and Cherry, J. P. (1981). Functional properties of co-precipitated protein isolates from cottonseed, soybean, and peanut flours, *Can. Inst. Food Sci. Technol. J.*, *14(4)*: 283.
113. Chang, P. R. and Sosulski, F. W. (1985). Functional properties of dry milled fractions from wild oats (*Avenua fatua L.*), *J. Food Sci.*, *50*: 1143.
114. Ma, C.-Y. and Harwalkar, V. R. (1984). Chemical characterization and functionality assessment of oat protein fraction, *J. Agric. Food Chem.*, *32*: 144.
115. Paredes-Lopez, O. and Ordorica-Falomir, C. (1986). Functional properties of safflower protein isolates: water absorption, whipping and emulsifying characteristics, *J. Sci. Food Agric.*, *37*: 1104.
116. Madhusudhan, K. T. and Singh, N. (1985). Effect of heat treatment on the functional properties of linseed meal, *J. Agric. Food Chem.*, *33*: 1222.
117. Dev. D. K. and Quensel, E. (1988). Preparation and functional properties of linseed protein products containing differing levels of mucilage, *J. Food Sci.*, *53*: 1834.

118. Wojnowska, I., Poznanski, S., and Bednarski, W. (1981). Processing of potato protein concentrates and their properties, *J. Food Sci.*, *47*: 167.
119. Devi, M. A. and Venkataraman, L. V. (1984). Functional properties of protein products of mass cultivated blue-green alga *Spirulina Platensis*, *J. Food Sci.*, *49*: 24.
120. Huang, Y. T. and Kinsella, J. E. (1986). Functional properties of phosphorylated yeast protein: Solubility, water-holding capacity, and viscosity, *J. Agric. Food Chem.*, *34*: 670.

Chapter 3

Emulsifying Properties of Proteins

3.1 Introduction

Emulsification is the most important process in the manufacturing of many formulated foods. Food emulsions are classified as macroemulsions with droplet size of 0.2 to 50 μm. Emulsion represents a heterogeneous mixture of fat globules. Food emulsions can be of the oil in water (O/W) or water in oil (W/O) type. The difference between O/W and W/O emulsions is that an O/W emulsion commonly exhibits a creamy texture, while a W/O system has greasy textural properties.

Protein emulsifying activity is the ability of the protein to participate in emulsion formation and to stabilize the newly created emulsion. The emulsifying capacity is the ability of the protein solution or suspension to emulsify oil. Emulsifying properties are useful functional characteristics which play an important role in the development of new sources of plant protein products for uses as foods. Proteins are the components that dominate in most food emulsions.

A significant number of foods are emulsions, dispersions, and foams, and in these systems, proteins, in combination with lipids and carbohydrates, are important stabilizers. The review of Halling [1] on the use of food proteins as stabilizers of emulsions and foams, provides extensive references to earlier reviews and reports. Walstra [2] presented the physical and colloidal aspects of emulsification and whipping for milk and milk products.

Required functional properties of proteins are affected by product utilization, for example in dairy foods, emulsifying properties are important, and milk substitutes should possess proper emulsifying capacity (EC), color, mouthfeel, flavor and solubility characteristics. In comminuted and other meats, limiting functional properties are: water holding capacity, emulsifying capacity, and emulsion stability (ES), fat binding and resistance of the functional properties to heat treatment. Comparative studies of the emulsifying properties of different proteins are difficult until all factors influencing protein emulsifying properties and methods of testing are standardized.

The characteristics used to describe emulsifying properties of proteins are EC, ES, and emulsifying activity (EA). They are used to describe the emulsifying properties of proteins in food emulsion systems. EC is presented as the amount of oil (ml) that is emulsified under specific conditions by 1 g protein. The emulsifying capacity of an emulsifier depends on its ability 1) to form the adsorption films around the globules, and 2) to lower the interfacial tension at the oil-water interface. Emulsion stability is the capacity of emulsion droplets to remain dispersed without separation by creaming, coalescing, and flocculation. Emulsifying activity is presented as the maximal interfacial area (cm^2) per 1 g of protein of a stabilized emulsion.

Many chemical and physical interrelated factors are involved in the formation, stability, and textural properties of protein-fat-water emulsions. EC and ES depend on the properties of proteins and conditions of emulsification and vary with the source of protein, its concentration, pH, ionic strength (salt type and concentration), and viscosity of the system. Nakai et al. [3] reported that solubility, surface hydrophobicity, and molecular flexibility influence the emulsification behavior of globular proteins. Emulsifying properties of proteins are influenced by equipment design, temperature of oil, and protein solution.

In certain foods, a natural protein ingredient is an effective stabilizer. Proteins are effective surface-active agents because they possess the capacity to lower the interfacial tension between hydrophobic and hydrophilic components in foods. Proteins participate in the formation of oil-in-water and water-in-oil emulsions and stabilize the emulsions that are formed. A stabilizing effect of proteins in the emulsion system results from the formation of a protective barrier around fat droplets, preventing emulsion coalescence.

The emulsifying capacity of proteins depends on the shape, charge, and hydrophobicity of the protein molecules, neutrality of dipoles, hydration of polar groups. Emulsion stability depends on the magnitude of these interactions [4]. To produce stable emulsions protein material should be selected that is soluble, has the ability to adsorb rapidly at the interface, has well-distributed charged groups, and has the ability to form a strong cohesive film. Stable emulsions are prepared with milk proteins because components of micellar casein and whey proteins possess the properties listed above.

3.2 Hydrophobic and Hydrophilic Properties of Proteins

The understanding of the emulsifying properties of plant and animal food proteins was enhanced by establishing a correlation between surface hydrophobicity and interfacial tension. Proteins with a large number of apolar amino acids, i.e. high hydrophobicity are surface active. The attempt was made to explain protein emulsifying properties by their surface hydrophobicity. The relationship between protein structure and functional properties (emulsifying and foaming capacity) was demonstrated by Kato and Nakai [5] as a result of determining protein surface hydrophobicity. They reported strong correlation between EC and hydrophobicity of proteins. Kato et al. [6] showed correlation between surface hydrophobicity and interfacial tension of the protein solutions and emulsifying properties of proteins.

Total hydrophobicity × surface hydrophobicity ($S \times S_o$) represents more than 71% of the variability in the emulsifying activity index [7]. A hypothesis was strongly supported that hydrophobicity plays a crucial role in determining other functional properties of proteins [8]. An effective hydrophobicity (S_o) measurement with *cis*-parinaric acid has been extended to the insoluble proteins, and the value of S_o was recommended as a predictor of some functional properties. Values of S_o have been used to predict emulsifying capacity of proteins. According to the hydrophobicity mechanism of protein emulsifying capacity, amphiphilic proteins having high surface hydrophobicity are adsorbed at the oil/water interface. Adsorbed proteins reduce the interfacial or surface tension facilitating the formation of emulsions. Proteins are surface active agents, and protein aqueous solution showed a surface tension about 25 mN/m, lower than that of pure water. At the oil/water interfaces, the surface pressures of protein vary in the range 15–25 mN/m, depending on the protein concentration, pH, ionic strength, temperature, etc. [9]. The concentration of protein at the interface and decrease of the interfacial tension is affected by protein hydrophobicity. The surface hydrophobicity of proteins correlated significantly with an increasing index of emulsifying activity and decreasing interfacial tension [5].

Globular proteins with a great surface hydrophobicity, such as lysozyme, ovalbumin and whey proteins improve their emulsifying capacity by moderate heating and partial unfolding. Proteins must possess a well-balanced distribution of hydrophilic and hydrophobic domains in the molecules. The higher EC of caseinates is related to their high solubility and to their dissociated and naturally unfolded structure. Additionally, they possess relatively high hydrophobicity and a separation of hydrophobic and hydrophilic regions of the polypeptide chains. Molecular hydrophobicity and surface active properties are correlated. Effective surface hydrophobicity (S_o) and interfacial tension influenced emulsifying activity.

A number of methods have been used to attach hydrophobic groups or change the hydrophobicity of proteins to improve emulsifying and foaming properties: attaching fatty acids, binding of hydrophobic amino acids by enzymatic or chemical reactions, and deamidation of gluten [10]. An effective method of surface hydrophobicity determination is the use of fluorescent probes which bind to the hydrophobic regions on the surface of the protein molecule [5].

The surface activity, and emulsifying properties have been improved by incorporation of the hydrophobic constituents to a hydrophilic protein. Alkyl esters have been incorporated by using papain as a catalyst under mild alkaline conditions (pH 9.0) [11]. In this treatment hydrolysis is carried out at a limited amount of susceptible peptide bonds, and ester groups are linked by peptide bonds. The EC increases gradually with alkyl chain length.

The side of the protein stabilizer, which is exposed to the discontinuous oil phase should be hydrophobic. Protein adsorption and orientation at the oil/water interface is affected mostly by its hydrophobicity [12]. The side of the protein stabilizer which is exposed to the continuous aqueous phase should be hydrophilic and should have most of the polar and charged groups exposed. Hydrophobic interactions stabilize the conformation of proteins in their native state in solution.

The hydrophobicity of proteins is related to their content of apolar amino acids. Bigelow [13] calculated the average hydrophobicity of several proteins by dividing the total hydrophobicity by the total number of residues. The average hydrophobicity of proteins ranged from 1000 to 12|000 cal/residue.

The amino acid composition influences the folding and reactivity of food proteins. Kato and Nakai [5] found that proteins with a high content of nonpolar amino acid residues (more than 30% of the total amino acids) exhibit good surface activity (emulsifying and foaming properties) and poor gelling ability. However, prediction of the EC of proteins from amino acid composition showed low correlation because it measured the total potential hydrophobicity of the protein rather than hydrophobic groups oriented at the surface after unfolding. Only a portion of nonpolar amino acid residues located in the interior of protein molecules in solutions participate in the emulsification of oil into an aqueous phase. Hydrophobicity of proteins influences protein solubility in water and indirectly emulsifying properties of proteins. Proteins interact with the oil surface more significantly if a large number of hydrophobic amino acids are at the surface. Surface hydrophobicity is not the only one factor determining emulsifying properties of proteins. β-lactoglobulin exhibited better emulsifying properties at pH 7 than at pH 3, although surface hydrophobicity of β-lacto-globulin is higher at pH < 3 than at pH > 3 [14].

The suitable hydrophilic-lipophilic balance (HLB) determines the emul-sifying properties of proteins. Aoki et al. [15] suggested that emulsifying properties of proteins are affected by a balance between the hydrophilic and lipophilic groups and did not necessarily increase as the protein became more lipophilic. A proper balance between hydrophilic and hydrophobic groups is necessary to lower the surface and interfacial tension. The hydrophilic-lipophilic properties of proteins enable them to orient at the oil-water interface with the lipophilic groups oriented towards the oil droplets and hydrophilic groups oriented toward the water phase. Despite the considerable lowering of the interfacial tension the emulsions are still thermodynamically unstable.

Emulsions of high stability can be prepared with a combination of different emulsifying agents exhibiting various HLB values. Maximum lowering of surface and interfacial tension was achieved when the water-oil absorption index, determined by spontaneous water and oil uptake, was nearly 2.0, i.e. protein absorbed twice as much water as oil [16]. If this ratio was higher than 2.0, proteins showed higher hydrophilic properties and lower EC. Elizalde et al. [16] supported observations of Kato and Nakai [5] that the greater the decrease in interfacial tensions, the higher the emulsifying activity of proteins. At the same time, Saito and Taira [17] found no correlation between surface hydrophobicity of plasma proteins and the emulsifying activity index. Poor correlation was found between EC and surface tension. Some proteins with low surface hydrophobicity exhibited good emulsifying properties.

An important property of proteins is sensitivity to surface denaturation at the oil-water and air-water interface. This sensitivity to denaturation may influence the functional properties of proteins. Since the surface hydrophobicity of protein increases with denaturation at the interface, the emulsifying properties may be improved. The sensitivity of flexible proteins to the action of protease was utilized as a method to detect the role of protein flexibility. Kato et al. [18] reported good correlations between the emulsifying activity ($r = 0.93$), and the digestion velocity of proteins determined by α-chymotrypsin. The flexibility of the protein structure detected by protease digestion was correlated with the emulsifying properties.

Protein solubility is an important factor determining emulsifying properties of proteins. As protein solubility increased the capacity of a protein to form and stabilize emulsion was improved. Undissolved proteins are poor stabilizers because proteins must dissolve and migrate to the interface. The emulsifying capacity of corn germ proteins was influenced by solubility especially in diluted emulsions. Correlation was established between EC and protein solubility and the effects of pH on emulsion formation [19]. Minimum EC was found in the isoelectric range.

A significant number of plant proteins are underutilized in food preparation because of their emulsifying properties. Utilization of plant proteins as emulsion stabilizers is limited by their poor solubility. Solubility and functionality might be improved by protein modification. Solubility and surface activity of modified proteins are improved by partial hydrolysis during enzymic modification.

Increased concentration of soluble protein may cause the formation of smaller particles during emulsification and may increase the stability of the emulsion to creaming and drainage [1]. However, data related to correlation between EC and protein aqueous solubility are controversial. The positive correlation between solubility of proteins and EC and ES was reported. However, other data showed evidence that emulsifying properties and protein solubility are not well correlated [20, 21]. At the same time, solubility of proteins appears to contribute more to the quality of emulsions formed than to quantities of oil emulsified. Paulson et al. [22] suggested that insoluble proteins participate in emulsion stabilization. Insoluble protein particles incorporated during emulsification can increase stability of emulsions. Solubility and hydrophobicity of proteins have been suggested as indicators of EC of proteins [7]. Combinations of effective hydrophobicity and solubility were more effective predictors of emulsifying properties than hydrophobicity and solubility alone. The importance of surface hydrophobicity (S_o) and solubility to the EC and ES of proteins was reported by Shimizu et al. [23].

Changes in protein conformation during heat treatment affected solubility and hydrophobicity [7]. Heating of meat proteins to 70 °C resulted in significant two- to threefold increase in S_o of proteins and decrease in solubility values to 25%. Consequently, heating effectively exposed the hydrophobic residues of the proteins. Results of this study indicated that both hydrophobicity and solubility influence the emulsifying properties of meat proteins. Solubility was the more important parameter for predicting emulsifying activity index (EAI) and EC of protein samples with low solubility, whereas hydrophobicity was critical for predicting EAI and EC of samples with high solubility. Combination of high hydrophobicity and high solubility values resulted in good emulsifying functionality. Capacities of proteins to emulsify fat were improved with increasing hydrophobicity values at solubility values above 30%, and with increasing solubility values at any given hydrophobicity value. Heating at 75°C slightly improved EAI of myosin, however, the emulsifying stability index decreased [24]. Emulsifying properties of globular proteins were improved by heating due to unfolding of protein polypeptide chains. Nakai et al. [3] demonstrated the importance of protein surface hydrophobicity and reported a correlation between bound linoleate and dispersibility of surfactant treated soy, sunflower and rapeseed proteins. Treatment with surfactants affected solubilization of these proteins. The dispersibility was improved with increase bound linoleate. High protein dispersibility was associated with high effective hydrophobicity. The data that the So of proteins increased with dispersibility is contrary to the known mechanism of hydrophobicity-solubility relationship of proteins.

Partial hydrolysis with proteinase-treated soy, sunflower, and rapeseed protein isolates showed that these proteins did not change their So, whereas protein solubility increased from 55 to 95% [3]. This evidence demonstrated the more significant effect of So on emulsifying properties than solubility of proteins.

3.3 Interfacial Film Formation and Properties

The interfacial films of proteins at an oil-water interface have been thoroughly studied. However, there is no extended information related to thickness, structure, textural properties of films and protein conformation in films. The surface activity of proteins is expressed as the protein ability to migrate, adsorb, unfold and form a layer at the interface as a result of rearrangement [25]. Proteins form membranes or films around the fat droplets and lower the interfacial tension between water and oil. As a result proteins retard coalescence of fat droplets. Surface film formation is a result of diffusion and adsorption of protein molecules at the interface. Formation of the protein film is facilitated if protein is solubilized, and is influenced by protein capacity to diffuse at the interface. The property of proteins to form films is a subsequent result of protein diffusion and adsorption at the interface, partial unfolding and reorientation at the interface [26]. These processes are affected by intrinsic and extrinsic factors. The important property of proteins for their surface activity is molecular flexibility and conformation. Molecular properties of the proteins influence the formation of an interfacial film which occurs in sequential steps. Formation of the interfacial film is a three-step process and includes:

1. Diffusion of proteins to the interface,
2. Protein adsorption at the interface,
3. Changes of protein conformation resulted from unfolding and reorientation of protein molecules.

The first step in surface film formation is protein diffusion to the interface.

During emulsification the soluble protein diffuses and concentrates at the interface. The diffusion is influenced by protein concentration, molecular size, temperature, pH, ionic strength, and solubility. Soy proteins diffuse slowly to the interface when compared to other food proteins (whey proteins and caseinates). The difference in diffusion rate is probably related to the large size of soy protein particles. Soy proteins, including mostly the 7S and 11S globulins, have a complex quaternary structure with a molecular weight of particles ranging from 180 to 363|000. Caseinate, a widely used stabilizer, is a multicomponent mixture of interacting protein molecules with different surface activities. Milk caseinates with complex quaternary structures, like the soy proteins, have different surface properties, including surface adsorption. The protein molecules must penetrate to the interface and unfold. For rapid migration and adsorption at the interface, the solubility of proteins is critical. Milk caseins possess high surface activity and diffuse rapidly to the interface and decrease significantly interfacial tension at the concentrations above 10^{-3} wt%. This effect was obtained especially when the caseinate was dispersed in 0.2 M NaCl solution. The significant decrease in diffusion rate and surface activity was observed at a caseinate concentration below 10^{-3} wt%. The whey proteins diffuse quickly to the interface because they consist mainly of small molecules and molecular complexes. High speed of diffusion of whey proteins is due to the formation of small molecular complexes.

The second step in interfacial film formation is protein adsorption.

Formation of stable interfacial films and emulsions is an effect of rapid adsorption of the surface active protein at the oil-water interface. The protein layer at the interface forms an effective barrier against the coalescence of oil droplets. The dislocation of the protein polypeptide chains at the interface is affected by the free surface energy of adsorption. The adsorption and reorien-tation of protein surfactants at the interface is controlled by polar and apolar sites of proteins. The apolar groups contribute to hydrophobicity of proteins and affinity of the proteins for the oil phase. Proteins with an excess of hydrophobic groups have a better potential for adsorption at the interface than hydrophilic proteins. To be adsorbed at the interface, the protein molecule must overcome one or more barriers. It must compress molecules already adsorbed at the interface.

The adsorption and layer formation at the surface of fat droplets is affected by the amino acid composition, conformation of the protein, pH, concentration of ions, viscosity, temperature, hydrophilic-lipophilic balance (HLB), and intensity of homogenization (energy input). The rate of interfacial film formation and its properties are mostly affected by the emulsification process, especially shearing forces during homogenization and mixing. In emulsions some form of shear is necessary for the creation of a new surface area.

In food emulsions, proteins can serve as stabilizers causing interaction and competition between them in adsorption at the interface. Proteins exhibit different surface activity, and proteins with low activity can be displaced by those with high surface activity [27]. For example, β-casein is more surface active than αs_1-casein and can replace the latter at the interface. Proteins possess surface active properties that lower the surface tension of liquids when adsorbed at the oil-water interface. This surfactant effect of casein is obtained at a low concentration, 10^{-5} g/dl [28]. The decrease in interfacial tension is proportional to the number of hydrophilic and hydrophobic groups penetrating the interface. As protein molecules adsorb at the interface, decrease in surface tension indicates low surface free energy.

Various proteins are adsorbed at the interface differently; β-casein forms dilute monolayers, while lysozyme forms concentrated films. In the dilute protein solution monolayers protein molecules are completely unfolded. It was shown for polypeptides that α-helix is stable at the interface, and this property of the protein native structure remained in the surface-denatured state [12]. The rate of the protein adsorption at the interface is affected by the flexibility of the native protein.

Soy proteins form a thick, cohesive interfacial layer, with good rheological properties. Soy proteins are rapidly adsorbed at the interface. However, thickness of the soy protein layer decreases with a decrease in the size of fat droplets and increased interfacial surface [29]. The stability of the emulsion is significantly increased with increased protein concentration, reflecting the formation of a thicker film around the emulsified droplets. Emulsion formation from concentrated solutions of proteins, such as lysozyme, caused formation of multilayers which are formed by adsorption of monomeric or dimeric lysozyme molecules. The structure of adsorbed films surrounding fat droplets is influenced by the native structure of the protein.

The dispersivity of emulsions is characterized by fat particle size distribution and the stability by amount of protein adsorbed per unit area of fat surface (protein load). The quantity of adsorbed protein at the fat globule is largely determined by the fat surface area created and by the type of protein adsorbed. The final fat surface area of the emulsions is mostly determined by the emulsification conditions. This area increases more as a function of power input than as a function of homogenization.

Protein adsorption at the interface was determined by measuring droplet size and calculating the total surface area [30]. Protein concentration at the interface varied between 0.5 and 2.5 mg/m^2 for whey protein, and 20 mg/m^2 for casein. Sodium caseinate which does not exist in micellar form was adsorbed at a low level comparable to whey proteins. The type and thickness of the film depends upon the concentration and structure of the protein molecule. During protein migration and adsorption some surface denaturation may occur because of conformational unfolding and spreading. The adsorption is enhanced by salt addition as a result of a decrease in surface repulsions. Electron microscopic and chemical techniques have been used to determine the presence of an adsorbed layer surrounding fat particles in emulsion.

At the third step of interfacial film formation there is an unfolding of protein molecules.

Protein unfolding is important for formation of an interfacial layer around fat droplets. Protein molecules already adsorbed at the interface are exposed to conformational changes, rearrangement, and expansion. Proteins tend to unfold to establish a new thermodynamic equilibrium. Proteins unfold more easily at an oil-water than at an air-water interface because of more significant hydrophobic interactions [25]. Protein molecules unfold at the interface to varying degrees depending on their molecular properties and spread to cover the surface area. As a result, an extended film is formed. Protein molecules adsorbed at the interface interact between polypeptide chains and undergo rearrangement in order to achieve the state of lowest free energy. Protein rearrangement is affected by the nature of protein, molecular flexibility, and protein concentration at the interface.

In unfolded protein film, increase in the number of contacts per molecule is observed, making it more resistant to desorption. The level of protein unfolding at the interface is affected by amino acid composition. Degree of unfolding is affected by distribution of the nonpolar residues in a protein chain. Protein unfolding at the oil-water interface is affected by the conformational stability of the tertiary structure of protein molecule. Partial unfolding of proteins at the oil/water interface is facilitated by a decrease in the stability of the tertiary structure of globular proteins, i.e. by partial denaturation or a change in pH away from the isoelectric point [31]. Protein concentration in emulsion influences unfolding of proteins [32]. Extensive unfolding of protein at the interface may occur at low protein concentration insufficient for monolayer formation. Conformation changes lead to a loss of the tertiary and some of the secondary structure. The adsorbed layer of proteins is usually monomolecular, and multilayers may be formed at high protein concentration. The flexibility of native proteins and capacity to adsorb at the interface determine the packing of protein molecules in a film.

Molecules of lysozyme do not unfold at interface as readily as β-casein and form the film much more slowly. BSA has intermediate properties, i.e. it unfolds to a limited extent. High surface activity of BSA is related to the significant number of hydrophobic domains. BSA is more effective as an emulsion stabilizer than lysozyme and β-casein. Surrounding layer formed by BSA is stronger at lower protein concentrations [33]. Surface hydrophobicity of BSA, K-casein and β-lactoglobulin showed correlations with surface tension, interfacial tension and emulsifying activity of these proteins.

Strong protein films are formed by large molecules. Larger protein molecules with a certain amount of tertiary structure at an interface exhibit better surface active properties because of more significant intermolecular interactions [34]. Interfacial films with a more condensed structure are formed by globular proteins, probably because these proteins, e.g. BSA and β-lactoglobulin, retain more residual secondary and tertiary structure during adsorption at the interface. In the formed surface films of globular proteins, loss of tertiary structure may be greater than for extended proteins [25]. Globular proteins possess a high emulsifying capacity because of sufficiently flexible domains in their structure, the ability to form interfacial films rapidly, and a decrease in surface tension. The tertiary structure of globular proteins facilitates molecular interactions, i.e. hydrophobic, electrostatic, disulfide, and Van der Waals forces. These forces contribute to the rheological properties of the film which influence film stability. Whey proteins, β-lactoglobulin, BSA, and α-lactalbumin are globular and they form condensed films with a relatively high surface viscosity.

Rheological Properties of Protein Films

The integrity of the protein film and its stability to mechanical and thermal treatment, such as blending, homogenization, shearing, and heating, is influenced by its rheological properties, elasticity, viscosity, flexibility, and cohesiveness. Stable emulsions can be formed if the interfacial protein film is as thick as possible and possesses good rheological properties, if it is well hydrated, possesses a net negative charge, and contains exposed polar and charged groups. The physical properties of the protein interfacial film and its surface activity determine the ability of proteins to form and stabilize emulsions. When two fat droplets in emulsion approach each other, their stability to thermal and mechanical shocks is influenced by thickness of the protein film, its rheological properties, restorative ability, and other factors. The most important rheological property is surface elasticity of the protein film, i.e. ability of the film to be deformed upon the application of the stress or force with recovery of natural orientation after the stress is released.

An important indicator of protein film stability is surface viscosity which is the characteristic of mechanical strength and is determined by protein-protein and protein-water interactions. The maximal viscosity was found in protein films with the greatest degree of protein-protein interactions. There is only limited knowledge about the film structure and protein-protein, protein-lipid interactions of the protein film. A basic factor contributing to good emulsifying properties is the capacity of the protein to form a film with high shear moduli as the result of extensive interactions between protein molecules in the film. This ability of proteins reaches maximum at the pI of the protein, when electrostatic repulsions are minimal. Rheological properties of surface films are influenced by protein concentration, pH, salt concentration, and other factors. Rheological properties of the protein interfacial film and a rate of formation are determined to a great extent by conditions of emulsification, e.g. blending, mixing, shearing, and denaturation. These factors are more important than the properties of the proteins themselves and they determine the stability of the emulsion formed [35]. Textural, especially viscoelastic properties of protein film, are affected by interaction between protein molecules during film formation. Halling [1] found that cohesive films are more resistant to deformation and they enhanced emulsion stability. The physical properties of the continuous phase, e.g. solution of proteins and other components influence emulsion stability. Increasing viscosity of continuous phase reduces the mobility of fat droplets; as a result, contacts between droplets are limited and emulsion stability is enhanced [1]. High surface viscosity of protein films indicates strong interactions between protein molecules. Generally more viscous surface layers are formed from globular proteins than disordered proteins. Surface films formed in emulsions from whey proteins are more viscous than films formed from casein. Caseins form surface films at the oil-water interface with different viscous properties [36]. The surface viscosity of different caseins was in the order β-casein « αs_1-casein « κ-casein. High surface viscosity of κ-casein was related to polymerization via disulfide linkage.

Different proteins do not behave in the same way at interfaces. Looping of the apolar regions of protein stabilizer into the oil phase occurs as the surface concentration of protein increases. As a result of compression of these loops, the condensed film is formed. Dickinson et al. [36] when measuring interfacial tension observed an increase in the viscoelasticity of a protein film for several days after adsorption. This phenomenon was demonstrated with gelatin adsorbing at the *n*-hexadecane-water interface. Strengthening of the protein film might be due to the increase in number of cross-links and/or further protein accumulation at the interface. The emulsifying capacity of proteins is related to the proportion of non-polar amino acid residues on the surface of the protein molecule [37]. An increased number of non-polar residues on the protein surface will lower the energy barrier to adsorption which depends on the structure of proteins. An additional factor is that textural properties of the interfacial layer are affected by intermolecular hydrophobic interactions between non-polar residues on the surface of the protein.

3.4 Factors Affecting Emulsifying Properties of Proteins

3.4.1 Protein Concentration

Stability of emulsions is influenced by content of protein in preparation. Because of that, soy protein isolates are superior to soy protein concentrates. At low protein concentration thin film may be formed with low strength. While EC (g oil emulsified/g protein) decreases with protein concentration, ES markedly increases. ES is important in commercial conditions of emulsion preparation. Generally, the concentration of protein stabilizer in emulsion is low in the range 1.5–3%. Commercial protein stabilizers of higher concentration (5–10%) are also utilized. However, stable protein-oil-water emulsions can be obtained at low protein concentration, 0.2–1%.

The process of soy isolate extraction with incorporation of sodium bisulfite in extraction medium resulted in an improved EC of soy protein isolate. Presumably, film-forming properties of soy isolate were enhanced by cleavage of disulfide bonds. Excellent ECs were shown by soya and other oilseed proteins isolated by ultrafiltration [38]. Protein isolates obtained by this method were nonsignificantly denatured. Surface concentration of protein influences their adsorption. At low BSA interfacial pressure, the entire peptide backbone of β-casein is located at the interface with few loops and tails at the interface [32]. At higher surface BSA concentration, there is a significant increase in the number of loops with formation of multilayers. The formation of multilayers occurs at high protein concentration and proteins adsorbed on the first layer are mostly native because of the high rate of diffusion. The higher emulsion stability was related to the larger number of points of attachment of the unfolded protein at the interface.

The adsorption profile of gelatin was not affected by changes in pH or salt concentration [31]. Protein concentration at the interface increased with gelatin concentration in solution and reached a plateau at 0.625 mg/m^2. Sharp increase in protein adsorption was obtained when the concentration of gelatin exceeded 1.5% (w/v). The effect of disperse phase volume fraction (5 to 50%) and protein concentration in the aqueous phase (0.01 to 2%) on the size of droplets in emulsions prepared with a homogenizer was reported [30]. In milk-fat emulsions, diameter of fat droplets was not influenced by protein concentration in aqueous phase and increased slightly for disperse volume phase above 35%.

3.4.2 pH of Medium

Some proteins have optimal emulsifying properties at the isoelectric point (egg white, gelatin), whereas others perform better at Ph values away from the isoelectric point (soybean and peanut proteins). pH influences the EC of proteins indirectly by affecting their solubility, conformation and surface properties [1]. The pH-emulsifying property profile of various proteins resembles the pH-solubility profile. Definite stabilizing films of emulsions were established at all pH values below that of the isoelectric point of the protein and up to pH 6.5. The EC of peanut flour was poorest at pH 4.0, a level which is near the isoelectric point of the predominate native peanut protein (arachin). At this point, proteins have a net electrical charge of zero and minimum solubility and reactivity [39].

Improvement of the EC of BSA is probably related to facilitating molecular unfolding at the interface and enhanced electrostatic repulsion between emulsified droplets. The effects of pH and structural modifications on the emulsifying capacity of various proteins was reported [35]. A significant increase in the EC of BSA was found between pH 3 and 4. The EC of BSA progressively increased between pH 4 and 9, indicating that as net charge increased, the ability to form a film was enhanced. At pH 9.0 and higher, the EC of BSA decreased reflecting the altered tertiary structure as disulfide bonds were broken. The decrease in EC at pH values higher than 9.0 may reflect the alkali-induced hydrolysis of disulfide bonds in BSA. The adsorption isotherms at the oil/water interface as a function of equilibrium BSA concentration at pH 4, 5, and 6, were obtained [31]. The maximum level of BSA adsorption (2.5 mg/m^2) was obtained at pH 5.0, the isoelectric point of BSA. Protein adsorption decreased at pH values above or below the pI of BSA. Emulsifying properties of soy proteins are influenced by pH and ionic strength of the aqueous phase [20]. Maximum EC of soy isolate and soy 11S and 7S globulins was obtained at around pH 10.

Emulsion stability is affected by the pH of the protein solution by influencing charge. In viscous emulsions at the isoelectric pH, repulsions between protein molecules are minimal and rigid protein films are formed. As a result of intermolecular interactions in the protein film, emulsion stability is enhanced [1]. The effect of the isoelectric pH in diluted emulsions is different; there is decrease in repulsions between fat droplets that cause formation of the floccules and emulsion destabilization. At pH values far from the isoelectric point, zeta potential and repulsion is increased and flocculation is minimal. Electrical charges on fat droplets can arise by ionization, absorption, or frictional electricity produced during emulsification as a result of the large shearing forces.

3.4.3 Ionic Strength

Ionic strength influences the EC of proteins. The EC of soy protein preparations was enhanced by addition of NaCl [40]. The enhanced EC of soy proteins at 0.1 M NaCl is due to the formation of a more cohesive interfacial protein layer. A more cohesive layer is formed due to the influence of the ionic strength on the solubility and association-dissociation of soy protein subunits. Addition of salt can impair electrostatic interactions and affect protein diffusion and unfolding at the interface. However, the presence of NaCl decreases the electrostatic repulsion potential and may reduce emulsion stability. In milk emulsions, the level of free ionized Ca influences stability because the main components of caseinate are precipitated by small amount of ionic Ca. Ca binder, sodium citrate, is added to control level of ionic Ca in cream liqueurs. In sausage batters emulsifying capacity of proteins is increased by the presence of NaCl (0.5–1 M). Chloride and sulfate anions at concentrations of zero to 0.1 M significantly enhanced the EC of BSA [35]. Chloride increased EC of BSA slightly up to 0.6 M, but the "salt in" effect of sodium chloride reduced the rate of absorption at the interface at salt concentrations above 0.6 M. The BSA adsorption was not affected by ionic strength (NaCl) from 0.01 to 0.1 M, but was influenced by the species (anion/cation) of inorganic salts [31]. The BSA adsorption varied in the following order: $CaCl_2$ > LiCl > NaCl > Na_2SO_4 > KSCN.

3.4.4 Heat Treatment and Other Factors

Moderate increase in temperature is a critical factor in the formation of emulsions, causing an increase in the stabilizer adsorption at the interface. Penetration of the protein macromolecules is possible because of melting of the fat and decrease in the oil and fat viscosity. The weakening of the molecular structure of water will enhance hydrophobic interactions. Surface activity of proteins may increase as the result of denaturation, however, aggregation of proteins may reduce the concentration of effective adsorbing components of protein molecules. Emulsifying properties of proteins were improved by heat denaturation, correlating linearly with increasing surface hydrophobicity [6]. However, authors suggested that the improvement in surface properties by heat denaturation was dependent on the fact that no coagulation or loss in solubility occurred.

Heating caused increase in apparent viscosity of some of the proteins, which improved emulsifying properties of these proteins. Heat treatment of protein solutions causes protein denaturation and increases viscosity of the continuous phase or may result in gelation [1]. The improvement of protein functionality by heating some of the proteins is probably due to the exposure of the hydrophobic groups of amino acids and increasing total surface activity. Improvement in emulsifying properties was mainly observed when moderate heating was applied. The BSA adsorption at the interface was higher at lower temperature [31]. Adsorption of BSA was affected by the stability of globular protein. Decrease in values of BSA adsorption was found for destabilized globular structure as a result of pH change.

Excessive heating of whey protein caused decrease of surface hydrophobicity (So) probably due to hydrophobic interactions. Samples with the same solubility showed an increase in EC and ES with an increase in So [24]. Heating of soy proteins increased the EC of most of them probably due to increased So values. The ECs of BSA, gluten, and whey protein were not significantly affected by heating, while ES was slightly improved. Soy proteins and myosin showed increases in EC after heating, while ES decreased. An adverse heating effect on emulsifying properties was found for proteins: β-lactoglobulin, pea, canola, and casein. Considerable improvement in emulsifying properties was found for gelatin and ovalbumin. This was probably related to an increase in gelatin solubility. Excessive heating and increased emulsifying temperature lowered the EC of proteins probably because of excessive unfolding of BSA with destabilization of the film at the oil/water interface and coalescence of the emulsion [35]. The EC of soy proteins and the viscosity of their emulsions decreases as soy protein is thermally denatured [39]. The EC of soy 7S and 11S protein was impaired by heating. Consequently, unfolding of protein polypeptide chains and subsequent aggregation resulted in reduced EC. This effect of heating may be due to decreased solubility of the proteins. Prolonged heating of peanut proteins reduced EC.

3.5 Emulsion Stability

Emulsion stability is the ability of emulsion droplets to remain dispersed without coalescing, flocculation, or creaming. ES is not a characteristic of maximum oil addition, but rather the ability of the emulsion to remain stable and unchanged. Emulsions with low stability will appear visually as fat separation or creaming, which is caused by flocculation and coalescence. Protein emulsifiers possess the ability to reduce the tendency of the droplets in an emulsion system to coalesce and tendency of the emulsion to break down.

Emulsion stability is affected by the nature, and properties of the interfacial film. Protein stabilized emulsions have thick, hydrated interfacial films and possess a net charge. Thickness, viscosity, cohesiveness, and charge of the layer formed is determined by the nature of the protein and the conditions of emulsification. Stability of emulsion particles is enhanced by hydration of the proteins adsorbed in surrounding film [1]. In emulsions, proteins provide an effective barrier to coalescence. The stabilizing effect of proteins in emulsions results from the protective barrier they form around fat droplets, which further prevents their coalescence [41]. Long-term stability of emulsions depends basically on the thickness and strength of adsorbed protein films at the oil-water interface. Protein film possesses resistance to desorption when two droplets make contact and are capable of withstanding physical and thermal shocks and maintaining stability [1]. When repulsion exceeds the absolute value of the attractive forces by a value greater than a critical level, the fat globules will be stable in the emulsion. The protein adsorption and protein-protein interactions in the film are retarded by a high net charge, causing the formation of films with limited cohesiveness.

Fluctuations in storage temperatures are the common cause of emulsion destabilization. Low temperature could be an important factor in the stabilization of an otherwise poorly emulsified system. An increase in emulsion temperature accelerates the destabilization process because of increasing thermal energy and a reduction in viscosity and surface rigidity. Aoki et al. [15] reported lower emulsion stability when excessively denatured soy protein treated by 1-propanol was used as emulsifier. Temperature increased the protein adsorption as long as the protein remained soluble. The physical properties of emulsions are influenced by the composition of the continuous and discontinuous phase, the composition of surface layer around the droplets, and the size distribution of droplets [42]. Reduction in emulsion coalescence and separation results from reduction in the mobility of the fat droplets because of higher viscosity. An increase in emulsion stability is observed with an increase in viscosity of the continuous phase that determines the viscosity of the surface layer.

The emulsion properties are dependent not only on the protein stabilizer but also on the properties of the oil/fat, oil type and its volume. As the concentration of the discontinuous phase in the emulsion system increases, the more closely crowded the fat droplets become. A high concentration of fat droplets reduces their motion and tendency to settle while imparting a "creamed" appearance to the emulsion. Liquid vegetable oils are more acceptable for model testing of EC and ES because they do not develop crystalline phases at ambient temperatures. However, the chemical composition of vegetable oils varies depending on the conditions of cultivation and the cultivar of the oilseed. The composition of animal fats is influenced by the feed of the animal. Tarasevich et al. [43] reported that the adsorption density of protein decreased with increasing chain length of the hydrocarbon oil. Coalescence is more likely in concentrated emulsions with a high content of the oil (fat) phase, and when in the emulsion, the oil is almost all liquid, and the system is highly flocculated. In such emulsions there are multiple Brownian encounters between droplets, and a single fat droplet will have multiple areas of contact with other droplets.

High water and oil absorption capacity and viscosity of protein products have been associated with optimum emulsion stabilizing properties [44]. Kanterewicz [45] suggested a water-oil absorption index (WOAI) to characterize the relative importance of these factors in emulsion stability. Elizalde [46] derived a regression model for predicting emulsion instability from water- and oil-absorption capacities of the proteins and the composition of emulsion. In general combinations of high water and oil ratio (WR/OR) resulted in good emulsion stability, and these factors are highly interrelated. Proteins with low water-and oil-absorption capacity formed emulsions with low stability. Several processes cause emulsion instability. Instability of emulsion can be caused by aggregation, coalescence, flocculation, creaming, and/or phase inversion of the droplets. Coalescence reduces the net surface area of the emulsion system and causes emulsion instabilization. Uncontrolled coalescence shortens the shelf-life of salad-cream, cream liqueurs, etc.

The stability of emulsions to coalescence is markedly affected by the size of the droplets. Larger droplets coalesce faster than small ones, even when they have a thick adsorbed protein film [47]. Film thickness is an important factor of emulsion stability. Bigger droplets are expected to have thicker interfacial films than smaller droplets if protein concentration is constant. The suggestion that smallest fat droplets have the highest resistance to coalescence has not been proved experimentally. However, an increase in the degree of homogenization, i.e. obtaining an emulsion with a low average droplet size, enhanced stability of the emulsion. A strong correlation between surface hydrophobicity and coalescence rate constant was reported [47]. The monolayer adsorption density of β-lacto-globulin initially decreased from 2.8 to 1.4 mg/m^2 with an increase in S_o and a sharp increase to about 5 mg/m^2 with multilayer formation was observed. In the multilayers, proteins were loosely bound and did not increase stability to coalescence. For the stability to coalescence, the conformation stability of β-lacto-globulin was more important than film thickness.

Emulsion stability is influenced by the rheological properties of the adsorbed protein film – its elastic, cohesive, and restorative properties. Dickinson et al. [36] showed that formation of the film with stable rheological properties at the oil/water interface is a slow process and production of the stable emulsion in a homogenizer requires less than a second. In this study, pure proteins were used as the stabilizers. The same prolonged process of the development of rheological properties was found for egg yolk proteins. Low molecular weight stabilizers had a negative effect on the rheological properties of films, and as a result reduced emulsion stability. Various caseins are effective stabilizers because of extremely high flexibility, little secondary structure, and rapid unfolding at interfaces. The rate of coalescence de-creases with surface viscosity, and elasticity increases with higher protein load and

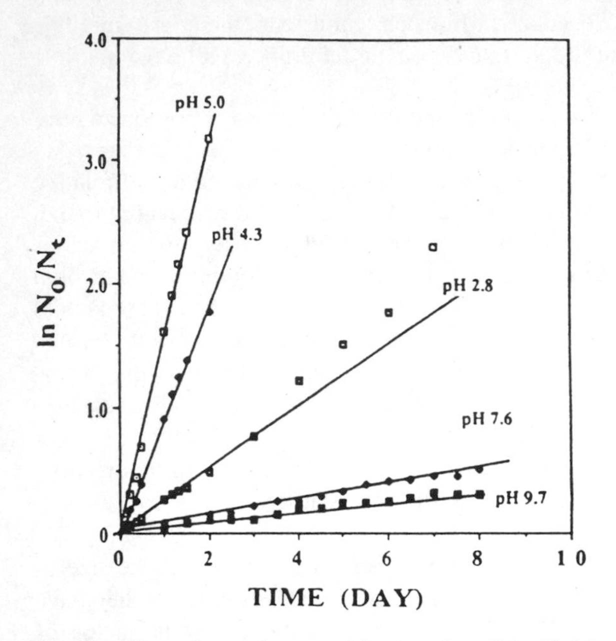

Fig. 3.1. First-order coalescence kinetics plot of 1n No/Nt versus time for ß-lactoglobulin-stabilized emulsions at different pH. From Ref. 50

greater compactness of the protein film [48]. However, viscoelastic properties of surface film are not the only factor responsible for emulsion stability. Graham and Phillips [49] did not find positive correlation between surface viscosity and coalescence stability.

Emulsion stability is influenced by the pH of the protein solution. The rheological properties of interfacial film and emulsion stability are influenced by the pH of the protein solution which affects protein solubility, net charge, and conformation of protein molecules at the oil-water interface [49]. Emulsions with a thicker adsorbed film exhibited the lowest stability to coalescence and maximum rate of coalescence was around pH 5.0 in emulsions stabilized with β-lactoglobulin (Fig. 3.1) [50]. In Fig. 3.1, the relationship between the coalescence rate constant and time of incubation is presented. pH 5 is close to the IEP of β-lactoglobulin. The coalescence rate constant increases at the pI range. The most stable emulsions were obtained at pH 9.7 with minimum surface protein concentration. Extensive flocculation of emulsion droplets in milk emulsions is near the isoelectric point, and coalescence decreases with pH far from the pI [51]. The effect of pH on ES is due to the alteration of the charge on the protein film.

Modified soy protein showed high emulsion stability in the slightly acidic region where the protein solubility remained at low level [15]. The ES of unmodified soy protein decreased with decrease in protein solubility. The HLB number of soy protein at pH 4.5 decreased and its ES increased as a result of modification by alcohol treatment, acetylation, and partial hydrolysis. Lipohilization of soy protein enhanced ES in slightly acidic region.

Processing treatment influence emulsion stability. Freezing of food emulsion increased coalescence and emulsion destabilization when the ionic strength was low [52]. The rate of coalescence was not affected by freezing when the ionic strength increased to 0.2 M. The rate of coalescence can be increased by refrigeration as a result of fat/oil crystallization and disruption of the protein film [53]. Emulsion destabilization by fat crystallization during refrigeration and water crystals during freezing is the result of rupturing of the interfacial film. Emulsions prepared with stabilizer "mannoprotein" containing 44% carbohydrate (mannose) and 17% protein at a pH range of 2 to 12 were stable to three cycles of freezing and thawing [54]. Coalescence causes an increase in average droplet size and the volume of the separated phase, and changes in turbidity and viscosity. The rate of coalescence in emulsions was determined by measuring the viscosity of emulsions [55]. Coalescence of emulsion causes a decrease in the emulsion viscosity. This method is effective for concentrated and viscous emulsions. To determine the rate of coalescence, the rate of the particle size distribution has to be measured as a function of holding time, although in most cases, only the amount of free oil needs to be measured.

Flocculation and coagulation are affected by the same phenomenon of contact between fat droplets. Interaction between fat droplets in emulsion causes flocculation. Flocculation is a kinetic phenomenon and is observed in emulsions with relatively large droplets. In food emulsions, larger droplets (>2 μm) flocculate rapidly. During development of flocculation when two colliding fat droplets approach close enough, the layers may merely compress without interpenetration, or they may interpenetrate without compression. In real systems there is a blend of both processes, probably biased in favor of compression. The theoretical interpretation of this process was reported by Fleer and Scheutjens [56]. Flocculation can result in an increase in emulsion viscosity as a result of the interconnected network formation. Flocculation is influenced by the pH, the ionic strength of the protein solution, and the nature of the protein stabilizer. Emulsion flocculation varied with type of protein and increased with increases in ionic strength [52]. Flocculation is facilitated since most emulsions contain salts, and the electrical barrier is decreased. In most food emulsions the ionic strength is between 10 and 100 mol m^{-3}, so that electrostatic repulsion is not a major stabilizing factor. Small-molecule surfactants reduce flocculation of protein-stabilized emulsions because they displace proteins from the interface [57]. The flocculation rate can be determined by a microscopic method and using the Coulter counter. Emulsions have been classified into the arbitrary scale according to their degree of flocculation [58]. Darling [59] suggested the method of flocculation measuring. Flocculation characterized by the degree of clustering in the emulsion was measured as the change of viscosity at constant shear with time.

Creaming is caused by the difference in density between the oil and water phases. Very small droplets, 0.5 μm or less, are not separated by creaming; and only larger droplets are involved in the creaming. Creaming is reversible and the cream layer can be redispersed, however, extensive creaming may not be reversible. Creaming is influenced by the droplet size in the emulsion, and a reduction in the size of droplets may reduce creaming. Creaming stability of emulsions prepared with whole milk protein, β-lactoglobulin, whey protein and micellar casein was improved with increasing energy input [60]. Creaming stability was markedly increased with increasing protein concentration. Heating of the emulsions to 70–80 °C improved creaming stability. An increase in pH from 6 to 9 decreased the rate of creaming. Creaming of food emulsions can be effectively reduced by addition of hydrocolloids, gums and other polysaccharides that cause an increase in the viscosity of the continuous phase. Emulsion gravitational separation can be accelerated by centrifugation at low speed to avoid breaking the interfacial film. The ultrasonic method of measuring creaming stability was presented by Howe et al. [61]. The principle of emulsion testing by ultrasound is based on the difference in ultrasound velocity in the oil phase and continuous phase. If the droplet size is smaller than the wavelength of ultrasound, the ultrasound velocity is affected only by the volume fraction of dispersed oil [62].

3.6 Measuring Emulsifying Properties

A variety of experimental methods have been applied to measure EC and ES. For this reason, there are difficulties in comparing the measurements carried out by different authors. Functionality of proteins in emulsions is influenced by the procedure of emulsion preparation, and there is no accepted standard procedure. Testing of protein EC in model systems is widely utilized and there seems to be a correlation between EC obtained in model systems and in commercial conditions. Significance of model tests of EC may be increased if they simulate food processing conditions at the maximal level.

Emulsifying properties of various proteins are difficult to compare because of the absence of the standard method for determination of emulsifying capacity and emulsion stability. Small variations in the design of emulsifying equipment, technique, speed of blender, rate of oil addition, type and properties of oil, temperature of continuous and discontinuous phase, source and properties of proteins influence emulsifying capacity of proteins. EC and ES may be measured by various methods: change of the particle size distribution by microscopy, change in the separated oil fraction, and change in drained water phase over time. Centrifugation is usually used to accelerate emulsion separation. Pulsed MNR of dilute emulsions and measuring of the dielectric constants of the medium have been tested. The EC test establishes the maximum amount of oil emulsified under chosen conditions. Electrical conductivity or resistance is measured for determining EC. After a certain volume of oil has been added, the properties of emulsion in the mixer have been changed and are referred to as either inversion or breaking. The endpoint is measured as an increase in the electrical resistance, a drop in viscosity, or a change in the visual appearance of the emulsion.

The best method of measuring emulsion stability is dispersion analysis to register the size distribution of the fat particles as a function of different factors. The most reliable method of emulsion dispersivity measuring is microscopic observation to assess changes in the dimensions of the individual droplets. Emulsion breakdown can be determined by the increase in the number of large oil particles (more than 20 μm). The disadvantage of this method is that oil particles of less than 0.8 μm are not detected. These globules can present a significant percentage especially in homogenized milk, soft drink and flavor emulsions. Microscopic measurements of the diameter of fat particles during emulsification showed that as the time of treatment in the emulsifying unit (blender, homogenizer, sonicator) increases, the droplet size decreases at first, but then approaches a limiting value. The spectroturbidity method also provides data about droplet size distribution and average diameter. The optical density of diluted emulsions is related to the interfacial area, i.e. the surface area of all droplets.

The EAI indicates the area of interface stabilized per unit weight of protein. EAI is also influenced by the conditions of emulsification and emulsifying equipment. The emulsifying stability index (ESI) includes measuring emulsion turbidity after heating in a boiling water bath. The emulsifying activity test includes emulsion centrifugation and the height of the unseparated layer is expressed as a percentage of the total.

3.7 Emulsifying Properties of Meat Proteins and Proteins Utilized as Extenders in Meat Products

3.7.1 Protein Functionality in Comminuted Meats

Meat proteins have been considered in a modern diet as a main protein source for humans, contributing about 35% of the protein intake. The batters of comminuted meat products (CMP) are complex colloidal suspensions of meat and fat particles extended with solubilized proteins and partially emulsified fat [63]. The quality of finished CMP is mainly determined by functionality of protein and lipid components. Manufacturing of CMP with the proper textural properties is related to the functionality of protein in the three-dimensional matrix of sausage batters. Formation of this matrix in sausage batter is due to the interactions between protein-water, protein-protein, and protein-lipid. Proteins are the major structural components in the system; they combine and develop the structure by binding water and fat. Various proteins are added to emulsion-type sausage batter to balance the quality and quantity of protein with processing functionality, nutritional value, and cost. Three groups of proteins serve as emulsifiers and stabilizers in comminuted meats: meat, milk, and various plant proteins. Nonmeat proteins from a variety of plant and animal sources are used extensively as binders, fillers, and extenders in CMP. Those plant and animal proteins can reduce cooking loss and formulation costs as well as improve CMP quality characteristics:

emulsifying capacity, emulsion stability, water retention, flavor, texture, and nutritive value. These nonmeat proteins assist the meat myosin and actin in stabilizing the emulsion system. Nonmeat proteins act as emulsion stabilizers to reduce the possibility of breaking of the emulsion and decreasing moisture and fat losses during cooking [64, 65].

Different protein sources have unique individual functionalities, such as emulsifying capacity, water and fat binding, viscosity, and protein solubility. Several workers have shown that different protein sources can improve physical and chemical properties in emulsion-type sausage, such as soy protein, wheat flour, milk proteins, cottonseed flour, sunflower protein, and wheat germ protein flour [65–67]. Stability of sausage batter is the result of gel and emulsion formation. For this reason, the term "meat or sausage emulsion" has been replaced by "meat batter" to demonstrate the more complex structure and effect of gelation and heat-setting properties of the meat proteins.

The basic operation in sausage batter preparation is comminution. During comminution, the muscle cells are broken down and the proteins are extracted. Insufficient extraction of myosin or protein-protein interaction results in excessive exudation and a mushy texture. Increase protein-protein and protein-water interactions in CMP may improve firmness and yield of the product. The level of protein extraction is affected by salt concentration, temperature, and the amount of added water before and during comminution. During comminution in sausage production, the fat cells are disrupted, the fat is released, and the fat dispersion is formed. The salt-soluble proteins form a surrounding layer at the surface of released fat droplets and a heat stable emulsion is formed. The capacity to form interfacial protein layer at the fat/water interface is one of the important functional properties of muscle proteins in CMP. The fat that is not emulsified is unstable and can form fat cups in CMP. Added milk and plant proteins participate in free fat emulsification and stabilization of the structure. The stabilized fat globules are retained in the network of swollen fibers of the protein sol matrix and prevent fat and water separation and product shrinkage.

Sausage batters are considered to be colloidal dispersions that include homogenized particles of muscular, connective and adipose tissues, solubilized proteins (salt- and water-soluble), salt, water, and other ingredients added during formulation. The physical and chemical stabilization of sausage batters is obtained as a result of both partial emulsification of lipid components, formation of a protein matrix and gelation of muscle proteins [68]. Although the sausage batters do not possess the properties of o/w emulsion, they exhibit some characteristics that are related to emulsions. Heat treatment stabilizes the emulsion as a result of protein denaturation, coagulation, and aggregation. As a result CMP with low moisture loss without fat coalescence is formed.

The theory of emulsion formation and stabilization in meat batters is indirectly transferred from the structure and properties of other food emulsions. The emulsification theory describing processes during comminution was widely accepted in the 1960s–1970s as being responsible for sausage batter stability in CMP. Formation of a stable protein-fat-water system was due to the emulsification of fat particles by forming water- and salt-soluble protein film on their surfaces and protein matrix. The most important functional properties of animal proteins and protein extenders of animal and plant origin are:

1. Water retention resulted from protein-water interactions;
2. Fat binding resulted from protein-lipid interactions and emulsification; and
3. Formation of the CMP structure as a result of protein-protein and protein-water interactions.

Parks and Carpenter [65] determined the binding value constants using the Saffle binding system: NFDM, 27.5; soy isolate, 17.6; milk protein hydrolyzate, 15.8; soy concentrate, 13.0; autolyzed yeast, 8.3; soy flour, 7.6. The binding value constants determined mainly by binding and emulsifying properties are affected by the processing of protein manufacturing. Saffle [69] recommended that meat raw material be classified based on emulsifying (binding) constants, where the percentage of salt extractable proteins was multiplied by the emulsifying efficiency (ml oil emulsified/100 mg of salt soluble protein) and constants have been obtained. Binding constants were suggested for predicting minimum cost formulations for sausage mixes stable upon cooking.

Emulsions of CMP have been considered as a gel-type emulsion with fat dispersed uniformly in the protein-gel matrix and emulsion stability is affected by distribution of fat particles and rigidity of the gel [70]. Meat proteins and water form a matrix or lattice which binds and partially emulsifies fat particles. The protein matrix structure is built up from previously dissolved water- and salt-soluble proteins. The fat ingredient in meat emulsions is a dispersed or discontinuous phase. In the continuous phase, proteins, especially salt-soluble proteins, and other components are dissolved. The scanning electron micrographs show protein encapsulating the fat droplets and forming the surrounding matrix [71]. In cooked sausage batters, these membranes have been disrupted with the formation of a number of pores or openings. The protein matrix had been disrupted with the protein being coagulated. However, there was no fat separation caused by fat particles coming to the surface. The fat separation could be possible if the protein membrane was broken before the protein matrix set up. Formation of the typical protein-fat-water emulsion was demonstrated during batter comminution with fat globules stabilized by protein membranes. A scanning electron micrograph of a fat droplet surrounded by a protein coat in a sausage emulsion has been presented [72]. Protein film surrounding fat droplets is formed mainly from myosin.

The following factors influence the properties of meat emulsions: temperature, salt concentration, size of muscular and connective tissue particles, size of fat droplets, pH, added milk and/or vegetable proteins, conditions of comminution (shear force), autolytic state of the meat (pre- and postrigor), use of frozen and fresh meat and others. The preparation of meat emulsions requires a certain amount of shear force. The comminution is performed in a silent cutter and emulsitator. Many processors use the silent cutter for comminution, reduction of the particle size, and as a mixer. Under these conditions, stable sausage emulsions are prepared in the emulsitator. With an increased time of treatment in silent cutter, more soluble proteins can be extracted and more stable emulsion can be prepared. Many sausage manufacturers, after grinding and blending of the meat with all the ingredients, perform comminution in the emulsitator without using a silent cutter. Properties of sausage emulsion and amount of fat emulsified were affected by the rate of mixing.

3.7.2 Emulsifying Properties of Various Muscular Proteins

Emulsifying properties of meat proteins are influenced by species of animals, sex, age, morphological structure, and processing treatment. Boneless cow and bull meat exhibits a higher EC than pork cheek meat. Turgut [73] reported no significant effect of age on the EC of the meat of four different species of animals. However, salt-soluble proteins from sheep and goat muscle showed higher EC than from cattle and waterbuffalo. Differences in EC were explained by differences in protein extractability. Turgut and Sink [74] and Turgut [73] reported the effect of age and sex on the EC of bovine muscle. An emulsion stability test showed that meat emulsions from old animals released a lower amount of oil and liquid than meat from younger animals. The EC of salt-soluble proteins from older animals was higher than that from younger animals. There was no significant effect of sex, regardless of age (between 6 and 20 months), on the EC of semitendinosus muscle.

Proteins of meat obtained from by-products of slaughtering possess different EC and ES. For example, proteins obtained from beef heart showed high EC, but the emulsion obtained had low stability. However, in a sausage batter more sarcoplasmic than myofibrillar proteins will be dissolved and their role in stabilization of the muscle protein-fat-water system will increase. The insoluble protein fraction (connective tissue and unextracted myofibrillar proteins) plays an important role in emulsion formation and stabilization particularly in the presence of salts.

Frozen and thawed meat does not have the emulsifying properties of fresh meat. The effect of frozen storage at –20 °C for 8 months and thawing on protein solubility, EC, and viscosity in meat and fish was reported [75]. Chicken myofibrillar proteins showed the highest EC, both initially and throughout thesto-rage period. The EC decreased during frozen storage, and the storage effect was more significant for fish than for pork proteins. The effect of frozen storage on EC could be related to the loss of protein solubility. Irradiated muscle proteins soluble in water were used for the determination of emulsifying and gelling properties [76]. Physicochemical properties of muscular proteins affected by irradiation treatment influenced emulsifying and gelling capacity. Surface and emulsifying properties of modified gelatins had more in common with low-molecular-weight food emulsifiers than with proteins [77]. When used at high concentration (5 wt%) these surfactants retarded the freezing of water. They can be recommended for stabilization of emulsions and protection against breakdown during frozen storage.

Table 3.1. Lard in water emulsifying activities and emulsion stabilities of EMG products at different oil volume frations

	Oil volumes fraction					
Surfactant	0.2	0.3	0.4	0.5	0.6	0.7
	Emulsifying activity (m^2 ml)					
EMG-6	0.129	0.225	0.253	0.280	0.244	0.190
EMG-8	0.229	0.281	0.467	0.475	0.553	0.562
EMG-12	0.178	Ne[a]	NE	NE	NE	NE
	Emulsion stability					
EMG-6	0.79	0.91	0.84	0.69	0.68	0.55
EMG-8	0.89	0.92	0.75	0.67	0.58	0.58
EMG-12	0.00	Nd[b]	ND	ND	ND	ND

[a] NE = no emulsion formed.
[b] ND = not determined because no stable emulsion was formed. From Ref. 79

Functionality of gelatin highly hydrophilic protein was considerably improved by covalently attached L-leucine *n*-alkyl ester, a highly hydrophobic group [78]. Enzymatically modified gelatin with an amphiphilic structure was produced. Modified gelatin showed desirable emulsifying, and antifreezing properties [79]. The modified gelatins with covalent attachment of L-leucine *n*-hexyl ester (EMG-6), *n*-octyl ester (EMG-8) or *n*-dodecyl ester (EMG-12) were prepared. EMG-8 showed the highest emulsifying activity in lard-EMG-8-water emulsion system at oil volume fraction from 0.2 to 0.7 (Table 3.1) [79]. EMG-8 and EMG-6 showed almost the same values for emulsion stability. EMG-8 dispersed lard more evenly than the other ingredients in the structure of meat emulsions. Fat globules were more homogeneously distributed in the structure of sausage with EMG-8. Because of desirable emulsifying properties EMG-8 was added in the formulation of CMP.

Salt- and Water-Soluble Proteins

Generally the role of protein solubility in emulsifying properties has been stressed and correlation between the soluble protein level and the EC and ES of meat extracts has been found [80]. Soluble myofibrillar proteins have been considered as having higher emulsifying properties. Emulsifying properties of myofibrillar and sarcoplasmic proteins are important in processing of meat emulsions. Within muscle proteins, myofibrillar proteins are the major emulsifiers. Some of these proteins in meat emulsions will be dissolved, particularly in the presence of salt. The emulsifying effect of these proteins includes formation of a thin protein film around the fat particles.

The capacity of meat proteins to bind and emulsify fat during comminution is influenced by the amount of water- and salt-soluble proteins potentially available and the capacity of these proteins to emulsify fat. Water-soluble meat proteins exhibited very low emulsifying properties compared to emulsifying capacity of salt-soluble proteins. Gaska and Regenstein [81] have reported that high-salt insoluble proteins may play an important role in emulsion formation during comminution. Low- and high-salt washed muscle, resuspended in either 0.15 or 0.6 M NaCl, pH 7.0 showed good EC. Correlation has been established between emulsifying properties of salt-soluble meat proteins and surface hydrophobicity, solubility, and content of sulfhydryl groups [82]. Hydrophobicity, solubility and sulfhydryl group content of salt extractable proteins from beef and fish influenced their emulsifying and fat binding properties [7].

The Effect of Postmortem Changes

Emulsifying properties of meat proteins are influenced by meat postmortem changes and pH. Prerigor meat because of high protein solubility, capacity to bind water, and excellent emulsifying properties was widely utilized in the U.S. up until the 1930s or early 1940s. Prerigor beef showed more salt-soluble protein extracted than postrigor meat. Proteins of prerigor meat showed more superior emulsifying properties than proteins of postrigor meat. The high EC of prerigor meat could be preserved by meat preblending with ice and salt. In rigor meat, formation of crossbridges between the thin and thick filaments caused decreased extractability of myofibrillar proteins, myosin and actomyosin. During rigor mortis as pH declines to about 5.4, the pI of myosin is approached with a loss in the ability to bind water. Postmortem pH fall is strongly affected by temperature with higher temperature enhancing the rate of pH drop.

Protein extracts prepared from beef samples with pH 6.0–6.3 had lower hydrophobicity and higher EC and fat binding properties than proteins extracted from meat samples with pH 5.5–5.7 [83]. Lower emulsifying properties of the samples with pH 5.5–5.7 were related to protein denaturation. The pI of the meat proteins is in the range pH 4.5–5.5. Samples in the prerigor state had higher extractability and EC of protein component than samples in the postrigor state. The effect of pH on functional properties is related to the decrease of the net charge density and zeta potential when the pH of the tissue approaches the isoelectric point.

The effect of prerigor pressure treatment on the EC of meat proteins in sausage batters was reported [84]. Before the onset of rigor mortis, the EC of muscle homogenate from the control samples was higher than the pressure treated samples. Partial denaturation of proteins was probably responsible for the decrease in EC. The EC of pressure-treated and control samples was not different when tested at 24 and 168 h postmortem. The EC of myofibrillar and sarcoplasmic protein extracts from control samples was slightly higher than pressure-treated samples. However, overall, the EC of muscle proteins was not detrimentally affected by pressure treatment.

The Effect of Temperature

Practical experience in the sausage industry has shown that if the temperature of the sausage emulsion in the chopper exceeds 15–22 °C, emulsion breakdown will occur. Emulsion breakdown could be related to protein denaturation and/or to the mechanical action of the chopper knives which increases contact of the fat particles. The temperature effect was related to a decrease in viscosity. The emulsion at high temperature is less viscous and less stable. At higher temperatures there is an increase in the coalescence of fat droplets. Lee [85] suggested that the protein film which covers the solid fat particles during comminution increases the solid fat dispersing in the continuous phase without formation of a true emulsion. Overchopping and overheating may destabilize a raw batter as a result of protein denaturation due to heating and shearing, disruption of the protein matrix, fat liquefaction and separation.

The chopping temperature affected the microstructure of meat emulsions with a standard frankfurter formulation (25% fat) [71]. The effect of endpoint chopping temperature at 10, 16, 22, and 28 °C on the ultrastructure of meat emulsions with a grind-mix-emulsifying production system was evaluated. Scanning electron micrographs suggested that emulsion stability was attained as a result of two functions: first, related to the interfacial protein film thickness, and the second related to the integrity and density of the emulsion matrix and its ability to retain that integrity during thermal processing. These functions appeared to be related to the fat- and water-binding capacity of the meat emulsion. Overheating of CMP during heat treatment at 80 °C causes collagen gelatinization and formation of undesirable pockets of a heat-reversible gel [86].

The Effect of Ionic Strength

During production of sausage batters, addition of 1.5–3.0% NaCl increases the extractability of myofibrillar proteins from the meat and improves the emulsifying properties and flavor of the final product. Emulsifying properties of meat proteins are influenced by the concentration of salts and pH of water soluble proteins. Ionic strength enhanced the emulsifying properties of proteins by increasing the amount of protein in solution. When added during formulation, sodium chloride markedly increases the surface and interfacial tension of water. However, protein ingredients in the emulsion significantly decrease interfacial tension. In the presence of salt, myosin showed the greatest binding capacity, and in the absence of salt, the binding capacity of myosin is enhanced by sarcoplasmic proteins.

The Effect of Fat Properties

In meat emulsions an insignificant amount of potential fat is emulsified, and available salt-soluble proteins are not utilized. The dispersivity of the emulsion and the amount of fat emulsified in meat emulsions is affected by the properties of the fat. The size of the fat droplets affects emulsion stability. The smaller the fat particles, the more stable the emulsion if a sufficient amount of emulsifier covered the fat droplets. Significant discrepancies in the size of the fat particles in meat emulsions probably influence stability of the system during subsequent heat treatment.

The Emulsifying Properties of Proteins in Poultry Meats

In processed chicken products as comminuted meats (frankfurters and bologna), chicken loaves and rolls, the functional properties of chicken meat proteins such as water and fat binding, emulsifying and gelling capacity mainly affect the quality of the finished product. Myosin and actomyosin had the highest EC but lacked ES when compared to the other fractions. Myosin from chicken muscle performed as an effective emulsifier when added at low concentrations, but the emulsions prepared showed insufficient stability [87]. At a myosin concentration below 2 mg/ml, an oil layer in addition to the cream layer was observed in the emulsion. The exhaustively washed myosin below 4 mg/ml produced emulsions with low stability. The need for a higher concentration of exhaustively washed myosin to stabilize the emulsion might be related to its insolubility. In this case, a larger concentration of myosin will be required to stabilize the emulsion. Stability of emulsions prepared with heat-treated (40 °C to 75 °C) myosin was related to the gelling ability of the myosin molecules [87]. Maximum gelation of myosin occurs between 60 °C and 70 °C. Proteins of broiler meat had a higher EC than found in hen, turkey and duck meat [88]. The EC of myofibrillar proteins was considerably higher than sarcoplasmic ones and EC was affected by the pH of the solution in a model system. Myosin, actin and actomyosin demonstrate a higher EC while stroma proteins decrease the EC of the system.

The optimum EC for sarcoplasmic proteins was obtained at pH 6.0. In myofibrillar proteins the optimum EC was at pH 7. The EC of these proteins decreased in the pH range 5 to 6. A positive correlation (linear relationship) was observed between pH and EC in the mixture of both protein fractions. Myofibrillar proteins showed a considerably higher ES than sarcoplasmic proteins when measured by the amount of water and oil released during heating of the emulsion sample. The mixture of sarcoplasmic and myofibrillar proteins showed intermediate EC and ES values. The EC and ES of meat mixtures prepared from broiler breast muscles with pH range from pH 5.7 to 6.2 and 6.5 was not different [88].

The emulsifying capacity of fresh extracted chicken protein was lower than nonextracted protein paste, but there was no difference in EC between extracted and nonextracted protein [89]. The frozen samples of proteins showed lower EC.

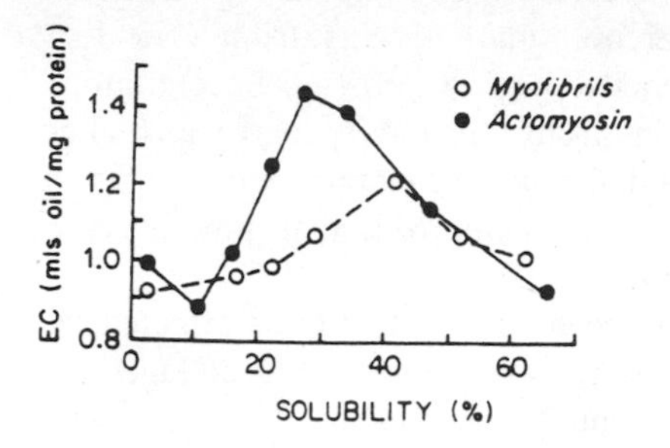

Fig. 3.2. Effect of duration of enzyme treatment (solubility) on the emulsifying capacity of 1.0% (w/v) MDF protein suspensions in 0.1 M NaCl, 0.05 M K phosphate, pH 7.0. From Ref. 90

The extracted chicken protein gave a considerably higher emulsion stability than unextracted. This was probably related to a much lower fat to protein ratio (0.8) in the extracted protein as compared to unextracted (2.26). The EC of mechanically deboned chicken myofibrillar proteins was improved by partial proteolysis with acid protease [90]. The maximum EC of myofibrils was observed at 42% solubility, while the maximum EC of actomyosin occurred at 28% solubility (Fig. 3.2) [90]. Actomyosin showed a higher EC than the myofibrils. Proteolysis increased the EC of samples compared to controls in the pH region of low solubility.

Improved EC of modified meat proteins was demonstrated also by an emulsified meat cooking test, indicating an increase in cooked yields, and less cooked-out fat. The effect of potassium phosphate and sodium acetate washing on the emulsifying properties of the proteins in turkey dark meat has been reported [91]. Washing of dark meat resulted in bleaching and loss of specific flavor. The EC of turkey dark meat proteins was not affected by the extraction process. These data were in contrast to those of Jimenez-Colmenero and Garcia [92] who reported a reduction in the EC of mechanically deboned pork meat due to water washing. This was related to the low EC of sarcoplasmic proteins extracted during the washing procedure and tested in the presence or absence of myofibrillar proteins. The breast tissue more effectively stabilized the emulsion than the unextracted and extracted thigh tissues. Phosphate-washed tissue showed significantly higher emulsion stability than acetate/phosphate washed tissues and a control of unwashed thigh tissue.

The Emulsifying Properties of Fish Proteins

Salt-soluble fish proteins are considered to have emulsifying properties superior to water soluble proteins. Borderias et al. [93] reported the EC of myofibrillar and sarcoplasmic proteins of various fish species and found a correlation between the EC and the concentration of soluble proteins. The EC decreased with an increase in protein concentration, and this may be due to a greater degree of unfolding of proteins during the shearing involved in the emulsification. Another factor is that, as protein concentration increases, the emulsion system is over-saturated with proteins and the EC no longer depends on either the concentration or on the type of protein. Variations among the fish species were related to different concentrations of soluble proteins. The sarcoplasmic proteins from four fish species exhibited the same emulsifying capacity. At low protein concentrations (5 mg/ml), the EC of the sarcoplasmic proteins from four fish species was higher than the EC of the myofibrillar proteins and the muscle homogenates. This is controverts the statement from Gaska and Regenstein [81], that sarcoplasmic proteins did not take part in emulsion formation.

The relationship between fish collagen concentration and EC as affected by pH and sodium chloride levels has been reported [94]. EC of the collagenous material from hake increased as collagen concentration increased. The EC values were higher for the collagenous material from the muscle than those from the skin. When the EC was expressed in g oil/100 mg soluble protein, EC values remained stable at 1 and 2% NaCl, and increased notably at 3% NaCl. EC values remained low and stable in the pH range 1 to 5 and increased markedly at pH 5–6. This increase in EC is probably due to a reduction in the number of intermolecular cross-links. The EC was generally higher for the collagenous material from hake than that from trout. The EC of minced fish meat decreased during ice storage and subsequent frozen storage [95]. A marked decrease in EC occurred during the first 90 and 120 days of frozen storage with fresh and 3-day ice-stored samples, respectively. The decrease in EC might be explained by a concurrent decrease in muscle protein solubility due to protein denaturation induced by freezing and frozen storage.

Grinding of fish muscle in the frozen state at –29 °C to –23 °C resulted in an extensive loss of functional properties when compared to grinding in the thawed state [96]. All tested functional properties, level of salt-soluble proteins, viscosity, emulsifying capacity and elasticity were affected by grinding in the frozen state. The functionalities of freshly frozen samples were affected less than those frozen and stored for 8 and 18 weeks at –29°C. The fish sausage prepared from muscle ground while frozen had a soft, crumbly cooked structure. The sausage prepared from fish ground while thawed had an acceptable cooked texture. Fish proteins are sensitive to temperature, pH, salts and contact with polar solvents and this may result in a loss of functional properties during preparation of dry fish protein concentrates and isolates.

A combination of radiation (250 Krad) and heating (63 °C for 10 min) resulted in the precipitation of 75% of the total proteins [97]. Gamma-irradiation and heating in alkaline pH (8.0–9.0) has been utilized for the preparation of fish protein concentrate. The procedure caused the precipitation of 75% of fish proteins which accounted for 80% of myofibrillar proteins. The EC of fish concentrate obtained by irradiation and heat treatment was influenced by its concentration. The EC (ml oil/mg protein) decreased as the concentration of protein in solution increased.

Collagen as an Emulsion Stabilizer

The stroma proteins exhibit deleterious effects on meat quality, reduce tenderness, decrease the emulsifying capacity of meat and the water holding capacity. Sausage manufacturers use large amounts of connective tissue to reduce processing costs. However, extended replacement of meat proteins by collagen can decrease emulsifying capacity and WHC with formation of gel pockets, gel caps and poor peelability. The emulsion stability of the raw bologna was not different between bologna with or without collagen. Collagen-containing bologna showed a firmer texture even at higher fat levels. The added collagen was of a fibrous nature and had a low soluble fraction which is probably responsible for the absence of gel pockets in the product. The replacement of 20% lean meat with hide collagen did not alter raw sausage emulsion stability, cook yield, water activity, and expressible juice [98].

Collagenic proteins of bones can be utilized in human food since they exhibit good functional properties. The satisfactory protein can be extracted using low temperature and alkaline solutions [99]. Emulsifying capacities of bone extracts and plasma proteins were lower than the EC of meat proteins. However, EC values for the salt soluble proteins were approximately the same for beef muscle protein and bone protein extracts. The mixtures of bone protein extract and meat proteins exhibited EC values between those of the components. Mixtures with 30% bone protein extract showed better EC than plasma proteins considered as an effective emulsifier.

3.7.3 Emulsifying Properties of Blood Proteins

Production of white emulsifier with an acceptable EC from bovine blood proteins requires the separation of whole blood into a serum fraction and red blood cell concentrate. Emulsifying activity of globin increased with increased globin concentration when the protein concentration was below 0.5% and did not change with further increase in globin concentration [100]. Emulsifying activity (EA) of blood globin, measured by optical density, was greater than that of hemoglobin and ovalbumin. EA decreased in the isoelectric zone owing to the low solubility of globin. The EA of globin in the isoelectric zone was increased by acetylation, but in this case, the decrease of EA was found in the acidic pH region. The EA of hemoglobin continued to increase gradually with protein concentration up to 2%. Globin showed an extremely low EAI near pH 7, probably due to the low protein solubility. Bovine serum albumin (BSA) showed good EC in the pH range 4–9 and a rapid drop in EC was found at pH below 4 [35]. BSA is a relatively hydrophobic protein and forms stable protein films at the oil-water interface. Unfolding of the tertiary structure of BSA at the interface improves the properties of the interfacial layer. BSA has 17 disulfide bonds which impart a certain degree of structural stability. Native BSA formed a stronger layer at the interface and formed more stable emulsions. Modification of BSA by succinylation considerably increased EA in the pH range 4–7 when compared to native BSA. Emulsifying activity (EA) was influenced by increase in protein concentration and temperature. Protein concentration influenced formation of a stable interfacial protein film. In emulsions stabilized with BSA, maximum EA was obtained at ~ 1% protein level, and a further increase in protein increased emulsion stability. MacRitchie [25] reported that < 8 mg of protein was sufficient to form complete coverage for 1 m^2 of an oil-water interface.

Emulsifying activity is influenced by temperature of emulsification. At the lower temperatures (5–8 °C) the rates of protein diffusion and possibly denaturation were reduced because thermal energy of the dispersion was reduced [35]. More stable interfacial film and emulsion was formed at 10–25 °C when BSA was adsorbed more rapidly at the interface. Excessive homogenization resulted in unfolding of BSA and protein denaturation with destabilization of the film at the oil-water interface and coalescence of fat droplets. Sodium chloride and sodium sulfate affected EA of BSA solutions [25]. At low ionic strength (< 0.02 M), low EA of BSA was observed because the increased charge repulsion at the oil-water interface reduced the amount of BSA adsorbed at the interface. Optimum EA was found at intermediate ionic strength levels (0.1–0.6 M), and the decrease of EA was observed at high ionic strength of NaCl (>0.75 M). Irradiation and fumigation treatment did not affect EAI of plasma protein, 52.1–61.6 m^2/g, whereas heating at 63 °C for 20 min in solution significantly decreased EAI (44.1%) [101]. Hydrophobicity of irradiated plasma protein was slightly enhanced, while heating slightly lowered hydrophobicity.

3.8 The Functionality of Nonmeat Proteins in Comminuted Meats

A wide variety of protein additives of plant and animal origin are available for use in CMP to improve emulsifying capacity. Studies of the functionality of nonmeat proteins in CMP have focused on such functional properties as water retention, fat binding, emulsifying capacity, sensory properties, textural properties, storage stability, and nutritional quality. In the future, the use of plant proteins as additives in foods will increase significantly. Use of high levels of plant proteins in CMP with protein, fat and moisture content equivalent to those of an all-meat product may cause problems in emulsion stability, texture, flavor and color [102]. Stable sausage emulsions, without fat separation during the cooking process, are of paramount economic importance to the processed meat industry. At low meat replacement levels in CMP containing 20–30% fat, a variety of oilseed proteins could produce acceptable products. However, when higher replacement levels are used, a combination of protein types might be required in order to produce acceptable products. Incorporation of plant and animal proteins in sausage batters can be considered advantageous especially in batters with a low myosin content. Muscle proteins are better emulsion stabilizers in comminuted meats than most of the plant proteins. Commercial use of soybean proteins as binders in comminuted meats has shown them to be compatible with chopped meat systems.

Contradictory hypotheses of the protein-fat structural relationships have been proposed for CMP. In a sausage emulsion, the amount of emulsified fat is rather low and this emulsion is not stable and is sensitive to coalescence during heat treatment. The concept that the batter of comminuted meat products is related to true emulsions was not accepted. Meat batters are more similar to materials with visco-elastico-plastic properties with fat globules distributed comparatively homogeneously in the structure. In batters of frankfurters, fat particles were mechanically fixed within the structures of the protein filaments. Carroll and Lee [103] examined the thermal stability of meat emulsions before and after the cooking process by light, scanning and transmission electron microscopy. Microstructures changed as a result of increasing chopping temperatures. The fat droplets changed from spherical to oval and irregular shapes, accompanied by the formation of channels within the protein matrix, resulting from a weakening of the matrix. After thermal processing, the globule membranes were disrupted and the protein of the continuous phase was coagulated into dense, irregular zones [103]. Jones and Mandigo [71] observed small pores in the protein membrane surrounding fat globules in the finished product. These factors appear to be directly related to the fat-binding and water-holding ability of the sausage emulsions.

Protein substances are optimal stabilizers for fat emulsion, i.e. pre-emulsified fat. Preemulsified fat may play an important role in improving the sausage processing by stabilizing the fat component for CMP. Preemulsified fat can be prepared by using water, fat and protein which are components of sausage ingredients. Preemulsified fat was prepared by ultrasonic emulsification of melted fat and water with protein additives, NFDM, sodium caseinate, and blood plasma proteins, which gave a superior quality to the finished product [104]. The superior quality was approved by higher WHC, juiciness, tenderness and more uniform distribution of fat particles of finished product.

3.8.1 Milk Proteins

Dairy proteins have been considered in a modern diet as a main protein source for humans, contributing about 25% of the protein intake. Various milk protein ingredients can be utilized to improve the flavor, texture, appearance, and nutritional value of CMP, as well as providing a greater flexibility of formulation to minimize processing losses. Such ingredients are especially useful for cases of limited salt-soluble myofibrillar proteins in low meat or cheap meat products.

Nonfat dry milk (NFDM) improves yield, color, flavor, and stability of sausage batter upon cooking. NFDM has a very high nitrogen solubility index, high emulsifying capacity, and moderate water holding capacity compared with other nonmeat proteins. Whey protein concentrate (WPC) is highly nutritious, with a protein efficiency ratio (PER) of 3.0–3.2 (2.5 for casein), and contains signifi-

Table 3.2. Water holding capcity (WHC), viscosity, and cooking losses of sausage batters containing milk proteins and corn germ protein flour

Treatment[1]	Added water[2] (%)	WHC	Viscosity x 10^5 (cPs)	Cooking losses (%)
Control	25.0	0.375^{c}	3.40^{a}	11.75^{a}
CGPF, 3.5%	28.5	0.559^{a,b}	3.32^{a,b}	8.57^{b,c}
NFDM, 3.5%	28.5	0.586^{a}	2.26^{c}	7.17^{c}
WPC, 3.5%	28.5	0.507^{b}	2.77^{c}	9.66^{b}
SC, 2.0%	27.0	0.594^{a}	2.99^{b,c}	9.51^{b}

^{a,b,c,d} Means in the same column with similar superscripts are not significantly different ($P > 0.05$).

[1] CGPF, NFDM, WPC, and SC are corn germ protein flour, nonfat dry mild, whey protein concentrate, and sodium caseinate, respectively.

[2] Sausage batter were formulated with 25% added water for control samples; 1.0% water was added with each 1.0% additive used. From Ref. 107

cant amounts of essential amino acids such as lysine, tryptophan, methionine, and cystine. Whey protein is an excellent emulsifier at all pH values in comminuted meats. Whey proteins have many S-S bonds and are globular, strongly folded structures. Whey proteins are sensitive to heat; they unfold and form -S-S bonds, resulting in gelation. WPC could replace lean beef in the fat stabilization of meat emulsions.

Caseinates not only stabilize the fat emulsions but also contribute indirectly to better water binding and consistency. They save part of the soluble meat proteins from denaturation in the interface, leaving them available for gel formation. With the incorporation of caseinates, the stability of CMP becomes less sensitive to chopping temperatures, resulting in a more reliable and flexible process [105]. This occurs because caseins are rather unique in having very little, if any, secondary structure, because of high levels of proline, resulting in extremely high heat stability [106]. Caseinates have a random coil structure because of the high proline and the low sulfuric amino acids content. As a consequence, caseinates show no heat gelation and stability to denaturation. High electrical charge of caseinates makes them soluble in water and good emulsifiers. Caseinates may be applied in dry or presolubilized form at the beginning of the comminution process, but optimum stability is obtained when caseinates are incorporated as the stabilizers of preemulsified fat.

The functionality of various milk proteins and corn germ protein flour (CGPF) in comminuted meats has been reported [107]. Sausage batter properties of WHC, viscosity, and cooking losses are presented in Table 3.2. WHC of sausage batters with milk proteins and CGPF added were higher than that of the control batter. There were no differences in WHC among samples containing NFDM, sodium caseinate (SC) and CGPF. The WHC of sausage batters affects important quality characteristics of the finished products. The stability of sausage batter and water absorption by milk proteins and CGPF decreased the amount of free water and increased bound water. Stability of sausage emulsion and WHC affected the cooking losses similarly. The control batter with lower WHC had higher cooking losses of 11.75% than all four treatment batters (from 7.17% to 9.66%). The sausage batters with NFDM added showed the lowest cooking loss as a result of binding fat and water. This corresponds to the report of Holland [108] that NFDM significantly improves sausage batter stability upon cooking. Viscosity was lower for batters with milk proteins than for batter of the all-meat control (Table 3.2). This low viscosity may have been due to the high protein solubility of milk proteins that would not increase viscosity of the batter and/or to the dilution effect of the additional water (2.0–3.5% more) in the extended batters. Payne and Rizvi [109] reported that, as the percentage of added water increased in sausage batters at 10°C, the apparent viscosity decreased.

3.8.2 Soy Proteins

Soybeans are potentially the most abundant and economical sources of food proteins. The range of soy protein utilization will be determined by functional and physical properties of the proteins rather than the nutritional value. Soy proteins are used in CMP mainly for enhancing the emulsifying capacity and stabilizing emulsions. Stability is the main factor determining cooking yields. Soy proteins incorporated in sausage batters can increase emulsifying properties, viscosity, gel forming, and water holding in CMP. The role of protein gelation in the stabilization of emulsion-type sausages has been emphasized. These functional properties contribute to the formation of stable meat emulsions and improve moisture retention, appearance, and yield of the finished products. The effect of nonmeat proteins at 20, 40, and 60% substitution on sausage emulsion stability has been studied by recording released fat, released liquid, and cooking yield [65]. Soy protein isolate and NFDM were most effective, while soy flour and autolyzed yeast were least effective in emulsion stabilization. Soy concentrate and milk protein hydrolyzate showed intermediate stabilizing properties.

Soy proteins incorporated in CMP can be considered as heat-induced gels, and presumably soy-meat protein interactions are responsible for the textural properties of CMP. Soy protein isolates improved emulsion stability in CMP, cooking yields and binding strength in restructured hams. Peng et al. [110] obser-ved 11S protein-myosin interactions at heating temperatures between 85 and 100 °C. Experiments showed that basic subunits of 11S soybean interacted with myosin heavy chain subunits, and no interaction was observed for acid 11S subunits. The initial interaction temperature for either native or preheated myosin with native 11S protein was between 80 and 90 °C. In contrast preheated 11S protein interacted with myosin at a temperature range of 90–100 °C probably due to reduced reactivity of the acidic subunit. Peng et al. [110] suggested that hydrophobic bonds favored at high temperatures have been involved in 11S protein-myosin interactions at 85–100 °C. These temperatures are above the end-point

Table 3.3. Water holding capacity (WHC), viscosity, and cooking losses of sausage batters containing soy and corn germ proteins

Treatment[1]	Added water[2] (%)	WHC	Viscosity x 10^{-5} (cPs)	Cooking losses (%)
Control, 0%	25.0	0.61[a]	3.66[a]	10.09[a]
SF, 3.5%	28.5	0.66[b]	3.41[a]	7.68[b]
SI, 3.5%	28.5	0.69[b]	3.47[a]	7.94[b]
CGPF, 3.5%	28.5	0.66[b]	3.57[a]	8.10[b]
LSD		0.044	0.322	0.681

[a,b] Means within the same column with different superscripts are different (P > 0.05).
[1] SF, SC, SI, and CGPF are soy flour, soy concentrate, soy isolate and corn germ protein flour, respectively.
[2] Sausage batters were formulated with 25% added water for the control sample; in experimental samples; 1% water was added for each 1% plant protein source. From Ref. 111

temperatures in processed cooked meats (68–70 °). The textural properties of these products result from myosin-myosin or myosin-other meat protein interactions. This phenomenon explains the softer texture in soy-containing frankfurters [111].

Formation of stable meat batters with added soy flour (SF), soy concentrate (SC), soy isolate (SI), and CGPF is presented in Table 3.3 [111]. Meat batters extended with four plant proteins exhibited higher ES and WHC than those without protein addition. There were no significant differences among samples containing SF, SC, SI and CGPF. No significant differences were found in viscosity of meat batters in control and experimental treatments (Table 3.3). Highest cooking losses of meat batters were observed in the control all-meat treatment containing 25% water. No difference in cooking losses was found among samples containing SF, SC, SI, and CGPF. The excess of water added with these additives was absorbed by soy and corn proteins. High cooking losses and low cook yield could result from low emulsion stability in meat batter.

However, several researchers have reported that SI had greater effect in emulsion stabilizing than SC and SF [65]. Consequently, an increased addition of plant protein can stabilize sausage emulsion, increase WHC, and decrease cooking losses, thereby increasing product yield.

In order to develop new food items based on soy protein or to continue using soy proteins as additives, their objectionable flavor and aroma should be eliminated or reduced. Two major off-flavor and off-aroma notes occur in soy protein products: the grassy or beany note and the bitter and astringent characteristic [112]. The bitter flavors are formed from nonvolatile, oxygenated fatty acids. Phosphatidylcholine bound to soy protein develops a bitter flavor upon oxidation, lipoxygenase-mediated or autoxidation [112]. The characteristic be any flavor or aroma are partly derived from the raw materials and partly formed by

Table 3.4. Cook stability, water-holding capacity (WHC), and yield of frankfurters containing soy proteins as powder and preemulsified fat (PEF)

Sample[1]	Cook Stability (%)[2]	WHC[2]	Yield (%)[3]
Control	89.12[a]	0.595[a]	79.57[a]
SFp	91.43[a,b]	0.708[b]	81.37[a]
SFe	94.38[c,d]	0.765[c,d]	85.90[b]
SCp	91.10[a,b]	0.704[b]	80.98[a]
SCe	92.94[b,c]	0.750[c]	84.62[b]
SIp	95.85[d]	0.795[d,e]	84.31[b]
SIe	95.48[d]	0.805[e]	89.62[c]
LSD	2.51	^0.399	2.84

[a,b,c,d] Means in same column with different letters are significantly different (P >0.05).
[1] SFp = soya flour added as powder, SFe = soya flour added as PEF, SCp = soya concentrate added as powder, SCe = soya concentrate added as PEF, SIp = soya isolate added as powder, SIe = soya isolate added as PEF.
[2] Means from 8 replications.
[3] Means from 3 replications. From Ref. 114

oxidation of residual lipids [113]. Alcohols and carbonyl compounds were found to be the main components of the grassy or beany notes of raw soybean. Among them, *n*-hexanol and *n*-hexanal seemed to be important flavor components in terms of effect and quantity. Many other compounds have been found to be responsible for off-flavor and off-aroma, and interactions between various compounds also may contribute to the off-flavors [112, 113]. Beany off-flavor is extremely difficult to eliminate or mask. Most chemical methods and heat treatments used caused protein denaturation or possibly generated a cooked or toasted off-flavor. Suggested chemical methods of extraction of lipids, aromatic and flavor compounds are noneffective. They reduce protein functionality and may possibly promote the formation of toxic substances during chemical treatment.

An effective physical method was developed to mask specific off-flavor of soy proteins and to increase their utilization in the production of comminuted meats. The functional and sensory characteristics of frankfurters made with soy flour, concentrate, or isolate incorporated in formulations either as a powder or in a form of preemulsified fat (PEF) have been evaluated. Experimental samples with PEF exhibited high emulsion stability. The highest means of cooking stability were obtained with the SFe, SIp and SIe, whereas the control samples showed highest losses upon cooking (Table 3.4) [114]. SFe had a higher cooking stability than the SFp samples, but cooking stability of SIe and SCe frankfurter was not significantly higher than with SIp and SCp frankfurters. The control frankfurters, containing no added soy proteins, had the lowest WHC (Table 3.4) and hence the lowest yield.

Table 3.5. Effect of preemulsification of soya proteins on the flavor of frankfurters at 1 day of storage.

	Flavor characteristics[1,2]			
Sample[3]	Meaty	Bitter	Beany	Acceptability
Control	41.83^{d}	8.80^{a}	12.50^{a}	44.16^{c}
SFp	25.33^{b}	26.33^{d}	36.20^{d}	22.60^{a}
SFe	43.50^{d}	$16.80^{b,c}$	$18.33^{b,c}$	$40.60^{b,c}$
SCp	19.80^{a}	30.60^{d}	39.00^{d}	26.25^{a}
SCe	41.87^{d}	19.60^{c}	21.00^{c}	$40.83^{b,c}$
SIp	36.80^{c}	26.27^{d}	20.40^{c}	38.80^{b}
SIe	42.80^{d}	13.50^{b}	$14.60^{a,b}$	44.00^{c}

[a,b,c,d] Means in same column with different letters are significantly different ($P<0.05$).
[1] Means from three replications.
[2] Data were collected from a 60-point scale, 0 = weak, 60 = strong.
[3] SFp = soy flour added as powder, SFe = soya flour added as preemulsified fat (PEF), SCp = soya concentrated added as powder, SCe = soya concentrate added as PEF, SIp = soya isolate added as powder, SIe = soya isolate added as PEF. From Ref. 114

Treatments containing soy proteins in the PEF form (SFe, SCe) had higher WHC than those containing proteins added as a powder. Similar results were reported by Lin and Zayas [115] and Zayas [104] where different proteins such as corn germ protein, blood plasma, sodium caseinate, and nonfat dry milk incorporated as PEF increased the WHC. Increased cooking stability and WHC for frankfurters with SFe and SCe resulted from higher batter stability and water binding when proteins were incorporated as PEF. The yield reflects the batter stability, i.e. ability of a frankfurter to retain water and fat during the cooking and smoking process (Table 3.4). Frankfurters with emulsified soy isolate (SIe) had the highest yield, whereas the control, SFp, and SCp ranked the lowest. The three frankfurters (SFe, SCe, SIe) containing soy proteins in the PEF form had higher yield than their counterparts (SFp, SCp, and SIp), containing powdered forms of soy proteins. This indicated better water and fat binding properties in products containing PEF.

The sensory data for flavor analysis are given in Table 3.5 [114]. The control frankfurters were rated the same for meatiness as the PEF samples (SFe, SCe, and SIe). The samples containing soy proteins added as PEF were more meaty, less bitter, and less beany than those with soy proteins incorporated as the powder (SFp, SCp, and SIp). The all-meat control was the least bitter, and the control and SIe samples were rated lowest in beaniness. The treatments containing soy proteins as PEF were rated highly acceptable and were not significantly different from the all-meat control. The PEF samples (SFe, SCe, and SIe) were superior to samples with soy added as the powder (SFp, SCp, and SIp) indicating a reduction of the off-flavor as a result of a physical masking phenomenon. Also, the PEF samples were rated as high as the all meat control for meatiness and soy acceptability.

Finding new methods to decrease specific off-flavors of soy proteins in meat products that are non-destructive to protein functionality is of utmost importance in the food industry. Soy proteins incorporated as PEF increased the WHC, cooking stability, and yield of products and had no detrimental effect on the color of frankfurters. The incorporation of soy proteins (soy flour and concentrate) as a preemulsified fat in comminuted meat products decreased the specific soybean and bitter notes which are the two most offensive characteristics of meat products containing soy protein additives. Utilization of preemulsified soy proteins resulted in finished products with improved sensory characteristics and enhanced functional properties.

3.8.3 Corn and Wheat Germ Proteins

Defatted corn germ protein flour (CGPF) produced by a hexane extraction method had good nutritional quality [116], high water retention, fat binding, emulsifying capacity, and emulsion stability [64, 117]. The level of fat emulsified during sausage batter processing affected the textural and water binding properties of the finished product [104]. The structure of frankfurters was significantly altered by the incorporation of a corn germ protein-lard-water emulsion, resulting in a homogeneous structure [64]. The influence of the CGPF pretreatment (blending and emulsification) before incorporation in the formulations, on its functionality in comminuted meats was reported [118]. The new approaches were pretreatment of CGPF by blending and emulsification of CGPF with fat and water and partial emulsification of fatty tissue with CGPF and water.

The emulsification process and production of an emulsion with highly dispersed fat droplets affected the textural properties of sausage batters. Changes in water binding of the batters affected their viscous properties, WHC and yield of the finished product (Table 3.6) [118]. Incorporation of CGPF and fat in the sausage batters as an emulsion significantly enhanced WHC of the system, presumably because of water retention in the emulsion around the fat droplets. Our previous studies showed that as the result of emulsification and dislocation of the protein stabilizer at the interface protein-fat-water, hydrophilic properties of the protein and batter system increased [119]. Batter viscosity was higher for the samples containing CGPF-fat-water emulsion and preemulsified fatty tissue (PEFT) than for the samples containing defatted CGPF as a powder and for control samples without CGPF incorporation (Table 3.6). Firmness values of the three experimental and control samples were not significantly different. However, the shear force value was significantly higher in the group with CGPF used as the stabilizer of PEFT comparing with control. An increase in the shear force value of the samples with PEFT indicated that structure modification had occurred with the emulsifying of adipose pork tissue with CGPF.

Table 3.6. Textural characteristics, color, water holding capacity (WHC) and yield of frankfurters containing corn germ protein (CGP) incorporated after different pretreatment techniques[d]

Treatment	Batter Viscosity, x 10^{-5} cPs	Firmness, kg	Shear Force, kg	Color, Redness	WHC	Yield, (%)
Control[e]	2.60[a]	0.61[a]	5.37[a]	2.10[a]	0.59[a]	118.05[a]
CGP powder[f]	2.63[a]	0.59[a]	6.06[a,b]	2.08[a,b]	0.88[b]	122.87[c]
CGP-fat-water emulsion[g]	3.32[b]	0.58[a]	5.61[a,b]	2.04[b]	0.83[b]	126.52[b]
CGP as the stabilizer of PEFT[h]	3.01[c]	0.57[a]	6.32[b]	2.09[a]	0.80[b]	122.40[c]

[a,b,c] Means in the same column with different superscripts are significantly different ($P > 0.05$).
[d] In all experimental samples 3% CGP was utilized; control sausage batters were formulated with 23% added water and experimental with 26% added water.
[e] Control, without CGP.
[f] CGP utilized as a powder additive.
[g] CGP utilized as the stabilizer of preliminary prepared CGP-fat-water emulsion.
[h] CGP utilized as the stabilizer of CGP – adipose prok tissue – water (preemulsified fatty tissue = PEFT), prepared by comminution. From Ref. 118

There were no significant differences in redness of frankfurters, except the sample containing the CGPF-fat-water emulsion (Table 3.6). Because of the incorporation of the white emulsion in the batter, the redness values slightly decreased. However, these results indicated no significant color changes of frankfurters with the incorporation of defatted CGPF. Defatted CGPF used in the formulations of frankfurters significantly increased WHC for all three experimental samples (Table 3.6), decreased the amount of free water and increased bound water in the system. The processed yields of frankfurters of the three experimental groups were significantly higher than that of control groups (Table 3.6). Amino acid profiles were not significantly different between CGPF preparation techniques and control product. Lysine content in frankfurters was in the range 4.91–7.37 g per 100 g protein, compared with the FAO/WHO amino acid pattern of 4.2 g per 100 g protein. On the basis of grams of amino acid per 100 g protein, defatted CGPF and frankfurters processed by different CGPF pretreatment techniques met or exceeded the provisional amino acid pattern for human consumption in all essential amino acids (except isoleucine). Lower isoleucine content in frankfurters was not the result of CGPF incorporation. Slight differences were observed for the sulfur-containing amino acid (cystine).

Significantly lower off-flavor was found for batches with CGPF incorporated as a stabilizer of CGPF-fat-water emulsion. No unacceptable flavor notes were detected. Lower off-aroma and off-flavor scores were obtained as a result of the emulsification. Conclusion can be made that this was related to the formation of surrounding layers of defatted CGPF around fat droplets and absorption of the aromatic and flavor substances in the system. A dilution effect also contributed to lower off-aroma and off-flavor scores for frankfurters containing CGPF as a stabilizer of CGPF-fat-water emulsion and PEFT.

There are three methods of CGPF incorporation – as a powder, as a stabilizer of CGPF-fat-water emulsion, and PEFT – and they have been used to modify some of the quality characteristics of sausage batters and finished products. All these methods of incorporating CGPF in the frankfurter formulations increased WHC and viscosity of the batters and the yield of finished products. The results of sensory, chemical, and quality tests showed that CGPF utilization with the three methods of pretreatment had no adverse effect on the finished products.

Corn protein was an effective stabilizer of PEFT. A significant increase in fat binding and WHC and the stability of sausage batter was established (Table 3.7) (Zayas, J. F. and Lin, C. S., unpublished data). Insufficient time and temperature for protein to be hydrated may explain the insignificant effect of a 2% protein powder additive. However, increased thermal stability (less water loss) was obtained by addition of 2% protein as a powder additive and as a stabilizer of PEFT. Increase in yield of the finished product corresponded to higher thermal stability of sausage emulsion.

The effect of CGPF concentration on the microstructure of comminuted meats has been reported [115]. At a 2% level of CGPF, there are small particles with a globular or elongated shape, which may represent the homogeneous structure of fat emulsion or the fat globules confined locally within the protein matrix. These protein micellar structures seem to have a thin discontinuous membrane and flow into a long shape. The coarse structures of the large fat globule do have a thick membrane which has an irregular shape resulting from coalescence. Differences in the protein matrix and fat globules were found in commercial frankfurters [120]. The structures ranged from coarse to fine in terms of fat droplets and muscle components. The structure had small fat droplets distributed in the protein matrix. Theno and Schmidt [120] observed a true emulsion formed in the CMP. The level of fat emulsification depends on the properties of protein and fat components, temperature of emulsification, equipment for sausage batter comminution, and other factors.

Wheat Germ Proteins

In flour milling operations, the wheat germ is separated as a milling by-product. Wheat germ is a unique source of concentrated nutrients not utilized much for human consumption. Wheat germ protein flour (WGPF) has a high concentration of protein, rich in essential amino acids, such as lysine, methionine, and threonine, which are absent in many other cereal proteins [121]. The amino acid content of wheat germ is above the FAO/WHO pattern for essential amino acids. Wheat germ is a good source of macro- and micro-elements, the vitamins, thiamine, riboflavin, pantothenic acid, and niacin,apart from being the richest source of tocopherol from

Table 3.7. Batter water holding capacity, thermal stability and cook yield with corn protein additive

Corn Protein %	Water Holding Capacity	Thermal Stability		Cook Yield %
		Fat loss	H_2O loss	
		ml/30g		
0	0.6586[a]	1.43[a]	6.13[a,b]	125.09[a]
Powder Additive				
1	0.7083[a,b]	2.57[a]	6.30[a,b]	126.78[a,b]
2	0.6972[a,b]	1.37[a]	4.97[b]	128.13[b]
Stabilizer of Emulsion				
1	0.7052[a,b]	2.70[a]	7.03[a]	127.39[a,b]
2	0.7773[b]	2.20[a]	4.89[b]	128.19[b]

[a,b] Means in the same column with different superscripts are different ($P < 0.05$). Zayas, J.F. and Lin, C.S. unpublished data.

Table 3.8. Viscosity, adhesiveness, water holding capacity (WHC), batter stability, and cooking loss

Treatment	Added water (%)[d]	Viscosity (10^{-5} cps)	Adhesive ness, kg	WHC	Batter stability[f]	Cooking loss[f]
Control 1	28.0	2.32[a]	2.55[b,c]	0.60[b]	9.06[b]	15.15[b]
Control 2	31.5	1.93[c]	2.38[c]	0.48[c]	11.79[a]	17.63[a]
WGPF[e]	31.5	2.24[b]	2.91[a]	0.66[a]	5.93[c]	12.81[c]
Sf[e]	31.5	2.20[b]	2.69[a,b]	0.69[a]	5.80[c]	10.65[c]
CGPF[e]	31.5	2.19[b]	2.64[b]	0.66[a]	6.61[c]	13.06[c]

[a,b,c] Mean values in the same column with different superscripts are signifanctly different ($P < 0.05$).
[d] Water added during batter formulation.
[e] WGPF = wheat germ protein flour, SF = soy flour, CGPF = corn germ protein flour.
[f] Batter stability and cooking loss in ml. From Ref. 125

sources of plant origin. Baldini et al. [122] found wheat germ to be as stable as maize germ, upon storage and antioxidant treatment. Incorporation of defatted WGPF in bread formulations improved the quality and nutritive value of bread, biscuits, rolls, and other flour-based products, and microwave-baked muffins [123]. An interesting antiatherosclerotic effect was reported from wheat germ in diets of male quails [124].

Functional properties of WGPF, soy flour (SF), and CGPF in batters of comminuted meats are presented in Table 3.8 [125]. Viscosity of batters containing tested protein extenders was lower than that of control 1. No difference was found in viscosity among treatment samples. However, viscosity of experimental batters was higher than that of control 2, though they had the same level of added water. Adhesiveness of meat batters containing WGPF was higher than those of control 1 and 2 and the treatment containing CGPF. WHC and stability of meat batters were significantly affected by incorporated protein additives (Table 3.8). Samples containing WGPF, SF, and CGPF showed higher WHC and higher batter stability than controls 1 and 2. Lowest values for WHC and batter stability were obtained for control 2. No differences in WHC and batter stability were found among treatments containing added extenders. Consequently, experimental batters showed higher thermal stability. Cooking losses of WGPF, SF, and CGPF samples were lower than those of controls 1 and 2. However, no difference was found between treatment samples.

3.9 Milk Proteins as Emulsifiers in Food Systems

Proteins in milk are widely recognized for their superior functional properties, high sensory characteristics, and nutritional quality. Milk proteins, especially caseins, are very effective emulsifying agents in many foods. The EC of milk proteins is affected by their molecular flexibility, i.e., the ability to unfold at the interface, and to form an interfacial film. The emulsifying properties of milk proteins are widely utilized in the dairy industry in the manufacturing of various foods. The function of milk proteins in dairy emulsions is stabilization of the system against separation, coalescing, and flocculation. Large amounts of caseinates, whey, and milk protein concentrates and isolates are produced as functional ingredients added to various formulated foods. NFDM was the main emulsifier in sausage processing until Na caseinate and WPC became available within the last two to three decades. The manipulation of milk proteins in the formulations of foods affects the extent of milk protein utilization as food ingredients. The functionality of milk protein ingredients can be markedly influenced by the order of incorporation during formulation.

Some milk proteins exhibit inconsistent and poor functionality as a consequence of the method of production. The inconsistence of milk protein functionality is related to fluctuations in milk composition (breed, seasonality), processing (heat) treatment history, and type of whey (acid or sweet). World annual whey production is estimated in excess of 224 billion lb (more than 100 million metric tons) [126]. WPCs with >50% protein can be produced by ultrafiltration and WPIs with >90% protein by ion-exchange adsorption technology. NFDM and WPC powders are major milk protein ingredients in formulated foods. NFDM is produced by pasteurization of skim milk, vacuum concentration and spray drying under low and high temperature processing conditions. As a result two NFDM products with different properties are obtained. NFDM obtained by low temperature treatment with high soluble protein is an effective emulsifier. Because of mild heating conditions there is minimal level of protein denaturation and formation of the complexes with casein micelles. More highly concentrated milk proteins such as caseins and WPCs have good EC and emulsion stability.

Milk proteins, because of high surface activity, rapidly adsorb at the oil-water interface, producing the surface layer that stabilize oil/fat droplets against flocculation or coalescence [127]. The important properties of the milk film in emulsions is its stability to heating and its ability to form a gel on cooling. The caseinates have extremely good emulsifying properties, and since the whey proteins have been removed, they are heat stable in sterilized liquid products. However, it is necessary to improve the emulsifying properties of WPC. The caseinates are excellent surfactants because of their ability to envelope oil droplets or air bubbles with a protein membrane with high elasticity and resistance to mechanical disruption. The EC of milk proteins is determined by the rate of protein emulsifier diffusion to the interface, a decrease in interfacial tension, and formation of a film surrounding the fat droplets. Milk proteins exhibit surface activity, i.e. they adsorb at the interface of liquids at low concentrations and reduce the surface tension [128]. Determination of the milk proteins' emulsifying properties is limited by variation in composition of caseinates, WPC, and NFDM in the extent of denaturation during processing, conditions and duration of storage, and absence of the standard methods for testing emulsifying properties. The surface activity of milk proteins has been extensively studied because of their availability in pure form and good surfactant activity.

Functionality of milk proteins in cream liqueurs is of practical importance. Stable emulsions with 15–20 wt% alcohol are produced. The lower interfacial tension during homogenization in the presence of alcohol leads to the formation of emulsions with high dispersivity [129]. The effect of ethanol on emulsion stability may be related to all macromolecular food emulsifiers or limited to dairy proteins which possess high surface activity. When milk proteins (Na caseinate and whey protein isolate) were tested together with ethanol, the average droplet size at first decreased and increased at high alcohol concentrations (30%) [130]. A significant reduction of surface tension was observed when sodium caseinate, ethanol and *n*-tetradecane were used for emulsification [129].

3.9.1 Emulsifying Properties of Caseins and Caseinates

Caseins and caseinates are widely utilized in the food industry because of their emulsifying properties and enhancement of flavor of various foods. A number of substitutes for ice cream have appeared in the U.S. and most of them contained Na caseinate. Na caseinate is an emulsifying and foaming agent in ice cream and frozen desserts; it contributes to the smoothness of body and texture. In sour cream products based on vegetable oil, the Na caseinate acts as an emulsifier and stabilizer of fat emulsion, improves flavor and water binding. Sodium caseinate is one of the best non-meat protein emulsifiers. It contains approximately 90% protein and is completely soluble in water. The increase in concentration of Na caseinate resulted in an improved emulsion stability. Na caseinate was a more effective emulsifier than soy-sodium-proteinate. Waniska et al. [35] reported a greater EC of β-casein than soy isolate and β-lactoglobulin, but it was inferior to ovalbumin, the 11S fraction of soy protein and BSA. β-casein adsorbs and spreads rapidly at the interface and occupies a greater area per molecule than BSA or lysozyme [12]. A high level of original Ca in casein significantly diminishes its EC. Na caseinate with very high solubility in water forms viscous solutions. Casein subunits consist of a nonuniform distribution of polar and hydrophobic residues along their polypeptide chains. The sodium and potassium caseinates are good emulsifiers because they are soluble, show resistance to thermal denaturation and coagulation, and rapidly form interfacial films. These caseinates form stable emulsions over a wide range of temperature, pH, and salt concentrations [128].

The cryocasein with high water solubility, water retention and the ability of the native casein to form curd was prepared by ultrafiltration and cryodestabilization [131]. The cryocasein exhibited lower emulsifying activity (EA) than Na and Ca caseinates [132]. The EA of cryocasein, Na and Ca caseinates increased with the increase of protein concentration. At 0.5 and 1%, cryocasein showed lower EA than Na and Ca caseinates, however, there was no difference found at 2% protein levels. Cryocasein showed lower EA than caseinates at neutral and alkaline pH and lowest EA was recorded at pH 4. Cryocasein exists as micelles in the medium and exhibits lower migration and adsorption at the interface than both caseinates. The stability of emulsions prepared with two tested caseinates and cryocasein increased with the increase of pH from 6 to 12. Protein-stabilized emulsions usually exhibit their maximum stability in the isoelectric pH range [1]. Emulsions of lower stability were prepared with cryocasein than with Na and Ca caseinates when the ionic strength was adjusted in 0–1 M NaCl.

Surface active properties of the caseins and caseinates are useful in a variety of applications and are related to unique amphiphilic conformation and their ability to form aggregates. Casein micelles rapidly form a surface layer on the surface of freshly formed fat droplets and stabilize the emulsion. Caseinates rapidly migrate and concentrate at the oil/water interface. Caseins can be utilized for studying the relationship between structure and their functionality in foods, because the structures of caseins are known. In milk substitutes, caseins are utilized for nutrition and for physical and functional properties such as EC and ES. A number of beverages have been developed with casein products as emulsion stabilizing agents.

It is known that the natural milk fat globule membrane is fragile and is mostly replaced by a layer of casein in milk used in food manufacturing. In natural milk, fat particles are stabilized by a third phase oriented at the fat globule surface which is called the milk fat globule membrane (MFGM). It is an important surface active substance consisting mainly of protein and lipid [133]. Removal of the protein of

MFGM by proteolytic digestion and phospholipid by phospholipase C led to the destabilization of fat globules in milk. The protein membrane in milk emulsion consists of caseins, lipoproteins, immunoglobulins, and phospholipids. Extremely tightly bound proteins in the milk fat globules membrane are highly associated hydrophobic lipoproteins. Micelles of casein, adsorbed from the aqueous phase, contribute significantly to the stability of milk fat globules in homogenized milk. Casein micelles are disrupted during homogenization as a result of the high pressure treatment.

Caseins are excellent emulsifiers because they are extremely flexible, possess no tertiary structure and contain little secondary structure, and they easily unfold at the interface. For this reason, nonsignificant denaturation of caseins was found during emulsification, and they were effective emulsifiers in systems with a prolonged or repeated emulsification process. In casein molecules, hydrophobic and hydrophilic polarities are present that give the caseins excellent functional properties, particularly EC and ES. The HLB of the caseins and physical properties are determined by primary and secondary structure and the sites of phosphorylation and glycosylation [134]. Increased phosphorylation of caseins might increase the thermal stability of casein micelles.

The four major fractions of casein, αs-, K-, β-, and γ-caseins contain subfractions and genetic polymorphs. Each of the fractions and subfractions possesses a unique primary structure and physicochemical properties. There are three hydrophobic regions in αs_1-casein that make this casein highly surface active [135]. αs-casein contains one major subfraction (αs_1-Cn) and several minor components. αs_1-Cn subfraction of casein has four genetic variants. The β-, K-, and whole α-caseins have also been fractionated and their subfractions characterized. In emulsions prepared with αs_1-casein, the hydrophobic N-terminal peptide (1–23) is adsorbed rapidly at the oil/water interface and is responsible for emulsion stability [136]. Emulsifying properties of αs_1-casein and two peptides, αs_1 (24–199) and αs_1 (1–23) measured as EAI and EC were reported [137]. The EAI of αs_1 (1–23) decreased significantly when the concentration was lower than 2% while that of αs_1-casein was not concentration dependent. The adsorption of αs_1 (1–23) at the oil globule was extremely high, indicating that this peptide is interactive. The EAI of αs_1-casein was at the minimum at the isoelectric point (pH 5.0) and αs_1 (1–23) showed high EAI in the acidic pH region. This property of αs_1 (1–23) can be utilized for emulsion preparation in foods with acidic pHs.

A synergistic effect in the emulsifying activity was suggested by Shimizu et al. [23] between αs_1-casein and other peptides. Enzymatic hydrolysis is frequently an effective means of improving functional properties of the proteins, such as emulsifying and foaming properties. Shimizu et al. [23] separated additional fraction of small peptide αs_1-Cn (fl 54–199) which significantly contributed to the EC of αs_1 (1–23). The EAI was increased by pepsin hydrolysis in which αs_1-casein was shown to be split into two peptides [137]. No significant effect on EAI was found for hydrolysis with plasmin and chymosin.

β-Casein is a most effective stabilizer because it decreases surface tension more significantly than other caseins. The highly amphiphilic and flexible molecules of β-casein rapidly form films while the globular milk proteins such as serum albumin or β-lactoglobulin require a much longer time for film formation [138]. In β-casein molecules, there are large portions of the protein without the presence of charged groups. The sections of molecules could be found that contain at least six nonpolar amino acids and no charged groups [139]. Surface-active properties of β-casein were compared with the more highly structured BSA and lysozyme. β-casein exhibited higher rate of adsorption at the interface than BSA and lysozyme [12]. The C-terminal region of β-casein is highly hydrophobic and imparts surface activity to β-casein [135]. At low β-casein concentrations, molecules of β-casein were completely unfolded at the interface [140]. If the β-casein concentration approached the monolayer level, the β-casein molecules ceased to exist in a totally spread layer and started to form a series of loops. The authors demonstrated the effect of protein structure on the formation of an interfacial layer in emulsions [140]. Multilayers of β-casein are not formed at the low β-casein concentration; they can be formed above a saturated surface concentration of ~2.5 mg m^{-2}.

Protein molecules with disulfide bonds have a lower ability to unfold because of their rigidity. The EC of complex mixtures of proteins was affected by the presence of disulfide bonds. Reimerders et al. [141] reported controversial data that emulsions of low stability have been prepared with β-casein, probably because of lack of tertiary structure. α-Casein was a more effective stabilizer than β-casein and the highest emulsion stability was observed for K-casein. Replacement of αs_1-casein by β-casein during incubation increased as the concentration of β-casein in the emulsion increased. However, even at high concentrations of β-casein (3-4 ml/ml), αs_1-casein was not replaced completely by β-casein. Therefore, probably some of the αs_1-casein has been adsorbed irreversibly. Dickinson et al. [142] reported that more hydrophobic β-casein is more surface active than αs_1-casein and Na and Ca caseinates have intermediate surface properties. The various milk proteins showed the next order in decreasing surface tension: K-casein > αs_1-casein > K-casein > β-lactoglobulin > α-lactal-bumin > BSA.

Proteins possess the property to replace each other at the oil/water interface, i.e. β-casein displaced αs_1-casein at the surface of fat globules stabilized completely by αs_1-casein [143]. The reverse exchange process was also found to occur when β-casein was replaced by αs_1-casein. There was no preference found for either αs_1- or β-casein adsorbed at the oil/water surface during homogenization. However, during emulsion incubation surface αs_1-casein was partially replaced by β-casein. The molar surface ratio of β-casein to αs_1-casein in emulsion was higher than the ratio of β-casein in the original Na-caseinate [28].

Emulsifying properties of dephosphorylated casein were dependent on the amount of casein in solution [144]. The pH of the solution and solubility of the proteins were major factors in the EC of the protein. Emulsions made with dephosphorylated caseins were similar to native casein emulsions but tended to be less stable. The palmitoyl residues with a hydrophobic nature can be covalently attached to αs_1-casein, and isopeptide bonds are formed with the E-amino groups

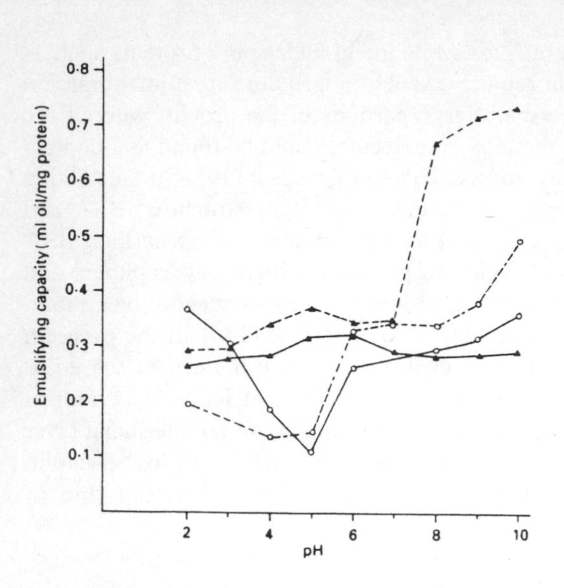

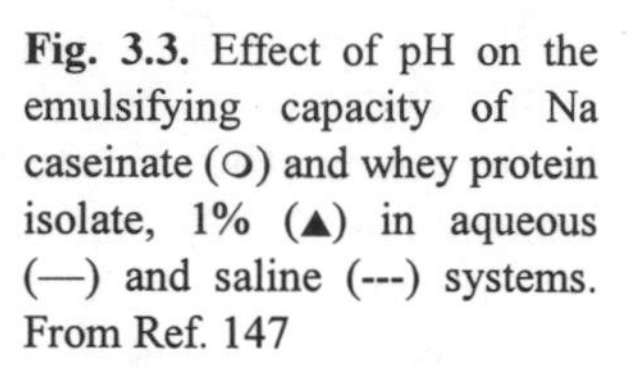
Fig. 3.3. Effect of pH on the emulsifying capacity of Na caseinate (○) and whey protein isolate, 1% (▲) in aqueous (—) and saline (---) systems. From Ref. 147

of lysine residues [145]. The palmitoyl-αs_1-casein molecules were able to form micelles and improve the functionality of αs_1-casein. The incorporation of palmitoyl residues improved the emulsifying properties of αs_1-casein. The palmitoyl-αs_1-casein highly incorporated samples gave stable emulsion without visible flocculation and coalescence. High ES was observed when there was a high level of palmitoyl residues incorporation due to the improved amphipathic nature of the protein. Only 0.1% (w/v) of this stabilizer was enough to form and stabilize an emulsion when the oil:water ratio was 3.

Na caseinate is highly soluble and can be dispersed rapidly in water. The solubility of casein is reduced by interaction between phosphoserine groups of caseins and calcium. The caseinates and casein micelles exhibited higher interfacial activity than whey proteins, native β-lactoglobulin, or the albumins. Sodium caseinate differed from other proteins in the kinetics of its emulsification.

Na caseinate and WPC diffused to the interface at a higher rate than soy protein. Na caseinate exhibited markedly higher surface activity than gelatin and it was preferentially adsorbed from the mixture of casein and gelatin [146]. Na caseinate and WPI gave similar EC-concentration profiles at pH 7 (Fig 3.3) [147]. As the emulsifier concentration increased, the EC decreased. The EC of aqueous solutions of WPI remained relatively constant at pH range 2–10 and saline solutions of WPI showed greater EC than aqueous solutions [147]. The EC of Na caseinate was minimal in the isoelectric pH range, and EC of WPI was relatively independent of pH (Fig. 3.3). The EC increased at pH's above 6. In the pH range 2–4 aqueous solution of Na caseinate gave greater EC than saline solutions. At low level (<1%) WPI exhibited higher EC than Na caseinate, and at above 1% Na caseinate was marginally better than WPI. The ES was improved with an increase in protein concentration from 0.1 to 5%. Emulsions of higher stability were prepared with WPI than Na caseinate in the range 0.25–2.5%. The emulsions con-

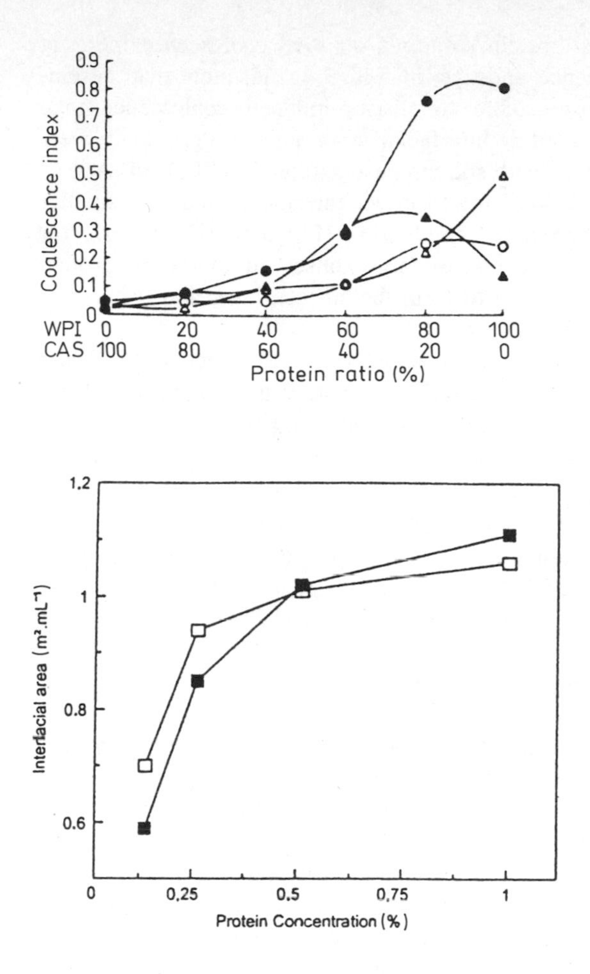

Fig. 3.4. Emulsion coalescence index: unheated protein solution (○); heated protein solution (Δ); monogly-cerides added (.5% in the oil phase) (•, ▲). From Ref. 149

Fig. 3.5. Effect of protein concentration on interfacial area of milk protein-stabilized emulsions. (■) Casein-stabili-zed emulsions; (□) whey pro-tein-stabilized emul-sions. From Ref. 150

containing less than 1% protein stabilizer were poorly dispersed and lacked stability. Similar high emulsifying activity of sodium caseinate and WPI were reported [148]. Sodium caseinate exhibited excellent emulsifying activity at 25, 55, and 65 °C and pH 6, 7, and 8. Sodium caseinate and WPI, despite major differences in surface activity, exhibited similar values for protein flexibility and emulsion activity.

Complementary or synergistic effects of recombination of casein and whey protein fractions were reported [149]. Preferential adsorption of casein over whey protein was observed at the various ratios of WPI to sodium caseinate (CAS). Presumably, the first layer in contact with the oil phase is formed from pure casein. Caseins are adsorbed rapidly and spread out onto fat droplets, forming a thin layer and blocking adsorption sites for whey proteins. The coalescence index was determined as a function of protein composition (Fig. 3.4) [149]. Emulsions from unheated protein solutions showed maximum coalescence index (minimum stability) around 80:20 WPI:CAS ratio. The addition of monoglycerides in emulsions from unheated mixed protein solutions doubled coalescence index at any ratio tested. The coalescence indexes in Fig. 3.4 indicated that casein-containing emulsions were more stable to stirring-induced coalescence than emulsions from pure whey proteins. Interfacial area and protein load were determined on emulsions stabilized with sodium caseinate and WPI [150]. At low protein concentrations, WPI emulsion showed larger interfacial area with smaller oil droplets than sodium caseinate emulsions (Fig. 3.5). However, for concentrations higher than 0.5% the reverse trend was observed. Increasing protein concentration reduced the time required to form the surface layer and resulted in increased surface area.

Emulsifying properties of acidic caseins at pH <3.5 and effect of NaCl and $CaCl_2$ have been reported [151]. The EC of Na caseinate was affected by pH and decreased from 325 to 264 g oil/g protein as pH increased from 1.5 to 3.5 and increased from 251 to 268 g oil/g protein between pH 5.5 and 7.0. The EC increased markedly above pH 7.0 to ~700 g oil/g protein at pH 8.5. Low concentrations (<5 mM) of NaCl or $CaCl_2$ increased the EC of acidic casein (pH 2.5) and the higher concentration showed the opposite effect. The EC of Na caseinate increased with addition of $CaCl_2$ up to 5 mM $CaCl_2$.

3.9.2 Emulsifying Properties of Whey Proteins

Whey proteins remain in the milk after precipitation of caseins by rennet or acid pH (at pH 4.6). Whey proteins are highly valuable because of a good essential amino acids pattern, high in lysine, tryptophan, methionine and cystine. In milk, the ratio of whey proteins to casein micelles is about 1500:1. The high content of salt and lactose makes whey powder unsuitable for many food applications. To increase the potential as a functional ingredient in foods, it must be demineralized and delactosed. The industrial production of food WPC with up to 80% protein is mainly based on ultrafiltration. WPC with 35% protein are common and cheaper to produce.

WPCs have a significant potential for application in food manufacturing because of high functional properties, nutritional quality, excellent sensory properties, and low cost. Whey proteins possess antimicrobial properties, e.g. lysozyme, lactotransferrin, and lactoperoxidase. Whey proteins are suitable for designing new food products. They have a bland flavor, acceptable emulsifying properties, heat-induced gelling properties, and high nutritional value. However, whey is disposed as waste at the 40% level, resulting in a loss of a potential food and an environmental and economic burden [152]. Whey proteins have been incorporated in cream cheeses to increase the binding of water and fat without affecting the taste of the product. Stable, non-synerising emulsions were obtained by homogenization of heat-denatured whey proteins and fat. A pre-emulsified protein-fat-water system was incorporated in cream cheeses and cheese spread formulations [153]. Formulation of cream cheese spreads has been developed containing 59% fat, cultured cream blended with up to 60% whey proteins (on a protein basis). The quality of the cheese spread was not different from a commercial product. Yoghurt was fortified with WPC to replace skim milk solids. Addition of WPC improved the viscous properties of yoghurt and reduced level of syneresis. Whey protein products were recommended for replacement of egg yolk as effective emulsifying and stabilizing agents at low pH values. WPC with 75% protein is used as a replacement for egg proteins in a variety of foods.

One of the main problems that has limited utilization of WPC as a food ingredient has been the absence of constant functional properties. In complex food systems, WPC has demonstrated a lack of consistency [154]. Functional properties of whey proteins are influenced by parameters of heating, the pH of the medium, salt concentration, and mechanical treatment during production of WPC. The functionality of proteins is influenced not only by environmental factors but also by the kind of protein, and the way by which it has been produced, or the presence of nonprotein components. Emulsifying properties of WPC are affected by lipid, ash and sulfhydryl content and use of SH-group content of WPC to predict emulsifying properties has been proposed [155]. The variations in emulsifying properties of commercial whey proteins is mainly related to the processing conditions that cause protein denaturation and loss of solubility.

As a result of heat-induced interaction between whey proteins and casein micelles at pH around 6.5, new textured dairy products can be developed. Textured milk proteins with two levels of fat content (11 and 15%) were characterized by similar emulsifying capacities [30]. Variation in fat content in textured milk proteins influenced functionality. The EC and ES of protein preparation with 15% fat content was higher than for 11% fat. Addition of NaCl markedly increased EC of textured milk proteins.

Whey proteins are not considered to be effective emulsifiers because of the absence of the appropriate balance of hydrophobic and hydrophilic groups [156]. These groups are distributed uniformly in the primary structure of whey proteins. Whey proteins' adsorption at the interface is affected by pH and heat stability of whey protein stabilized emulsions decreased with addition of $CaCl_2$. WPCs prepared from whole milk whey had poor emulsifying properties [157]. WPCs prepared from skim milk whey had lower surface hydrophobicity than with other WPCs. Emulsions prepared with WPCs obtained from skim milk whey, whole milk whey, and buttermilk-enriched skim milk whey showed variations in EC. The lower EC was found for WPCs with higher free fat content and lower denaturation temperature. The stability of emulsion prepared with WPCs from whole milk whey was lower than other WPCs. The highest ES was found for skim milk whey, probably due to higher protein and lower fat content.

Emulsifying Properties of Individual Whey Proteins

WPCs are heterogeneous entities and the major whey proteins are β-lactoglobulin (54%), α-lactalbumin (21%), BSA, the immunoglobulins, and various polypeptides [128]. Methods and techniques have been developed for the production of commercial preparations of relatively pure α-lactalbumin and β-lactoglobulin,

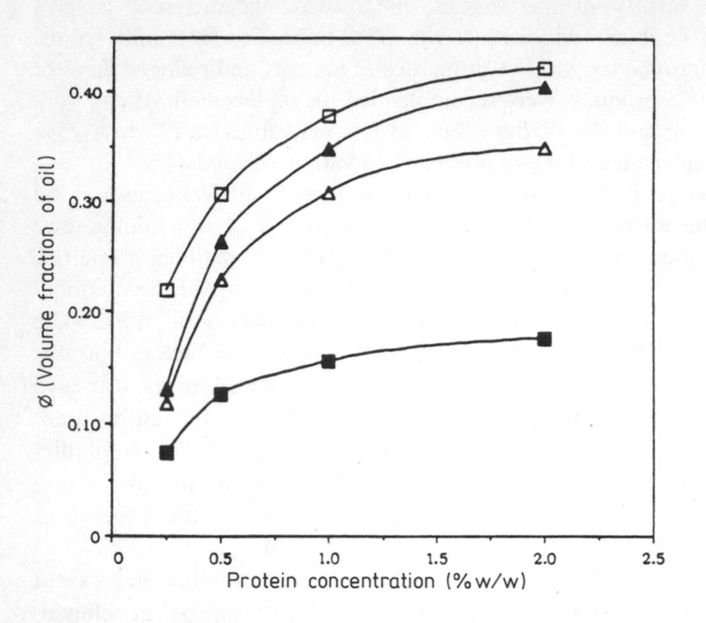

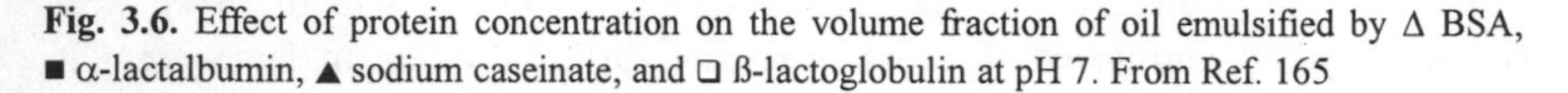
Fig. 3.6. Effect of protein concentration on the volume fraction of oil emulsified by Δ BSA, ■ α-lactalbumin, ▲ sodium caseinate, and □ ß-lactoglobulin at pH 7. From Ref. 165

which may be utilized as standards for testing their functional properties [158]. Functionality of whey proteins should be explored on the basis of their structural properties. There are few studies that relate functionality to the structure of food proteins [159]. Functional properties of acid WPCs have been related to the physicochemical properties and composition [160]. The functional properties of whey proteins are affected by changes in the globular folded structures of the molecules [161]. The conformation and physicochemical properties of whey proteins can be modified by heat treatment.

The type of protein stabilizer affects ES, for example, proteins with bulky tertiary structures, e.g., BSA, stabilize emulsions more effectively than caseinates [1]. Proteins with large molecules and residual tertiary structure form more stable emulsions than proteins with smaller molecules and limited ordered structure, i.e. higher rigidity [12]. Despite a rigid structure, whey proteins readily adsorb at the interface and form stable emulsions. Large proteins form interfacial films with higher viscosity, thickness and strength. Native BSA with a greater degree of tertiary structure, formed a stronger and more cohesive film [35]. When tertiary and secondary structure was disrupted by urea treatment of BSA, the emulsifying activity was lost.

β-Lactoglobulin formed more stable and homogeneous emulsions than the more hydrophobic caseins [162]. Importance of the β-lactoglobulin α-helical conformation for β-lactoglobulin hydrophobic interactions with lipid residues was suggested by Brown et al. [163]. β-Lactoglobulin forms the most viscous films at the interface and viscosity increases on aging probably due to the formation of disulfide bonds [164]. Films formed by β-lactoglobulin at relatively high surface concentration showed viscoelastic properties, and at low concentration films were purely viscous. BSA and β-lactoglobulin are sensitive to protease digestion and exhibit high EAI compared with ovalbumin and lysozyme, which are resistant to protease digestion. Ovalbumin and lysozyme exhibited inferior emulsifying properties [18]. The EAI is expressed as surface area created per unit mass of protein.

Different milk proteins exhibited various capacities to emulsify oil as determined by density measurements (Fig. 3.6) [165]. Nearly all of the oil present was emulsified by β-lactoglobulin and Na caseinate at concentration of 2.0% (w/w) and pH 7. BSA emulsified slightly less oil than β-lactoglobulin or Na caseinate while α-lactalbumin emulsified the least amount of oil. Tested milk proteins emulsified less oil as the pH was lowered, and highest volume fractions occurred at pH 7. A decrease in EC at low pH range (pH 5.0) was related to minimal protein solubility and increased protein aggregation.

Sufficiently stable emulsions were prepared with the WPC and β-lacto-globulin enriched fractions and they exhibited comparable emulsifying proper-ties. α-lactalbumin-enriched WPCs exhibited average EC but poor emulsion stability [166]. WPC fractionated into β-lactoglobulin and α-lactalbumin enriched fractions showed improved emulsifying properties. Emulsions made with β-lactoglobulin and α-lactalbumin had an average EC of 185 ml oil/g protein and 120 ml oil/g protein, respectively. β-lactoglobulin enriched fractions showed similar emulsifying properties to WPC samples.

Protein Hydrophobicity and Emulsifying Properties

Surface hydrophobicity is related to the level of hydrophobic amino acids exposed at the surface of protein molecules. The surface hydrophobicity is also affected by protein unfolding during denaturation. Emulsifying activity and water retention is affected by the distribution of hydrophobic groups in proteins like β-lacto-globulin. In the molecule of most proteins, some hydrophobic groups are exposed at the surface of molecules and can participate in hydrophobic interactions with other molecules. This level of hydrophobic groups in proteins is called the "effective hydrophobicity". This hydrophobicity is an important factor because hydrophobic interactions play a major role in the adsorption of proteins on the surface of fat globules. During the whey protein adsorption, initially there is hydrophobic interaction between the protein and the fat surface in which hydrophobic groups are involved. Bigelow [13] calculated the average hydrophobicities of milk proteins: lactoferrin (1096), serum albumin (1120), immunoglobulin (1095), β-lactoglobulin (1217), and α-lactalbumin (1119). The EAI and coalescence stability generally increased with increasing protein hydrophobicity and solubility [166]. The linear correlations between protein hydrophobicity and the emulsified oil volume fraction (r^2 =0.59) and initial surface area in emulsion ($r^2 = 0.61$) has been established. The emulsifying properties of K-casein and β-lactoglobulin did not correlate well with their surface hydrophobicities [167]. However, correlation was found between emulsifying properties and calculated molecular hydrophobicity [165]. Much more research is required to determine the correlation between protein functionality and structure.

The β-lactoglobulin modified by fatty acid attachment showed higher EAI, especially at very low levels of incorporation of stearic acid and decreased with the extent of incorporation [168]. The improvement in emulsifying property might be related to the surfactant-like property of β-lactoglobulin modified with fatty acid. The most significant improvement of amphiphilic properties and EAI was obtained at 0.3 mol of fatty acid/mol of protein. At this level of fatty acid incorporation, the highest solubility of β-lactoglobulin was observed. An increase in EAI was related to improved alignment of the protein at the oil-water interface. β-Lactoglobulin lipophilization also increased stability of emulsions, and the maximal ES was obtained at 3 and 4 mol of fatty acid attached to 1 molof β-lactoglobulin [168]. Further increase in fatty acid attachment exhibited negative effects on emulsion stability. Whey proteins modified with fatty long-chain *N*-acylamino acids added at level 0.1-0.5% exhibited enhanced emulsifying properties [169]. Increase in ES was obtained by attachment of ethyl residues to β-lactoglobulin through esterification [170]. During emulsification they found that over 40% of the ethyl esterified protein was adsorbed at the oil-water interface, i.e. four times more than the native protein.

Surface hydrophobicity (So) is not always correlated with EA, and native β-lactoglobulin was a more effective emulsifier than chemically modified forms of β-lactoglobulin, although it had the lowest So [170]. Surface active properties of proteins cannot be attributed to surface hydrophobicity alone. This was demonstrated by α-lactalbumin that showed average emulsifying and foaming properties and low surface hydrophobicity [18]. Low correlation between the average hydrophobicities and adsorbability of protein at any pH has been reported [171]. The surface hydrophobicities are not the most important factors influencing selectivity of adsorption at the fat-water interface. The strong dependence of protein adsorption on pH suggests that average hydrophobicity is not critical in adsorption of proteins.

Effect of pH on Emulsifying Properties

Structure and intermolecular associations of β-lactoglobulin changed with pH [134]. At pH 7 increased flexibility of β-lactoglobulin resulted in increased unfolding of the protein molecule and formation of a stronger film. As a result more stable emulsions with smaller fat droplets were formed. Conformational changes of proteins are pH-dependent and they influence "effective hydrophobicity" of protein molecules. Electrostatic properties of proteins influence their adsorption at the interface. Isoelectric points of milk proteins are in the pH range 4.2 to 5.4. The minimal emulsion stability was at pH 5, close to the isoelectric

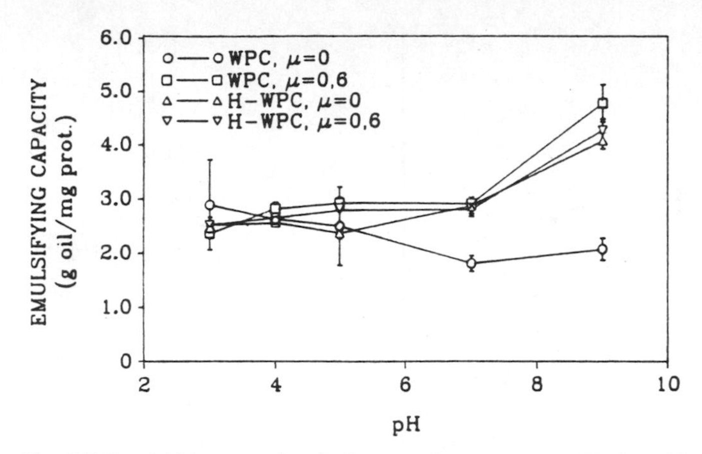

Fig. 3.7. Emulsifying capacity of whey protein concentrate (WPC) and heated WPC over a range of Ph (3 to 9) at two ionic strengths (μ=0 and 0.6). From Ref. 173

point of whey proteins. A decrease in stability is caused by aggregation and coalescence at this pH range.

EAI values for WPC samples were higher at pH 7 than at pH 6 and 8, and emulsion stability was observed at pH 7 [172]. An optimum pH value of 6.31 for heat-treated WPC for EAI was reported. The EC of WPC dispersed in water (μ=0) decreased gradually in the pH range from 3 to 9 (Fig. 3.7) [173]. In the presence of salts (μ=0.6), EC was stable at acidic and neutral pH, and increased at pH 9. There was no correlation between solubility and EC. Greater EC values were also obtained for WPI in saline solutions at pH above 7. The presence of salt may have induced some change of conformation with release of the hydrophobic sites that interact more effectively with oil. Shimizu et al. [14] reported the relationship between whey proteins hydrophobicity and emulsifying properties at different pH values. Protein adsorption at the surface of fat globules was affected by pH. β-lactoglobulin exhibited low adsorption in interfacial film and low EAI at pH 3, and increase in EAI was observed with an increase in pH to 9. β-Lactoglobulin has a relatively rigid conformation at pH 3 and resists surface denaturation. This explains the reduced protein adsorption and low EA at pH 3. The effect of pH on emulsifying properties of β-lactoglobulin was due to conformational changes. The low film-forming capacity at acidic pH was due to higher resistance to unfolding because of β-lactoglobulin rigidity and more stable structure. Stable film formation by β-lactoglobulin at alkaline pH values is caused by higher flexibility, higher surface activity, and expansion of β-lactoglobulin.

β-Lactoglobulin adsorption at the interface was most rapid at pH 5.3, less rapid at pH 6.3, and slow at pH 3.3 [174]. The rate of β-lactoglobulin adsorption measured by surface pressure was rapid during the first minute, more gradual for the next 17 min, and reached equilibrium after 6 h. This was reflected in equilibration of surface pressure after 6 h with no packing of the β-lactoglobulin in the

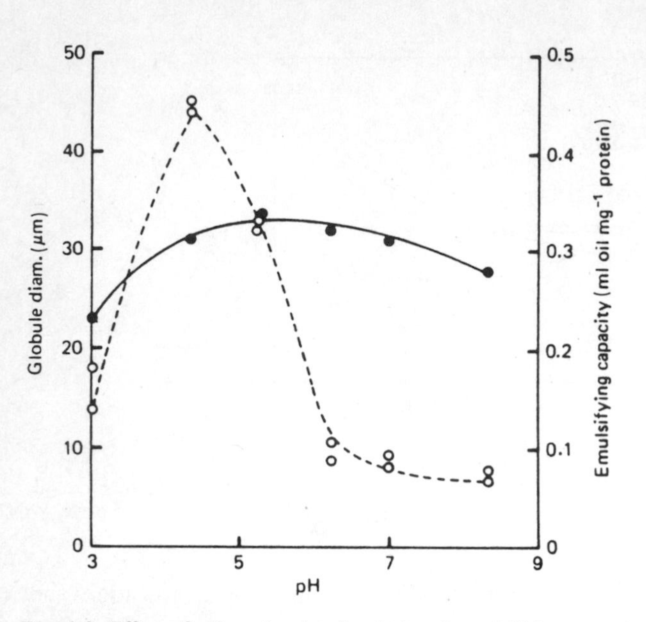

Fig. 3.8. Effect of pH on droplet size (○) and emulsifying capacity (●) of a solution containing 0.5% (w/v) protein powder. From Ref. 180

surface film. The charge properties of β-lactoglobulin affect its surface activity, which is at a maximum around pH 4.9 (isoelectric point of β-lactoglobulin is 5.2). The maximum rate of protein adsorption is near the isoelectric point. The β-lactoglobulin and other milk proteins exhibited the highest rate of adsorption and rearrangement at the lowest net charge [175].

The lowest volume fraction of oil emulsified by 2% BSA occurred at pH 4.0 with formation of a highly viscous emulsion [176]. The adsorption of BSA at the interface was minimal at pH 4. Conformational changes of α-lactalbumin at pH lower than its isoelectric point exposed the hydrophobic domains of the α-lactalbumin molecules [156]. These conformational changes induced by pH affected the adsorption of proteins and showed the aggregation of proteins at the interface. Whey proteins form the interface film with a granular structure. The interfacial film formed from whey protein solutions at pH 9 and neutral pHs consisted mainly of β-lactoglobulin. Adsorption of β-lactoglobulin was reduced at lower pH. Film formed at pH 5 was more heterogeneous. The adsorption of individual whey proteins was highly selective. In emulsions stabilized by WPI, selective adsorption of β-lactoglobulin and α-lactalbumin over BSA and immunoglobulins occurred at the oil-water interface [177]. The formation of the high molecular weight protein polymers with time following emulsion preparation correlated with the disappearance of β-lactoglobulin and α-lactal-bumin.

Adsorption of the individual whey proteins at the fat globule surface of emulsions at various pH values was reported by Shimizu et al. [178]. Whey protein films contained 2–3 mg polymerized protein per m^2 of film surface. The amount of whey protein in the interfacial film was 2.0, 7.6, 3.0, and 2.7 mg m^{-2} at pH 3, 5, 7, and 9, respectively. Whey proteins mainly adsorbed in the film were α-lactalbumin, β-lactoglobulin, and BSA. Gel electrophoresis (SDS-PAGE) of emulsions stabilized with WPI showed formation of disulfide bonds involving β-lactoglobulin and α-lactalbumin at the emulsion droplet surface [179]. The extent of disulfide bonds formation increased with storage time. Aggregation of the emulsion droplets occurred with time and the formation of disulfide bonds between proteins occurred at the oil-water interfaces. Formation of such bonds increased viscoelastic properties of the interfacial film and stability to coalescence. Langley et al. [180] found that the EC of WPCs and droplet fat size in emulsion were dependent on pH (Fig. 3.8). The maximum droplet size was found at pH 4.0, whereas EC had a maximum value at pH 5-6 that corresponds to isoelectric point of α-lactalbumin, β-lactoglobulin, and the caseins. Shimizu et al. [178] reported maximum coconut oil adsorption in the whey protein solution at pH 4-6. Emulsions contained three times more protein associated with interfacial film at pH 5 than at pH 7. β-lactoglobulin was selectively adsorbed at pH 9 as a main protein of interfacial film.

The stability of emulsions prepared with milk proteins was studied with light scattering methods, and a decrease in surface area was observed with time of incubation (50 h) [165]. The rates of coalescence in emulsions stabilized by casein and α-lactalbumin were similar at pH 7 and pH 5; and for BSA it was not affected by pH values. The rate of coalescence in emulsions stabilized by β-lactoglobulin was similar at pH 4 and 5 but slower at pH 7. Coalescence of fat droplets is considered the most severe form of emulsion degradation as it requires more energy to reverse than the other mechanisms.

The Effect of Ionic Strength

The influence of salt on emulsifying properties of proteins is due to the effect of salt on protein conformation and solubility. Changes in ionic strength in the range 0-1.0 M NaCl had little effect on the stability of emulsions oil-water-WPC [156]. The intensity of diffusion and adsorption of whey protein decreased at concentrations below 1 mg/dL. The volume fraction of oil emulsified by 1% solution of β-lactoglobulin decreased from $v = 0.36$ to 0.33 as the ionic strength increased from 0.1 to 0.2. A similar effect of ionic strength was found for the other milk proteins.

The Effect of Protein Solubility

The emulsifying properties of whey proteins are influenced by the solubility of WPC prepared from Cheddar cheese wheys. The WPC from buttermilk-enriched skim milk whey had lower solubility than WPC from skim milk and whole milk whey [157]. The strong correlation between solubility and emulsifying properties of whey proteins (β-lactoglobulin) and sodium caseinate was found, especially at protein concentrations less than 0.4% [181]. Bech [182] reported a relationship between the solubility of whey proteins concentrated by ultrafiltration and their EC. Emulsifying properties of WPC were influenced by protein solubility [155]. The higher the percent of soluble protein, the lower was the emulsion rating after 40 min of centrifugation.

The Effect of Heating

The important functional requirement of WPC is heat stability in the formulated products during sterilization. Whey proteins are thermally denatured when heated in solution at 60–75 °C [183]. The objective in the production of infant formulas is to prevent whey protein self-aggregation. Generally, heat treatment is detri-mental to protein functionality, especially heating at temperatures above 60–70°C. Heating of WPC dispersions at temperature above 70 °C either before or after emulsification caused aggregation of whey proteins and was detrimental to their emulsifying properties [161]. Heating of emulsions stabilized by whey proteins at concentrations higher than 6% causes gelation and occlusion of the fat. According to Foley and O'Connel [147] heat treatment (70–100 °C) of aqueous solution of milk serum proteins before emulsification decreased the EC and ES. However, heat treatment after emulsification did not affect emulsion stability. Emulsions stabilized with Na caseinate were not affected by heating used before or after emulsion formation. Limited denaturation improved emulsifying properties of whey proteins. Improvement in emulsifying properties is probably due to unfolding of protein molecules and exposition of hydrophobic groups. Soluble β-lactoglobulin is an effective emulsifier, however its EC is sensitive to denaturation. α-lactalbumin exhibited good emulsifying properties that are probably related to the sensitivity of α-lactalbumin to surface denaturation at oil-water interfaces [18].

Stability of whey proteins to heat treatment can be enhanced if at least part of the globular protein is incorporated as a complex with an anionic polysaccharide such as carrageenan, and sodium alginate. Turgeon et al. [184] reported that improvements in emulsifying properties of heated WPCs were affected by the degree of protein denaturation. They found correlation between EC and rate of adsorption of protein and peptide components at an interface. Emulsifying properties of whey proteins were improved by limited heat denaturation, however, they were still less active than β-casein [135]. The highest EAI was found for WPC samples heat-treated to yield 41% soluble protein, and a decrease in EAI was observed if solubility increased or decreased from 41%. Heating influenced not only the total amount of soluble protein, but also the ratio of β-lactoglobulin to α-lactalbumin. The WPC samples with 41% soluble protein produced more stable emulsions than with other treatments.

The EC of WPCs as measured by absorbance values at 500 nm increased from 0.58 to 0.70 after 6 months of storage of WPCs at temperatures of 5 and 20 °C [185]. Storage time, temperature and water activity showed significant effect on EC of WPCs. More significant increase in EC was found for WPCs stored in air

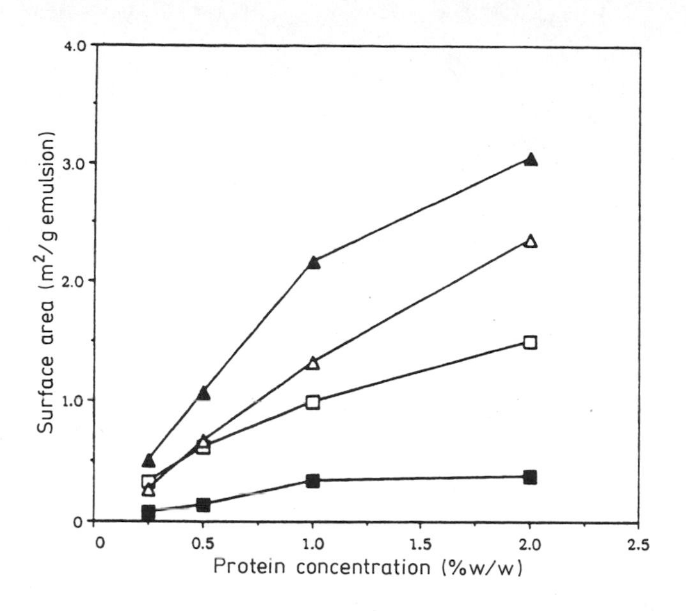

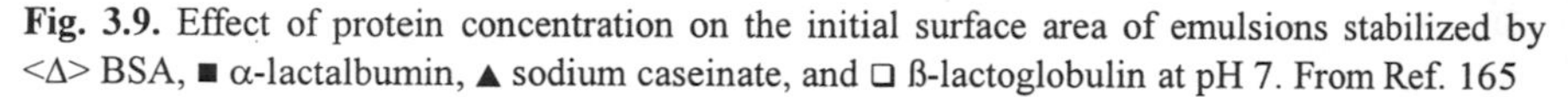

Fig. 3.9. Effect of protein concentration on the initial surface area of emulsions stabilized by <Δ> BSA, ■ α-lactalbumin, ▲ sodium caseinate, and □ ß-lactoglobulin at pH 7. From Ref. 165

than in nitrogen at 20 °C. However, if storage temperature was too high (40 °C), the EC of WPCs increased during the first 3 months of storage, then decreased to almost the initial value after 6 months storage.

The Effect of Protein Concentration

The quantity of any protein adsorbed during emulsification is affected by protein concentration in the solution or dispersion. The stability of an emulsion is enhanced by increase in protein concentration that facilitates more protein adsorption with higher protein load [156]. Protein adsorption at the interface is influenced by the protein/fat ratio in emulsion [30]. Fat droplets aggregation was established with increasing fat to milk protein mass ratios [186]. The fraction of adsorbed protein increased with increased fat to protein mass ratios. In the emulsions with a fixed fat content, the fraction of adsorbed protein increased when the protein present decreased. Fat to protein mass ratios determined droplet size, surface area in emulsion, and protein surface coverage. Stability of emulsions prepared from WPC, coconut oil, and distilled water was positively correlated with protein concentration. The stability of emulsions prepared with 1 and 2% WPC was very low and a stable emulsion was formed at 6% WPC level in dispersion [156]. Concentration of protein stabilizer affected the viscosity of the emulsion.

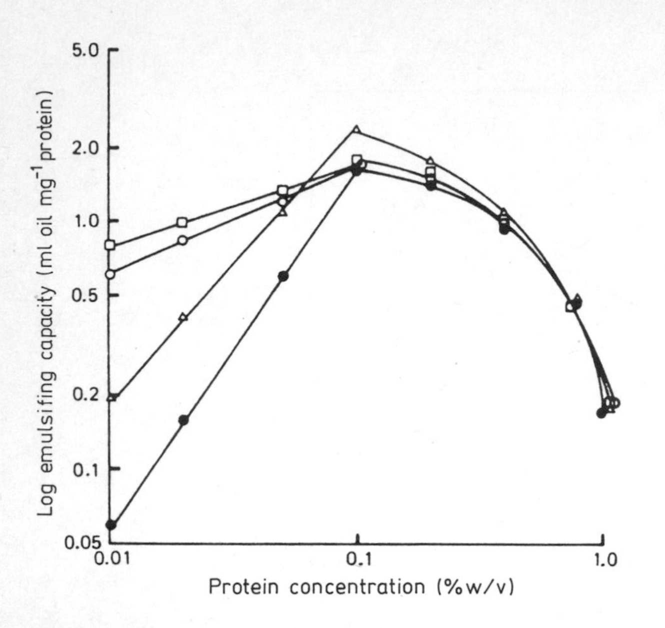

Fig. 3.10. Emulsifying capacity of whey protein powders at pH 7.0. ○ powder 1; ● powder 7; Δ powder 8; □ powder 11. From Ref. 180

The initial surface areas of the emulsions stabilized by sodium caseinate, BSA, and α-lactalbumin were directly proportional to increasing protein concentration for 0.25, 0.5, and 1% protein (Fig. 3.9) [165]. At 2% protein concentration the surface area was less than predicted. When the protein concentration increased from 0.25 to 0.5%, the emulsions prepared with β-lactoglobulin doubled in surface area. However, this trend was not observed when the protein concentration was increased from 1 to 2%. The EC of ion-exchange fractionated WPC increased with an increase in protein concentration up to 0.1% (w/v) at pH 7, and a decrease in EC was observed at protein concentrations higher than ~ 0.2% because of packing effects (Fig. 3.10) [180]. The size of fat droplets was found to be independent of protein concentration. At a protein concentration of 0.01%, most of the oil was stabilized (1.5–2.5 ml oil/mg protein). Powders of α-lactalbumin showed similar stabilizing properties while powders 7 and 8, high in β-lactoglobulin, had the least effect.

3.10 Emulsifying Properties of Egg Proteins

Egg yolks are widely utilized as emulsifiers. In salad dressings, the most important emulsifying agent is egg yolk, which also influences the stiffness and stability of the emulsion. For this reason, knowledge of the emulsifying properties of individual yolk proteins is of practical importance. There is limited information about functional properties, particularly the EC of individual egg yolk proteins. Egg yolk is a natural emulsion containing phospholipids, of which ~79% is lecithin and lysolecithin. Components of yolk, lipoproteins, lipovitellin, lipovitellenin, and phospholipids act as emulsifiers. Yolk components such as low-density lipoproteins (LDL), high-density lipoproteins (HDL) form a protective layer on the oil droplets. The most important proteins in yolk granules are: α- and β-lipovitellins (70%), phosvitin (16%), and LDL (11%). Phosvitin is a phosphoprotein and contains about 10% phosphorus and represents about 80% of the protein phosphorus in yolk. Phosvitin is separated into two components, α- and β-phosvitin.

Emulsifying properties of egg yolk mostly depend on the lipoprotein fraction, especially LDL emulsifying properties. The emulsifying activity (EA) of LDL was little affected by the changes in pH or ionic strength [187]. Authors suggested that the protein component of LDL is the main contributor to its emulsifying activity. A slight increase in EA (absorbance at 500 nm) was observed when lecithin was added to all tested concentrations of LDL. The interfacial tension of LDL did not decrease with the addition of lecithin. The mean droplet size in emulsions prepared with LDL was much smaller than that for emulsions prepared with BSA. An increase in low molecular weight fractions of the LDL polypeptide moiety was observed with treatment with protease. The EA of protease-treated LDL decreased as the amount of polypeptide moiety decreased. This decrease in EA was related to the amount of the polypeptide moiety and not affected by the kind of proteases used in the experiment [188]. The same trend in the decrease in EA was found for protease-treated BSA, and the EA of LDL-treated was much higher than BSA-treated samples. The size of fat globules in emulsions prepared with LDL or BSA gradually increased with the decrease in concentration. However, this increase in globule size for protease-treated LDL emulsions was much smaller than that of protease-treated BSA emulsions. The micelle-like structure of LDL might influence the high emulsion stability of LDL. The soluble complex in yolk including LDL and phospholipid exhibited much better emulsifying properties than either lecithin vesicles or lecithin suspensions. Spray-dried plasma fraction of yolk exhibited the highest emulsion activity of all yolk fractions in distilled water [189]. The nonspray-dried plasma fraction of yolk showed higher emulsion activity than nonspray-dried whole yolk and the nonspray-dried granule fraction.

The emulsifying properties of phosvitin as a function of protein concentration, oil volume, mixing speed and emulsification time in comparison with BSA have been reported [190]. The EA and ES increased with phosvitin concentration, oil volume fraction and mixing time. The increase in the EA of phosvitin was about 17% when the protein concentration increased from 0.1 to 0.5%. No further increase in EA was found with concentrations above 0.5%. The EA increased by 274% with an oil volume fraction increase from 0.17 to 0.67 with phosvitin as stabilizer, and 231% for BSA. The EA increased with increased mixing speed from 10|000 to 22|000 rpm by 149% for phosvitin and 129% for BSA [190]. The EA and ES of phosvitin increased with protein concentration, mixing speed and time [191]. Phosvitin exhibited better emulsifying properties than BSA at pH 7

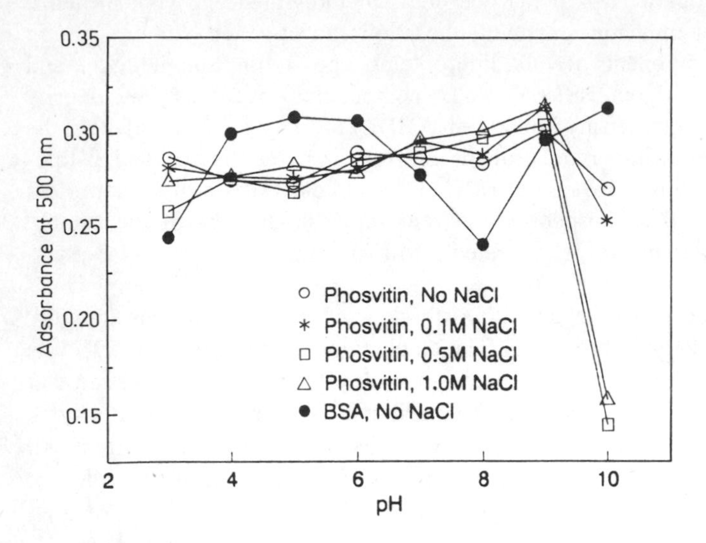

Fig. 3.11. Emulsifying activity of phosvitin and BSA. From Ref. 191

(Fig. 3.11). The EA of phosvitin increased slightly in the pH range from 3 to 9 followed by a sharp decrease at pH 10. The BSA exhibited an increased EA in the pH range from 3 to 6, and then decreased to a minimum at pH 8 and increased again at pH 9 and 10. Phosvitin had a higher ES than BSA at all pH levels except pH 5 and 10. This might be related to the alteration of the charge of the protein molecules. The ES of phosvitin was lowered by NaCl (0.1–1 M) and the effect of NaCl was greater in the acidic pH range. The emulsions most stable to salt were prepared at pH 7–8 which is the pH normally found in eggs. The oil separation in emulsion increased with concentration of NaCl.

A decrease in LDL emulsifying activity with an increase in the treatment temperature has been reported (Fig. 3.12) [192]. However, the EC and ES index decreased considerably when heating temperature was above 75 °C. Morphological changes of LDL were observed at 75 °C heating temperature. The relationship was suggested between the morphological change of LDL and its functionality [192]. The percentage of extracted lipid was considered as an index of morphological change of LDL, because of its large decrease at temperatures above 75°C. The percentage of extracted lipid correlated significantly with EC ($r = 0.984$).

The development of the low-fat egg yolk products by extraction is an effective way of decreasing the high fat content in egg yolk products. The influence of the extraction process on the functionality of the protein component is the most important indicator of acceptability of this treatment method. Yolk was fractionated into oil, protein and aqueous components using solvent extraction [193]. The native egg yolk formed more stable emulsions than one of the fractions, or these fractions in combinations. Partly defatted egg yolk powder has been pre-

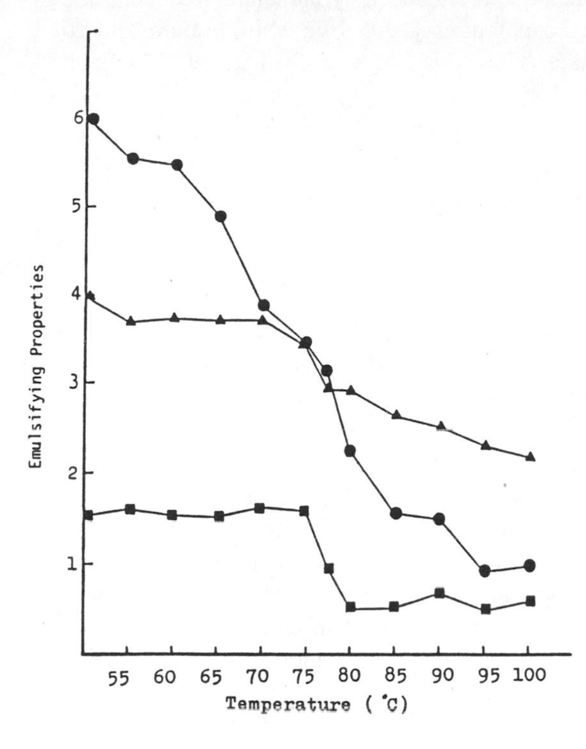

Fig. 3.12. Emulsifying properties of LDL as a function of heating temperature. [● Emulsifying activity index x 10^{-1} (m^2/gl); ▲ emulsifying capacity (ml); ■ emulsifying stability index (min)]. From Ref. 192

pared using various solvents, and the most efficient was hexane-isopropanol, which extracted more than 50% of the lipid [190]. After serial solvent extraction a decrease in protein solubility and the EA of the egg yolk protein concentrates was observed. Hexane-isopropanol reduced the EA to about 14% and protein solubility to 13.8% from 15.7%. Extraction with other solvents resulted in a loss of EA of up to 41%. This showed that the soluble protein was an important contributor to EA. The serial extractions removed triglycerides from egg yolk, and residual lipids contained an increasing proportion of phospholipids. A water-soluble protein fraction was prepared from egg yolk powder by centrifugation and filtration [190]. The EA of this fraction did not change significantly after each extraction with hexane, except chloroform-methanol extraction where the EA of the water soluble fraction increased from the first to third extractions, and decreased at the fourth extraction. The water soluble fraction that resisted denaturation had excellent EA and ES. Poor emulsifying properties of ovalbumin were found by Cheftel et al. [194]. Egg white should be removed from egg yolk before drying if egg yolk is utilized as an emulsifier. In commercial liquid egg yolk, albumen adheres to the yolk. Ovalbumin exhibited excellent emulsifying properties when it formed a conjugate with dextrin [195]. The EC was enhanced by ultrasonic treatment of egg yolk lecithin with ovalbumin and conalbumin [196]. Egg white proteins modified with fatty long chain *N*-acylamino acids added at a level of 0.1–0.5% exhibited enhanced emulsifying properties [169].

3.11 Emulsifying Properties of Plant Proteins

3.11.1 Soybean Proteins

Soybean and other legume proteins have a good balance of essential amino acids, in particular, they are a good source of lysine. Soybean proteins have reached the highest degree of refinement and are utilized as binders and extenders in a variety of processed foods. The most important functionality of soy proteins in processed meats and other foods is their emulsifying property,together with fat and water retention. The EC and ES of soy proteins are useful functional characteristics which play an important role in the development of soy protein-containing products for use as a food. The emulsifying activity of soy concentrate and isolate is important in various food applications especially when they are used in formulations of sausage batters. Use of soy proteins in emulsified food systems could promote fat binding to reduce cooking losses, improve EC, and maintain stability of the emulsion system.

Soy proteins play two roles in emulsification, i.e., forming and stabilizing the oil-in-water emulsion. The increasing utilization of soy protein products is due to several factors, including abundance, low cost, desirable textural and good nutritional properties. Soy proteins have several specific functions that participate in the formation of fat emulsions in foods such as control of viscosity, stabilization of fat droplets, fat and water absorption. The stabilizing effect of soy proteins in emulsions is related to high electrical charge and more hydrophilic-lipophilic groups within protein structures which increase the protein-lipid and protein-water interactions. The protein-lipid and protein-water interactions are the major factors of emulsion formation and affect the appearance, color, texture, and yield of finished products.

The emulsifying properties of soy proteins are improved by heat denaturation because of the exposure of hydrophobic groups and increased surface hydrophobicity and decreased surface tension. Surface properties of proteins are influenced by heat treatment, and Kato et al. [197] found that surface hydrophobicity increased with increasing temperature. The increase in surface hydrophobicity correlated linearly with EC and ES. The freshly made emulsions composed of heat-pretreated soy proteins had a small droplet size (average 1.84 μm) and a narrow size distribution compared to emulsions stabilized by crude isolate Purina 500E (average size 3.07 μm) [198]. Emulsion stability was also improved with the heat-pretreated soy protein. The emulsion stabilizing properties appear to be positively correlated with the percentage of protein adsorption into the cream phase. The soy protein formed a protective film around fat globules as observed by Flint and Johnson [199] and this process was strongly pH-dependent. Formation of the stable film was observed at the acidic side of the isoelectric point, down to pH 1.0.

Studies have been carried out on the emulsifying properties of 7S and 11S proteins; these proteins differ very much in their gelling, texturizing and film forming properties. The 11S protein is the main protein in soya in terms of quantity and functionality. Sulfur amino acid content in 7S globulin is very low and sulfhydryl groups are probably not present. Fractionation of soy isolate to 7S and 11S fractions showed that the 7S protein-rich fraction exhibited better emulsifying properties than the 11S protein-rich fraction [20]. The EC and ES of the 7S protein fraction was higher than the 11S protein fraction in the pH range 2–10. Better emulsifying properties of 7S globulins can be related to a higher rate of diffusion to the interface, and also possibly that disulfide bonding in the 11S globulin may inhibit unfolding and decrease interactions at the oil/water interface. Both the 7S and 11S protein fractions showed their lowest EC in the isoelectric region and increased at pH's above and below this region. The ES of the 7S fraction was also lowest in the isoelectric region, but thc ES for the 11S fraction was lowest at neutral pH and did not correlate with protein solubility. The EC and ES of 7S and 11S fractions in the pH range 3 to 8 remarkably increased when treatment with HCl for partial hydrolysis was applied. This increase was especially significant in the isoelectric and neutral region. Partially hydrolyzed proteins are commercially utilized to improve their functionality. The effect of hydrolysis on emulsifying properties differs depending on the kind of protein and the conditions of hydrolysis.

Emulsions stabilized with the 11S fraction showed a markedly increased breaking stress as the protein concentration increased [20]. Heating of the protein solution at 95 °C for 5 min prior to emulsification increased breaking stress from 2 to 4 times. The difference in breaking stress between heated and unheated samples became significant starting at a protein concentration of around 0.5%.

The ES of emulsions prepared with soy isolate from mature beans was higher than those from immature ones [200]. This was due to the increased 7S fraction in the mature soybeans. The major factor affecting the ES of the emulsions stabilized with soy isolate was the 7S/11S protein ratio. ES increased with an increasing ratio of 7S/11S protein fractions in the mixture [200]. The correlation for the relationship between ES and 7S/11S ratio was 0.846. Storage time did not affect soy protein solubility. The ES of soy isolates from 3 and 6 months maturation stages remained similar and the presence of soy protein was the most important factor in preparing a stable emulsion. Addition of 2% NaCl at neutral pH decreased soy protein solubility and impaired ES [200]. Addition of 2% NaCl impaired the stability of 7S and 11S protein stabilized oil-in-water emulsions. Different functional properties of 7S- and 11S globulins are related to varying molecular weight, carbohydrate content, disulfide bond content, and isoelectric points. The mixing of soy concentrate and isolate influenced emulsifying properties and the percentage of protein adsorption onto the oil globule surface

[201]. The high emulsion stabilizing properties (77%) of soy isolate was decreased to a low level 15%, when the emulsion was prepared with a mixture of soy concentrate and isolate. The main reason for this effect of mixing two soy proteins was the initial coating of the oil droplets with a thin layer of soy concentrate which prevented adsorption of soy isolate.

Emulsifying properties of purified soy protein produced by ultrafiltration was reported [202]. Tested protein had an EC and ES comparable to commercial soy isolate (Promine-D) in the acidic and isoelectric pH range. High emulsifying properties of soy protein obtained by ultrafiltration are probably due to its high phospholipid (lecithin) content and high solubility. A mayonnaise analog with use of full-fat soy protein in the formulation showed good sensory properties. Protein was incorporated after 6–8 months of storage. Full-fat soy protein with 33% fat can impair sensory properties of mayonnaise because of fat oxidation.

Most of the soy proteins are processed into finished foods with other protein ingredients, such as meat, milk, fish proteins, starches or fats. Preparing blends of soy proteins and caseins or whey protein diminished the undesirable off-flavor of soy protein [201]. Emulsion stabilizing properties have been determined by centrifugation of emulsions for heated soy isolates (SI), sodium caseinates, and mixtures of these proteins. In emulsions prepared using the shearing force method, the stabilizing properties of heated SI decreased to a low level when the mixture was tested. The heated SI dispersions with high viscosity (protein concentration 4%) showed high emulsion stabilizing properties. However, a mixture of heated SI and sodium caseinate with viscosity similar to heated SI showed lower stabilizing properties than those of the sodium caseinate. The main reason for low stabilizing properties when mixtures of heated SI and sodium caseinate are applied is the initial coating of the oil globules with a thin layer of sodium caseinate which prevents adsorption of heated SI. Tornberg [29] reported that the caseinates disperse in the aqueous phase as smaller particles than that of the heated SI. For this reason, sodium caseinate can diffuse to the oil/water interface more easily than heated SI.

The ECs of soybean and peanut co-isolates, containing cottonseed protein were between 142.8 and 150 ml oil/g co-isolate, which is greater than the EC of cottonseed isolate (73 ml oil/g) [203]. The ECs of co-isolates and isolates were greatest at pH 9.0 and 2.5 and lowest at pH 5.0.

The effects of different processing conditions such as pH, temperature of incubation, and protein concentration on the EC and ES of soy and corn germ proteins were reported. The factors influencing the functionality of soy flour (SF), soy concentrate (SC), SI, and corn germ protein flour (CGPF) were related to the conditions of their utilization as supplements of meat proteins in the production of comminuted meat products [204]. The results of experiments are shown in Fig. 3.13 and Fig. 3.14. In pH-temperature interaction (Fig. 3.13), the response surface of EC for the four samples (SF, SC, SI, and CGPF) reflected the interdependence of pH and temperature effects. The EC of the four samples tended to increase with increasing pH from 6 to 8. However, temperatures of protein solu-

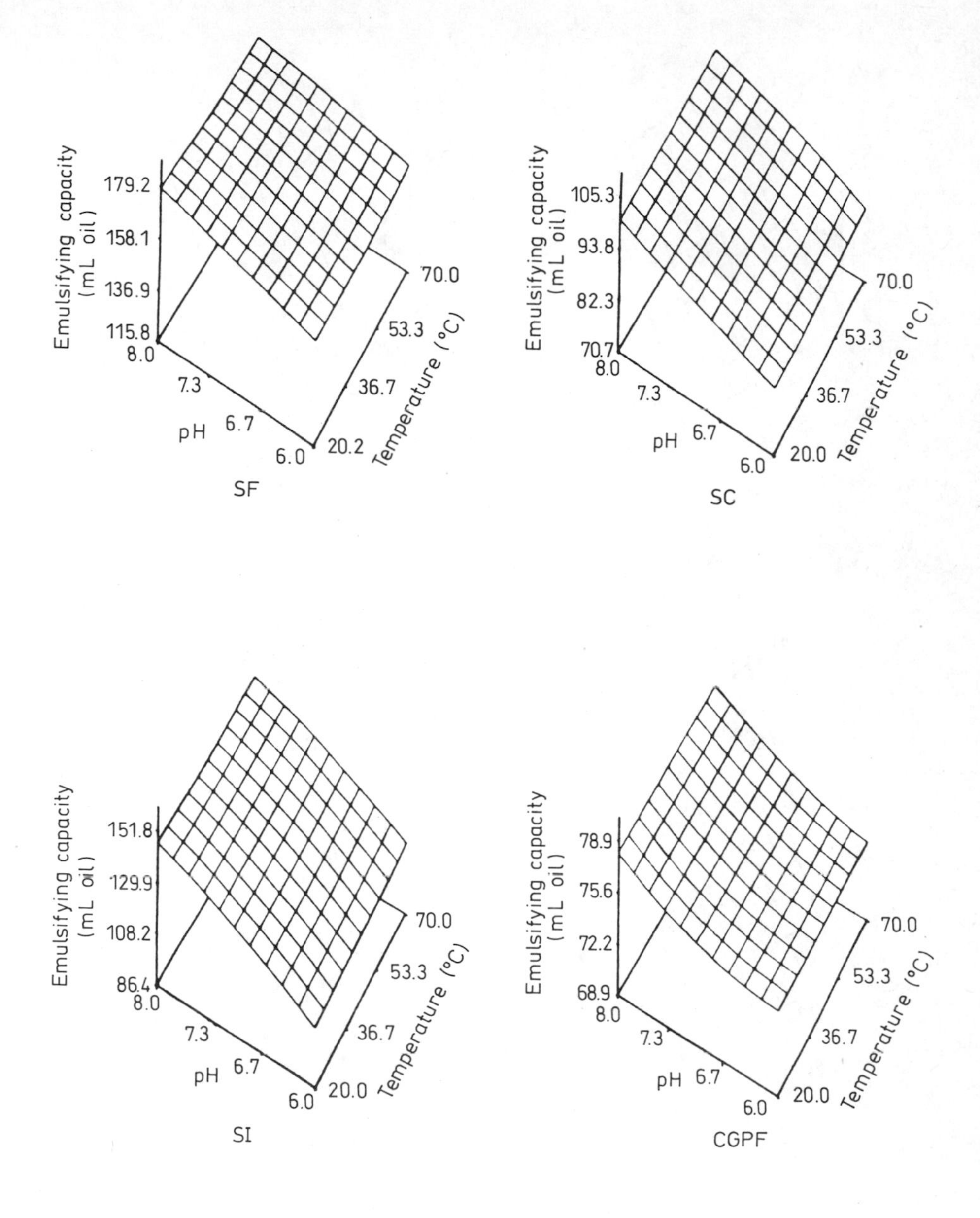

Fig. 3.13. Emulsifying capacity of soy flour (SF), soy isolate (SI), sodium caseinate (SC), and corn germ protein flour (CGPF) as a function of pH and temperature of incubation. From Ref. 204

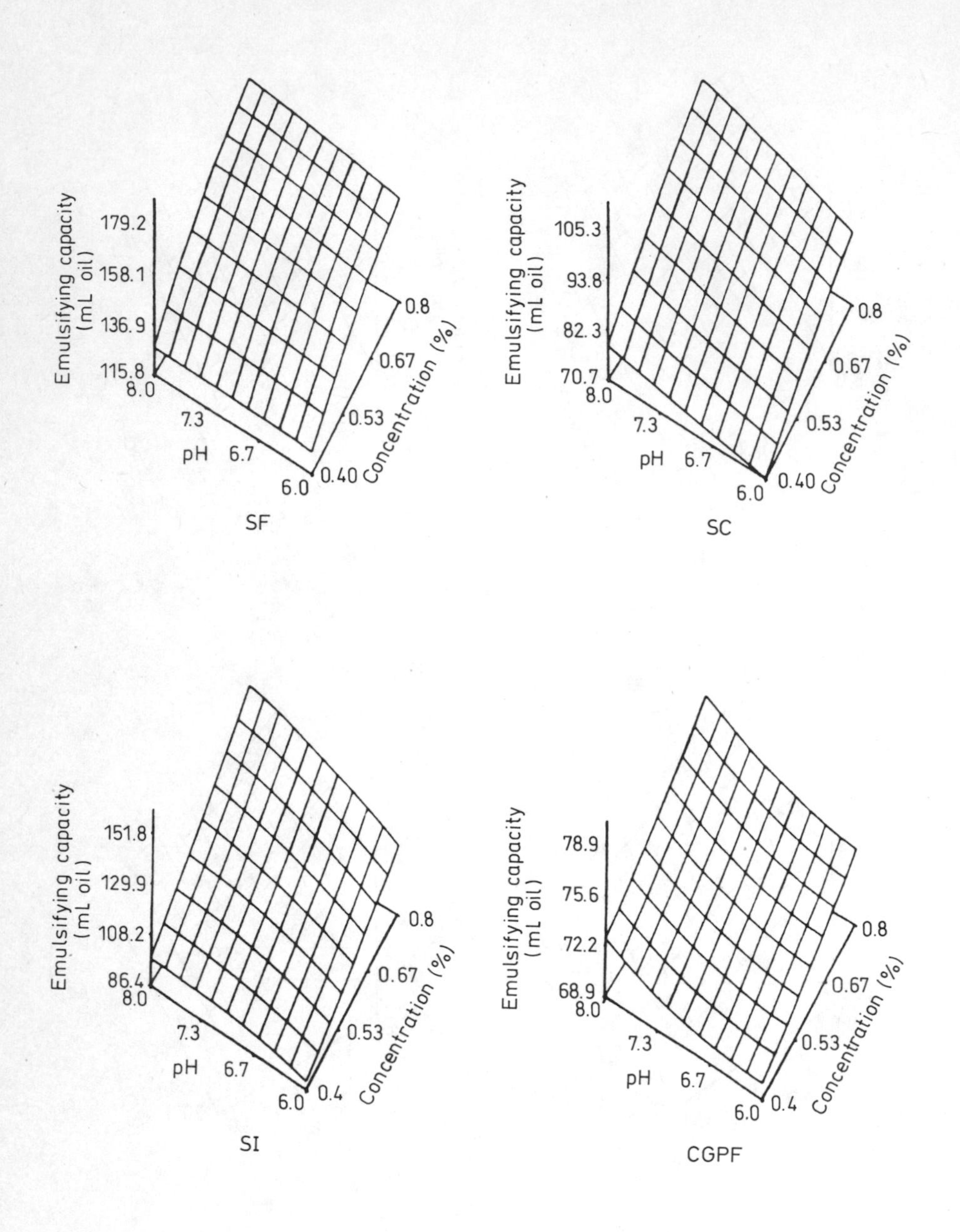

Fig. 3.14. Emulsifying capacity of soy flour (SF), soy isolate (SI), sodium caseinate (SC), and corn germ flour (CGPF) as a function of pH and sample concentration. From Ref. 204

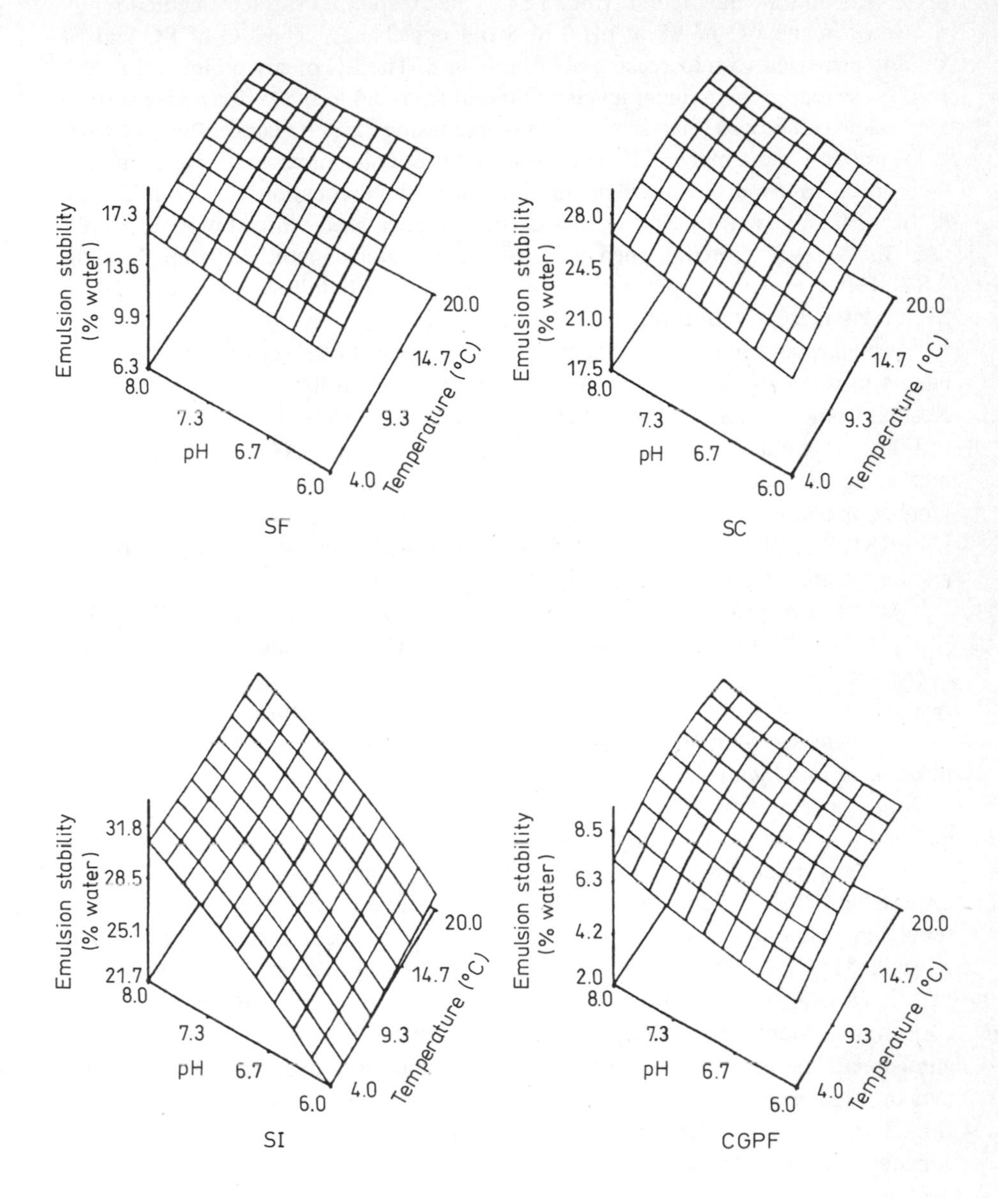

Fig. 3.15. Emulsion stability of soy flour (SF), soy isolate (SI), sodium caseinate (SC), and corn germ flour (CGPF) as a function of pH and temperature of incubation. From Ref. 204

tions from 20 to 70 °C did not affect the EC of these four samples. In regard to pH-concentration interaction (Fig. 3.14), the response surface methodo-logy showed that the EC of SF at pH 6 to 8 did not change. The EC of SC and SI slightly increased with increasing pH from 6 to 8. The EC of soy proteins (SF, SC, and SI) increased with higher levels of protein from 0.4 to 0.8%. Response surface methodology showed that samples with increasing protein concentrations from 0.4% to 0.8% had a higher EC. The surfactant properties of proteins was related to their ability to lower the interfacial tension during emulsification. Soy flour had the highest EC, followed by SI, SC, and CGPF. EC increased more slowly for CGPF than for the soy proteins when protein concentrations increased from 0.4% to 0.8%. However, CGPF appeared to have the highest EC followed by SF, SC, and SI in relation to protein content.

Proteinaceous emulsifiers, such as soy proteins, can act as surface-active agents in water-in-oil emulsions and are widely used in the food industry [205]. Results from response surface methodology in Figs 3.15 and 3.16 show that the stability of emulsions obtained with SF, SC, SI, and CGPF increased with increasing pH from 6 to 8 [204]. However, ES was not affected by increased incubation temperatures from 4 to 20 °C at pH 6–8. In pH-temperature interaction, ES of SF, SC, SI, and CGPF tended to increase with increase in pH from 6 to 8 and temperature from 4 to 20 °C (Fig. 3.15).

An effect of protein concentration on ES was found in all samples (Fig. 3.16) [204]. ES of SF and SC was not influenced by pH in the range 6–8. SI showed decreased ES with increasing pH from 7 to 8. Comparison of the results from response surface methodology showed that all samples had the same tendency:

ES increased with an increase in protein concentration from 0.4 to 0.8% and pH from 6 to 8, with the exception that ES did not increase with an increase in protein concentration from 0.6 to 0.8% at pH 6 for SI. Higher stability of emulsions in food products improves their appearance and storage stability.

Proteins aided formation of emulsions and helped to stabilize them during processing [206]. Proteins form a charge layer around fat droplets causing mutual repulsion, reducing interfacial tension, and preventing coalescence. Leman and Kinsella [171] stated that the amount of protein adsorbed on the fat globule is correlated with the protein concentration during emulsification. Increased protein concentration facilitated adsorption of protein, resulting in enhanced stability of emulsions. Therefore, increasing protein concentration lowered the interfacial tension, and stabilized the emulsion in all samples (SF, SC, SI, and CGPF) (Fig. 3.16). The highest ES was found for SI at pH 8 and 0.8% protein concentration. In relation to protein content, SC had the highest ES at pH 8 and 0.8% level and the SF had the lowest ES at pH 6 with 0.4% protein concentration. Carbohydrates in these plant proteins would stabilize the emulsion systems. The results of this study showed that three soy protein products had different EC and ES. Different functionality of these protein products is related to their different proximate composition (protein, fat, minerals and carbohydrates content). Soya and corn germ protein differed in their structure, particularly arrangement of hydrophobic and hydrophilic groups in their conformational structure.

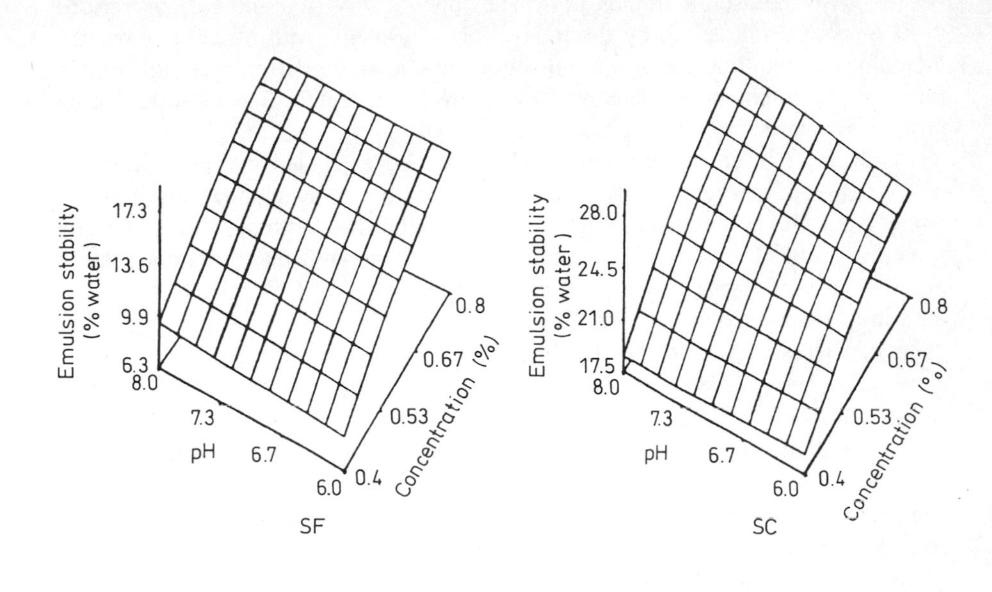

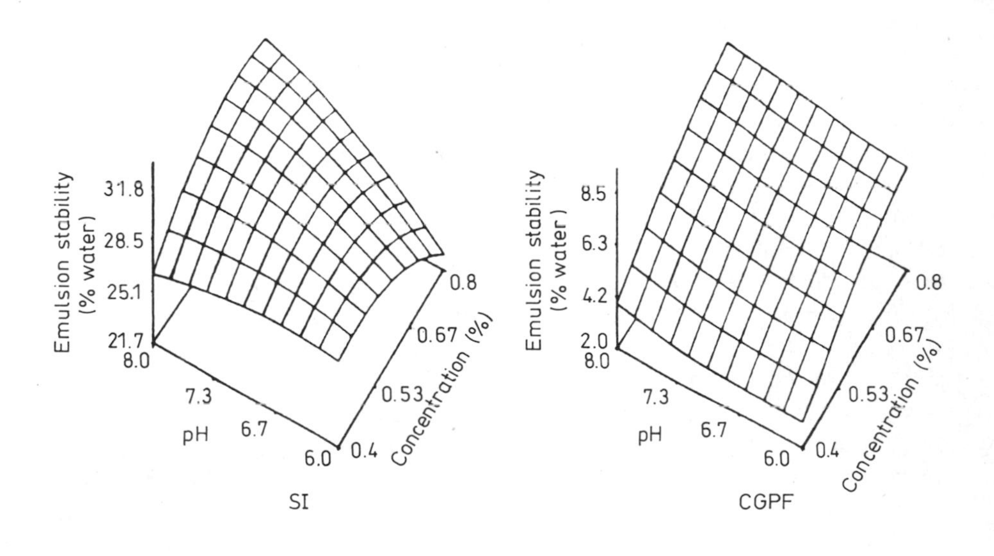

Fig. 3.16. Emulsion stability of soy flour (SF), soy isolate (SI), sodium caseinate (SC), and corn germ flour (CGPF) as a function of pH and temperature of incubation. From Ref. 204

These structural properties can reduce the surface and interfacial tension between two immiscible liquids [24]. The functionality (EC and ES) of various proteins is also influenced by the interaction of proteins with other components, including protein-lipid and protein-water interactions. Differences in protein functionality might be caused by various levels of protein denaturation during processing, particularly during hexane fat extraction.

Utilization of soy proteins in foods might be expanded by fortification of emulsifying properties. Many methods have been developed to improve emulsifying properties of food proteins. Soy lecithin has a strong affinity for soy proteins and has been utilized to obtain a soy lecithin-soy isolate (SI) complex to improve its emulsifying properties. The EC of the lecithin-SI complex increased as the ratio of soy lecithin to soy isolate increased [207]. The EA of the lecithin-SI complex increased considerbly (5 times) after treatment with 50% ethanol. The untreated lecithin-SI complex showed 2.5 times higher EA than SI. The most significant increase in EA was obtained at an ethanol concentration of 40–60% at 25 °C for 30 min and the ratio of lecithin to SI in the complex 1:4. Conformational changes of SI in the ethanol treated lecithin-11S and lecithin-7S soy protein complexes were established. The EC of the lecithin-SI complex increased considerably after heating in boiling water for 1 min. Heating equally increased the EC of 11S and a lecithin-11S protein complex. Heat treatment affected only the aggregation of 11S protein to increase the EC, while aggregation of 7S protein did not increase its EC. However, Aoki et al. [20] indicated that the EC of the 7S and 11S protein-rich fractions decreased as a result of heating at temperatures between 50 °C and 95 °C, with the lowest value of EC at 85 °C.

The EC of soy proteins increased with increasing pH of the dispersions from which emulsions were prepared [208]. At pH 4.5, the ECs of heat-treated samples were much lower than those of the unheated samples. Arteaga et al. [209] reported emulsifying properties of blends consisting of three different commercial soy concentrates (SC): low-heat treated, protease-treated, and alcohol-treated. High EA and ES exhibited protease-treated SC, and low emulsifying properties were found for alcohol-treated SC.

Incorporation of a lipophilic component into the hydrophilic soybean glycinin molecule affected the surface active (emulsifying) properties of these palmitoyl proteins. Emulsifying activity of soybean glycinin increased 2.5 times with the incorporation of palmitic acyl residue in the presence of 8 M urea at pH 9.0, at 25°C, with and without 2-mercaptoethanol [210]. At higher protein concentrations, it was observed that the amount of oil present was the limiting factor for the palmitoyl proteins. These proteins remain soluble in spite of their increased hydrophobicity.

The effect of enzymatic deamidation with alcalase on the emulsifying properties of soy proteins was measured as the absorbance of diluted emulsions at 500 nm [211]. The EA of the heat-treated hydrolysates and the deamidated, heat-treated hydrolysates was greater than control, intact proteins. An increase in EA was obtained under mildly acid (pH 4–6) and alkaline conditions. The EA of protein hydrolysates increased with the increase in the rate of deamidation probably due to a greater protein solubility, surface hydrophobicity and negative charges. Increased solubility of modified soy protein caused the formation of a strong interfacial layer [212]. The pH-emulsifying activity profiles were similar to the pH-solubility profiles in the pH range 2–9. Waniska et al. [35] suggested that for emulsion formation, electrostatic and hydrophobic interactions which influence EC and a certain degree of the stability of tertiary structure are necessary. Deamidated soy protein with increased ES can be utilized in coffee whiteners, beverages and other foods.

3.11.2 Pea and Bean Proteins

Pea protein concentrate (83% protein) extracted in water at pH 5, dialyzed and then centrifuged, exhibited excellent emulsifying properties [213]. The pea protein preparations had good emulsifying properties, comparable to that of soy isolate and higher than soy concentrate and gluten. More stable emulsions were prepared with pea vicilin than with the pea legumin, however legumin showed a higher EC [214]. The vicilin had a greater EC at pH 4, where it was least soluble. ES values were similar for legumin, vicilin, and mixed globulin fractions. The high ES at pH 3–4 may be related to protein denaturation that increases cohesive forces and stabilizes the interfacial film. The EC of chickpea protein isolates was higher for isoelectric protein (72.9%) than for micelle protein (63.7%) and soy isolate (50.8%) [215]. Generally the sodium pea proteinates showed better functionality than isoelectric pea isolates. Samples of pea protein isolates and soy isolate showed a similar EC (33–38%) and sodium pea proteinate showed the highest EC (52%).

The functionality of freeze-dried protein powders of cowpea, peanut and soya was reported [216]. The protein powders were prepared from unfermented and fermented extracts by freeze-drying. The EC of freeze-dried protein powders increased as the concentration of powder in 25 ml solution increased from 0.5 to 3 g. Protein powders prepared from extracts fermented with *S. thermophilus* and unfermented peanut and soybean showed a greater EC than cowpea powders. Metabolic activities of *L. bulgaricus* and *S. thermophilus* may influence the EC of tested legume protein powders. Pea protein isolates have been modified by acylation with succinic and acetic anhydride at 1.0, 3.0, and 5.0 mmol anhydride/g protein [217]. Acylation increased the solubility of the pea protein isolates in the pH range of 4.5–6.0 compared to the nonmodified proteins. The EC and ES of acylated pea protein isolate increased with the extent of acylation up to 3.0 mmol anhydride/g protein. In acylated pea protein, the hydrophilic groups were exposed in unfolded protein chains.

The use of lupin seed protein is limited by the presence of toxic quinolizidine alkaloids. However, lupin protein meal unlike soybeans do not contain trypsin inhibitor or hemaglutinins. Functional properties of the protein from various varieties of lupin might be different because the ratio of the different globulin fractions is not constant. Stable emulsions were prepared at low lupin protein isolate concentration (0.04%) [218]. The EC increased on both the basic and acidic side of the isoelectric point (pH 5.0) due to increased solubilization. The EC of lupin protein isolate was similar to soy isolate.

Two potential products from dry beans are high protein flour and protein concentrate. The protein-rich fractions of dry legume flours exhibited good EC [219]. The EC of the pin-milled flour of pinto-bean, navy bean and chick-pea was 17.50, 33.75, and 36.25 ml oil/g dry sample, respectively. There were differences among pinto, navy bean, and chick-pea flour in emulsion stability. Pinto bean exhibited the lowest EC and had the lowest ES (64.38% water separated) followed by navy bean (37.9%) and chick-pea (25.34%). Navy and kidney bean protein concentrates and commercial soy product exhibited a similar EC [220]. Navy bean protein flour exhibited a higher EC (0.12 ml oil/mg protein) than soy flour (0.08 ml oil/mg protein), while navy, kidney, and soy protein concentrate displayed a similar EC (0.12 ml oil/mg protein). A positive correlation was found between the EC and the protein content of a mixture of navy and pinto bean flour [221]. The EC increased from 19.2 ml/g for the high starch fraction of pinto bean to 42.5 ml/g for the blend containing 30% high protein fraction.

Field pea and faba bean protein products had a similar emulsifying capacity (34.6–38.6 ml oil/0.1 g sample) as soy protein products (37.2–45.1 ml oil/0.1 g sample) and can be recommended as fillers and extenders in sausage batters or bakery products [222].

Succinylation of winged bean protein rapidly increased the EC beyond 54% succinylation. At the 82% succinylation level, the EC increased 2 times, and then decreased [223]. The EC of 80% acylated winged bean protein increased from 35 to 47 ml oil/g and then decreased. Generally, since the nitrogen solubility of acylated protein flours is higher, an increase in EC is expected. The decrease in EC at higher degrees of acylation can result from the decreased protein solubility. The improved EC of acylated protein can be utilized in milk-like beverages, coffee whiteners and milk products.

3.11.3 Corn Proteins

The emulsifying capacity and emulsion stability of defatted corn germ protein flour (CGPF) can be used to define how these proteins can be added to existing foods and how they can replace more expensive proteins traditionally used. A knowledge of the emulsifying properties of defatted CGPF is necessary to evaluate their potential use as food additives. We have conducted studies on the basic factors affecting the functional properties of plant and animal proteins, including CGPF as an additive in comminuted meats. The solubility of CGPF preparations for pH values between 6 and 8 was determined to relate protein solubility to EC and ES [117]. Corn germ protein isolate had good protein solubility at neutral and low pH and the ability to stabilize oil-in-water emul sions. Defatted CGPF contains largely globulins and albumins which possess lower viscoelastic properties than wheat proteins (gliadin and glutenin). The individual functionali-

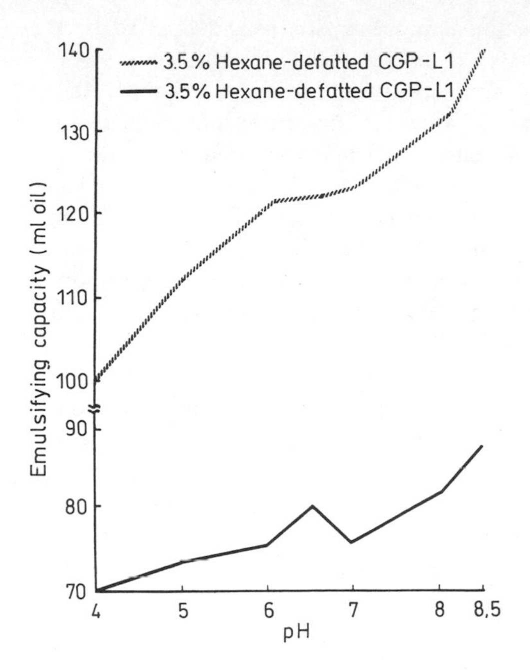

Fig. 3.17. Emulsifying capacity of corn germ protein defatted by conventional and modified hexane extraction method as a function of pH. From Ref. 224

ties of several kinds of corn protein and emulsifying properties of supercritical and two hexane extracted CGPF preparations have been reported [117]. One CGPF was processed by the conventional hexane extraction and the other by a modified technique. The EC was measured by using a model system to test how much oil could be emulsified into an aqueous phase containing the test protein. ES was measured by using a model system to test the stability of an emulsion prepared under specific conditions. The emulsions were prepared from corn oil (20 to 40%), defatted CGPF preparations (3 to 7%), and water (total 150 g).

At present, conventional oil extraction by hexane leaves a certain amount of residual lipids in the defatted CGPF, which reduces its quality. The modified hexane extraction method tested here markedly reduced the amount of residue oil in CGPF. As the result of the utilization of modified method, the flavor characteristics and color of CGPF were notably improved. Hexane defatted CGPF had a food-grade quality and considerable potential for use as a supplement in a variety of foods. Figure 3.17 shows the effect of pH on the EC of two hexane-defatted CGPFs [224]. Both preparations had the lowest levels of EC in their isoelectric region, and EC increased at pH values above this region. As shown in Fig. 3.17, both hexane-defatted CGPFs indicated minimum values of EC at pH 4.0–4.5. The relationship between solubility and EC was found, i.e., the emulsion capacity of defatted CGPFs correlates with their solubility. At a higher pH, protein had higher solubility, which certainly increased the EC of defatted CGPF. A significant increase in EC was obtained at a pH range of 7.0–8.5. A hexane-defatted CGPF obtained by the modified extraction process had a higher

EC than one obtained by the conventional process. This was supported by the evidence of higher protein solubility.

An optimal composition of the components is needed to obtain a stable emulsion, with little fat or water separated after holding or centrifugation. The EC and ES were affected by the CGPF concentration. The ES was determined for both defatted CGPF samples at concentrations from 3 to 7%. For each sample of defatted CGPF, several protein and fat concentrations and emulsifying time combinations resulted in relatively effective emulsification. An increase in protein concentration resulted in increased ES. There was no measurable fat separation in these emulsions. High ES was obtained as the result of defatted CGPF specific properties and carbohydrate component of CGPF (about 34–38%), and formation of a system with modified textural properties. Defatted CGPF obtained by the modified extraction process had the highest ES, i.e., the least water separation at 7% protein concentration and 40% fat in emulsion. At lower protein concentrations (3–3.5%), emulsions of lower stability were obtained. Centrifugation of these emulsions resulted in higher amounts of separated water. Holding temperature also influenced the ES of the emulsions [224].

Lower values of ES were obtained for the conventionally hexane-defatted CGPF. Increasing the defatted CGPF content increased the ES of the system [224]. An important factor influencing the EC and ES of conventionally defatted CGPF was its higher protein denaturation compared with samples defatted by the modified process. Differences in protein denaturation of samples defatted by modified and conventional processes was established by differential scanning calorimeter studies (Zayas and Lin, unpublished data). Emulsifying time in the range 2–5 min had no significant effect on the ES of emulsions at protein concentrations from 3 to 7%. McWatters and Holmes [39] showed that large concentrations of soluble proteins were not necessarily related to maximum emulsifying capacity and emulsion stability. Undissolved particles of proteins in the aqueous phase of emulsion can also stabilize the system by serving as a physical barrier to coalescence of oil droplets. Protein particles are absorbed at the oil-water interface more quickly than individual protein molecules. Both fat concentration and emulsifying time contributed to the increased ES of conventional defatted CGPF, which was not shown with other factors. Emulsions produced with defatted CGPF obtained by the modified extraction procedure had consistently less water and emulsified oil separated after centrifugation than did those containing defatted CGPF extracted by the conventional procedure. Increasing the emulsifying time slightly increased the ES. However, the predominant factor was protein concentration, which contributed most to the increased ES. The general trend was observed that hexane-defatted CGPF obtained by a modified process had higher EC and ES than that obtained by the conventional extraction process. Because of better functional properties in the model system, hexane-defatted CGPF obtained by the modified process was recommended for utilization as a food additive.

Functional properties have been tested for two CGPF preparations, supercritical carbon dioxide (SC-CO_2)-defatted and hexane-defatted in relation to their protein solubility, emulsifying capacity and emulsion stability in a model system that corresponded to the conditions for sausage batter processing [117]. Within a pH range of 7–9 of the protein solution, higher protein solubility was observed, which certainly increased the EC of that protein. A significant increase in EC was obtained with a pH range of 7.0–8.5. SC-CO_2-defatted CGPF had a higher EC and ES than hexane-defatted CGPF. The difference in ES between the SC-CO_2- and hexane-defatted CGPF could be explained by the higher water retention, fat binding and higher protein solubility of the SC-CO_2 CGPF preparation. Another important factor could be the higher level of fat of the hexane-defatted CGPF (11.1%) than of the SC-CO_2-defatted CGPF (1.6%). A stable emulsion was obtained with 7.0% SC-CO_2-defatted CGPF and 40% corn oil. The ES of SC-CO_2-defatted CGPF decreased as incubation temperature increased and increased as emulsifying times increased. The temperature of incubation was an important factor for ES of SC-CO_2-defatted CGPF, when low emulsifying times were used. A more stable emulsion was produced with low incubation temperatures. Overall results showed that SC-CO_2-defatted CGPF had better functional properties over hexane-defatted CGPF.

3.11.4 Miscellaneous Proteins

Peanut Protein

The highest EC for partially defatted nonfermented peanut flour was observed at pH 6.0 when steam had been applied for 15 min [225]. Increasing the heating time to 45 min caused a decrease in EC. Fungul fermentation of the peanut flour increased the EC at pH 2.0–10.0.

Sunflower Proteins

Sunflower proteins have a significant potential as a source of food protein because of the absence of any antinutritive factor.

Sunflower protein flour exhibited a better EC than soy protein products. Acetylated sunflower protein concentrate showed greater values of EC and ES than the succinylated samples [226]. The pH-EC profiles of acylated and control sunflower concentrate followed the pattern of their pH-solubility profiles. The minimum EC was in the acidic pH range and a remarkable increase in EC was observed until pH 7 and a sharp decrease in EC at alkaline pH values. Modified sunflower concentrate at a 10% succinylation level exhibited a more significant increase in EC. An increase in sunflower concentrate emulsifying properties was probably related to the increased protein solubility of the modified sunflower concentrate. Succinylation improved the emulsifying properties of sunflower protein isolate in the pH range 4 to 10. Sunflower isolate succinylated at a level of 27% showed the greatest EC [227]. Acetylation had little effect on emulsifying properties of sunflower isolate, although EA increased at pH 3–4. In the presence of NaCl emulsifying properties of succinylated sunflower isolate remained superior to the control and the acetylated isolate. Succinylated (87%) sunflower isolate incorporated in meat proteins at different ratios increased EC and maintained a similar level of ES. Succinylated sunflower isolate with improved emulsifying properties has a potential for application in comminuted meats, salad dressings and mayonnaise.

Rice Proteins

The EC of rice bran protein concentrate suspension was significantly affected by pH [228]. Full-fat and defatted concentrate showed positive correlation of EC with the pH and nitrogen solubility. The ES was also affected by changes in pH. The ES (85%) of full-fat samples was higher than the defatted concentrate (74%), which contained more carbohydrates (50.8%). The full-fat concentrate contained 38.2% carbohydrates. The EC of rice bran protein concentrate was influenced by ionic strength (0.1 and 1.0 M NaCl). The effect of NaCl on the EC of rice protein concentrate was related to the salting out effect.

Rapeseed and Canola Proteins

Rapeseed protein concentrate, prepared according to Thompson et al. [229] as a 1% dispersion, emulsified more oil (108.0 ml oil/g sample) than rapeseed flour (89.8 ml oil/g sample). Emulsions stabilized with rapeseed concentrate were more stable (27 ml) than with rapeseed flour (44 ml) but less stable than with soy isolate (14 ml). The high EC, water and fat binding suggest the possible use of rapeseed concentrate in processed meats. Rapeseed meal and rapeseed protein isolate exhibited a lower EC than those of soy meal and isolate [230]. The EC of rapeseed isolate was in the range 25.0–30.3 ml oil/100 mg sample and soy isolate 43.5 ml oil/100 mg sample. Rapeseed isolate and soy isolate gave stable emulsions with no aqueous phase separation during 1–6 days of incubation at 21 °C. The emulsifying properties of rapeseed isolate were improved with increased protein solubility. Rapeseed meal had a lower EC and ES than soybean meal. Emulsion activity (EA) of canola protein increased with the extent of succinylation and was affected by pH [231]. Addition of NaCl at a 0.35 M level increased the EA of canola protein. Succinylation increased the molecular surface area of the protein, apparently as a result of electrical repulsion forces between identical charges. Increases in EA and ES of succinylated canola oil was related to the protein solubility, hydrophobicity, zeta potential and rheological properties of protein dispersions.

Cottonseed Proteins

The EC of cottonseed protein meal was slightly higher in 5% NaCl solutions (71–74 ml oil/g meal) than in water (63–67 ml oil/g meal) [232]. The increased EC of cottonseed protein meal in 5% NaCl solution was related to higher protein solubility. The EC of cottonseed protein isolate increased markedly at levels of succinylation above 60%, and no effect has been observed at levels of succinylation below 60% [233]. The increase in EC could result from unfolding or swelling of protein molecules at high levels of succinylation. Secondary changes of protein molecules may alter the emulsifying properties of protein. The effect of protein water solubility on EC was tested and a positive, but not linear correlation was found. There was no change in EC of cottonseed protein isolate up to 20–40% protein solubility, however EC increased markedly in the range of protein solubility from 50 to 80%.

Guar Proteins

Guar protein isolates prepared by various methods exhibited an EC in the range 56–72 ml oil/g sample [234]. The 2-propanol extracted meal isolate had the highest EC (72 ml oil/g sample). Guar protein meal showed a greater EC than soybean meal, but the EC of guar protein isolate was similar to soy isolate at all pH's except around 4.0 [235]. The minimum values of EC were around pH 4.5 (minimum protein solubility) and EC values increased on either side of pH 4.5. The EC of guar protein was higher on the acidic than the alkaline side of the isoelectric point. Guar protein isolate did not form a stable emulsion in the pH range 4.4–5.2. The EC of guar protein isolate was slightly higher than soy isolate. The EC of guar proteins increased with NaCl concentration up to a maximum and then decreased. The pH-emulsion capacity profile (pH range 2–11) of guar protein meal extracted by different methods have a similar trend to that of the pH-nitrogen solubility profile [236]. The EC for methanol-extracted guar protein was lower (54 ml oil/g protein) than that of extracted by isopropanol or ethanol (75 and 82 ml oil/g protein, respectively). Other guar proteins tested had an EC in the range 87–90 ml oil/g protein. Protein solubility was decreased by autoclaving, however the EC was not affected.

Sesame and Safflower Proteins

Sesame protein isolates formed stable cream-colored emulsions, however emulsion stability was lower than for soy isolate. The EC of sesame α-globulin increased significantly in the alkaline pH range [237]. This demonstrated the suggestion that the dissociated and partly unfolded polypeptide chains formed a viscous and cohesive interfacial film around the emulsion particles more effectively. The EC of sesame proteins was approximately 22% lower than the EC of the typical egg yolk emulsion [238]. When the sesame protein dispersion was prepared with lecithin, the EC increased to the values of egg-yolk-stabilized emulsions. The sesame protein exhibited higher ES than soy isolate.

Safflower protein isolates obtained by a micellization procedure (dilution in cold water) showed no difference in EC to those obtained by isoelectric precipitation [239]. However, the ES was higher for the isolates obtained by the micellization procedure than those of the isoelectric isolates.

Linseed Proteins

Emulsions of high stability were produced with high-mucilage linseed protein concentrates [240]. The low-mucilage linseed protein isolate exhibited lower EC and ES than high-mucilage concentrates, indicating the role of mucilage in enhancing emulsifying properties. The effect of heating and pH in the range pH 1–12 on the EC of linseed and soy protein flour was reported [241]. The U-shaped curved relationship was obtained for the EC vs pH profile with linseed and soy protein flours with the minimum at pH 4.5. Heated linseed protein showed a lower EC than unheated, and the EC had a broad minimum in the pH range 3 to 6. The EC vs pH profile for tested proteins resembled the nitrogen solubility vs pH profile curve. Linseed protein flour showed better EC values at acidic pH compared to soy protein meal, however the soy protein exhibited better EC values at alkaline and neutral pH. The better EC in soy protein is due to higher content of protein in the sample. A gradual increase in the EC of linseed protein was found with an increase in NaCl concentration from 0 to 0.2 M NaCl.

Oat Proteins

Oat protein concentrate showed an emulsion activity index (EAI) comparable to gluten [242]. Emulsifying properties were directly related to oat protein solubility. The defatted oat concentrate had a lower EAI than the nondefatted samples. The ES of emulsions with oat protein concentrate was considerably lower than soy isolate and gluten. The pH-EAI curves of globulin, prolamin and glutelin fractions followed the pH-solubility curves. The processed wild and domestic oat products gave a comparable EC to wheat and potato protein flours [243]. Bran fractions of oat gave a slightly higher EC than those of the meals and flours. However, the processed oat bran and flour samples exhibited poor ES. The defatted wild oat bran also exhibited a high EC, however the emulsions were unstable.

The use of oat proteins is restricted because of poor functional properties. Solubility, in particular, is poor at neutral and slightly acidic pH. Mild acid hydrolysis has been used to deamidate oat protein isolate to improve functional properties [244]. The EC, EAI, and emulsion stability index (ESI) of unmodified and deamidated oat protein isolate progressively increased with increasing levels of modification, and the ESI was significantly higher in the deamidated samples. The surface hydrophobicity of oat protein isolate was increased by deamidation. Deamidation caused a significant increase in the EAI over a wide range of pH. The effect of deamidation on the EC of oat protein isolate can be related to an increase in protein solubility and surface hydrophobicity. Oat protein extracted from defatted oat (variety Sentinel) was acylated with acetic or succinic anhydride. Acylation increased the EC, EAI, and ESI of oat proteins considerably, especially at higher levels of modification [245]. The ES was significantly increased even with a low degree of acylation. The EC of the acylated oat protein blended with meat proteins increased markedly and the oat proteins showed a higher EC than meat proteins. The EC of the mixtures decreased progressively with an increase in the ratio of meat in the blends. Ma [245] suggested that interaction between acyla-

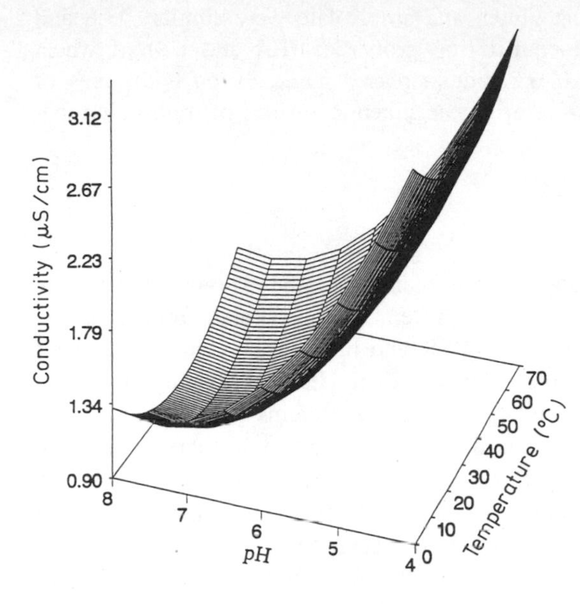

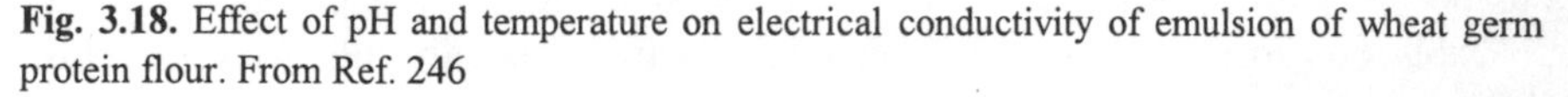

Fig. 3.18. Effect of pH and temperature on electrical conductivity of emulsion of wheat germ protein flour. From Ref. 246

ted oat protein and meat proteins had a synergistic effect on the EC. The pH-EAI profile of acylated oat protein resembled the pH-solubility profile with minimum EAI at pH 4.5. The EAI was increased by acylation at all pH's except pH 1.5. Succinylation increased the solubility of oat proteins markedly especially at acidic and near neutral pH [245]. Succinylated oat proteins showed a higher EAI than acylated, particularly at alkaline pH's. The modified oat protein could find application in the production of comminuted meats because of its good EC and fat binding.

Wheat Proteins

The emulsifying capacity of wheat germ protein flour (WGPF) at pH 4–8 and temperature of incubation 5–70 °C is presented in Fig. 3.18 [246]. The highest emulsifying capacity of WGPF was at pH 4–5 and 70 °C while the lowest was at 30°C. The high emulsifying capacity of WGPF at 70 °C was probably due to strong interactions between the proteins and gelatinized carbohydrates and maximum rigidity of the proteins in the isoelectric pH range. Denaturation of proteins of WGPF by enhancing macromolecular flexibility and surface hydrophobicity might have improved emulsifying properties of WGPF. In food ingredients such as WGPF, containing both protein and polysaccharide, complications may arise due to the limited thermodynamic compatibility of two or more polymers in a common solvent. The rheological properties of the interfacial layers are influenced by interactions of protein with high molecular weight polysaccharide. The EA of wheat gluten and soy isolate were similar, 43.8 and 40.8%, respectively [247]. The emulsifying properties (EA and ES) of wheat gluten were the lowest at pH 4.0–6.0, but increased markedly on both sides of pH 4.0–6.0. The emulsifying profile of wheat gluten at various pH resembled that of NFDM.

Potato Proteins

Functional properties of whey-potato protein blends over the pH range 4 to 8 were affected by the proportion of two tested protein sources. Potato protein is the byproduct of starch manufacture and its PER and biological value were greater than that of casein and comparable to whole egg. The functionality of potato proteins may be improved by blending with dairy proteins [248]. Emulsifying properties were improved as pH increased and the proportion of potato proteins in the blend decreased. The lowest EAI was observed at the pI of two proteins, i.e., between pH 4.6–5.5 for whey protein and pH 5 for potato protein. The effect of proportion of potato: whey protein on the EAI changed with pH 4–8. The ES of protein composites increased linearly within the pH range 4 to 8. The ES of composites with 25% potato protein was markedly reduced, especially at pH 8, probably due to protein-protein interactions between potato and whey proteins. Wojnowska et al. [249] reported the EC of 100 mg spray-dried potato proteins obtained by ultrafiltration and concentrated to 1:15. The potato protein concentrates prior to drying exhibited larger EC. Emulsions stabilized with potato protein did not break down after heating at 80 °C for 30 min.

Alfalfa Leaf Proteins

Alfalfa leaf proteins showed high emulsifying properties despite low solubility values. The EC of leaf protein concentrate (LPC) decreased with an increase in LPC concentration [250]. The ECS of alfalfa LPC at 0.5, 1.0, and 2.0% were 700, 521, and 246 ml oil/g LPC, respectively. LPC (2% w/v) formed very stable emulsions with a high viscosity like that of mayonnaise. The EA and ES of succinylated LPC increased to 31.8 and 36.3%, respectively. The emulsifying properties of edible alfalfa protein concentrate are affected by the pH with a maximum EA and heat stability of the emulsion at pH 6 [251]. The ES is more or less constant between pH 4 and pH 10. The addition of NaCl, 1.0 M slightly reduced the EA and ES, and addition of sucrose improved the EA and ES of alfalfa protein concentrate.

Yeast Proteins

The EAI of phosphorylated yeast protein with 73% protein content was low at pH 5 but exceeded EAI of BSA at pH 7.0. The yeast protein with high charge and low nucleic acid was prepared by phosphorylation with phosphorus oxychloride. Phosphorylated yeast protein had solubilities of 3.0 and 9.5 mg/ml at pH 5 and 6–8, respectively. Phosphorylation of yeast nucleoprotein improved the EA, especially at pH 6–7 probably due to an increase in solubility and viscosity [252]. These factors and an increased negative charge on the phosphorylated protein may facilitate formation of the strong interfacial film and reduce coalescence of fat droplets [1]. The EA of phosphorylated yeast protein was greater than several food proteins at neutral pH. The EA was decreased by addition of NaCl up to 0.2 M. The EA of phosphorylated yeast protein progressively increased with an increase in protein concentration from 0.2 to 1% due to the formation of thicker and more stable surface films. Thicker films around fat droplets produce more stable emulsions. The ES of phosphorylated yeast protein was slightly greater when heated at 90 °C and incubated for 60 min than emulsions prepared with BSA and egg albumin. In these emulsions the negatively charged phosphoryl groups in the protein probably reduced the coalescence of fat droplets. The emulsifying properties of the two yeast proteins were compared with soy protein concentrate [253]. The EC of *C. utilis* protein concentrate was two times higher than soy concentrate, 30.0 and 15.0 ml/g protein, respectively.

Egusi Seed Proteins

Partially defatted egusi seed meal is made into patties that serve as a meat substitute. The egusi seed protein exhibited excellent emulsifying properties and formed thin emulsions at pH 2.5 and 5.0 [254]. Thicker mayonnaise-type emulsions were formed at pH 6.5–10.5. Positive correlation was found between protein solubility and EC.

Tepary Flour Proteins

The ECs of tepary flour, albumin, globulin and soy protein isolate were similar and consisted of 32.5, 31.5, 27.0, and 31.2 ml oil/100 mg protein, respectively [255]. The EC of the sample can be influenced by other components, especially carbohydrates. The EC data of those proteins are comparable with those of other plant proteins such as faba bean (34 to 37 ml oil/100 mg protein).

References

1. Halling, P. J. (1981). Protein-stabilized foams and emulsions, *Crit. Rev. Food Sci. Nutr.*, *15*: 155.
2. Walstra, P. (1983). In *Encyclopedia of Emulsion Technology – 1* (P. Becher, ed.), Marcel Dekker, New York, p. 57.

3. Nakai, S., Ho, L., Helbig, A., Kato, A., and Tung, M. A. (1980). Relationship between hydrophobicity and emulsifying properties of some plant proteins, *Can. Inst. Food Sci. Technol. J.*, *13(1)*: 23.
4. Nakai, S. and Powerie, W. D. (1981). *Cereals: A Renewable Resource, Theory and Practice* (Y. Pomeranz and L. Munch, eds.), Amer. Assoc. Cereal Chem. Symposium, p. 736.
5. Kato, A. and Nakai, S. (1980). Hydrophobicity determined by a fluorescent probe method and its correlation with surface properties of proteins, *Biochem, Biophys. Acta.*, *624*: 13.
6. Kato, A., Tsutsui, N., Kobayashi, K. and Nakai, S. (1981). Effects of partial denaturation on surface properties of ovalbumin and lysozyme, *Agr. Biol. Chem.*, *45*: 2755.
7. Li-Chan, E., Nakai, S. and Wood, D. F. (1984). Hydrophobicity and solubility of meat proteins and their relationship to emulsifying properties, *J. Food Sci.*, *49*: 345.
8. Shimada, K. and Matsushita, S. (1980). Relationship between thermocoagulation of proteins and amino acid compositions, *J. Agric. Food Chem.*, *28*: 413.
9. Das, K. P. and Kinsella, J. E. (1990). Stability of food emulsions: Physicochemical role of protein and nonprotein emulsifiers, *Adv. in Food And Nutr. Res.*, *34*: 81.
10. Matsudomi, N., Kato, A., and Kobayashi, K. (1982). Conformation and surface properties of deamidated gluten, *Agric. Biol. Chem.*, *46*: 1583.
11. Yamashito, M., Arai, S., Amano, Y., and Fujimaki, M. (1979). A novel one-step process for enzymic incorporation of amino acids into proteins: application to soy protein and flour for enhancing their methionine levels, *Agric. Biol. Chem.*, *43*: 1065.
12. Phillips, M. C. (1981). Protein conformation at liquid interfaces and its role in stabilizing foams and emulsions, *Food Technol.*, *35(1)*: 50.
13. Bigelow, C. (1967). On the average hydrophobicity of proteins and the relation between it and protein structure, *J. Theor. Biol.*, *16*: 187.
14. Shimizu, M., Saito, M., and Yamauchi, K. (1985). Emulsifying and structural properties of β-lactoglobulin at different pHs, *Agric. Biol. Chem.*, *49*: 189.
15. Aoki, H., Taneyama, O., Orimo, N., and Kitagawa, I. (1981). Effect of lipophilization of soy protein on its emulsion stability properties, *J. Food Sci.*, *46*: 1192.
16. Elizalde, B. E., Pilosof, A. M. R., and Bartholomai, G. B. (1991). Relationship of absorptive and interfacial behavior of some food proteins to their emulsifying properties, *J. Food Sci.*, *56*: 253.
17. Saito, M. and Taira, H. (1987). Heat denaturation and emulsifying properties of plasma protein, *Agric. Biol. Chem.*, *51*: 2787.
18. Kato, A., Komatsu, K., Fujimoto, K., and Kobayashi, K. (1985). Relationship between surface functional properties and flexibility of proteins detected by the protease susceptibility, *J. Agric. Food Chem.*, *33*: 931.
19. Lin, C. S. and Zayas, J. F. (1987). Protein solubility, emulsifying stability, and capacity of two defatted corn germ proteins, *J. Food Sci.*, *52*: 1615.
20. Aoki, H., Taneyama, O., and Inami, M. (1980). Emulsifying properties of soy protein: Characteristics of 7S and 11S proteins, *J. Food Sci.*, *45*: 534.
21. Flint, F. O. and Johnson, R. F. P. (1981). A study of film formation by soy protein isolate, *J. Food Sci.*, *46*: 1351.
22. Paulson, A. T., Tung, M. A., Garland, M. R., and Nakai, S. (1984). Functionality of modified plant proteins in model food system, *Can. Inst. Food Sci. Technol. J.*, *17*: 202.
23. Shimizu, M., Lee, S. W., Kaminogawa, S., and Yamauchi, K. (1986). Functional properties of a peptide of 23 residues purified from the peptic hydrolyzate of αs_1-casein: Changes in the emulsifying activity during purification of the peptide, *J. Food Sci.*, *51*: 1248.
24. Voutsinas, L. P., Cheung, E., and Nakai, S. (1983). Relationships of hydrophobicity to emulsifying properties of heat denatured proteins, *J. Food Sci.*, *48*: 26.
25. MacRitchie, F. (1978). Proteins at interfaces, *Adv. Protein Chem.*, *32*: 283.

26. Kinsella, J. E. and Whitehead, D. M. (1987). In *Proteins at Interfaces: Physicochemical and Biochemical Studies* (J. L. Brash and T. A. Horbett, eds.), ACS Symp. Series No. 343, Amer. Chem. Soc., Washington, D. C., p. 629.
27. Dickinson, E. and Stainsby, G. (1988). Emulsion stability, In *Advances in Food Emulsions and Foams* (E. Dickinson and G. Stainsby, eds.), Elsevier, Amsterdam, p. 1.
28. Kinsella, J. E. (1981). Functional properties of proteins: possible relationships between structure and function in foams, *Food Chem.*, *7*: 273.
29. Tornberg, E. (1979). The adsorption behavior of proteins at an interface as related to their emulsifying properties, *ACS Symp. Ser.*, *92*: 6.
30. Oortwijn, H. and Walstra, P. (1979). The membranes of recombined fat globules. 2. Composition, *Neth. Milk Dairy J.*, *33*: 134.
31. Das, K. P. and Chattoraj, D. K. (1980). Adsorption of proteins at the polar oil-water interface, *J. Colloid Interface Sci.*, *78*: 422.
32. Graham, D. E. and Phillips, M. C. (1979). Proteins at liquid interfaces. I. Kinetics of adsorption and surface denaturaiton, *J. Colloid Interface Sci.*, *70*: 403.
33. Kinsella, J. E. (1982). Protein structure and functional properties: Emulsification and flavor binding effects, In *Food Protein Deterioration, Mechanisms and Functionality* (J. P. Cherry, ed.), Amer. Chem. Soc., Washington, D. C., p. 301.
34. Song, K. B. and Damodaran, S. (1987). Structure-function relationship of proteins: Adsorption of structural intermediates of bovine serum albumin at the air-water interface, *J. Agric. Food Chem.*, *35*: 236.
35. Waniska, R., Shetty, J. K., and Kinsella, J. E. (1981). Protein-stabilized emulsions: effects of modification on the emulsifying activity of bovine serum albumin in a model system, *J. Agric. Food Chem.*, *29*: 826.
36. Dickinson, E., Murray, B. S., and Stainsby, G. (1985). Time-dependent surface viscosity of adsorbed films of casein and gelatin at the oil-water interface, *J. Colloid Interface Sci.*, *106*: 259.
37. Nakai, S. (1983). Structure-function relationships of food proteins with an emphasis on the importance of protein hydrophobicity, *J. Agric. Food Chem.*, *31*: 676.
38. Manak, L., Lawhon, J., and Lusas, E. (1980). Functioning potential of soy, cottonseed, and peanut protein isolates produced by industrial membrane systems, *J. Food Sci.*, *45*: 236.
39. McWatters, K. H. and Holmes, M. R. (1979). Influence of moist heat on solubility and emulsification properties of soy and peanut flours, *J. Food Sci.*, *44*: 774.
40. Stone, M. and Campbell, A. M. (1980). Emulsification in systems containing soy protein isolates, salt and starch, *J. Food Sci.*, *49*: 1713.
41. Kinsella, J. E. (1979). Functional properties of soy proteins, *J.A.O.C.S.*, *56*: 242.
42. Walstra, P. (1984). *Food Science and Technology: Present Status and Future Direction* (B. McLaughlin, ed.), Boole Press, Dublin, p. 323.
43. Tarasevich, B. N., Izmailova, V. N., Morozova, L. Z., and Novoselova, M. A. (1984). Effect of nonpolar phase on adsorption of proteins at water-hydrocarbon liquid interface, *Colloid J.*, *84*: 1028. (English translation from *Kolloidn. Zh.*, *84*: 1191.)
44. Elizalde, B. E., Kanterewicz, R. J., Pilosof, A. M. R., and Bartholomai, G. B. (1988). Physicochemical properties of food proteins related to their ability to stabilize oil-in-water emulsions, *J. Food Sci.*, *53*: 845.
45. Kanterewicz, R. J., Elizalde, B. E., Pilosof, A. M. R., and Bartholomai, G. B. (1987). Water oil absorption index (WOAI): A simple method for predicting the emulsifying capacity of food protein, *J. Food Sci.*, *52*: 1381.
46. Elizalde, B. E., Pilosof, A. M. R., and Bartholomai, G. B. (1991). Prediction of emulsion instability from emulsion composition and physicochemical properties of proteins, *J. Food Sci.*, *56*: 116.
47. Das, K. P. and Kinsella, J. E. (1990). Stability of food emulsions: Physicochemical role of protein and nonprotein emulsifiers, *Adv. in Food and Nutr. Res.* (J. E. Kinsella, ed.), *34*: 81.

48. Rivas, H. J. and Sherman, P. (1984). Soy and meat proteins as emulsion stabilizers. 4. The stability and interfacial rheology of o/w emulsions stabilized by soy and meat protein fractions, *Colloids Surf.*, *11*: 155.
49. Graham, D.E. and Phillips, M.C. (1980). Proteins at liquid interfaces. V. Shear properties. *J. Colloid Interface Sci.*, *76*:240.
50. Das, K. P. and Kinsella, J. E. (1989). pH dependent emulsifying properties of β-lactoglobulin, *J. Dispersion Sci. Technol.*, *10*: 77.
51. Fox, P. T. and Mulvihill, D. M. (1982). Milk proteins: molecular, colloidal and functional properties, *J. Dairy Res.*, *49*: 679.
52. Tornberg, E. and Ediriweera, N. (1987). Coalescence stability of protein stabilized emulsions, *Spec. Publ.-Soc. Chem.*, *58*: 52.
53. Walstra, P. (1987). Physical principles of emulsion science, In *Food Structure and Behavior* (J. M. V. Blanshard and P. Lillford, eds.), Academic Press, Orlando, Florida, p. 87.
54. Cameron, D. R., Cooper, D. G., and Neufeld, R. J. (1988). The mannoprotein of *Saccharomyces Cerevisiae* is an effective bioemulsifier, *Appl. Environ. Microbiol.*, *54*: 1420.
55. Shioya, T., Kako, M., Taneya, T., and Sone, T. (1981). Influence of the thickening on the whippability of creams, *J. Texture Stud.*, *12*: 185.
56. Fleer, G. J. and Scheutjens, J. M. H. M. (1982). *Adv. Colloid Interface Sci.*, 16, *341*: 360.
57. Darling, D. F. and Birkett, R. J. (1987). Food colloids in practice, *Spec. Publ.-R. Soc. Chem.*, *58*: 1.
58. Tornberg, E. and Ediriweera, N, (1988). Factors that influence the coalescence stability of protein stabilized emulsions estimated from the proportion of oil extracted by hexane, *J. Sci. Food Agric.*, *46*: 93.
59. Darling, D. F. (1982). Recent advances in the destabilization of dairy emulsions, *J. Dairy Sci.*, *49*: 695.
60. Leman, J., Haque, Z., and Kinsella, J. E. (1988). Creaming stability of fluid emulsions containing different milk protein preparations, *Milchwissenschaft*, *43*: 286.
61. Howe, A.M., Mackie, A.R., and Robbins,M.M. (1986). Technique to measure emulsion creaming by velocity of ultrasound. *J. Dispersion Sci. Technol.*, *7*:231.
62. Hibberd, D. J., Howe, A. M., Mackie, A. R., Purdy, P. W., and Robbins, M. M. (1987). Measurement of creaming profiles in oil-in-water emulsions, *Spec. Publ. – R. Soc. Chem.*, *58*: 219.
63. Hermus, R. and Albers, H. (1986). Meat and meat products in nutrition. 32nd European Meeting of Meat Research Workers, Gent, Belgium.
64. Lin, C. S. and Zayas, J. F. (1987). Influence of corn germ protein on yield and quality characteristics of comminuted meat products in a model system, *J. Food Sci.*, *52*: 545.
65. Parks, L. L. and Carpenter, J. A. (1987). Functionality of six nonmeat proteins in meat emulsion system, *J. Food Sci.*, *52*: 271.
66. Wills, R. B. and Kabirullah, M. (1981). Use of sunflower protein in sausages, *J. Food Sci.*, *46*: 1657.
67. Gnanasambandam, R. and Zayas, J. F. (1992). Functionality of wheat germ protein in comminuted meat products as compared with corn germ and soy proteins, *J. Food Sci.*, *57*: 829.
68. Ziegler, G. R. and Acton, J. C. (1984). Mechanisms of gel formation by proteins of muscle tissue, *Food Technol*, *38(5)*: 77.
69. Saffle, R. L. (1966). Linear Programming – Meat Blending, IBM Publication No. E20-0161–0, White Plains, New York.
70. Lee, C. M., Carroll, R. J., and Abdollahi, A. (1981). A microscopical study of the structure of meat emulsions and its relationship to thermal stability, *J. Food Sci.*, *46*: 1789.

71. Jones, K. W. and Mandigo, R. W. (1982). Effect of chopping temperature on the microstructure of meat emulsions, *J. Food Sci.*, *47*: 1930.
72. Smith, D. M. (1988). Meat proteins: Functional properties in comminuted meat products, *Food Technol.*, *4*: 116.
73. Turgut, H. (1984). Emulsifying capacity and stability of goat, waterbuffalo, sheep and cattle muscle proteins, *J. Food Sci.*, *49*: 168.
74. Turgut, H. and Sink, J. D. (1983). Factors affecting the emulsifying capacity of bovine muscle and muscle protein, *J. Food Sci.*, *48*: 841.
75. Colmenero, F. J. and Borderias, A. J. (1983). A study of the effects of frozen storage on certain functional properties of meat and fish protein, *J. Food Technol.*, *18*: 731.
76. Zabielski, J., Kijowski, J., Fiszer, W., and Niewiarowicz, A. (1984). The effect of irradiation on technological properties and protein solubility of broiler chicken meat, *J. Sci. Food Agric.*, *35*: 662.
77. Shimada, A., Yamamoto, I., Sase, H., Yamazaki, Y., Watanabe, M., and Arai, S. (1984). Surface properties of enzymatically modified proteins in aqueous solution, *Agric. Biol. Chem.*, *48*: 2689.
78. Watanabc, M., Toyokawa, H., Shimada, A., and Arai, S. (1981). Proteinaceous surfactants produced from gelatin by enzymatic modification: Evaluation for their functionality, *J. Food Sci.*, *46*: 1467.
79. Sase, H., Watanabe, M., Arai, S., and Ogawa, Y. (1987). Functional and sensory properties of meat emulsions produced by using enzymatically modified gelatin, *J. Food Sci.*, *52*: 893.
80. Gillet, T. A., Meiburg, D. E., Brown, C. L., and Simon, S. (1977). Parameters affecting meat protein extraction and interpretation of model system data for meat emulsion formation, *J. Food Sci.*, *42*: 1606.
81. Gaska, M. T. and Regenstein, J. M. (1982). Timed emulsification studies with chicken breast muscle: soluble and insoluble myofibrillar protiens, *J. Food Sci.*, *47*: 1438.
82. Nakai, S., Li-Chan, E., and Hayakawa, S. (1986). Contribution of protein hydrophobicity to its functionality, *Die Nahrung*, *30*: 327.
83. Li-Chan, E., Nakai, S., and Wood, D. F. (1985). Relationship between functional (fat binding, emulsifying) and physicochemical properties of muscle proteins. Effects of heating, freezing, pH and species, *J. Food Sci.*, *50*: 1034.
84. Elgasim, E. A., Kennick, W. H., Anglemier, A. F., Elkhalifa, E. A., and Koohmaraie, M. (1982). Effects of prerigor pressurization on the emulsifying capacity of muscle protein, *J. Food Sci.*, *47*: 861.
85. Lee, C. (1985). Microstructure of meat emulsions in relation to fat stabilization, *Food Microstructure*, *4*: 63.
86. Wu, F. Y., Dutson, T. R., and Smith, S. B. (1985). A scanning electron microscopic study of heat-induced alterations in bovine connective tissue, *J. Food Sci.*, *50*: 1041.
87. Huber, D. G. and Regenstein, J. M. (1988). Emulsion stability studies of myosin and exhaustively washed muscle from adult chicken breast muscle, *J. Food Sci.*, *53*: 1282.
88. Kijowski, J. and Niewiarowicz, A. (1978). Emulsifying properties of proteins and meat from broiler breast muscles as affected by their initial pH values, *J. Food Technol.*, *13*: 451.
89. Ozimek, G., Jelen, P., Ozimek, L., Sauer, W., and McCurdy, S. M. (1986). A comparison of mechanically separated and alkali extracted chicken protein for functional and nutritional characteristics, *J. Food Sci.*, *51*: 748.
90. Smith, D. M. and Brekke, C. J. (1985). Enzymatic modification of the structure and functional properties of mechanically deboned fowl proteins, *J. Agric. Food Chem.*, *33*: 631.
91. Elkhalifa, E. A., Graham, P. P., Marriott, N. G., and Phelps, S. K. (1988). Color characteristics and functional properties of flaked turkey dark meat as influenced by washing treatments, *J. Food Sci.*, *53*: 1068.

92. Jimenez-Colmenero, F. and Garcia, M. E. (1981). Effect of washing on the properties of mechanically deboned meat, In *Proceedings of the 27th European Meeting of Meat Res. Workers* (O. Pranld, ed.), Vienna, p. 359.
93. Borderias, A. J., Jimenez-Colmenero, F. and Tejada, M. (1985). Viscosity and emulsifying ability of fish and chicken muscle protein, *J. Food Technol.*, *20*: 31.
94. Montero, P. and Borderias, J. (1991). Emulsifying capacity of collagenous material from the muscle and skin of hake (*Merluccius L.*) and trout (*Salmoirideus Gibb*): Effect of pH and NaCl concentration, *Food Chemistry*, *41*: 251.
95. Vidya Sagar Reddy, G. and Srikar, L. N. (1991). Preprocessing ice storage effects on functional properties of fish mince protein, *J. Food Sci.*, *56*: 965.
96. Groninger, H., Hawkes, J. W., and Babbitt, J. K. (1983). Functional and morphological changes in processed frozen fish muscle, *J. Food Sci.*, *48*: 1388.
97. Warrier, S. B. and Ninjoor, V. (1981). Fish protein concentrate (FPC) from Bombay duck isolated by radiation-heat combination procedure: Functional and nutritional properties, *J. Food Sci.*, *46*: 234.
98. Rao, B. R. and Henrickson, R. L. (1983). Food grade hide collagen in bologna effect on functional properties, texture and color, *J. Food Quality*, *6*: 1.
99. Caldironi, H. A. and Ockerman, H. W. (1982). Bone and plasma protein extracts in sausages, *J. Food Sci.*, *47*: 1622.
100. Nakamura, R., Hayakawa, S., Yasuda, K., and Sato, Y. (184). Emulsifying properties of bovine blood globin: A comparison with some proteins and their improvement, *J. Food Sci.*, *49*: 102.
101. Hayashi, T., Biagio, R., Saito, M., Todoriki, S., and Tajima, M. (1991). Effect of ionizing radiation on sterility and functional qualities of dehydrated blood plasma, *J. Food Sci.*, *56*: 168.
102. Mittal, G. S. and Usborne, W. R. (1985). Meat emulsion extenders, *Food Technol.*, *33*: 121.
103. Carroll, R. J. and Lee, C. M. (1981). Meat emulsions-fine structure relationships and stability, *Scanning Electron Microscopy*, *111*: 447,
104. Zayas, J. F. (1985). Structural and water binding properties of meat emulsions prepared with emulsified and unemulsified fat, *J. Food Sci.*, *50*: 689.
105. Visser, F. M. W. (1985). Milk proteins in meat products and soups or sauces, In *Milk Proteins 84* (T. E. Galesloot and B. J. Tinbergen, eds.), Proc. International Congress on Milk Proteins, Pudoc, Wageningen, the Netherlands, p. 206.
106. Schmidt, R. H. and Morris, H. A. (1984). Gelation properties of milk proteins, soy proteins, and blended protein systems, *Food Technol.*, *38(5)*: 85.
107. Hung, S. C. and Zayas, J. F. (1992). Functionality of milk proteins and corn germ protein flour in comminuted meat products, *J. Food Quality*, *15*: 139.
108. Holland, G. C. (1984). "A meat industry perspective on the use of dairy ingredients," Proc. Conf. Canadian Dairy Ingredients in the Food Industry, Ottawa.
109. Payne, N. N. and Rizvi, S. S. H. (1988). Rheological behavior of comminuted meat batters, *J. Food Sci.*, *53*: 70.
110. Peng, I. C., Dayton, W. R., Quass, D. W., and Allen, C. E. (1982). Investigations of soybean 11S protein and myosin interaction by solubility, turbidity and titration studies, *J. Food Sci.*, *47*: 1976.
111. Wang, C. R. and Zayas, J. F. (1992). Comparative study of corn germ and soy protein utilization in comminuted meat products, *J. Food Quality*, *15*: 156.
112. Kinsella, J. E. and Damodaran, S. (1980). Flavor problems in soy proteins, In *Analysis and Control of Less Desirable Flavors in Food and Beverage*, Academic Press, New York, p. 95.
113. Oliver, A., Hseieh, A., Hsuang, S., and Chang, S. (1981). Isolation and identification of objectionable volatile flavor compounds in defatted soybean flour, *J. Food Sci.*, *47*: 16.
114. Lecomte, N. B., Zayas, J. F., and Kastner, C. L. (1993). Soya proteins functional and sensory characteristics improved in comminuted meats. *J. Food Sci.*, *58*: 464.

115. Lin, C. S. and Zayas, J. F. (1987). Microstructural comparisons of meat emulsion prepared with corn protein emulsified and unemulsified fat, *J. Food Sci.*, *52*: 267.
116. Barbieri, R. and Casiraghi, E. M. (1983). Production of a food grade flour from defatted corn germ meal, *J. Food Technol.*, *18*: 35.
117. Lin, C. S. and Zayas, J. F. (1987). Protein solubility, emulsifying stability and capacity of two defatted corn germ proteins, *J. Food Sci.*, *52*: 1615.
118. Zayas, J. F. and Lin, C. S. (1989). Effect of pretreatment of corn germ protein on the quality characteristics of frankfurters, *J. Food Sci.*, *54*: 1452.
119. Gorbatov, V.M. and Zayas, J. F. (1973). "Technological role of animal fats in the production of minced meat products," Proceedings of XIX European Congress of Meat Research Workers, Paris, V. *3*: 785.
120. Theno, D. M. and Schmidt, G. R. (1978). Microstructural comparisons of three commercial frankfurters, *J. Food Sci.*, *43*: 845.
121. Zaitsev, V. I and Khomets, V. G. (1982). Nitrogen compounds of wheat and its products, *FSTA*, vol. 14, 11 M 1370.
122. Baldini, V. L. S., Iaderoza, M., and Draetta, I. S. (1982). Chemical and biochemical characterization of wheat germ and maize germ, *Coletanea do Instituto de Technolagia de Alimentos*, *12*: 1.
123. Turnbough, J. M. and Baldwin, R. E. (1986). Enhancing the nutritive value and appearance of microwave baked muffin, *Microwave World*, *7(4)*: 7.
124. Li Shaochen (1990). Experimental study on antiatherosclerosis effect of wheat germ, *Acta Academiae Midicinae Shandong*, 28(1) 18. *FSTA* vol. 22, 6 M 24.
125. Gnanasambandam, R. and Zayas, J. F. (1992). Functionality of wheat germ protein in comminuted meat products as compared with corn germ and soy proteins, *J. Food Sci.*, *57*: 829.
126. van Hoogstraten, J. J. (1987). The marketing of whey products: A view from Europe "J'ai deux amours," In *Bulletin 212, Trends in Whey Utilization*, Int. Dairy Fed., Brussels, Belgium, p. 17.
127. Dickinson, E. (1989). Surface and emulsifying properties of caseins, *J. Dairy Res.*, *56*: 471.
128. Kinsella, J. E. (1984). Milk proteins: Physicochemical and functional properties, *CRC Crit. Rev. Food Sci. Nutr.*, *20–21(3)*: 197.
129. Dickinson, E. and Woskett, C. M. (1988). Effect of alcohol on adsorption of casein at the oil-water interface, *Food Hydrocolloids*, *2*: 187.
130. Burgand, I. and Dickinson, E. (1990). Emulsifying effects of food macromolecules in presence of ethanol, *J. Food Sci.*, *55*: 875.
131. Lonergan, D. A. (1983). Isolation of casein by ultrafiltration and cryodestabilization, *J. Food Sci.*, *48*: 1817.
132. Moon, T. W., Peng, I. C. and Lonergan, D. A. (1989). Functional properties of cryocasein, *J. Dairy Sci.*, *72*: 815.
133. Kanno, C. (1989). Emulsifying properties of bovine milk fat globule membrane in milk fat emulsion: conditions for the reconstitution of milk fat globules, *J. Food Sci.*, *54*: 1534.
134. Swaisgood, H. E. (1982). Chemistry of milk protein, In *Developments in Dairy Chemistry*, v. 1 (P. F. Fox, ed.), Applied Sci. Publ., New York, p. 1.
135. Modler, H. W. (1985). Functional properties of nonfat dairy ingredients – a review. Modification of products containing casein, *J. Dairy Sci.*, *68*: 2195.
136. Shimizu, M., Takahashi, T., Kaminogawa, S., and Yamauchi, K. (1983). Adsorption onto an oil surface and emulsifying properties of bovine α_{S1}-casein in relation to its molecular structure, *J. Agric. Food Chem.*, *31*: 1214.
137. Shimizu, M., Lee, S. W., Kaminogawa, S., and Yamauchi, K. (1984). Emulsifying properties of an N-terminal peptide obtained from the peptic hydrolyzate of α_{S1}-casein, *J. Food Sci.*, *49*: 1117.

138. Ruegg, M. and Moor, U. (1984). Effect of calcium on the hydration of casein: I. Water vapor sorption and fine structure of calcium caseinates compared with sodium caseinates in the pH range 4.6–8.0, *J. Dairy Res.*, *51*: 103.
139. Eigel, W. N., Butler, J. E., Ernstrom, C. A., Farrell, H. A., Harwalker, V. R., Jennes, R., and Whitney, R. M. (1984). Nomenclature of proteins of cow's milk: fifth revision, *J. Dairy Sci.*, *67*: 1599.
140. Graham, D. E. and Phillips, M. C. (1979). Proteins at liquid interfaces. III. Molecular structures of adsorbed films, *J. Coll. Int. Sci.*, *70*: 427.
141. Reimerders, E., Lorenzen, P., Precht, D., and Mehrens, H. (1986). The determination of the size distribution of oil-water emulsions with regard to stability, *Lebensm.-Wiss. u.-Technol.*, *19*: 82.
142. Dickinson, E., Robson, E. W., and Stansby, G. (1983). Colloid stability of casein-coated polystyrene particles, *J. Chem. Soc. Faraday Trans.*, *12*: 2937.
143. Dickinson, E., Whyman, R. H., and Dalgleish, D. G. (1987). Colloidal properties of model oil-in-water food emulsions stabilized separately by αs_1-casein, β-casein and K-casein, In *Food Emulsions and Foams* (E. Dickinson, ed.), Royal Society of Chemistry Special Publication No. 58, London, p. 40.
144. Van Hekken, D.L. and Strange, E.D. (1993). Functional properties of dephosphorylated bovine whole casein. *J. Dairy Sci.*, *76*: 3384.
145. Haque, Z. and Kito, M. (1983). Lipophilization of α_{s1}-casein. 2. Conformational and functional effects, *J. Agric. Food Chem.*, *31*: 1231.
146. Dickinson, E., Pogson, D. J., Robson, E. W., and Stainsby, G. (1985). Colloids and surfaces, *14*: 135.
147. Foley, J. and O'Connell, C. (1990). Comparative emulsifying properties of sodium caseinate and whey protein isolate in 18% oil in aqueous systems, *J. Dairy Res.*, *57*: 377.
148. Lee, S. Y., Morr, C. V., and Ha, E. Y. W. (1992). Structural and functional properties of caseinate and whey protein isolate as affected by temperature and pH, *J. Food Sci.*, *57*: 1210.
149. Britten, M. and Giroux, H. J. (1991). Emulsifying properties of whey protein and casein composite blends, *J. Dairy Sci.*, *74*: 3318.
150. Britten, M. and Giroux, H. (1993). Interfacial properties of milk protein-stabilized emulsions as influenced by protein concentration. *J. Agric. Food Chem.* *41*: 1187.
151. Monhanty, B., Mulvihill, D. M., and Fox, P. F. (1988). Emulsifying and foaming properties of acidic caseins and sodium caseinate, *Food Chem.*, *28*: 17.
152. Liao, S. Y. and Mangino, M. E. (1987). Characterization of the composition, physicochemical and functional properties of acid whey protein concentrates, *J. Food Sci.*, *52*: 1033.
153. Modler, W. H., Poste, L. M., and Butler, G. (1985). Sensory evaluation of an all-dairy formulated cream-type cheese produced by a new method, *J. Dairy Sci.*, *68*: 2835.
154. Harper, W. J. (1984). Model food system approaches for evaluating whey protein functionality, *J. Dairy Sci.*, *67*: 2745.
155. Peltonen-Shalaby, R. and Mangino, M. E. (1986). Compositional factors that affect the emulsifying and foaming properties of whey protein concentrates, *J. Food Sci.*, *51*: 91.
156. Yamauchi, K., Schimizu, M., and Kamiya, T. (1980). Emulsifying properties of whey proteins, *J. Food Sci.*, *45*: 1237.
157. Patel, M. T. and Kilara, A. (1990). Studies on whey protein concentrates. 2. Foaming and emulsifying properties and their relationships with physicochemical properties, *J. Dairy Sci.*, *73*: 2731.
158. Modler, H. W. and Jones, J. D. (1987). Selected processes to improve the functionality of dairy ingredients, *Food Technol.*, *10*: 114.
159. Nakai, S. and Li Chan, E. (1985). Structure modification and functionality of proteins: quantitative structure-activity relationship approach, *J. Dairy Sci.*, *68*: 2763.

160. Mangino, M. E. (1987). Characterization of the composition, physicochemical and functional properties of acid whey protein concentrates, *J. Food Sci.*, *52*: 1033.
161. De Witt, J. N. (1989). Functional properties of whey proteins, In *Developments of Dairy Chemistry*, v. 4, p. 285.
162. Haque, Z. and Kinsella, J. E. (1988). Emulsifying properties of food proteins. II. Bovine serum albumin, *J. Food Sci.*, *53*: 416.
163. Brown, E. M., Carrol, R. J., Pfeiffer, P. E., and Sampugna, J. (1983). Complex formation in sonicated mixtures of β-lactoglobulin and phosphatidylcholine, *Lipids*, *18*: 111.
164. Castle, J., Dickinson, E., Murray, B. S., and Stainsby, G. (1987). In *Proteins at Interfaces: Physicochemical and Biochemical Studies* (J. L. Brash and T. A. Horbett, eds.), ACS Symp. Series No 343, Amer. Chem. Soc., Washington, D.C., p. 118.
165. Klemaszewski, J. L., Das, K. P., and Kinsella, J. E. (1992). Formation and coalescence stability of emulsions stabilized by different milk proteins, *J. Food Sci.*, *57*: 366.
166. Slack, A. W., Amundson, C. H., and Hill, Jr., C. G. (1986). Foaming and emulsifying characteristics of fractionated whey protein, *J. Food Proc. and Preserv.*, *10*: 81.
167. Nakai, S. (1983). Structure-function relationships of food proteins with an emphasis on the importance of protein hydrophobicity, *J. Agric. Food Chem.* , *31*: 676.
168. Akita, E. M. and Nakai, S. (1990). Lipophilization of β-lactoglobulin: Effect on hydrophobicity, conformation and surface functional properties, *J. Food Sci.*, *55*: 711.
169. Ma, C.Y., Paquet, A., and McKellar, R.C. (1993). Effect of fatty N-acylamino acids on some functional properties of two food proteins. *J. Agric. Food Chem.*, *41*: 1182.
170. Matarella, N. L. and Richardson, T. (1983). Physicochemical and functional properties of positively-charged derivatives of bovine β-lactoglobulin, *J. Agric. Food Chem.*, *31*: 972.
171. Leman, J. and Kinsella, J. E. (1989). Surface activity, film formation, and emulsifying properties of milk proteins, *Crit. Rev. Food Sci. Nutr.*, *28(2)*: 115.
172. Beuschel, B. C., Culbertson, J. D., Partridge, J. A., and Smith, D. M. (1992). Gelation and emulsification properties of partially insolubilized whey protein concentrates, *J. Food Sci.*, *57*: 605.
173. Turgeon, S. L., Gauthier, S. F., and Paquin, P. (1992). Emulsifying property of whey peptide fractions as a function of pH and ionic strength, *J. Food Sci.*, *57*: 601.
174. Waniska, R. D. and Kinsella, J. E. (1985). Properties of β-lactoglobulin: Adsorption and arrangement during film formation, *J. Agric. Food Chem.*, *33*: 1143.
175. Tornberg, E., Granfeldt, Y., and Hakansson, C. (1982). A comparison of the interfacial behavior of three food proteins adsorbed at air-water and oil-water interfaces, *J. Sci. Food Agric.*, *33*: 904.
176. Das, K. P. and Chattoraj, D. K. (1981). Nen-Newtonian viscosity of protein stabilized emulsion, *J. Coll. Interface Sci.*, *78*: 422.
177. Monahan, F.J., McClements, D.J., and Kinsella, J.E. (1993). Polymerization of whey proteins in whey-protein-stabilized emulsions. *J.Agric. Food Chem.*, *41*: 1826.
178. Shimizu, M., Kamiya, T., and Yamauchi, K. (1981). The adsorption of whey proteins on the surface of emulsified fat, *Agric. Biol. Chem.*, *45*: 2491.
179. McClements, D.J., Monahan, F.J.,and Kinsella, J.E. (1993). Disulfide bond formation affects stability of whey protein isolate emulsions. *J. Food Sci.*, *58*:1036.
180. Langley, K. R., Millard, D., and Evans, E. W. (1988). Emulsifying capacity of whey proteins produced by ion-exchange chromatography, *J. Dairy Res.*, *55*: 197.
181. Dumay, F. and Cheftel, I. C. (1986). The emulsifying properties of whey protein concentrates: Correlations between different methods of evaluation, *Sci. Aliment.*, *6*: 147.
182. Bech, A. M. (1981). The physical and chemical properties of whey proteins, *Dairy Industries International*, *46(11)*: 25.
183. de Witt, J. N. (1984). Functional properties of whey proteins in food systems, *Neth. Milk Dairy J.*, *38*: 71.

184. Turgeon, S. L., Gauthier, S. F., and Paquin, P. (1991). Interfacial and emulsifying properties of whey peptide fractions obtained with a two-step ultrafiltration process, *J. Agric. Food Chem.*, *39*: 673.
185. Hsu, K. H. and Fennema, O. (1989). Changes in the functionality of dry whey protein concentrate during storage, *J. Dairy Sci.*, *72*: 829.
186. Tomas, A. and Paquet, D. (1994). Effect of fat and protein contents on droplet size and surface protein coverage in dairy emulsions. *J. Dairy Sci.*, *77*: 413.
187. Mizutani, R. and Nakamura, R. (1984). Emulsifying properties of egg yolk low density lipoprotein (LDL): Comparison with bovine serum albumin and egg lecithin, *Lebensm.-Wiss. u.-Technol.*, *17*: 213.
188. Mizutani, R. and Nakamura, R. (1985). The contribution of polypeptide moiety on the emulsifying properties of egg yolk low density lipoprotein (LDL), *Lebensm.-Wiss. u.-Technol.*, *18*: 60.
189. Dyer-Hurdon, J.N. and Nnanna, I.A. (1993). Cholesterol content and functionality of plasma and granules fractionated from egg yolk. *J. Food Sci.*, *58*: 1277.
190. Chung, S. L. and Ferrier, L. K. (1991). Conditions affecting emulsifying properties of egg yolk phosvitin, *J. Food Sci.*, *56*: 1259.
191. Chung, S. L. and Ferrier, L. K. (1992). pH and sodium chloride effects on emulsifying properties of egg yolk phosvitin, *J. Food Sci.*, *57*: 40.
192. Tsutsui, T. (1988). Functional properties of heat-treated egg yolk low density lipoprotein, *J. Food Sci.*, *53*: 1103.
193. Larsen, J. E. and Froning, G. W. (1981). Extraction and processing of various components from egg yolk, *Poultry Sci.*, *60*: 160.
194. Cheftel, J. C., Cuq, J., and Lorient, D. (1985). Amino acids, peptides, and proteins, In *Food Chemistry*, 2nd edit. (O. R. Fennema, ed.), Marcel Dekker, New York, p. 301.
195. Kato, A., Murata, K., and Kobayashi, K. (1988). Preparation and characterization of ovalbumin-dextrin conjugate having excellent emulsifying properties, *J. Agric. Food Chem.*, *36*: 421.
196. Nakamura, R., Mizutani, R., Yano, M., and Hagakawa, S. (1988). Enhancement of emulsifying properties of protein by sonicating with egg yolk lecithin, *J. Agric. Food Chem.*, *36*: 729.
197. Kato, A., Osako, Y., Matsudomi, N., and Kobayashi, K. (1983). Changes in the emulsifying and foaming properties of proteins during heat denaturation, *Agric. Biol. Chem.*, *47*: 33.
198. Rir, I., Feldman, I., Aserin, A., and Garti, N. (1994). Surface properties and emulsification behavior of denatured soy proteins.*J. Food Sci.*, *59*: 606.
199. Flint, F. O. and Johnson, R. F. P. (1981). A study of film formation by soy protein isolate, *J. Food Sci.*, *42*: 1351.
200. Yao, J. J., Tanteeratarm, K., and Wei, L. S. (1990). Effects of maturation and storage on solubility, emulsion stability and gelation properties of isolated soy proteins, *J.A.O.C.S.*, *67(12)*: 974.
201. Aoki, H., Shirase, Y., Kato, J., and Watanabe, Y. (1984). Emulsion stabilizing properties of soy protein isolates mixed with sodium caseinate, *J. Food Sci.*, *49*: 212.
202. Lah, C. L. and Cheryan, M. (1980). Emulsifying properties of a full-fat soy protein product produced by ultrafiltration, *Lebensm.-Wiss. u.-Technol.*, *13*: 259.
203. Barardi, L. C. and Cherry, J. P. (1981). Functional properties of co-precipitated protein isolates from cottonseed, soybean and peanut flours, *Can. Inst. Food Sci. Technol. J.*, *14(4)*: 283.
204. Wang, C. R. and Zayas, J. F. (1992). Emulsifying capacity and emulsion stability of soy proteins compared with corn germ protein flour, *J. Food Sci.*, *57*: 726.
205. Dickinson, E. and Stainsby, G. (1987). Progress in the formulation of food emulsions and foams, *Food Technol.*, *41(9)*: 74.

206. Szuhaj, B. F. and Sipos, E. F. (1989). Food emulsifiers from the soybean: soy protein products, In *Food Emulsifiers: Chemistry, Technology, Functional Properties and Application* (G. Charalambous and G. Doxastakis, eds.), Elsevier Science Publishing Co., New York, p. 113.
207. Hirotsuka, M., Taniguchi, H., Narita, H., and Kito, M. (1984). Increase in emulsification activity of soy lecithin-soy protein complex by ethanol and heat treatments, *J. Food Sci.*, *49*: 1105.
208. Ashraf, H. L. (1986). Emulsifying properties of ethanol soaked soybean flour, *J. Food Sci.*, *51*: 193.
209. Arteaga, G. E., Li-Chan, E., and Nakai, S. (1992). A mixture design approach for studying emulsifying properties of soy protein concentrates, *Book of Abstracts*, IFT Annual Meeting, New Orleans, p. 157.
210. Haque, Z., Matoba, T., and Kito, M. (1982). Incorporation of fatty acid into food protein: Palmitoyl soybean glycinin, *J. Agric. Food Chem.*, *30*: 481.
211. Hamada, J. S. and Marshall, W. E (1989). Preparation and functional properties of enzymatically deamidated soy proteins, *J. Food Sci.*, *54*: 598.
212. Huang, Y. T. and Kinsella, J. E. (1987). Effects of phosphorylation on emulsifying and foaming properties and digestibiity of yeast protein, *J. Food Sci.*, *52*: 1684.
213. Swanson, B. G. (1990). Pea and lentil protein extraction and functionality, *J.A.O.C.S.*, *67(5)*: 276.
214. Koyoro, H. and Powers, J. R. (1987). Functional properties of pea globulin fractions, *Cereal Chem.*, *64(2)*: 97.
215. Paredes-Lopez, O., Ordorica-Falomir, C., and Olivares-Vazquez, M. R. (1991). Chickpea protein isolates: Physicochemical, functional and nutritional characterization, *J. Food Sci.*, *56*: 726.
216. Schaffner, D. W. and Beuchat, L. R. (1986). Functional properties of freeze-dried powders of unfermented and fermented aqueous extracts of legume seeds, *J. Food Sci.*, *51*: 629.
217. Johnson, E. A. and Brekke, C. J. (1983). Functional properties of acylated pea protein isolates, *J. Food Sci.*, *48*: 722.
218. King, J., Aguirre, C., and de Pablo, S. (1985). Functional properties of lupin protein isolates (*Lupinus albus cv Multolupa*), *J. Food Sci.*, *50*: 82.
219. Han, J. Y. and Khan, K. (1990). Functional properties of pin-milled and air-classified dry edible bean fractions, *Cereal Chem.*, *67(4)*: 390.
220. Kohnhorst, A. L., Smith, D. M., Uebersax, M. A., and Bennink, M. R. (1990). Compositional, nutritional and functional properties of meals, flours, and concentrates from navy and kidney beans (*Phaseolus vulgaris*), *J. Food Quality*, *13*: 435.
221. Gujska, E. and Khan, K. (1991). Functional properties of extrudates from high starch fractions of navy and pinto beans and corn meal blended with legume high protein fractions, *J. Food Sci.*, *56*: 431.
222. Sosulski, F. W. and McCurdy, A. R. (1987). Functionality of flours, protein fractions and isolates from field peas and faba bean, *J. Food Sci.*, *52*: 1010.
223. Narayana, K. and Narasinga Rao, M. S. (1984). Effect of acetylation and succinylation on the functional properties of winged bean flour, *J. Food Sci.*, *49*: 547.
224. Zayas, J. F. and Lin, C. S. (1989). Emulsifying properties of corn germ protein, *Cereal Chem.*, *4*: 263.
225. Prinyawiwatkul, W., Beuchat, L.R., and McWatters, K.H. (1993). Functional property changes in partially defatted peanut flour caused by fungal fermentation and heat treatment. *J. Food Sci.*, *58*: 1319.
226. Canella, M., Castriotta, G., and Bernardi, A. (1979). Functional and physicochemical properties of succinylated and acetylated sunflower protein, *Lebensm.Wiss. u.-Technol.*, *12*: 95.
227. Kabirullah, M. and Wills, R. B. H. (1982). Functional properties of acetylated and succinylated sunflower protein isolate, *J. Food Technol.*, *17*: 235.

228. Bera, M. B. and Mukherjee, R. K. (1989). Solubility, emulsifying, and foaming properties of rice bran protein concentrate, *J. Food Sci.*, *54*: 142.
229. Thompson, L. U., Liu, R. F. K., and Jones, J. D. (1982). Functional properties and food applications of rapeseed protein concentrate, *J. Food Sci.*, *47*: 1175.
230. Dev, D. K. and Mukherjee, K. D. (1986). Functional properties of rapeseed protein products with varying phytic acid contents, *J. Agric. Food Chem.*, *34*: 775.
231. Paulson, A. T. and Tung, M. A. (1988). Emulsification properties of succinylated canola protein isolate, *J. Food Sci.*, *53*; 817.
232. Rahma, E. H. and Narasinga Rao, M. S. (1984). Gossypol removal and functional properties of protein produced by extraction of glanded cottonseed with different solvents, *J. Food Sci.*, *49*: 1057.
233. Choi, Y. R., Lusas, E. W., and Rhee, K. C. (1983). Molecular structure and functionalities of protein isolates prepared from defatted cottonseed flour succinylated at various levels, *J. Food Sci.*, *48*: 1275.
234. Tasneem, R. and Subramanian, N. (1986). Functional properties of guar (*Cyamopsis tetragonoloba*) meal protein isolates, *J. Agric. Food Chem.*, *34*: 850.
235. Nath, J. P. and Narasinga Rao, M. S. (1981). Functional properties of guar proteins, *J. Food Sci.*, *46*: 1255.
236. Tasneem, R., Ramamani, S., and Subramanian, N. (1982). Functional properties of guar seed meal detoxified by different methods, *J. Food Sci.*, *47*: 1323.
237. Kinsella, J. E. and Mohite, R. R. (1982). Characteristics and properties of sesame proteins, *ACS Symp. Ser.*, *206*: 327
238. De Padua, M. R. (1983). Some functional and utilization characteristics of sesame flour and proteins, *J. Food Sci.*, *48*: 1145.
239. Paredes-Lopez, O. and Ordorica-Falomir, C. (1986). Functional properties of safflower protein isolates: water absorption, whipping and emulsifying characteristics, *J. Sci. Food Agric.*, *37*: 1104.
240. Dev. D. K. and Quensel, E. (1988). Preparation and functional properties of linseed protein products containing differing levels of mucilage, *J. Food Sci.*, *53*: 1834.
241. Madhusudhan, K. T. and Singh, N. (1985). Effect of heat treatment on the functional properties of linseed meal, *J. Food Sci.*, *33*: 1222.
242. Ma, C. Y. (1983). Chemical characterization and functionality assessment of protein concentrates from oats, *Cereal Chem.*, *60(1)*: 36.
243. Chang, P. R. and Sosulski, F. W. (1985). Functional properties of dry milled fractions from wild oats (*Avenua fatua L.*), *J. Food Sci.*, *50*: 1143.
244. Ma, C. Y. and Khanzada, G. (1987). Functional properties of deamidated oat protein isolates, *J. Food Sci.*, *52*: 1583.
245. Ma, C. Y. (1984). Functional properties of acylated oat protein, *J. Food Sci.*, *49*: 1128.
246. Bolnedi, V., and Zayas, J.F. Emulsifying properties and oil absorption of wheat germ protein flour. 1994 *Annual IFT Meeting, Book of abstracts, Atlanta*, p. 15.
247. Oomah, B. D. and Mathieu, J. J. (1987). Functional properties of commercially produced wheat flour solubles, *Can Inst. Food Sci. Technol. J.*, *20(2)*: 81.
248. Jackman, R. L. and Yada, R. Y. (1988). Functional properties of whey-potato protein composite blends in a model system, *J. Food Sci.*, *53*: 1427.
249. Wojnowska, I., Poznanski, S., and Bednarski, W. (1981). Processing of potato protein concentrates and their properties, *J. Food Sci.*, *47*: 167.
250. Knuckles, B. E. and Kohler, G. O. (1982). Functional properties of edible protein concentrates from alfalfa, *J. Agric. Food Chem.*, *30(4)*: 748.
251. Fiorentini, R. and Galoppini, C. (1981). Pilot plant production of an edible alfalfa protein concentrate, *J. Food Sci.*, *46*: 1514.
252. Huang, Y. T. and Kinsella, J. E. (1987). Effects of phosphorylation on emulsifying and foaming properties and digestibility of yeast protein, *J. Food Sci.*, *52*: 1684.
253. Okezie, B. O. and Kosikowski, F. V. (1981). Extractability and functionality of protein from yeast cells grown on cassava hydrolysate, *Food Chem.*, *6*: 7.

254. Akobundu, E. N. T., Cherry, J. P., and Simmons, J. G. (1982). Chemical, functional, and nutritional properties of egusi (*Colocynthis citrullus L.*) seed protein products, *J. Food Sci.*, *47*: 829.
255. Idouraine, A., Yensen, S. B., and Weber, C. W. (1991). Tepary bean flour, albumin and globulin fractions functional properties compared with soy protein, *J. Food Sci.*, *56*: 1316.

Chapter 4

Oil and Fat Binding Properties of Proteins

4.1 Introduction

The importance of the functional property of fat binding depends on the type of food,e.g., fluid emulsions, powders, dairy foods, sausage products, dough and bread. Binding of fat with food components, in particular proteins and carbohydrates, influences the textural and other food quality properties. The ability of proteins to absorb and retain fat and to interact with lipids in emulsions and other food systems is important in food formulations. Measurements of fat binding may provide data for selecting a protein raw material. Many properties of foods involve the interactions of proteins and lipids: formation of emulsions, fat emulsification in meats, fat entrapment in sausage batters, flavor absorption, and dough preparation. Fat absorption of proteins is affected by protein source, processing conditions, composition of additive, particle size and temperature. Proteins of plant origin, for example sunflower proteins, contain numerous nonpolar side chains that bind hydrocarbon chains, thereby contributing to increased oil absorption. Lipid absorption and binding have been studied much less than water absorption and binding. Although the mechanism of fat absorption or fat binding has not been completely explained, fat absorption is mainly attributed to the physical entrapment of oil or fat by proteins.

Fat binding is closely related to the protein content of the component, and this property is primarily the physical entrapment of oil due to chemical modifications of proteins that increase bulk density. In protein powder foods fat binding is influenced by the size of powder particles. Protein powders with a low-density and a small particle size adsorb and entrap more oil than high-density protein powders do. Kinsella [1] explained the mechanism of fat absorption as a phenomenon of physical oil entrapment. The mechanism of fat binding can be explained with knowledge of the protein material microstructure. Later Kinsella [2] and Sathe et al. [3] suggested that oil absorption is determined by fat binding by the nonpolar side chains of proteins. Nonpolar side chains of protein molecules are the primary sites of lipid-protein interactions. Oil absorption of plant protein blends is affected by protein concentration, the number of nonpolar sites and protein-lipid-carbohydrate interactions.

Insoluble and hydrophobic proteins have a high oil binding capacity. The fat binding capacity of proteins has been correlated with protein surface hydrophobicity (So) and with So×S where S = protein solubility [4]. Tightly bound oil absorbed to protein surfaces or to hydrophobic sites in proteins may be difficult to remove by solvent extraction. Usually, protein products with a low nitrogen solubility have a high fat binding capacity.

Formation of the protein-lipid complexes is involved in the process of emulsion stabilization. Interaction of serum proteins with fat globule membranes during homogenization of milk affects the properties of milk and dairy foods. During milk homogenization only caseins constituted the newly formed surface of fat globules [5]. The serum proteins, β-lactoglobulin and α-lactalbumin adsorb to the fat globule membranes only when the fat globules have been heated above 70 °C. During heating serum proteins are attached to cysteine-containing caseins on fat globule membranes. Fat binding of proteins is not caused by emulsification alone and should not be confused with their emulsifying properties when the concentration of protein, pH, temperature, ionic strength are more stringent factors with a curvilinear relationship between the protein concentration and the emulsifying capacity.

Complexes of proteins and lipids as natural components of foods, or formed during food processing, storage, and transportation influence the physical and chemical properties of foods. The presence of lipids in cell membranes and organelles show that protein-lipid complexes occur in biological systems. The first pure lipoprotein complex was isolated from horse serum. Lipoprotein compounds occurring widely in plant tissues have been isolated from plant tissues and isolated fractions had lipid contents and buoyant densities close to low density lipoproteins (LDL) (~ 75% lipid) and high density lipoproteins (HDL) (~ 50% lipid) of human blood serum [6]. It was found that at 2 °C phospholipid vesicles consisting of 155 molecules interacted with phospholipase A_2, i.e. the protein catalyzed the hydrolysis of phospholipids and formed a lipid-protein complex [7]. This complex contained 80 molecules of phospholipid and 2 molecules of the enzyme.

Oxidation of lipids and interactions between oxidized lipids and proteins are known to occur in vivo in aging tissues and in various foods, such as oilseeds, frozen meat, frozen and dehydrated fish, etc. This interaction influences the functionality of proteins in food systems. Lipid-protein interaction is accelerated by the presence of oxidized lipids in food systems, such as in frozen fish, fish meal, meat, and in oilseeds. Proteins may interact with oxidized lipids and they may act as a trap for lipid peroxides and secondary products of lipid oxidation. Products of fat oxidation, hydroperoxides and their breakdown products (i.e. ketones, aldehydes, alcohols, etc.) interact with the terminal functional groups of amino acids in proteins and enzymes, i.e. -SH, -OH, -COOH, and $-NH_2$ This interaction influences the functional and nutritional properties of proteins and flavor. The aldehydes from oxidized lipids interacted primarily with amino groups in carbonylamine reactions.

The capacity of proteins to retain lipids is affected by protein-lipid interactions and the spatial arrangement of the lipid phase which is determined by lipid-lipid interactions. Bonds involved in protein-lipid interactions are hydrophobic, electrostatic, hydrogen and non-covalent [2]. Protein-lipid complexes include hydrogen bonds as specific bonds between the lipid and protein moieties of a complex. Hydrophobic bonds are important in stabilizing protein-lipid complexes. Proteins have hydrophobic regions differing in size and number. The interaction between proteins and fatty acid anions alter protein structure by decreasing intramolecular hydrophobic bonds. Hydrophobic interactions influence protein aggregation. It is rather difficult to see how London-van der Waals forces between non-polar amino acid residues in protein chains and lipid chains could contribute significantly to the stability of protein-lipid complexes.

Damage to protein functionality and nutritional value has been studied in various model systems under different conditions of temperature, relative humidity, presence of antioxidants, oxygen, etc. Protein-lipid interactions markedly influence functional properties of proteins in food systems. Oxidized lipid-protein interactions cause denaturation of proteins and as a result a decrease in solubility and salt-extractability, a decrease in water holding capacity,and a hardening of texture.

Interactions with primary and secondary products of lipid oxidation lead to the polymerization of proteins with a significant increase in the molecular weight of proteins. At the same time, the high stability of unsaturated lipids in freeze-dried protein concentrates was related by Hudson and Karis [8] to the protective effect of their structural complexes with proteins. The damage of such amino acids as lysine, methionine and tryptophan as a result of interactions with oxidized lipids depends on processing conditions, temperature and time of treatment, and storage, water activity and oxygen tension. When casein or ovalbumin reacted with oxidizing ethyl linoleate at 50–6 °C in an aqueous medium or in the dry state, a significant decrease in nutritive value was found. Szebiotko et al. [9] found maximum lipid-protein interaction and protein degradation and losses of tryptophan at the maximum rate of lipid oxidation and hydroperoxide formation. Losses in available lysine appeared to take place in the initial induction period.

Losses of lysine and methionine nutritionally are most important. Lysine is the most sensitive amino acid, and several authors have shown extensive losses of lysine after lipid oxidation reactions. The ε-amino group of lysine reacts with aldehydes to form Schiff's bases. Mechanism of lysino-alanine (LAL) formation was studied by Finley et al. [10] using proton NMR to quantify the reaction products. The oxidation of cysteine residues in glutathione by linoleic acid hydroperoxide was observed. Four oxidation products were detected of which cystine monoxide and dioxide were the major products with smaller quantities of cysteine sulphinic acid and cysteic acid. They undergo β-elimination yielding dehydroalanine, the intermediate of lysino-alanine formation. Methionine is relatively easily oxidized to methionine sulphoxide during lipid oxidation.

4.2 Fat Binding Properties of Proteins Animal Origin

4.2.1 Muscle Proteins

In the animal products processing, the fat binding capacity (FBC) is important in comminuted meats and for applications such as meat protein substitutes and extenders. The FBC is critical in improving flavor carry-over in comminuted meats and simulated foods. The property of proteins to bind fat is of special importance in muscle foods because it improves flavor retention and mouthfeel. An increase in fat absorption and binding, and a decrease in cooking losses and maintainence of dimensional stability is important for the quality of coarse ground meats. The high capacity of animal and plant proteins to absorb oil was found by De Kanterewicz [11] using the Baumann apparatus. The equilibrium of oil absorption was obtained after 3 to 12 min of incubation (Fig. 4.1) [11]. The oil absorption capacity was expressed as the volume of oil/g protein at equilibrium. Li-Chan et al. [12, 13] reported the relationship between structure and functionality of meat proteins and demonstrated the importance of hydrophobicity, solubility, and SH group content of salt-extractable meat and fish proteins. They related these properties to fat binding and emulsifying properties. The fat binding capacity of protein extracts from various species of animals and poultry was affected by soluble protein concentration rather than by the source. Heat treatment of glass beads coated with bovine serum albumin (BSA) imparted a high-oil-binding capacity (3.4–3.6 ml oil/g BSA) to BSA samples [14]. The original nonheated BSA had no oil-binding capacity and was easily dissolved in water. The fat binding of blood globin was 2.0 g soybean oil/g protein product, that is slightly lower than that obtained for oilseed proteins. An increase in oil-binding ability was found for other proteins, such as ovalbumin, casein, gluten, and soybean protein.

The fat binding capacity of rockfish protein was positively correlated to surface hydrophobicity and negatively correlated to dispersibility. Rockfish proteins exhibited high FBC when prepared with higher hydrophobicity and lower dispersibility. The FBC of various proteins was differently affected by heating [15]. A positive effect of heating on FBC was found for BSA, β-lactoglobulin, ovalbumin, whey protein, and casein and a negative effect was found for soy and pea proteins. High protein solubility had a negative effect on the FBC. The low FBC of soluble proteins was related to the absence of available binding sites in protein molecules for interaction with oil or to the limited access of hydrocarbon chains to the binding sites of proteins.

Model studies with pure fish actin showed that increasing the ionic strength of the buffer solution caused polymerization of the soluble actin (G-actin) with its precipitation [16]. This process was accompanied by an increase in the ability of actin to bind neutral and polar lipids. Lipid-protein complexes was formed as a result of this interaction. A more significant increase in the binding of neutral lipids was found when the ionic strength increased to more than 0.1 M. Higher concentrations of salt induce changes in the actin molecule pertaining to a more hydro-

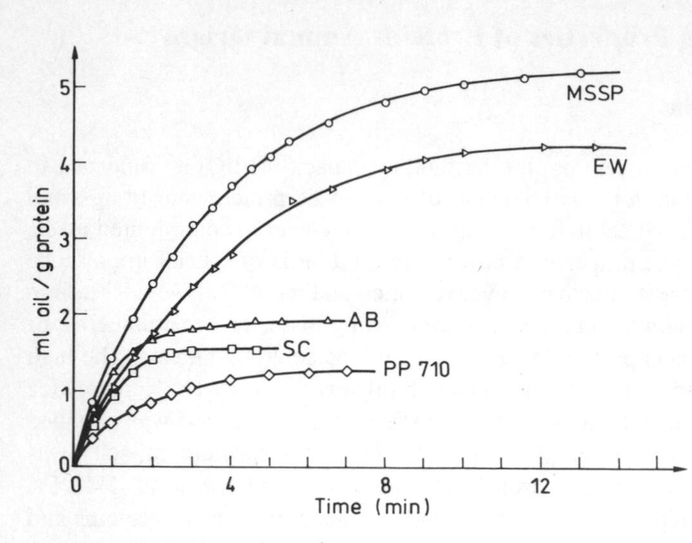

Fig. 4.1. Spontaneous oil uptake by different proteins. *AB*: bovine albumin; *EW*: egg white powder; *MSSP*: meat salt-soluble protein; *PP710*: Purina protein 710; *SC*: sodium caseinate. From Ref. 11

phobic nature. Consequently, these hydrophobic interactions between fish actin and neutral lipids are stronger than the binding forces between fish actin and polar lipids [16]. Shenouda and Pigott [17] reported studies with fish actin showing that the polymerization of actin activated by divalent cations gives a different binding pattern with lipids than that given in the absence of such cations. They found that excess of Ca^{2+} or Mg^{2+} cations increased the exposure of hydrophobic groups on the molecules of F-actin. As a result of that, binding of neutral lipids increased. Increase in Ca^{2+} or Mg^{2+} cations changes properties of actin which is less soluble in aqueous media. Experiments carried out in model conditions with fish actin showed the negative effect of increased salt concentration.

Fat Binding in Comminuted Meat Products

Meat proteins contribute to sausage batter stability and the textural properties of finished sausage products. The basic function of meat proteins in comminuted meat products (CMP) is binding and retaining the fat and water contained in the raw material and that added during blending or comminution. Fat binding is a result of encapsulation, structure formation, gelling, and partial emulsification. The stability of sausage batters and the binding of fat particles is determined by the ability of the proteins, mainly myofibrillar proteins, to form a three dimensional protein matrix with small capillaries. In this protein matrix particles of fat and water are dispersed, and as a result, a multiphase system is formed. The properties of proteins such as solubility, fat and water binding, swelling, emulsifying capacity and emulsion stability, and gelling characteristics are of major importance for the stability of sausage batter and formation of a stable heat-set protein matrix. These functional properties of meat proteins in CMP are influenced by factors such as the species, age, and sex of animals, pre-slaughter treatment of animals and postmortem treatment of the meat, the pH of the meat and the effect of added salts and protein extenders. The consistently high fat binding capacity of the plant and animal proteins added as extenders and replacers of meat proteins may affect the structural properties of the batter and binding of the fat in the finished meat products.

The objective of sausage processing is to create a specific structure of sausage batter with required characteristics. Sausage batter is a two-phase system, consisting of a matrix of salt-soluble proteins with suspended fat particles and water. During the comminution of meat the muscle fibers are separated, and disrupted into small segments. The sarcolemma is disintegrated, and the myofibrils and filaments are released. Some insoluble proteins, muscular and connective tissue particles are dispersed in a liquid water phase. In sausage batter, fat particles are enveloped by solubilized myofibrillar and sarcoplasmic proteins. However, at the fat/water interface, the myofibrillar proteins are preferentially absorbed.

Myosin which is most readily absorbed at the interface possesses a specific function. Myosin appears to be more surface-active than actomyosin or actin and readily migrates to the interface. Free myosin forms a monolayer at the lipid-water interface in raw sausage batter systems [18]. Myosin is the most important stabilizer in meat emulsions. It has been suggested that light meromyosin is attracted towards the aqueous phase in meat emulsions. However the hydrophobic region of myosin is located in the heavy meromyosin fragment and specifically in the globular head region. Hydrophobicity is generally regarded as having a significant correlation with interfacial tension and the emulsifying index. For this reason, hydrophobic proteins orient more readily at the interface and participate in fat binding. Deng et al. [19] related the phenomenon of muscle protein-protein (actomyosin) interactions in meat batters to the stabilizing effect on fat and water.

In sausage batters, fat droplets and fatty tissue particles are entrapped in the three dimensional network of the gel formed by actin, myosin, actomyosin, and other proteins. This system is stabilized by dissolved proteins surrounding fat particles as a film of protein that prevents fat coalescence during heat treatment. Encapsulation of the fat particles as a result of the formation of a surrounding matrix was demonstrated by Jones and Mandigo [20]. Sausage batter stability is affected by the strong protein layer surrounding the fat globules. The interfacial film of salt soluble proteins is a coating surface of the fat droplets. Thickening of the protein film releases melted fat in the form of fat pockets. The sarcoplasmic proteins form a weak gel and do not contribute significantly to the stability of sausage batter and the finished product.

The procedure of sausage batter manufacturing begins with the lean meats, those highest in myofibrillar proteins. During overchopping, the protein interface is broken by further comminution. Since this once used protein cannot participate in the interface formation again, the fat particles agglomerate and the system separates into a fat and water phase. If the interface formed is disrupted during the manufacturing process, the new proteins (of animal or plant origin) must be added to reform the interface. In the traditional process of sausage batter manufacturing, salt was added to lean meat at the optimal level for protein solubilization, and, after protein solubilization had been reached, fat meats were added to the formulation. In sausage batters, meat proteins must be solubilized and dispersed to function effectively. The important role of salt in sausage formulations is the solubilization of myofibrillar proteins. In the soluble form these proteins serve as stabilizers of the emulsion; additionally they envelope the fat particles. Proteins with good functionality in meat systems increase their solubility with increased temperature, showed a good gelling and water holding capability.

During sausage batter comminution, as soon as the cell wall is broken, the fat exudes and forms small droplets. Fat losses increase with more cell damage during comminution. For this reason, the time and degree of comminution influences fat losses. Various binders and extenders of animal and plant origin are added to sausage formulations for one or more of the following reasons: to increase fat and water binding; to increase the yields of finished products; to improve the stability of the sausage emulsion; to reduce production costs; and to improve sensory (flavor) characteristics. Gelatin is used as a binder for some loaf type meat products. The solubilization of salt-soluble myofibrillar proteins can be diminished if extenders and binders absorb water from the mixture prematurely. Milk proteins are well-accepted by consumers as meat additives; they improve the flavor characteristics of the products. The specific properties which sodium caseinate possesses are that it is preferentially absorbed at the fat/water interface in a meat emulsion and significantly improves fat and water binding characteristics. Sodium caseinate is an effective stabilizer of the emulsion in sausage batter.

The main functional property of milk proteins in CMP is fat and water binding and they are most effective in CMP with a high fat content. A high fat binding capacity is obtained if milk proteins are utilized as a stabilizer of preemulsified fat. Preparation of preemulsified fat includes blending of the mixture with a protein:fatty tissue:water ratio from 1:5:5 to 1:8:8 depending on the emulsifying capacity of the milk proteins [21]. The preemulsified fat after holding ~ 12–24 h is added to the formulation.

The quality of sausage batters is determined by fat and water binding during comminution and through heat processing. If sausage batters are manufactured incorrectly, instability of the sausage batters results in fat separation, aggregation of the matrix itself, and the separation of water. A frequent problem faced by sausage manufacturers is the formation of gelatin and fat pockets on the outside or in the interior of comminuted meats. The formation of gel pockets is a common problem in the production of liver sausages with a low level of salt-soluble myofibrillar proteins. To eliminate gel pocket formation, the collagen content in a formula should be limited to a maximum of 25% of the total protein. A certain amount of collagen, however, may even be desirable for formation of proper textural properties, especially in liver sausages. Sausage batter must retain an elastic texture through several cycles of solid-liquid fat transitions during cooking, chilling, storage, freezing, and reheating in the home.

Factors Influencing Fat Binding in Comminuted Meats

The stability of meat batters is affected by comminution temperature, and optimal stability of sausage batters is obtained at a final comminution temperature between 12 and 16 °C. With increased comminution time and a temperature up to 16 °C, the number of fat particles evenly dispersed in the protein matrix increa-ses. However, sausage batters with a temperature after comminution of 18–20 °C or more are unstable, and this is related to protein conformational changes and to the effect on the fat melting characteristics. The fat binding capacity of the protein component in sausage batters is sufficiently high if the comminution temperature has not reached 18 °C, which coincides with the start of the second melting range (higher melting point fraction) of the fat. High chopping temperature may lead to the formation of very small fat particles. In this case higher concentrations of proteins are necessary for stabilizing fat particles in the system. A low protein/fat ratio in sausage batter can cause destabilization of the dispersed phase. Thermal stability of sausage batters and binding of fat and water during comminution is related to the phase transition of fat in the temperature range of 14–30 °C.

The effect of time of comminution and batter temperature on fat binding in sausage batter has been reported [22]. The binding capacity of fat and water increased with increasing temperature to 10–21 °C after which it decreased. Satisfactory fat binding was observed after 6 min of comminution and a batter temperature of 10 °C, and remained acceptable with up to 14 min of comminution and temperature 23–24 °C. After that point, a steep increase in the amount of fat and moisture released was observed. The liquefaction of fat decreased rather then increased the fat binding capacity. The fat component which makes up to ~25% of sausage batter should play an active role in stabilizing the proteins/water/fat mixture. Fat and water binding during comminution are interrelated [22]. In the meat system it appears that there is a relationship between fat and water binding. The fat is released when water is released and the ordered structure disintegrates. When the batter temperature was lowered by the addition of dry ice and subsequent comminution for 5–6 min, the fat and water binding was gradually regained. No direct correlation was found between sausage batter temperature and fat and water binding in this study.

During chopping, if batter temperature was in the range 18–30 °C, pork fat was liquified and caused the interfacial protein film to assume an irregular, nonuniform shape [19]. As a result of the broken protein film, extensive fat separation was observed. The difference in oil separation between beef and mackerel batter systems was related to variation of available protein during the comminution process. There was no fat loss at pH 6.3 in mackerel batter as a result of low protein-protein interaction and an excess of protein to bind fat in the meat batter during chopping. In meat batters containing 2.1% NaCl chopped within a temperature range 6 to 8 °C, no fat and oil separation was observed [19].

The fat binding capacity in sausage batters remained constant until temperatures exceeded 20 °C in a beef-pork fat system and 28 °C in a beef-oil system [19]. Chilling of the sausage batter and rechopping recovered the fat binding capacity, but fat binding diminished when temperature exceeded 12 °C for the beef-pork fat system and 18 °C for the beef-oil system. Protein-protein interactions affected fat binding by decreasing the number of exposed hydrophobic moieties within the protein molecule that could interact with fat. A decrease in fat binding was found at a later stage of chopping than at the stage when a decrease in water binding was observed. The loss of binding capacity was partially reversible and cooling the batters to 0 °C by addition of dry ice, and rechopping partially recovered the fat and water binding capacity. A decrease in oil droplet size with prolonged chopping was reported by Deng et al. [19] and after 14 min chopping, the droplets had a very good protection from interfacial protein film. Low oil separation was found in the system. In overchopped batters, the protein film was disrupted and oil separation was observed.

Sausage batter deaeration can improve and stabilize binding of the fat component, i.e. decrease fat disintegration and formation of fat pockets. An increase in the fat binding by deaeration is obtained as a result of more proteins being available to cover the fat particles. The gel of protein is more compact and during heating is more resistant to stretching and shrinking. The higher level of fat tended to decrease the meat emulsion stability. Higher levels of fat probably caused greater fat and gel-water release during heating due to insufficient fat binding by the protein-fat-water network. More stable sausage batters have been prepared with a protein content of 12% than with 11%.

During heat treatment of sausage batters, the protein heat-set matrix formed from muscle proteins with a system of capillaries holds fat and water added during comminution. The system is stabilized during cooking as a result of protein denaturation and formation of a firm gel. A smokehouse cooking process can reach the endpoint temperature of 70–72 °C. During heating, a cohesive network structure is developed as a result of a rearrangement of the dissolved myofibrillar proteins. These proteins interact with insoluble meat proteins on the meat surface. An evenly distributed fat component in the protein structure prevents its shrinking when heat-denatured during heating. After comminution and before heating, a certain number of fat globules are surrounded by a distinct membrane. This protein membrane is disrupted after heating, and the previously dense protein phase is disrupted due to coagulationand forms dense irregular zones. Jones and Mandigo [20] suggested that fat expansion during heat treatment could cause disruption of the surrounding interfacial layer. During heating, structural changes occur because of protein denaturation and this will affect the protein matrix. For fat stability in sausage batters, the properties of the protein network entrapping the fat particles are of great importance.

The level of fat released is affected by the content and properties of connective tissue around the fat cells. The softness of fat is of less importance for fat loss during cooking. Generally, soft fats have a higher connective tissue contents and thicker cell walls than hard fats. Cell walls of hard fats are more easily broken down than those of softer fats causing higher losses of fat.

Methods for measuring fat binding are crude and only measure the released fat in the final stage and give no information about coalescence within the product. An important characteristic of fat binding is the loss of fat during cooking. For determining loss of fat due to heating, the temperature of the sample has to be kept above the melting temperature of the fat during centrifugation. The amount of fat released within the product was determined by extraction with hexane to determine the heat stability of the meat batters [23]. The correlation between the amount of extracted fat from the batter and the amount of fat separated during heating was established. Microscopic studies showed that fat remained after extraction was predominantly enclosed in intact fat cells. The heat stability of meat batters during processing was due to the integrity of fat cells rather than emulsified fat. However, microscopic studies of raw batters before heat treatment were not performed, and it was not established whether fat released from fat cells during comminution was in the form of fat conglomerates or in the form of highly dispersed droplets. Fat particle distribution in comminuted meats was studied qualitatively by microscopy and quantitatively by image analysis. Image analysis techniques made it possible to determine the perimeters and areas of dispersed fat particles, and fat particle size distribution in different meat products or depending on the various treatments. The effect of different treatments on fat component dispersion and stability may be evaluated by the area of fat particles.

4.2.2 Soy Proteins in Comminuted Meats

In comminuted meats, soy proteins added as extenders and replacers of meat proteins have been shown to improve binding of the structure and decrease fat and moisture losses during processing, particularly heat treatment. Soy proteins have been added to promote fat absorption or binding and maintain stability during processing. The stability of the sausage batters containing soy proteins is influenced by fat and water binding capacity. The fat and water binding ability of the batters is crucial to the production of comminuted meat products. Soy proteins are good binders functioning much like meat proteins in sausage batters. The functionality of soy proteins in sausage batters is to promote fat absorption or binding and decrease losses during heating. The gelling properties of soy proteins might accelerate fat binding.

Soy proteins have been incorporated in the formulation of frankfurters as a powder and preemulsified fat (PEF) [24]. Soy isolate emulsified (SIe) displayed the least liquid and fat separation, followed by soy concentrate emulsified (SCe) (Fig. 4.2). Other researchers have also reported that soy isolate had superior emulsion stabilizing properties over soy concentrate and flour [25]. The six experimental batters lost significantly less fat and water than the control. Rice et al. [26] reported similar findings for ground beef patties. All soy protein samples incorporated as PEF, i.e. soy flour emulsified (SFe), SCe, and SIe retained more moisture than their counterpart samples added as the powder, soy flour powder (SFp), soy concentrate powder (SCp), and soy isolate powder (SIp) (Fig. 4.2) [24]. The control and SFp samples released the largest amount of fat. The samples containing soy as PEF were superior to the samples containing soy protein added

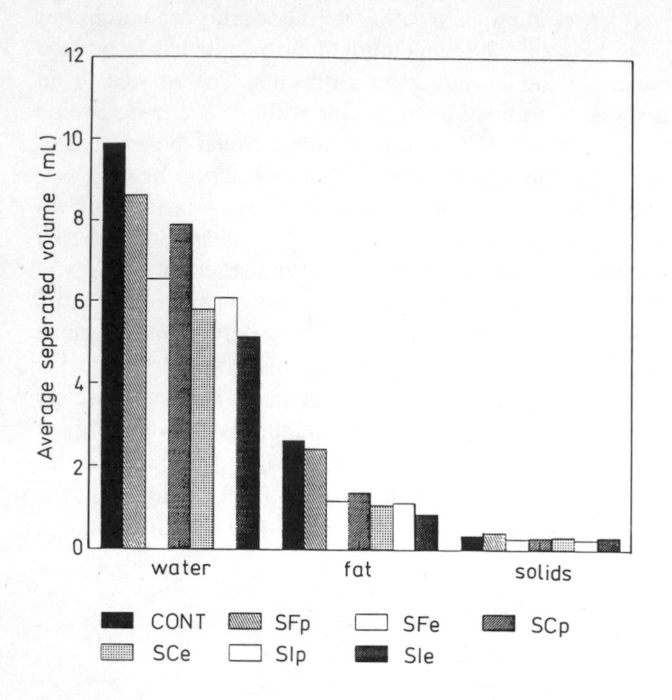

Fig. 4.2. The effect of preemulsification of sausage batters with soya proteins on sausage batter stability. *CONT*= control, *SFp*= soya flour added as powder, *SFe*= soya flour added as preemulsified fat (PEF), *SCp*= soya concentrate added as powder, *SCe*= soya concentrate added as PEF, *SIp*= soya isolate added as powder, *SIe*= soya isolate added as PEF. From Ref. 24

as the powder since less fat was released, indicating an elevated stability as a result of preemulsification of the soy proteins. Increased stability of the batter also increased yield and decreased cook losses of the frankfurters (Table 4.1) [24]. No differences existed between any of the treatments for solids released upon heating. Therefore, sausage batters formulated with soy proteins incorporated as PEF had significantly better emulsion stability than batters formulated with soy proteins added as the powder as exhibited by significantly lower cooking loss and fat released.

Increase in liquid and fat binding in pasteurized and sterilized luncheon meat as affected by incorporated 2 and 4% soy protein concentrate was established [27]. The pasteurized product retained more moisture than the sterilized one at all levels of soy protein concentrate added. Liquid retention increased markedly at 4% soy protein concentrate added to the pasteurized product. The increase in liquid retention was also observed in the sterilized products at both 2 and 4% extension levels as compared to the control. The pasteurized product retained more fat than did the sterilized product. The fat and liquid retention in the pasteurized product was markedly increased by the addition of 4% soy concentrate. Increasing soy concentrate from 2 to 4% resulted in a change in the percentage of canned contents

Table 4.1. Cooking stability, water-holding capacity (WHC), and yield of frankfurters containing soy proteins as powder and preemulsified fat (PEF)

Sample[1]	Cooking Stability (%)[2]	WHC[2]	Yield (%)[3]
Control	89.12^{a}	0.595^{a}	79.57^{a}
SFp	91.43^{a,b}	0.708^{b}	81.37^{a}
SFe	94.38^{c,d}	0.765^{c,d}	85.90^{b}
SCp	91.10^{a,b}	0.704^{b}	80.98^{a}
SCe	92.94^{b,c}	0.750^{c}	84.62^{b}
SIp	95.85^{d}	0.795^{d,e}	84.31^{b}
SIe	95.48^{d}	0.805^{e}	89.62^{c}
LSD	2.51	0.399	2.84

^{a,b,c,d} Means in same column with different letters are significantly different ($P < 0.05$).

[1] SFp = soya flour added as powder, SFe = soya flour added as PEF, SCp = soya concentrate added as powder, SCe = soya concentrate added as PEF, SIp = soya isolate added as powder, SIe = soya isolate added as PEF.

[2] Means from 8 replications.

[3] Means from 3 replications. From Ref. 24

contents minus separated liquid and fat from 69 to 100% in the pasteurized product and from 57 to 91% in the sterilized product [27].

4.2.3 Effect of Corn Germ Protein Flour on Fat Binding in Ground Beef Patties

Corn germ protein flour (CGPF) has been incorporated in CMP to increase fat absorption or binding and thus reduce heating losses of meat products [28]. A gel formed by a CGPF may interact with fat droplets and enhance fat absorption. The study was conducted to investigate fat binding of meat proteins and of hexane-defatted CGPF added to the formulations of beef patties at levels of 2.5, 5.0, and 7.5% [29, 30]. Fat retention is an important characteristic of beef patties extended with CGPF. Data of measurements of fat content and retention following heat treatment are presented in Table 4.2 [29,30]. The percentage of fat retention increased as the level of CGPF extension increased. Mean values for fat retention ranged from 52.74% in the control to 82.11% with the 30% added CGPF slurry (7.5% CGPF). However, there was no significant difference in fat retention between the raw all-meat control and beef patties extended with 10% CGPF slurry, or 2.5% CGPF dry powder. Extension with CGPF caused a decrease in the fat content of raw beef patties by the dilution effect of replacing meat fat and proteins with a defatted CGPF product. Following heat treatment, beef patties showed no significant differences of fat content among the various experimental samples, thereby illustrating the increased fat retention. Retention of fat was directly related to the percent of vegetable protein present in the patties. Results (Table 4.2) showed that CGPF has the ability to increase fat retention in conventionally heated beef patties, with a higher value for fat retention (82.11%) than for water retention (60.05%) at a 30% hydrated CGPF extension level. High fat binding and water retention will enhance the quality of the finished product, especially juiciness and textural properties and increased the yield of the products.

Nutritional guidelines recommend a reduction in the caloric intake from fat. Results indicate that manufacture of low-fat beef patties heated by microwave is possible with the utilization of plant proteins such as CGPF. Addition of a low-cost, defatted CGPF will not only lower production costs by replacing meat proteins and increasing yields, but will improve the nutritional quality of the product by lowering its fat content and, in turn, its cholesterol content. Following microwave heat treatment, beef patties showed significant differences for fat content. The fat content in microwave heated beef patties containing added CGPF was significantly lower than that of the control patties for all three extension levels (Table 4.2). At the same time, beef patties with 10 and 20% CGPF slurry contained more fat than those with 30% CGPF slurry. This was due to the dilution effect of adding defatted CGPF to ground beef during formulation. However, there were no significant differences in fat content between conventionally heated patties.
A desirable nutritional characteristic of CGPF is its low fat content. However, this positive quality would be partially negated if beef patties extended with CGPF retain a greater percentage of fat in cooking than all-meat patties. Compared with the all-meat control, fat retention increased at the 20% CGPF slurry extension level for conventionally heated and at 30% for microwave heated samples (Table 4.2). Fat retention accounted for some of the decrease in drip loss. However, there was no significant difference in fat retention between the all-meat control and beef patties extended with 10 and 20% CGPF slurry for microwave heated samples. Corn germ protein increased the degree of fat globule stabilization by forming a protein film on the surface of fat droplets, which prevented coalescence during heat treatment [31]. Data were collected to determine the effects of CGPF extension on the yield of beef patties (Table 4.2). Addition of 10, 20, and 30% CGPF slurry increased the yield of conventionally heated patties, however no differences were found between experimental patties. Addition of 20 and 30% CGPF slurry was found to increase cooking yields in microwave heated beef patties. There was no difference in mean values for percentage yield between the control patties and the 10% CGPF slurry treatment level for microwave heated samples. The yields with 20 and 30% CGPF slurry were 6.11% and 7.38% higher than that of the control, respectively. Although not significantly different, the trend was an increase in the percentage yield as the level of CGPF extension increased, because of higher water and fat retention.

Table 4.2. Content and retention of fat in raw, conventionally and microwave heated beef patties with and without added corn germ protein flour (CGPF)

Added	Fat Content (%)				Fat Retention (%)		Yield (%)	
CGPF Slurry^{d}	Conventionally heated		Microwave heated		Conventionally heated	Microwave heated	Conventionally heated	Microwave heated
(%)	Raw	Cooked	Raw	Cooked				
0	18.63^{a}	16.42^{a}	19.32^{a}	17.78^{a}	52.74^{a}	57.54^{a}	59.58^{a}	63.68^{a}
10	18.51^{a}	16.38^{a}	18.64^{a,b}	16.10^{b}	59.52^{a,b}	58.64^{a,b}	67.13^{b}	65.98^{a}
20	16.00^{b}	15.79^{a}	17.08^{b}	15.62^{b}	67.88^{b}	63.93^{a,b}	68.58^{b}	69.79^{b}
30	13.64^{c}	15.89^{a}	14.38^{c}	13.73^{c}	82.11^{c}	68.90^{b}	70.37^{b}	71.06^{b}

^{a,b,c} Means in the same column with the same superscirpt letters are not different ($P < 0.05$).
^{d} CGPF slurry = corn germ protein flour slurry; CGPF hydrated with distilled water in ratio of 1:3. From Ref. 29 and 30

4.2.4 Milk and Egg Proteins

The fat binding capacity of milk proteins affects the textural properties of foods. Fat binding by milk proteins is important for some applications as meat fillers and extenders, mainly because of the enhancement of flavor retention and improvement of mouthfeel, which are the key roles of fat functionality. De Kanterewicz et al. [11] reported that the oil absorption capacity of sodium caseinate is 1.5 ml oil/g. Variation in fat content (11 and 15%) in texturized milk proteins influenced functionality [32]. A lower fat absorption capacity of texturized milk protein was found for the product with a higher fat content.

Fat binding of milk proteins and corn germ protein flour (CGPF) was reported [33]. The fat binding index was defined as the percentage of fat bound by protein samples after thorough mixing and centrifuging. The fat binding capacity of sodium caseinate (SC) in the model system was highest, whereas that of whey protein concentrate (WPC) was the lowest (Fig. 4.3C and D), and those of CGPF and nonfat dry milk (NFDM) were similar (Fig. 4.3A and B) [33]. The superior fat binding of SC might be explained by differences in the oil binding capacity that are attributed to physical entrapment [1] and correlated to higher bulk density (porosity). Nevertheless, CGPF, the least porous powder, was still competitive with WPC in terms of fat binding capacity. This might be explained by the high hydrophobicity of corn protein, which contains numerous nonpolar side chains that can bind hydrocarbon chains [34]. Response surface analysis of fat binding by CGPF and WPC showed a maximum at the higher concentration (40%) and temperature (70 °C) after 30 min incubation. Voutsinas et al. [15] stated that the fat binding capacity of WPC increased with an increased heating time. A mild and slow heating process may make whey and corn proteins unfold, resulting in increa-

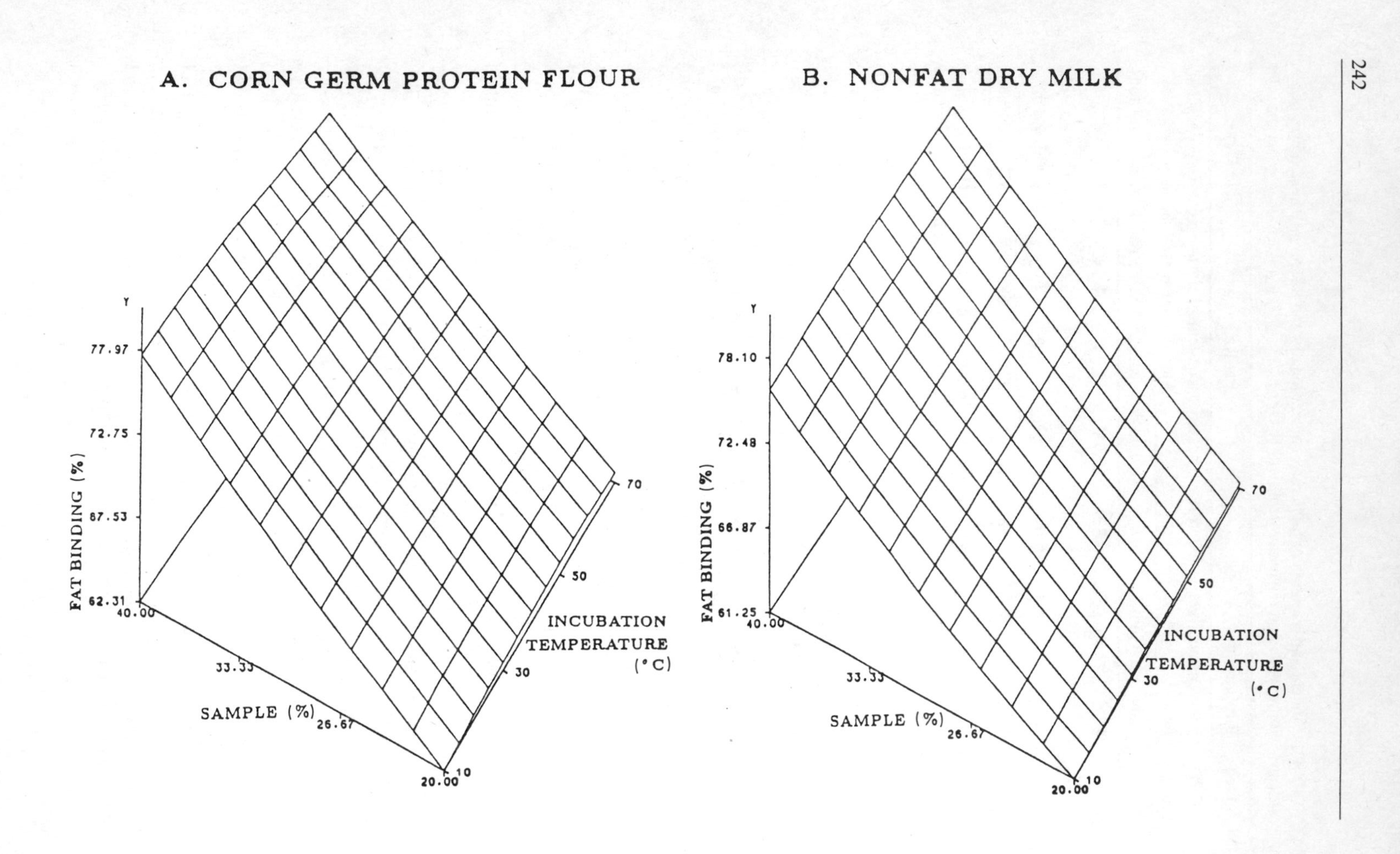

A. CORN GERM PROTEIN FLOUR
Y
77.97
72.75
67.53
62.31
FAT BINDING (%)
40.00
33.33
26.67
20.00
SAMPLE (%)
10
30
50
70
INCUBATION
TEMPERATURE
(°C)
B. NONFAT DRY MILK
Y
78.10
72.48
66.87
61.25
FAT BINDING (%)
40.00
33.33
26.67
20.00
SAMPLE (%)
10
30
50
70
INCUBATION
TEMPERATURE
(°C)

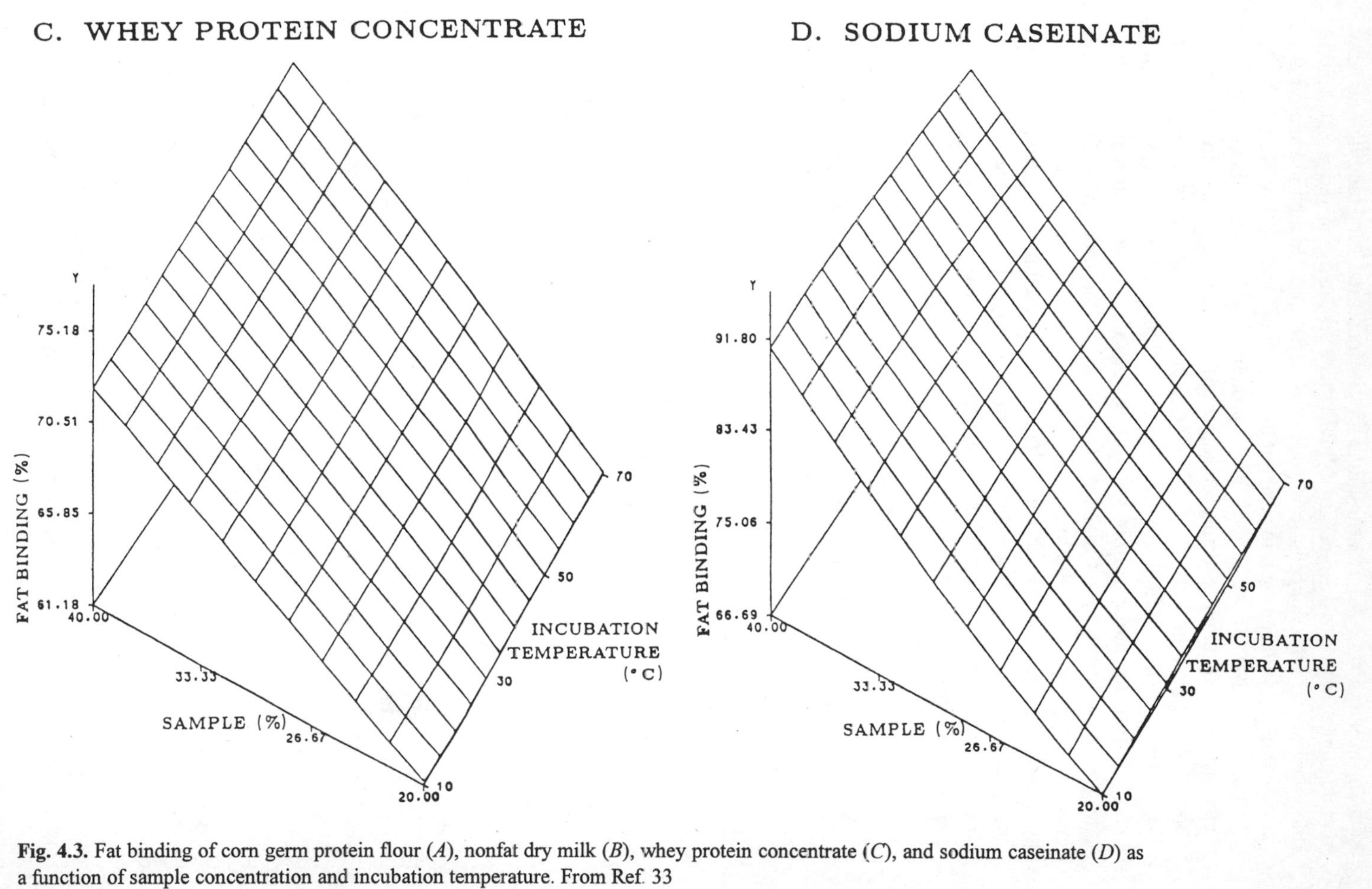

Fig. 4.3. Fat binding of corn germ protein flour (*A*), nonfat dry milk (*B*), whey protein concentrate (*C*), and sodium caseinate (*D*) as a function of sample concentration and incubation temperature. From Ref. 33

sing hydrophobicity and contributing to an increased fat binding capacity. Although the response surfaces of NFDM and SC also showed a fat binding maximum at the high concentration (40%), no effects of incubation temperature and time were found for them in terms of the ability to bind fat (Fig. 4.3).

The objective of egg and other proteins modification is to improve functionality. Extensive structural changes of acylated proteins in the lysine content have been reported [35]. Acylated egg proteins were about as digestible as native proteins when studied in vitro. However, chicks fed the acylated protein as their only protein source did not grow as well as chicks fed with native protein. The growth was restored when lysine was added to the diets. Modification with fatty acids suggested that the hydrophobic portions of fatty acids associated with hydrophobic pockets on the surface of the protein resulted in an increase in surface charge and some change in structure. Modification of egg proteins with oleic acid showed the possibility that protein molecules of the mixture of two or more egg proteins were surrounded by oleic acid [36].

4.3 Fat Binding Properties of Proteins of Plant Origin

4.3.1 Soy Proteins

The quantity of fat bound by soy proteins is affected by various factors: processing method, concentration of protein, size of protein particles, hydrophobic properties, and the liquid properties of oil. The fat binding capacity of soy proteins is enhanced by denaturation of the protein due to the exposure of apolar amino acids. However, denaturation might reduce fat binding capacity of soy proteins due to the destruction of hydrophobic domains. A lipoprotein complex of denatured, unfolded soy proteins with lecithin can be obtained by sonication. Triglycerides can subsequently be incorporated into these complexes. These protein-lipid complexes can be utilized as emulsifiers in cake mixes. It has been demonstrated that native soybean globulins formed lipoprotein complexes after denaturation and increased the surface area of protein molecules and gave a greater proportion of hydrophobic residues. Interaction between proteins and lipids in soy milk at elevated temperatures is responsible for the processing of „yuba," traditional food in the Orient.

The fat absorption capacity of soy protein concentrate and isolate was evaluated as a function of temperature by Hutton and Campbell [37]. Soy isolate showed greater fat absorption than soy concentrate. However, when fat absorption values were compared to the protein content in the samples, the difference between soy concentrate and isolate was significantly lower. Consequently, protein component of the samples was primarily responsible for fat absorption, and carbohydrates contributed significantly to fat absorption. In food systems, especially with high protein content, fat and water absorption are related primarily to the protein level. In food systems with a high content of polysaccharides they can play

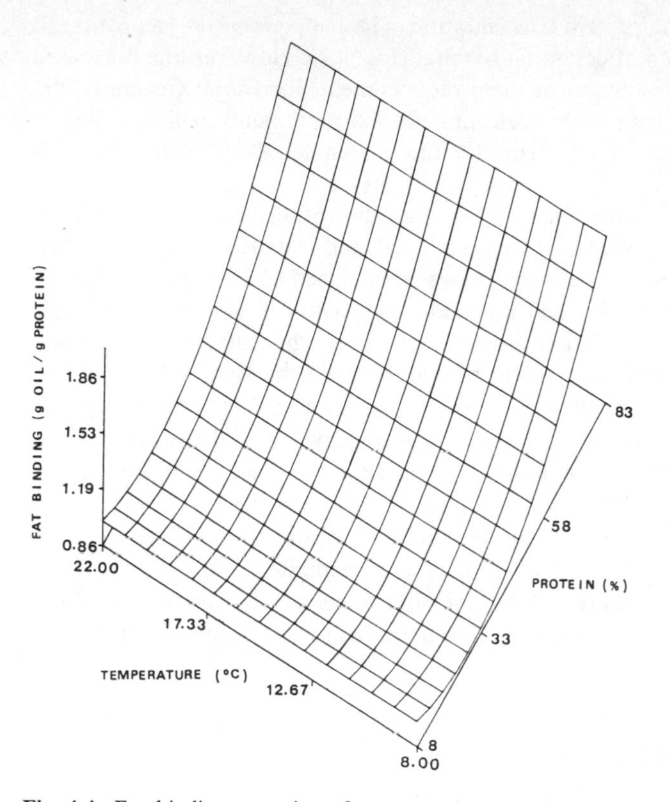

Fig. 4.4. Fat binding capacity of soy protein as a function of protein concentration and incubation temperature. From Zayas and Lin, unpublished data

a significant role in binding. The influence of other ingredients of food systems in addition to proteins should be investigated further.

Many studies have focused on the binding of lipids and solubility of proteins during dough mixing and baking [38]. Variations in the distribution of free and bound lipids with the type of supplemented soy protein isolate have been reported [39]. The supplementation of wheat flour dough was performed with 8% of various soy proteins. An approximately 60% decrease in free lipid content was measured for the unsupplemented dough and about 40% for the supplemented dough. A consistently higher degree of lipid binding was observed in doughs supplemented with isoelectric protein or Ca proteinate. Markedly higher lipid binding was observed in doughs supplemented with sodium proteinate heated at higher temperatures (drum dried) as compared with sodium proteinate heated at low temperatures after freeze- or spray drying.

Expression of the fat absorption values as a percentage of the weight of the protein (Fig. 4.4) brings the values for all soy samples into a relatively narrow range, further indicating that the protein was responsible for most of the fat absorption by the soy protein flour (Zayas and Lin, unpublished data). Fat binding increased with increasing protein concentration. The temperature of incubation in the range 8–22 °C had no effect on fat binding (Fig. 4.4). However, the responses were not proportionally increased as the protein contents increased. Generally, the less soluble protein isolates (isoelectric protein, Ca proteinate) exhibited higher binding capacity of free lipids than the more soluble alkali metal (K, Na) proteinates.

Defatted soy meal gave the highest fat absorption capacity (FAC) of 132 g/100 g sample compared to raw guar meal and defatted guar meal [40]. Correlation between FAC and bulk density was established by Dench et al. [41] for soy and sesame protein meals. The fat binding capacity of soy protein isolates produced by a membrane isolation process was higher than the commercial soy isolate [42]. The untoasted soy protein isolates showed the same oil absorption capacity as the glandless cottonseed nonstorage protein concentrate from a membrane isolation process. The lipophilic property and FAC of soy concentrate modified with sodium dodecyl sulfate (SDS) or sodium sulfate slightly increased (16–18%) when SDS treatment was carried out at pH 6 and 0.5% SDS [43]. FAC of soy concentrate markedly increased by combined treatment of soy concentrate with SDS and sodium sulfate. Soy concentrates exhibited FAC 5.9 ml oil/g concentrate comparing to untreated 4.1 ml oil/g concentrate. This effect of the combined treatment was due to exposition of more hydrophobic sites to interaction with oil.

4.3.2 Pea, Bean and Guar Proteins

Pea protein flours, concentrates and isolates have been recommended as protein sources in food preparation. Naczk et al. [44] showed that the oil absorption capacity (OAC) of pea protein preparations with 85–86% protein was comparable to that of gluten, 90.1–94.5% and 96.3%, respectively. However, the OAC was substantially lower than soy concentrate and isolate, 157 and 144%, respectively. The low OAC of pea protein is probably related to the low proportion of hydrophobic groups on the surface of the protein molecules. Oil absorption increased in proportion to the protein contents of the flour, protein fraction and isolate of field peas and faba bean [45]. Higher incubation temperatures (70 °C) had no adverse effect on the OAC.

The highest OACs were shown by freeze- and drum-dried sodium pea proteinates of 230 and 204%, respectively [46]. The OAC of other pea protein isolates was much lower: 90–104% for spray-dried, 122% for freeze-dried, and 127% for drum-dried. The mechanism of fat bindng is not fully understood, but formation of lipid-protein complexes is markedly responsible for oil retention. Paredes-Lopez et al. [47] reported that the OAC of chickpea protein isolates with 84.8–87.8% protein was higher for micelle protein (2.0 ml/g protein) than for isoelectric protein (1.7 ml/g protein) and was comparable to soy isolate (1.9 ml/g protein). Pea protein concentrate blended with cheese whey, heated at 63 °C for

30 min, concentrated and spray-dried compared favorably with NFDM in oil absorption capacity [48].

The oil absorption capacity for albumin was (2.58 g/g sample) and for globulin (1.67 g/g sample) of adzuki bean [49]. The OACs of black gram bean protein flour and concentrate were 1.1 and 3.4 g/g, respectively and were similar to the OAC of other bean proteins [50]. Acetylated protein (32% acetylation) showed an increased FAC from 190 to 270 g oil/100 g flour [51]. However, at 90% acetylation, a decrease in the FAC was found, and it was slightly lower than the control protein. There was no succinylation effect on the FAC of winged bean protein.

Fat absorption capacity of guar meal and soybean meal was 202 and 248 g/100 g protein, respectively [52]. The FAC for guar protein isolate and soy isolate was markedly higher, 338 and 268 g/100 g protein, respectively. Soy protein was more lipophilic and the FAC of soy proteins was greater than guar proteins. The FAC of guar protein isolate was much more than guar meal, 338 and 202 g/100 g protein, respectively. This suggests that, in this system, carbohydrates inhibited fat absorption. Guar protein isolates prepared by various methods exhibited an OAC in the range 105–133 g/100 g sample [53]. The values of the OAC for guar protein isolates were higher than for guar meals due to higher protein content. However, there was no strict correlation between guar protein content of the samples and their OAC values. The guar protein extracted by methanol showed a higher FAC (111 g/100 g) than proteins extracted by other methods (87–97 g/100 g) [40]. Raw guar meal showed a higher FAC than defatted and autoclaved meals on the protein basis. The level of the FAC of raw guar protein meal decreased after extraction with hexane, alcohols or acids, and after autoclaving. The inverse correlation was established between the FAC and the bulk density of the samples. Among the alcohol extracted proteins the methanol extracted had the highest FAC (119 g/100 g) and the lowest bulk density.

4.3.3 Corn Germ Proteins

The industry's selection of a protein raw material is based on data measuring its functional properties. The production of protein materials with desirable functional properties is of particular interest to food manufacturers. Nonmeat proteins derived from a variety of plant and animal sources are used extensively as fillers, binders, and extenders in meat systems. Intensive studies have been carried out on different proteins as additives in comminuted meats. Wills and Kabirullah [54] studied the effects of sunflower and soy proteins in wieners. Corn germ protein flour (CGPF) contains largely globulins and albumins, which possess lower visco-elastic properties than wheat proteins. Because CGPF is rich in starch, it stabilizes emulsions by absorbing or binding excess water, enabling more water to be added [55]. Some functional properties of CGPF have been reported, i.e., protein solubility, fat binding, emulsifying capacity, and emulsion stability [56, 57].

The degree to which the fat is extracted from corn germ meal is important not only for the yield of corn oil but also for the storage stability of the resulting CGPF. Lipids in CGPF will either enzymically or auto-oxidize with the formation of off-flavor compounds during storage, which will reduce the flavor and nutritional quality of the product. Conventional hexane extraction leaves residual lipids in CGPF [58]. Supercritical-CO_2 extraction is more efficient for the removal of triglycerides and bitter constituents (bound lipids), as well as for the inactivation of peroxidase enzymes, thereby maintaining the flavor and storage stability of CGPF. A modified process for hexane-extraction of oil from corn germ meal has been developed. This method significantly improved sensory characteristics of flavor and color of CGPF. A modified fat extraction technique was effective in producing CGPF with a low fat content (0.2% and less), and high storage stability [59]. This CGPF was tested as an extender in comminuted meats [56, 57].

The fat binding index was determined by the amount of oil adsorbed by one gram of CGPF. Fat binding was evaluated as a function of protein concentration, temperature of incubation and pH [56]. The functional properties of corn protein preparations reflect the composition of the sample; the nature and reactivity of proteins; their native structure and interactions with nonprotein components of CGPF preparations, as carbohydrates and lipids. They are also affected by environmental and processing conditions.

The ability to bind oil was significantly higher for SC-CO_2 CGPF than for hexane CGPF at a range of temperatures up to 85 °C (Fig. 4.5) [56]. The higher fat binding capacity of SC-CO_2 CGPF may be due to the lower fat content in the preparation. Less fat binding in the hexane defatted CGPF can be explained by higher original fat content in hexane-defatted CGPF than SC-CO_2 CGPF (Fig. 4.6) (Zayas and Lin unpublished data) and (Fig. 4.7) [56]. The lower fat binding va-

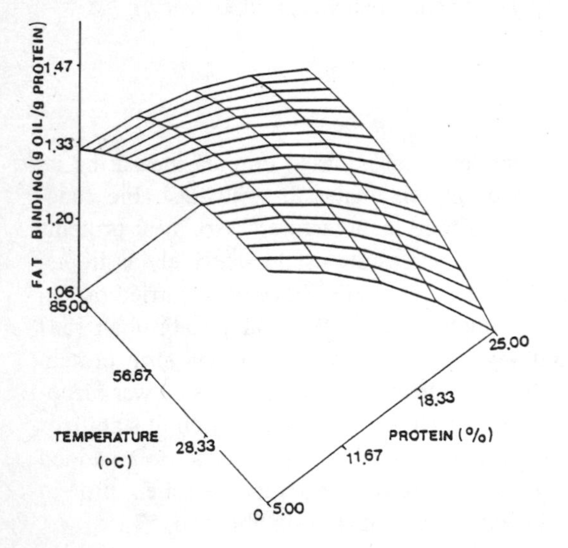

Fig. 4.5. Fat binding of SC-CO_2-defatted corn germ protein as a function of protein concentra-tion and incubation temperature. From Ref. 56

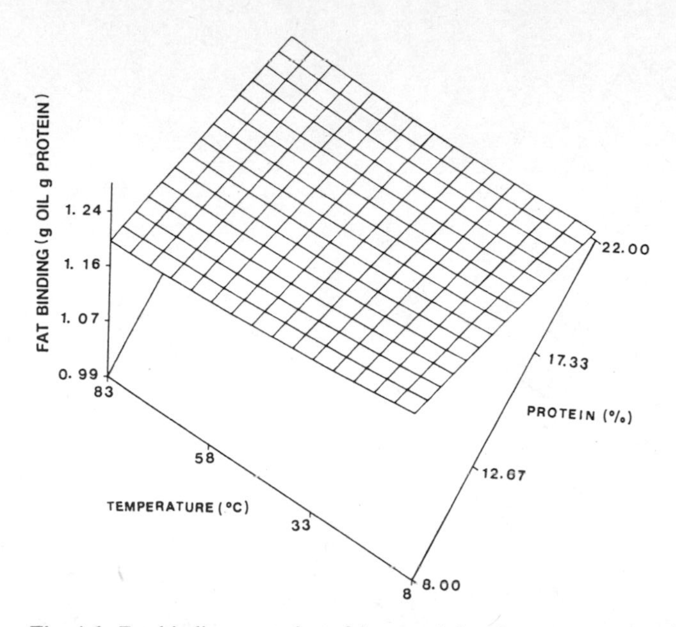

Fig. 4.6. Fat binding capacity of hexane-defatted corn germ protein processed by modified technique as a function of protein con-centration and incubation tem-perature. From Zayas and Lin, unpublished data

lues, when expressed on the basis of sample weight suggest that the protein was largely responsible for fat absorption.

Sausage batter is processed by comminution, with an end point temperature of 10–18 °C. Because of that, the fat binding capacity of CGPF at this range of temperatures had practical application. The higher fat binding capacity of the SC-CO_2 CGPF may affect the structural properties of the batter and binding of the fat in the finished meat products. The differences in fat binding attributable to temperature and the CGPF-temperature interactions were significant.

Response surface analysis of SC-CO_2 CGPF (Fig. 4.5) showed a maximum point at 0 °C incubation temperature and 5% protein concentration under tested conditions. Possibly, the high viscosity of the system at lower temperatures contributed to greater ease of fat entrapment. Fat binding (g of oil/g of protein) decreased as protein concentration increased and slightly decreased as temperatu-re increased at 5% protein concentration. The high concentration of CGPF meant less oil in the ratio, which decreased the availability of fat for CGPF to bind. However, it increased as temperature increased at 25% protein concentration.

Lower overall responses were established for hexane CGPF (Fig. 4.6) than SC-CO_2 CGPF. Hexane CGPF reached a maximum point at high temperature and high protein concentration. Temperature became a significant factor for fat bin-ding at high protein concentration. At low temperature, fat binding decreased as protein concentration increased. Overall responses of fat binding were slightly higher for modified hexane CGPF (Fig. 4.6) than for conventionally processed

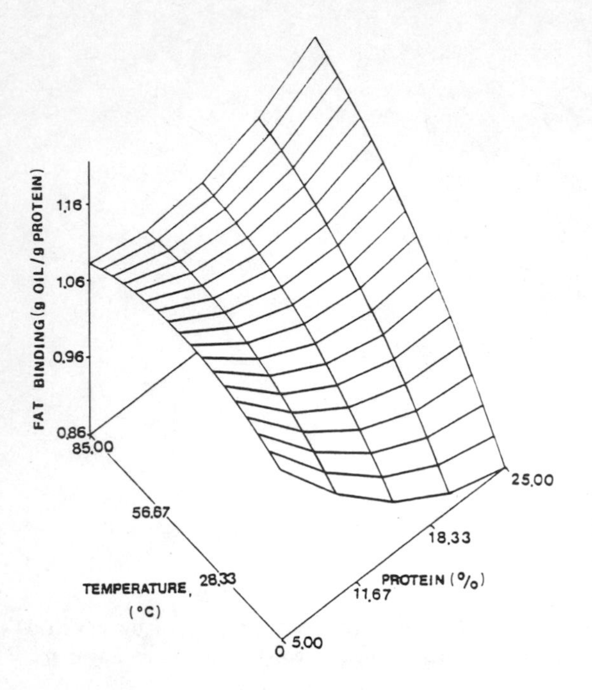

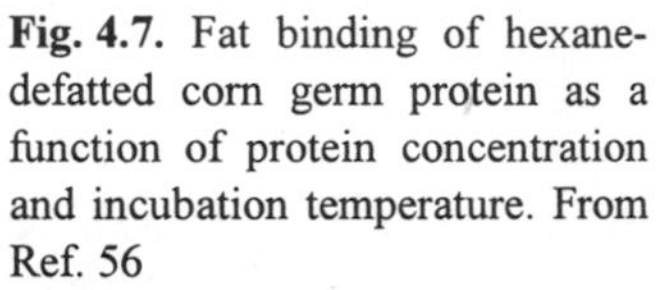

Fig. 4.7. Fat binding of hexane-defatted corn germ protein as a function of protein concentration and incubation temperature. From Ref. 56

CGPF (Fig. 4.7). Because of the better functional properties, particularly higher fat binding capacity and better sensory properties, blend flavor and white color, CGPF processed by modified procedure was recommended for utilization as an additive in comminuted and other meat products. CGPF processed by the modified procedure was more lipophilic and had a higher fat binding capacity. At the end of incubation at temperatures from 5 °C to 80 °C, both CGPFs had a similar fat bindng capacity. The fat binding capacity of CGPF processed by modified procedure was most efficient at a low protein percentage in the mixture. Soy protein flour had a 1.5 times higher fat binding capacity than these two CGPFs. The overall pattern showed that, in model systems, SC-CO_2 CGPF had better functional properties, such as fat binding capacity, than hexane CGPF. Hexane CGPF had a more yellow color, more fat residue and more protein denaturation by heat during sample preparation. The fat binding of SC-CO_2 CGPF was most efficient at low protein percentage in the mixture. Because of the better functional properties in a model system, particularly higher fat binding capacity, SC-CO_2 CGPF was recommended for utilization as an extender in comminuted meats.

4.3.4 Wheat Proteins

The breadmaking quality of wheat is related to quantity and quality of its protein. Acid-soluble proteins from good and poor breadmaking flours differ in hydrophobicity. This difference in hydrophobicity may explain, in part, differences in mixing times or mixing tolerances of the flours. The relationship between the amount of protein eluted and the amount of protein absorbed by the hydrophobic gel varied with protein concentration, showing interactions between proteins. Lipid-protein interactions during the dough preparation affected the loaf volume of bread. Fat absorption of wheat flour solubles obtained by ultrafiltration and spray-drying was similar [60]. Higher fat absorption of 0M-25 was related to its lower bulk density compared to other wheat flour solubles. Strong correlation ($r = 0.85$) was found between bulk density of wheat flour solubles and fat absorption. This may be due to physical entrapment of the oil.

In wheat gluten, natural fat is very tightly bound and difficult to separate, even by Soxhlet extraction with ethyl ether. Formation of the wheat protein-lipid complexes during bread making was proved by using ^{14}C-labeled lipid [61]. The greatest amount of lipid in the complex was not with a glutenin, a major determinant of bread quality, but a previously unknown protein fraction with low molecular weight and a tendency to form aggregates. The lipid fraction of wheat flour, comprising about 2% by weight, has a significant effect on the mixing and baking properties of dough. This fraction was identified as a protein and was named ligolin [61]. This protein fraction has been postulated as playing a key role in the formation of the gluten structure. The discovery of ligolin functionality in binding with lipids clarified the behavior of gluten in dough processing. Different lipid-binding proteins were recently purified from wheat flour. Some of these proteins are tightly bound to amphiphilic lipids. A low molecular weight basic protein, puroindoline with high lipid binding capacity was purified and sequenced [62]. Puroindoline bound approximately five palmitoyl molecules. Before the work of Frazier et al. [61], the only lipoproteins that had been isolated and studied were high sulphur-containing, low molecular weight proteins named purothionins. Lipo-purothionins were isolated from wheat flour, and after disruption of the lipoprotein complex, the proteins were identified as globulin-like proteins.

Oil absorption of wheat germ protein flour (WGPF) at concentrations of 5–25% and temperatures of incubation of 5–70 °C is presented in Fig. 4.8 [63]. With the increase in the concentration of WGPF, oil absorption increased significantly due to increased concentrations of proteins and carbohydrates. Oil binding was influenced by protein content, the amount of nonpolar amino acids, and the bulk density of the protein powder. Highest oil absorption was obtained at 5 and 15 °C at 25% WGPF and the lowest at all temperatures (5, 15, 30, and 70 °C) at 5% WGPF level. Two-way ANOVA showed a maximum oil absorption at 5 °C and 25% WGPF concentration. Probably the high viscosity of corn oil at 5 °C was responsible for better oil absorption of WGPF. Using response surface regression, a minimum oil absorption (5.1 g) was predicted to be at 5% concentration and 46 °C.

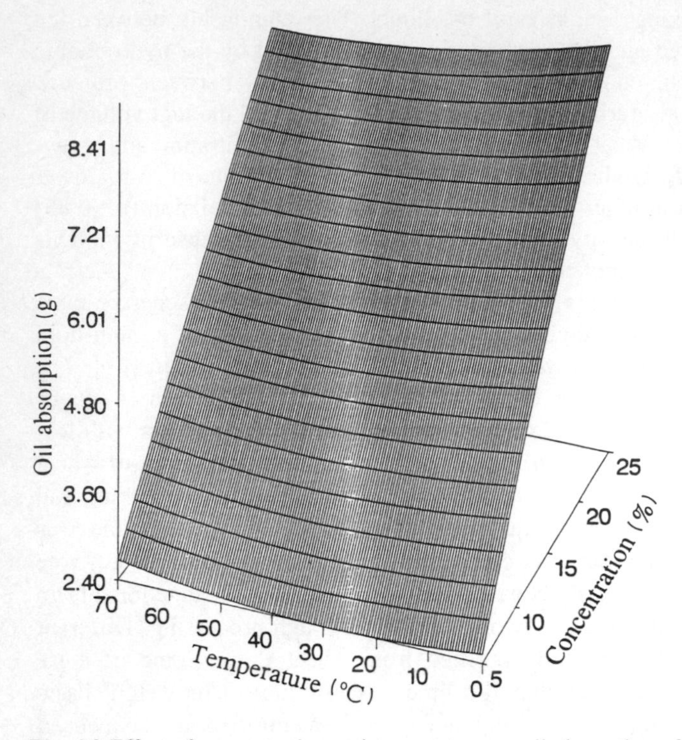

Fig. 4.8 Effect of concentration and temperature on oil absorption of wheat germ protein flour From Ref. 63

4.3.5 Cottonseed Proteins

Cottonseed proteins are easily separated into storage and non-storage protein fractions. These fractions have different solubilities, functional properties and nutritive values. Defatted cottonseed protein flour can be incorporated in the formulation of comminuted and coarse ground meat products. The use of cottonseed protein in meat products confers antioxidant protection, retarding the development of rancidity. The cottonseed flour is effective at levels as low as 3% in both the raw ground beef and the cooked beef products. Among the oilseed protein flours cottonseed protein flour showed higher OAC (1.7 ml oil/g product) than soybean and peanut, 0.9 and 1.3 ml oil/g product, respectively [64]. OAC of cottonseed, soy and peanut proteins was improved by their co-precipitation (3.0–3.1 ml oil/g product). OAC of both protein isolates and co-isolates was greater than respective flours.

The glandless cottonseed non-storage protein isolate exhibited a higher fat binding capacity (77.5 g oil/g protein sample) than storage protein isolate (43.7 g oil/g protein sample). However, fat binding of those two proteins was markedly lower than the fat binding capacity of soy isolate (132.7 g oil/g protein sample)

[65]. The fat-binding capacities of the glandless cottonseed protein isolates are comparable to soy isolate. Cottonseed protein obtained by the liquid cyclone process showed functional properties comparable to glandless cottonseed protein. Emulsifying properties, viscosity, and fat-binding of protein from the liquid cyclone process and glandless cottonseed protein are comparable with soy protein, Soya Fluff 200 W [66]. The fat binding capacity was 80.2 g oil/g of cottonseed flour obtained by the liquid cyclone process and 77.7 g oil/g of glandless cottonseed flour.

Oil absorption of modified cottonseed protein isolates was not significantly changed, except treatment with dimethylglutaric anhydride [67]. Oil absorption increased from 420 g/100 g protein isolate to 500 g/100 g of modified cottonseed isolate. Oil absorption was probably affected by the type and number of hydrophobic groups of protein molecules rather than the distribution of charges and disulfide bonds. Oil absorption of cottonseed protein isolate increased considerably with succinylation [68]. The mechanism of increasing oil absorption may be attributed to structural alteration of succinylated proteins. Enhanced oil absorption may be related to swelling and unfolding of protein. Acylation at levels above 60% caused a marked change in the OAC of cottonseed protein. The increase in the OAC is mostly attributed to a more significant exposure of hydrophobic groups in proteins unfolded by acylation. OAC is also influenced by the bulk density of the protein. The OAC of acylated proteins was higher than Promine-D, but less than defatted rapeseed protein flour.

4.3.6 Miscellaneous Proteins

All sunflower protein products bound more oil than the soy products. This is related to the more lipophilic properties of sunflower proteins. Sunflower proteins contain numerous nonpolar site chains that bind hydrocarbon chains, thereby, contributing to increased oil absorption. The OAC of sunflower albumin was considerably higher (310 ml/g sample) than those of globulins (130 ml/g sample) and control meal (213.3 ml/g sample) [69]. The high OAC of albumins is probably related to a low direct density value because fat absorption is attributed mainly to the physical entrapment of oil by the protein sample. The FAC was reduced by denaturation of albumins from 310.0 to 259.4 ml/g sample. However, it increased in the case of globulins fom 130.0 to 162.1 ml/g sample.

Rapeseed protein isolates prepared by countercurrent extraction of the rapeseed meal, followed by precipitations at pH 6.0 and 3.6, exhibited a higher OAC (1.92–1.96 g/g) than soybean isolate (1.56 g/g) [70]. However, low phytate rapeseed protein isolate (RPI) prepared by extraction of phytate at pH 4.0 showed a low OAC (0.95 g/g). The rapeseed meal showed a slightly lower OAC than soy meal. Oil absorption of protein samples is mainly affected by physical entrapment, particularly when OAC is determined by the centrifugation method. Rapeseed protein concentrate prepared according to Thompson et al. [71] showed a better OAC (263%) than rapeseed flour (240%) and soy isolate (200%). The increase in

OAC of rapeseed protein concentrate was probably related to a relatively high bulk density and a higher level of hydrophobic groups than in rapeseed flour. Beef patties containing rapeseed protein concentrate retained more of the initial fat and gave a higher yield than the control samples after cooking.

Succinylation did not change the fat binding capacity of rapeseed protein isolate but it increased slightly with acetylation [72]. This was probably related to different degrees of acylation.

Wild and domestic oat bran and flour exhibited similar oil absorption properties (bran, 120–137%, and flour, 56–62%, respectively) [73]. Defatted wild oat bran and flour showed the highest oil absorption, 178 and 80%, respectively. Domestic and wild oat meals were intermediate between the bran and flour fractions. Individual fractions of oat proteins (globulins, prolamins, glutelins) exhibited similar OAC except for albumins which showed a higher OAC. The OAC of oat protein concentrate was found to be slightly higher than soy isolate and much higher than gluten [74]. Fat binding of modified oat protein isolate was reported [75]. The maximum increase in FBC up to 3.56 ml/g was determined at 40.9% deamidation while the bulk density was not changed. Increases in FBC by deamidation might be related to changes in physicochemical and surface properties of the tested protein as a result of an increased surface area of the protein and surface hydrophobicity. Ma [76] reported that acylated oat protein a showed significant increase in FBC from 2.1 (control) ml/g to 4.95–6.35 ml/g, probably because of a marked decrease in bulk density. Succinylated oat protein exhibited an increase in FBC from 2.10 (control) ml/g to 5.25–6.30 ml/g.

The fat binding capacity of linseed protein was higher than soy proteins (334 g/100 g protein) [77]. There is a correlation between the FBC and apolar amino acid content in these proteins. The FBC of linseed proteins decreased considerably during heating. The FBCs of linseed protein flour and flour boiled for 15 min were 236 and 167 g/100 g of flour respectively [77]. The lower FBC of heat-processed proteins has been reported by other workers. Low mucilage linseed protein isolate showed a higher oil binding capacity than their high-mucilage counterparts [78]. Enhanced oil binding capacity was apparently due to a higher protein content. Peanut protein flour modified by controlled fungal fermentation exhibited increased oil retention capacity; it was more lipophilic than hydrophilic [79].

The oil absorption capacity of safflower protein isolate was found to be similar to soy concentrate and isolate. Oil absorption of safflower protein isolates is sufficiently high: 2.82 ml/g for protein isolate obtained by micellization procedure (dilution in cold water) and 2.44 ml/g for protein obtained by isoelectric precipitation [80].

Oil absorption of Nigerian conophor protein was comparable to soy protein, but the melon seed protein showed a much higher OAC than conophor and soy proteins [81].

Alfalfa protein concentrate exhibited a very high oil absorption capacity that was considerably higher than soy protein concentrate and isolate. Spirulina protein flour (18% protein) and concentrate (27% protein) of Spirulina cells and soybean

Table 4.3. Water and oil absorption capacities of tepary flour, albumin and globulin[a]

Sample	Water absorption		Oil absorption	
	g/g		ml/g	
	Sample	Protein	Sample	Protein
Tepary flour	0.73 ± 0.02	3.90 ± 0.08	0.55 ± 0.00	2.97 ± 0.03
Albumin	SW[2]	SW	3.10 ± 0.08	4.07 ± 0.1
Globulin	SW	SW	1.14 ± 0.29	1.26 ± 0.20
Soy protein isolate	3.86 ± 0.38	4.48 ± 0.46	0.83 ± 0.02	0.96 ± 0.02

[a] Means of duplicate determinations of defatted dry sample.
[b] *SW* = soluble in water. From Ref. 85

meal exhibited OACs of 190 g, 280 g, and 120 g of sample, respectively, i.e. 380, 373, and 240 g/100 g of protein, respectively [82]. Fat binding by freeze-dried leaf protein concentrate was significantly higher than by spray-dried at 85 and 140 °C outlet temperature [83]. The differences in fat binding of protein processed by different methods and temperature are probably caused by differences in bulk density.

The fat absorption capacity of the potato protein concentrate was affected by methods of precipitation, coagulation and drying methods [84]. The FAC was markedly higher for freeze-dried samples after heat coagulation (158%) than spray- (110%) and drum-dried (135%) samples. The FAC of spray-dried samples was higher (234%) with HCl as a coagulant at room temperature in comparison to HCl heat treatment (110%). The FAC of potato protein concentrate was significantly higher for spray-dried samples (188%) than freeze-dried samples (91%). FAC can also be affected by differences in protein content.

The albumin fraction of tepary was structurally more lipophilic and absorbed markedly more corn oil than tepary globulin, tepary flour or soy protein isolate (Table 4.3) [85]. The relatively high oil absorption capacity of tepary albumin (4.07 ml/g protein) and globulin is due to the presence of a large number of nonpolar side chains which might bind the hydrocarbon side chains of oil. These proteins appeared to be significantly more effective fat binders than soy protein isolate (0.96 ml/g protein).

The functional properties of jojoba protein concentrate have been reported [86]. Use of this protein for food purposes is restricted by the toxic component simmondsin, which must be removed. The jojoba protein was more soluble at alkaline pH than at acidic or neutral pH. The OAC of jojoba protein concentrate decreased with an increasing degree of purification. The oil absorption was lower than that of sunflower protein concentrate, but greater than that of soybean protein concentrate. The OACs of water and salt soluble protein extracts were higher (3 ml/g) than those washed with methanol/acetone solution (1.66 ml/g).

References

1. Kinsella, J. E. (1976). Functional properties of proteins in foods: A survey, *CRC Crit. Rev. Food Sci. Nutr.*, *7*: 219.
2. Kinsella, J. E. (1979]. Functional properties of soy proteins, *J.A.O.C.S.*, *56*: 242.
3. Sathe, S. K., Desphande, S. S., and Salunkhe, D. K. (1982). Functional properties of winged bean proteins, *J. Food Sci.*, *47*: 503.
4. Voutsinas, L. P. and Nakai, S. (1983). A simple turbidimetric method for determining the fat binding capacity of proteins, *J. Agric. Food Chem.*, *31*: 58.
5. Sharma, S. K., and Dalgleish, D. G. (1993). Interactions between milk serum proteins and synthetic fat globule membrane during heating of homogenized whole milk. *J. Agric. Food Chem.*, *41*: 1407.
6. Sitte, P. (1977). Functional organization of biomembranes, In *Lipids and Lipid Polymers in Higher Plants* (M. Tevini and H. K. Lichtenthaler, eds.), Springer-Verlag, Berlin, New York.
7. Soares de Araujo, P., Rosseneu, M. Y., Kremer, J. M. H., van Zoelen, E. J. J., and de Haas, G. H. (1979). Structure and thermodynamic properties of the complexes between phospholipase A_2 and lipid micelles, *Biochemistry*, *18*: 580.
8. Hudson, B. J. F. and Karis, I. G. (1976). Stability of lipids and proteins in leaf protein concentrates, *J. Sci. Food Agric.*, *27*: 443.
9. Szebiotko, K., Grzeskowiak, D., Walkowska, A., and Kopras, B. (1979). Changes in the content of tryptophan and available lysine during autoxidation of protein-lipid preparations, *Acta Aliment. Pol.*, *29*: 379.
10. Finley, J. W., Wheeler, E. L., and Witt, S. C. (1981). Oxidation of glutathione by hydrogen peroxide and other oxidizing agents, *J. Agric. Food Chem.*, *29*: 404.
11. De Kanterewicz, R. J., Elizalde, B. E., Pilosof, A. M. R., and Bartholomai, G. B. (1987). Water-oil absorption index (WOAI): A simple method for predicting the emulsifying capacity of food proteins, *J. Food Sci.*, *52*: 1381.
12. Li-Chan, E., Nakai, S., and Wood, D. F. (1984). Hydrophobicity and solubility of meat proteins and their relationship to emulsifying properties, *J. Food Sci.*, *49*: 345.
13. Li-Chan, E., Nakai, S., and Wood, D. F. (1985). Relationship between functional (fat binding, emulsifying) and physicochemical properties of muscle proteins. Effects of heating, freezing, pH and species, *J. Food Sci.*, *50*: 1034.
14. Seguchi, M. (1986). Lipid binding by protein films heated on glass beads and prime wheat starch, *Cereal Chem.*, *63(4)*: 311.
15. Voutsinas, L. P., Cheung, E., and Nakai, S. (1983). Relationship of hydrophobicity to emulsifying properties of heat denatured proteins, *J. Food Sci.*, *48*: 26.
16. Shenouda, S. Y. K. and Pigott, G. M. (1975). Lipid-protein interaction during aqueous extraction of fish protein: Actin-lipid interaction, *J. Food Sci.*, *40*: 523.
17. Shenouda, S. Y. K. and Pigott, G. M. (1977). Fish myofibrillar protein and lipid interaction in aqueous media as detected by isotope labeling, sucrose gradient centrifugation, polyacrylamide electrophoresis and electron paramagnetic resonance, In *Protein Crosslinking: Biochemical and Molecular Aspects* (M. Friedman, ed.), Ad. Exper. Med. Biol. 86-A, p. 657.
18. Jones, K. W. (1984). Protein lipid interactions in processed meats, In *Proceedings of 37th Annual Reciprocal Meat Conference*, National Live Stock and Meat Board, Chicago, IL, pp. 52–61.
19. Deng, J. C., Toledo, R. T., and Lillard, D. A. (1981). Protein-protein interaction and fat and water binding in comminuted flesh products, *J. Food Sci.*, *46*: 1117.
20. Jones, K. W. and Mandigo, R. W. (1982). Effects of chopping temperatures on the microstructure of meat emulsions, *J.Food Sci.*, *47*: 1930.
21. Wirth, F. (1985). Frankfurter-type sausages. Water binding, fat binding, development of structure, *Fleischwirtschaft*, *65(8)*: 937.

22. Brown, D. D. and Toledo, R. T. (1975). Relationship between chopping temperatures and fat and water binding in comminuted meat batters, *J. Food Sci.*, *40*: 1061.
23. Tinbergen, B. J. and Olsman, W. J. (1979). Fat cell rupture in a comminuted meat batter as a determinative factor of heat stability, *J. Food Sci.*, *44*: 693.
24. Lecomte, N. B. and Zayas, J. F. (1993). Soya proteins functional and sensory characteristics improved in comminuted meats, *J. Food Sci.*, *58*: 464.
25. Parks, L. L. and Carpenter, J. A. (1987). Functionality of six nonmeat proteins in meat emulsions, *J. Food Sci.*, *42*: 879.
26. Rice, D. R., Neufer, P. A., and Sipos, E. F. (1989). Effects of soy protein blends, levels and cooking methods on the nutrition retention of beef patties, *Food Technol.*, *43*: 88.
27. Schmidt, G. R., Means, W. J., Herriot, D. F., and Miller, B. F. (1983). The functionality of soy protein concentrate in canned luncheon meat, *Lebensm.-Wiss. u.-Technol.*, *16*: 55.
28. Lin, C. S. and Zayas, J. F. (1987). Influence of corn germ protein on yield and quality characteristics of comminuted meat products in model system, *J. Food Sci.*, *52*: 545.
29. Brown, L. M. and Zayas, J. F. (1990). Corn germ protein flour as an extender in broiled beef patties, *J. Food Sci.*, *55*: 888.
30. Brown, L. M. and Zayas, J. F. (1990). Effect of corn germ protein on quality characteristics of beef patties heated by microwave, *J. Food Proc. and Preserv.* , *14(2)*: 155.
31. Lin, C. S. and Zayas, J. F. (1987). Microstructural comparisons of meat emulsions prepared with corn protein emulsified and unemulsified fat, *J. Food Sci.*, *52*: 267.
32. Ozimek, G., Poznanski, S., and Cichon, R. (1981). Influence of selected factors on functional properties of textured milk proteins, *J. Food Technol.*, *16*: 575.
33. Hung, S. C. and Zayas, J. F. (1992). Protein solubility, water retention, and fat binding of corn germ protein flour compared with milk proteins, *J. Food Sci.*, *57*: 372.
34. Augustine, M. E. and Baianu, I. C. (1987). Basic studies of corn proteins for improved solubility and future utilization: a physicochemical approach, *J. Food Sci.*, *52*: 649.
35. King, A. J., Ball, Jr., H. R., and Garlich, J. D. (1981). A chemical and biological study of acylated egg white, *J. Food Sci.*, *46*: 1107.
36. King, A. J., Ball, Jr. H. R., Catignani, G. L., and Swaisgood, H. E. (1984). Modification of egg white proteins with oleic acid, *J. Food Sci.*, *49*: 1240.
37. Hutton, C. W. and Campbell, A. M. (1977). Functional properties of a soy concentrate and a soy isolate in simple system and in a food system. Emulsion properties, thickening function and fat absorption, *J. Food Sci.*, *42*: 457.
38. Chung, O. K., Tsen, C. C., and Robinson, F. J. (1981). Functional properties of surfactants in breadmaking. III. Effect of surfactants and soy flour on lipid binding in breads, *Cereal Chem.*, *58*: 220.
39. Chen, S. S. and Rasper, V. F. (1982). Functionality of soy proteins in wheat flour/soy isolate doughs. III. Protein and lipid binding during dough mixing, *Can. Inst. Food Sci. Technol. J.*, *15(4)*: 302.
40. Tasneem, R., Ramamani, S., and Subramanian, N. (1982). Functional properties of guar seed meal detoxified by different methods, *J. Food Sci.*, *47*: 1323.
41. Dench, J. E., Rivas, N. R., and Cyagill, J. C. (1981). Selected functional properties of sesame (*Sesamum indicum*) flour and two protein isolates, *J. Sci. Food Agric.*, *32*: 557.
42. Manak, L. J., Lawhon, J. T., and Lusas, E. W. (1980). Functioning potential of soy, cottonseed, and peanut protein isolates produced by industrial membrane systems, *J. Food Sci.*, *45*: 236.
43. Arce, C. B., Pilosof, A. M. R., and Bartholomai, G. B. (1991). Sodium dodecyl sulfate and sulfite improve some functional properties of soy protein concentrates, *J. Food Sci.*, *56*: 113.
44. Naczk, M., Rubin, L. J., and Shahidi, F. (1986). Functional properties and phytate content of pea protein preparations, *J. Food Sci.*, *51*: 1245.
45. Sosulski, F. W. and McCurdy, A. R. (1987). Functionality of flours, protein fractions and isolates from field peas and faba bean, *J. Food Sci.*, *52*: 1010.

46. Sumner, A. K., Nielsen, M. A., and Youngs, C. G. (1981). Production and evaluation of pea protein isolate, *J. Food Sci.*, *46*: 364.
47. Paredes-Lopez, O., Ordorica-Falomir, C., and Olivares-Vazquez, M. R. (1991). Chickpea protein isolates: Physicochemical, functional and nutritional characterization, *J. Food Sci.*, *56*: 726.
48. Patel, P. R., Youngs, C. G., and Grant, D. R. (1981). Preparation and properties of spray-dried pea protein concentrate – cheese whey blends, *Cereal Chem.*, *58*: 249.
49. Tjahjadi, C., Lin, S., and Breene, W. M. (1988). Isolation and characterization of adzuki bean (*Vigna angularis cv Takara*) proteins, *J. Food Sci.*, *53*: 1438.
50. Sathe, S. K., Deshpande, S. S., and Salunkhe, D. K. (1983). Functional properties of black gram (*Phaseolus Mungo L.*) proteins, *Lebensm.-Wiss. u.-Technol.*, *16*: 69.
51. Narayana, K. and Narasinga Rao, M. S. (1984). Effect of acetylation and succinylation on the functional properties of winged bean flour, *J. Food Sci.*, *49*: 547.
52. Nath, J. P. and Narasinga Rao, M. S. (1981). Functional properties of guar proteins, *J. Food Sci.*, *46*: 1255.
53. Tasneem, R. and Subramanian, N. (1986). Functional properties of guar (*Cyamopsis tetragonoloba*) meal protein isolates, *J. Agric. Food Chem.*, *34*: 850.
54. Wills, R. B. H. and Kabirullah, M. (1981). Use of sunflower protein in sausages, *J. Food Sci.*, *46*: 1657.
55. Bhattacharya, M. and Hanna, M. A. (1985). Extrusion processing of wet corn gluten meal, *J. Food Sci.*, *50*: 1508.
56. Lin, C. S. and Zayas, J. F. (1987). Functionality of defatted corn germ proteins in model system: fat binding capacity and water retention, *J. Food Sci.*, *52*: 1308.
57. Lin, C. S. and Zayas, J. F. (1987). Protein solubility, emulsifying stability and capacity of two defatted corn germ proteins, *J. Food Sci.*, *52*: 1615.
58. Christianson, D. D., Friedrich, J. P., List, G. R., Warner, K., Bagley, E. B., Stringfellow, A. C., and Inglett, G. E. (1984). Supercritical fluid extraction of dry-milled corn germ with carbon dioxide, *J. Food Sci.*, *49*: 229.
59. Zayas, J. F. and Lin, C. S. (1989). Frankfurters supplemented with corn germ protein: sensory characteristics, proximate analysis and amino acid composition, *J. Food Quality*, *11 (6)*: 461.
60. Oomah, B. D. and Mathieu, J. J. (1987). Functional properties of commercially produced wheat flour solubles, *Can. Inst. Food Sci. and Technol. J.*, *20 (2)*: 81.
61. Frazier, P. J., Daniels, N. W. R., and Russell Eggitt, P. W. (1981). Lipid-protein interactions during dough development, *J. Sci. Food Agric.*, *32*: 877.
62. Wilde, P. J., Clark, D. C., and Marion, D. (1993). Influence of competitive adsorption of a lysopalmitoyl-phosphatidylcholine on the functional properties of puroindoline, a lipid binding protein isolated from wheat flour. *J. Agric. Food Chem.*, *41*: 1570.
63. Bolnedi, Vani, and Zayas, J. F. Emulsifying properties and oil absorption of wheat germ protein flour. *Annual IFT Meeting, Book of Abstracts*, 1994, Atlanta, p. 15.
64. Berardi, L. C. and Cherry, J. P. (1981). Functional properties of co-precipitated protein isolates from cottonseed, soybean and peanut flours, *Can. Inst. Food Sci. Technol. J.*, *14 (4)*: 283.
65. Rhee, K. C. and Simmons, R. J. (1979). Evaluation of Selected Techniques to Determine the Functional Properties of Cottonseed Protein Derivatives. Final Report 12-14 –7001–850, Food Protein R & D Center, College Station, TX, p. 138.
66. Frank. A. W. (1987). Food uses of cottonseed protein, In *Developments in Food Proteins*, v. 5 (B. J. F. Hudson, ed.), Elsevier Applied Science, London and New York, p. 31.
67. Choi, Y. R., Lusas, E. W., and Rhee, K. C. (1982). Effects of acylation of defatted cottonseed flour with various acid anhydrides on protein extractability and functional properties of resulting protein isolates, *J. Food Sci.*, *47*: 1713.
68. Choi, Y. R., Lusas, E. W., and Rhee, K. C. (1983). Molecular structure and functionalities of protein isolates prepared from defatted cottonseed flour succinylated at various levels, *J. Food Sci.*, *48*: 1275.

69. Canella, M., Castriotta, G., Bernardi, A., and Boni, R. (1985). Functional properties of individual sunflower albumin and globulin, *Lebensm.-Wiss. u.-Technol.*, *18*: 288.
70. Dev, D. K. and Mukherjee, K. D. (1986). Functional properties of rapeseed protein products with varying phytic acid contents, *J. Agric. Food Chem.*, *34*: 775.
71. Thompson, L. U., Liu, R. F. K., and Jones, J. D. (1982). Functional properties and food applications of rapeseed protein concentrate, *J. Food Sci.*, *47*: 1175.
72. Thompson, L. U. and Cho, Y. S. (1984). Chemical composition and functional properties of acylated low phytate rapeseed protein isolate, *J. Food Sci.*, *49*: 1584.
73. Chang, P. R. and Sosulski, F. W. (1985). Functional properties of dry milled fractions from wild oats (*Avenua fatua L.*), *J. Food Sci.*, *50*: 1143.
74. Ma, C. Y. (1983). Chemical characterization and functionality assessment of protein concentrates from oats, *Cereal Chem.*, *60 (1)*: 36.
75. Ma, C. Y. and Khanzada, G. (1987). Functional properties of deamidated oat protein isolate, *J. Food Sci.*, *52*: 1583.
76. Ma, C. Y. (1984). Functional properties of acylated oat protein, *J. Food Sci.*, *49*: 1128.
77. Madhusudhan, K. T. and Singh, N. (1985). Effect of heat treatment on the functional properties of linseed meal, *J. Agric. Food Chem.*, *33*: 1222.
78. Dev, D. K. and Quensel, E. (1988). Preparation and functional properties of linseed protein products containing differing levels of mucilage, *J. Food Sci.*, *53*: 1834.
79. Prinyawiwatkul, W., Beuchat, L. R., and McWatters, K. H. (1993). Functional property changes in partially deffated peanut flour caused by fungal fermentation and heat treatment. *J. Food Sci.*, *58*: 1318.
80. Paredez-Lopez, O. and Ordorica-Falomir, C. (1986). Functional properties of safflower protein isolates: water absorption, whipping and emulsifying characteristics, *J. Sci. Food Agric.*, *37*: 1104.
81. Ige, M. M., Ogunsua, A. O., and Oke, O. L. (1984). Functional properties of the proteins of some Nigerian oilseeds: Conophor seeds and three varieties of melon seeds, *J. Agric. Food Chem.*, *32*: 822.
82. Devi, M. A. and Venkataraman, L. V. (1984). Functional properties of protein products of mass cultivated blue-green alga Spirulina platensis, *J. Food Sci.*, *49*: 24.
83. Knuckles, B. E. and Kohler, G. O. (1982). Functional properties of edible protein concentrates from alfalfa, *J. Agric. Food Chem.*, *30*: 748.
84. Knorr, D. (1980). Functional properties of potato protein concentrates, *Lebensm.-Wiss u.-Technol.*, *13*: 297.
85. Idouraine, A., Yensen, S. B., and Weber, C. W. (1991). Tepary bean flour, albumin and globulin fractions functional properties compared with soy protein isolate, *J. Food Sci.*, *56*: 1316.
86. Wiseman, M. O. and Price, R. L. (1987). Functional properties of protein concentrates from pressed Jojoba meal, *Cereal Chem.*, *64 (2)*: 94.

Chapter 5

Foaming Properties of Proteins

5.1 Introduction

The property of proteins to form stable foams is important in the production of a variety of foods. Foam can be defined as a two-phase system consisting of air cells separated by a thin continuous liquid layer called the lamellar phase. Food foams are usually very complex systems, including a mixture of gases, liquids, solids, and surfactants. The size distribution of air bubbles in foam influences the foam product's appearance and textural properties; foams with a uniform distribution of small air bubbles imparts body, smoothness, and lightness to the food. Proteins in foams contribute to the uniform distribution of fine air cells in the structure of foods. Body and smoothness of food foams is related to the formation of air bubbles that allow volatilization of flavors with enhanced palatability of the foods.

The most widely used protein foaming agents are: egg white, gelatins, casein, other milk proteins, soy proteins, and gluten. Proteins vary significantly in their foaming properties; for example, serum albumin is an excellent foaming agent while purified ovalbumin is poor. Protein foaming agents should possess the following properties. They should stabilize foams rapidly and effectively at low concentrations; perform as an effective foaming agent over the pH range which exists in various foods; perform effectively as a foaming agent in the medium with foam inhibitors such as fat, alcohol or flavor substances.

Our understanding of the functions of proteins in the formation of adsorbed protein films has been improved by studies in which radio-labelled proteins were used to obtain surface concentrations. Foamability and whippability of foaming agents are used interchangeably in the literature. The term whippability is applied when foam is obtained by a high blending or whipping treatment. The term foamability is applied when foam is prepared by injecting air or gas through the protein solution. The foaming properties of proteins are influenced by the source of the protein, methods and thermal parameters of processing, including protein isolation, temperature, pH, protein concentration, mixing time, method of foaming. Among many factors influencing foaming capacity (FC) of proteins the type of foaming equipment and method of agitation are important. Speed of whipping is important to foam properties and consumer acceptance.

5.2 The Mechanism of Foam Formation

In foam formation three stages are involved [1]. First, the soluble globular proteins diffuse to the air/water interface, concentrate and reduce surface tension. Second, proteins unfold at the interface with orientation of polar moieties toward the water; as a result of unfolding, there is orientation of hydrophilic and hydrophobic groups at the aqueous and nonaqueous phases. Third, polypeptides interact to form the film with possible partial denaturation and coagulation. Proteins rapidly adsorb at the interface and form a stabilizing film around bubbles which promote foam formation.

The basic function of proteins in foams is to decrease interfacial tension, to increase viscous and elastic properties of the liquid phase and to form strong films. Protein foamability is correlated with its capacity to decrease surface tension at the air-liquid interface. Proteins possess the ability to absorb at the interface and reduce surface tension. Surface tension at the gas-liquid interface is affected by temperature; the higher the temperature, the lower the surface tension due to changes in protein conformation [2] Different proteins possess different foaming powers, however, surface tension does not change proportionally to foaming power [3]. The surface tensions of soybean protein and egg albumin were low and close to hemoglobin, but the foaming power of hemoglobin was considerably higher. αs_1-casein had a higher surface tension than lysozyme, but the foaming power of αs_1-casein was significantly higher than lysozyme. Surface tensions of the protein solutions changed with time and almost reached a constant value within about 120 min after making a new surface [3]. The rate of surface tension change varied for various proteins and tested proteins needed different times to reach a constant value of surface tension.

Some proteins absorb at the gas/liquid interface and form a protective film around air bubbles. During the interfacial film formation at the first stage, the protein molecules are dislocated at the surface by diffusion and adsorption, and at the second stage they penetrate into the surface. Protein molecules must unfold to a certain degree and reorient at the interface with polar groups directed toward water and the nonpolar groups directed toward the air. A continuous film is formed as a result of polypeptides protein-protein interactions, and associations through electrostatic and hydrophobic interactions and hydrogen bonds. During foam formation, a monolayer of surface-denatured protein connected with the liquid is rapidly adsorbed at the interface of the colloidal system with air entrapment and bubbles formation. The limiting interfacial area occupied by one molecule of protein has been determined. In the monolayer at the air/water interface, BSA and ovalbumin were in extensively unfolded states while myoglobin, lysozyme, and β-lactoglobulin were only partially unfolded.

Protein adsorption is rapid in stirred conditions or at high protein concentration. The rates of adsorption were reflected in surface pressure of β-casein, BSA and lysozyme [4]. At similar concentrations, β-casein is more surface active than BSA and lysozyme. The rate of adsorption reflects the intensity of diffusion of the native protein to the interface. Different proteins create different surface pressure, however, protein concentration is the limiting factor. Because of the limited unfolding, lysozyme adsorption is lower than β-casein and surface pressure is lower than for β-casein [4]. At a bulk concentration above 1 mg/100 ml multilayers of lysozyme accumulate with a more ordered structure and greater viscosity.

The protein film can be compressed and may contain from 1 to 8 mg protein per square meter of the foam interface, depending on the tertiary and quaternary stability of the protein [5]. Unfolded at the interface, proteins retain some conformational structure and part of the native structure is retained. The interaction between the protein molecules in the film was measured, and physical properties of the film were established as was their importance in determining the stability of foams [6]. The protein film surrounding the air bubble should be resistant to moisture loss and withstand mechanical action and shearing forces during foam formation and storage. In foam surface layers, the protein molecules must interact with each other to form a surface film resistant to breaking.

The gas usually incorporated in foams is air, but carbon dioxide or nitrous oxide can be utilized. Whipping of protein solutions is most widely utilized method of foam production because it most closely resembles the commercial process of foaming. The volume of foam and its properties can be influenced by type of mixer used for foam formation. The relationship between the FC of proteins and their structure has been established.

Proteins with flexible molecules, such as β-casein rapidly reduce surface tension and have a good FC while globular proteins with highly ordered molecules, such as lysozyme have a poor foaming capacity. The low foaming power and k values were found in globular proteins such as lysozyme, and high foaming power was found in proteins with random coil flexible structure such as casein. These proteins easily penetrate at the interface. Soy proteins and egg albumin showed low foaming power, however they form extremely stable foams. Three proteins, β-casein, BSA and lysozyme, represent a wide range of protein structures, and their surface properties showed that structural differences were responsible for different behavior of these three proteins at the interfaces. To serve as an effective foam stabilizer, proteins must adsorb rapidly at the air/water interface. Graham and Phillips [6] showed different rates of foam formation on shaking protein solutions as related to the rates of increase in surface pressure on adsorption and suggested following order β-casein > BSA > lysozyme. The rate of surface denaturation and development of surface pressure can be related to the structural characteristics of the protein [7]. The rate of denaturation is faster for flexible random-structured proteins than for proteins with a tertiary structure.

The Effect of Protein Solubility

The most stable foams are formed with soluble proteins which can interact and form thick viscous films. In some studies, a correlation was found between the solubility of proteins and their foaming properties, FC and foam stability (FS). Protein solubility makes an important contribution to the foaming behavior of proteins. Theoretical studies suggested that when net proton charge increased, foaming capacity was enhanced [8]. This was established when the following proteins were utilized as foaming agents: egg white, fish protein concentrate, soy and leaf proteins. Cherry and McWatters [8] suggested that proteins in the liquid film should 1) be in the soluble state in aqueous media, 2) be concentrated at the liquid-air interface, and 3) be in the denatured state to possess high viscosity and strength. Protein particles in dispersion can stabilize foams as a result of being located at the air/water interface and serving as a physical barrier to bubble coalescence. The solid particles participated in foam stabilization and their removal reduced foam stability. However, foam was not formed with suspensions of the insoluble components.

Surface Flexibility of Proteins

The foaming properties of proteins depend on their ability to form flexible, elastic, cohesive interfacial film which is capable of entrapping and retaining air. This protein film should retain moisture and should be resistant to mechanical stress during foam formation and storage. High surface viscosity limits molecular flexibility of the film during foam formation when foaming agent solubility, mobility and migration to the interface are required. In foams stabilized with proteins, gas droplets are encapsulated by a viscoelastic protein film. The mechanism of foam formation and stabilization includes the capacity of proteins to form a cohesive protein film around gas/air bubbles. These protein films should be stable with sufficient mechanical strength to prevent rupture of foam during foaming and subsequent processing and storage. Viscoelastic protein films will respond to mechanical stress, because they are able to expand and compress. More stable foams are obtained if proteins form viscoelastic films than if they form highly viscous but rigid films. Foam stability is affected by protein intermolecular interactions that influence cohesiveness. Flexible proteins form foams with large bubbles and lower stability when ordered proteins form more stabile fine foams [9].

The relationship between molecular flexibility and FC of different proteins was reported [10]. Ratio of number of SS linkages/mol wt (SS/M) was used as an index of molecular flexibility. Proteins with high flexibility showed high FC when 11 model proteins were tested. More rigid proteins with low flexibility, i.e., lysozyme, ovomucoid and ribonuclease showed poor foaming properties [10]. Negative linear correlation ($r = -0.806$) was found between FC (ml) and SS/M. Experimental data showed that for protein molecules it is important to be flexible and to spread out at the air/water interface to stabilize foam from collapse. Graham and Phillips [7] determined that structures formed by proteins at the air/water interface reflect the intrinsic flexibility of their molecules.

The foaming properties of soy glycinin have been improved by reduction with 5 mM and 10 mM dithiothreitol protein (DTT) to reduce disulfide bonds and to increase protein hydrophobicity and film viscosity [11]. They determined the effects of cleavage of disulfide bonds on surface active and foaming activity of

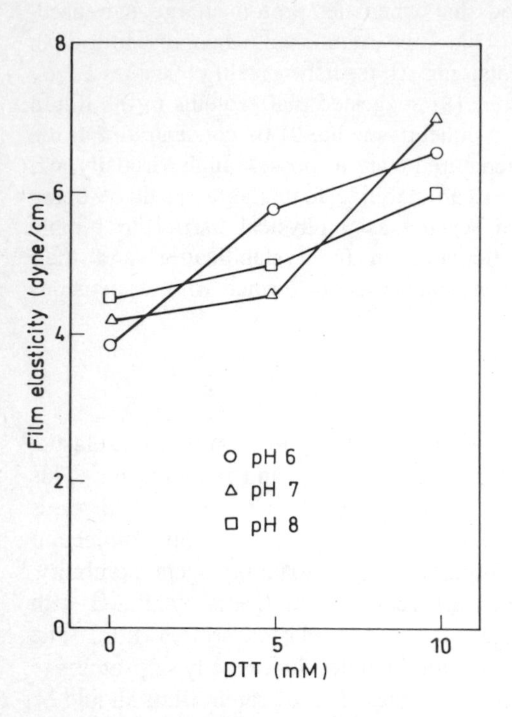

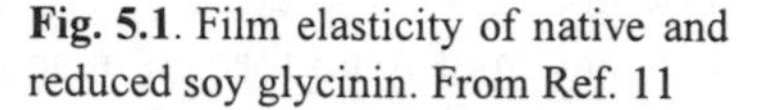
Fig. 5.1. Film elasticity of native and reduced soy glycinin. From Ref. 11

glycinin. Significant reorientation at the interface and protein-protein interactions produced stronger films. Reduction of all disulfide bonds and changes in molecular conformation was observed. Surface activity of glycinin was increased as a result of increase in surface hydrophobicity S_0 caused by significant exposure of nonpolar groups previously buried; unfolding of tertiary structure improved foaming properties. Protein diffusion at the interface depends upon the size and shape of the protein molecules. The rheological properties of interfacial film as surface yield stress and film elasticity increased as a result of reduction of disulfide bonds with extent reduction by DTT treatment (Fig. 5.1) [11]. Increase in yield stress is an indicator of higher intermolecular (protein-protein) interactions at the interface. The film elasticity of glycinin increased with the reduction of disulfide bonds and reflected greater molecular interactions to form a more cohesive film. The interfacial protein films in foams may be considered as layers of gel. The reduction of disulfide bonds of glycinin caused an increase in foam stability, measured as the half-life time of the liquid in the foam (Fig. 5.2) [11]. A significant increase in the half-life time of foam stability was found with a reduction between 5 mM and 10 mM DTT, i.e. a full reduction of disulfide bonds. This increase in foam stability was obtained especially at pH 6 and 7.

The effect of viscosity on foaming capacity and foam stability can be explained by the following factors [10]: 1) a lower rate of drainage of lamellae with increased viscosity; 2) a limitation for bubble movement through the solution; and 3) increased viscosity as an indicator of a more unfolded protein in solution. The vis-

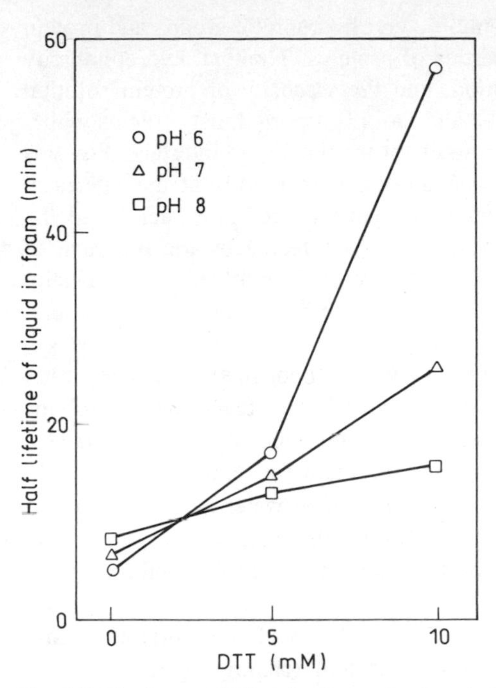

Fig. 5.2. The foam stability of native and reduced glycinin. From Ref. 11

cosity of protein dispersion influenced the foaming properties of proteins. Since the viscosity of a protein solution is a function of molecular size, shape, degree of hydration, and intermolecular interactions, are factors which influence FC. Extremely high viscosity can sometimes prevent air incorporation into the liquid protein solution. The increase in the viscosity effect presumably causes the decrease in overrun observed at high protein concentrations.

The Hydrophobic Properties of Proteins

The properties of proteins which enable them to form stable films in foams are affected by the molecular configuration of proteins, their intermolecular bonds, and the content and disposition of hydrophobic residues. Conformational changes of proteins at the air/water interface with unfolding expose hydrophobic regions and this accelerates the association of the polypeptides at the interface. Exposure of the hydrophobic groups at the interface facilitates the association of the polypeptides. As a result, a continuous cohesive film is formed around the air bubbles. The foams with high stability are produced when the hydrophobic region of proteins become situated at the interface, and molecules resist migration to the aqueous phase. Hayakawa and Nakai [12] suggested classifying the hydrophobicity of proteins into aliphatic and aromatic hydrophobicity. Aliphatic hydrophobicity is due to aliphatic amino acid residues, and aromatic hydrophobicity is due to aromatic amino acid residues.

The proper balance between hydrophobic and hydrophilic groups and protein solubility determines the foaming properties of proteins. The total hydrophobicity is obtained from amino acid composition, and the viscosity of protein solution determines the foam stability. Kitabatake and Doi [3] reported that proteins with a low polar to non-polar ratio unfold completely at the gas-liquid interface. Proteins with hydrophobic groups on their surface unfold more readily at the interface. There is a correlation between the content of hydrophobic amino acids and the surface properties of proteins. Protein unfolding is affected by the presence of hydrophobic groups, however some proteins remain unfolded with α-helix structure, and the number of unfolded structures depends on the surface pressure, i.e. the available space for unfolding.

A significant correlation ($r = 0.823$) was established between total hydrophobicity measured fluorometrically and the FC for eleven model proteins [7]. Total hydrophobicity rather than surface hydrophobicity S_o should have a more significant correlation with FC. The S_o, important for emulsification, could not explain the difference in the FC of proteins. At the air/water interface, proteins are more significantly uncoiled than at the oil/water interface. At the same time, measuring the exposed hydrophobicity by the fluorescence probe method can be used as an index of total hydrophobicity.

Various properties of proteins are required for foam formation and stabilization. The surface active properties of protein foaming agents are most important for foam formation. A direct correlation was established between the surface hydrophobicity of a protein and its ability to lower surface and interfacial tension [13]. An increase in FC was obtained with hydrophobic derivatives of caseins and other proteins as a result of better orientation at the air/water interface. Significant correlation ($r = 0.820$) was established between the hydrophobicity and FC for 10 different model proteins [10]. The optimum FC was established for hydrophobicity values 700 or above and dispersibility above 40%. Proteins with low hydrophobicity showed poor FC, and some proteins with low solubility showed a good FC.

Surface hydrophobicity was found to be less important in predicting foaming capacity than the emulsifying capacity of proteins. Measuring the foaming capacity of proteins defined as the volume of foam after aeration of the solution for 2 min showed no relationship between the foaming capacity and the surface hydrophobicity [9]. The foaming properties of proteins are enhanced by moderate heating and heat denaturation because of partial unfolding and exposure of hydrophobic groups and increased surface hydrophobicity and decreased surface tension. However, excessive heating and denaturation causes coagulation and aggregation and foam destabilization. With increasing hydrophobicity of proteins resulting from thermal denaturation, surface properties of proteins are improved with an improvement in FC and FS [14].

Absorption of the partially denatured protein molecules at the gas-liquid interface increases with time causing a decrease in surface tension as a result of changes in molecular arrangement. An increase in FC of denatured proteins when the proteins remain soluble may be explained by the more rapid lowering of surface tension on the adsorption for the unfolded molecule and stabilization by solid particles [15]. A curvilinear relationship between surface hydrophobicity and FC suggested that heat denaturation only enhances FC of proteins as a result of enhanced surface activity [15]. Surface hydrophobicity is generally increased by denaturation as the result of exposing more non-polar groups.

5.2.1 Factors Affecting Foam Formation

pH of Medium

The foaming capacity of proteins is influenced by environmental conditions such as pH. Protein food foams are mostly manufactured outside the isoelectric pH range of the proteins. If the pH of the food systems can be controlled it may improve the foaming properties of the proteins. Electrostatic attractions between proteins are at a maximum at the pI and more proteins adsorb at the interface reducing interfacial tension. Interfacial films are thickest, with high viscosity and elasticity because of extensive electrostatic bonding between molecules. Relationships between surface properties, surface pressure, surface viscosity, film elasticity and foam stability of BSA as a function of pH have been reported [16]. The surface pressures were measured at different pH values to determine their effect on foam stability. The maximum surface pressure was found at the near pI of BSA and decreased rapidly below pH 5 and above pH 6.

The size of the bubbles in foams formed in solutions of wheat proteins (glutenin) in 3 M urea was affected by pH and was larger in the isoelectric range. An acceleration in foaming power could be related to the denaturation of proteins at pH extremes. At the pI of proteins, the maximum values for viscosity, rigidity and stability of foam were obtained. Because of the maximum electrostatic attractions at the pI of proteins, the rheological properties of protein film, especially viscosity and rigidity, enhanced foam stability. At the isoelectric pH, proteins should adsorb easily at the interface because of minimal repulsion of the molecules. The surface elasticity of protein films critical for foam stability was at a maximum near the isoelectric point.

Comparatively high foam stability at the isoelectric point can be explained by the fact that the thickness and rigidity of the protein films adsorbed at the air/water interface is increased as a result of electrostatic intermolecular attractions at the isoelectric point of proteins. Electrostatic repulsion of the protein surface films is not important in the isoelectric pH region. The minimum in the „steady-state“ foam volume prepared with β-lactoglobulin was related to its rapid surface coagulation in the range of isoelectric point. A significant negative relationship was found between FS and charge density [10]. The FS increased considerably at low charge density. These data showed that foams can be destabilized as a result of intermolecular repulsion between protein molecules. The electrical properties of proteins influence the foam stability; the residual charge on the protein molecules is important for protein film stability. However, at the high electrical charge of protein molecules, the incomplete coverage with protein molecules is due to charge repulsion during film formation. The maximum FC of proteins was observed at pH's near the isoelectric point of soluble proteins. In the isoelectric pH range, the number of charges at the surface of protein molecules is minimal and surface viscosity is maximal.

The Effect of Protein Concentration

The foam volume and stability is influenced by protein concentration. Foams obtained with a higher concentration of proteins are more dense and more stable because of an increase in the thickness of interfacial films. Film formation on foam bubbles is influenced by protein concentration, and surface pressure is a function of protein conformation and concentration at the air/water interface. The rate of surface pressure change during foaming depends primarily on the rate of proteins adsorption. Fairly coherent films in the foam are formed at 0.1% protein in the liquid phase before foaming with a surface concentration of 1 mg/m^2 if completely adsorbed [17]. At this level of protein concentration, the surface rigidity of the film becomes substantial.

The maximum overrun of foam was obtained at the optimum protein concentration (2–8%) as a result of an appropriate viscosity of the liquid phase and thickness of the adsorbed film. At higher protein concentrations of foaming agent, thicker interfacial film is formed with a finer, more dense and stable foam. At high concentrations of protein or during blending, adsorption is so rapid that a bubble passing through a solution of protein is effectively enveloped by the protein film. The thickness of protein films increases at higher rates of adsorption and thick films are formed at 3–5 mg/m^2 [18]. The stabilizing effect of increased protein concentration is the increase in liquid viscosity at higher protein concentration. The foaming power and FS could be affected by a minor protein component that improves surface film stability.

The foaming capacity and foam stability of wheat germ protein flour (WGPF), corn germ protein flour (CGPF), soy flour (SF), nonfat dry milk (NFDM) and egg white powder (EWP) at protein concentrations from 1 to 8% are presented in Fig. 5.3 and Fig. 5.4 [19]. With the increase in concentration of EWP, NFDM, CGPF, and WGPF, foam expansion increased, while the SF decreased slightly at higher levels (Fig. 5.3). The highest foam expansion was obtained for EWP and other proteins at 8% levels and least at 1% levels among all proteins. This illustrates that various proteins exhibit different foaming properties due to differences in amino acid composition, sequence, molecular size of proteins, shape shape, surface polarity, charge, and hydrophobicity. An increase in overrun of SF and slight increase in WGPF foam at higher protein concentrations. The foam expansion of SF increased with concentrations up to 2%, above which it decreased. High viscosity of protein solutions during foam formation influenced the mobility and molecular flexibility of soy and wheat germ proteins.

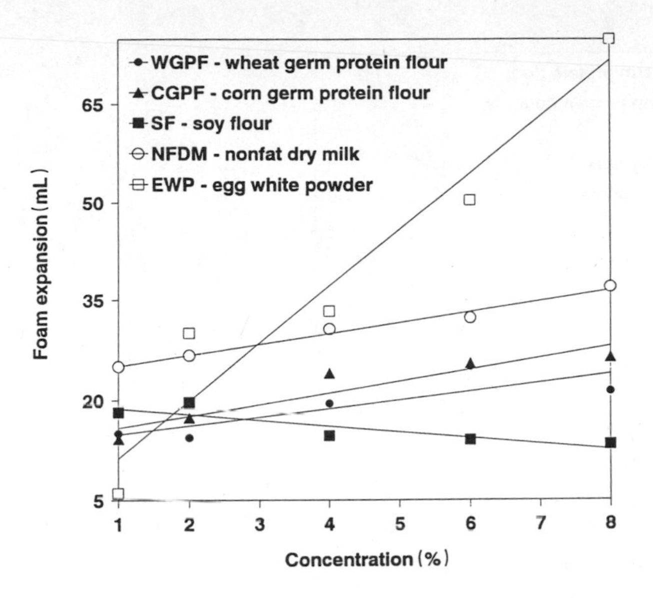

Fig. 5.3. Effect of protein concentration on foam expansion. From Ref. 19

The foam stability of proteins tested (Fig. 5.4), such as WGPF, CGPF, and EWP after 1 h of incubation, increased at higher levels of tested proteins (4–8%) [19]. SF exhibited the highest foam stability at the 2% level, and there was slight decrease at higher levels (4–8%). The foam stability of NFDM foams significantly increased at 6–8%. The highest foam stability was exhibited by EWP, especially at the 4–8% levels. The foam drainage was affected by the protein used to stabilize the foam. High viscosity at higher concentrations (4–8%) was effective in preventing gravity deformation of the film in WGPF, CGPF, NFDM, and EWP foams. Thicker and rigid interfacial films at higher protein concentrations might have reduced drainage within the lamella structure. The foam stability of SF decreased with increasing protein concentration, which was probably due to stronger protein-protein interactions at higher concentrations. At low bulk protein concentrations, adsorption was diffusion controlled, but at high concentrations, adsorp concentrations, adsorption was extremely rapid. The high surface concentrations of proteins might have affected the surface pressure. The magnitude of surface pressure is affected by molecular composition and the properties of proteins.

Whipping Aids

Whipping aids can be added to proteins to improve their FC and to improve deleterious changes of proteins caused by drying and heating. Organic solvents such as triethyl citrate and glyceryl triacetate are permitted whipping aids and are used commercially. The synergistic effect of organic solvents is obtained at low concentrations (0.01–0.03%). Ethanol is accepted in the brewing industry as a

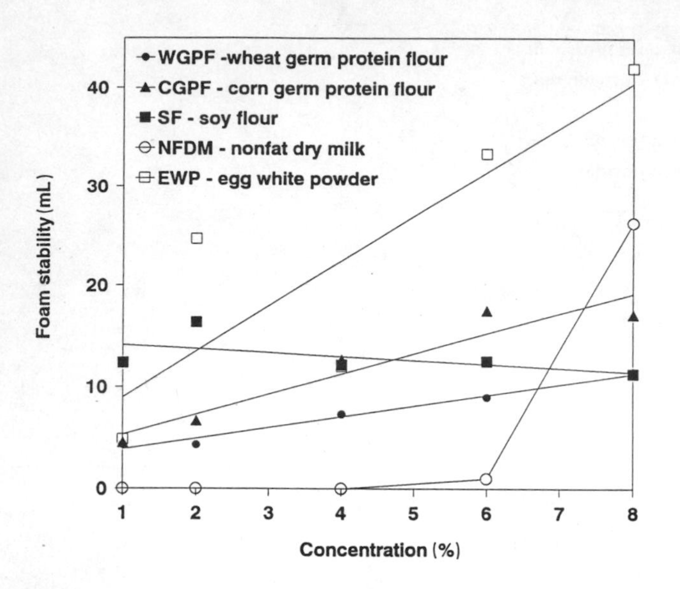

Fig. 5.4. Effect of protein concentration on foam stability; 1 h incubation. From Ref. 19

foaming aid for beer. The effect of solvents on the foaming capacity of proteins is related to the reduction of surface tension and protein insolubilization at the air/water interface [20].

The effect of sugars on the FC of proteins is related to the inhibition of heat denaturation. Sucrose at a level of 20% was an effective protective agent for egg white proteins during pasteurization and drying. Kinsella [21] suggested that the sucrose decreased the foaming capacity of proteins by inhibiting the incorporation of air bubbles into solution but increased foam stability as a result of increased bulk viscosity. The effect of sucrose on foam stability is proportional to the sucrose concentration. Enhancement of foaming properties of proteins with added sucrose is related not only to increased bulk viscosity, but sucrose also influences the surface properties of the protein films. Proteins adsorb faster to the air/water interface when sucrose is present. The level of sucrose required to increase protein adsorption was less than 50 g/100 g of water. It was found that sucrose retarded the thermal denaturation of proteins because a solution of sucrose is a poorer solvent for the protein than water [22]. Unfolding of protein macromolecules caused by thermal treatment is lower when the solvent is a sucrose solution comparing with water. Addition of sucrose in the protein solution increased surface tension at an air-water interface. From the other hand, the beneficial influence of sucrose on ovalbumin foams was explained in terms of the protein's low surface hydrophobicity and promotion of protein adsorption at the air-water interface [9].

Addition of NaCl influences the foaming capacity of proteins because salt affects the solubility, viscosity, unfolding and aggregation of proteins. Frequently, NaCl increases foam overrun and reduces foam stability. The effect of different ions on foaming properties depends on their influence on structure and protein conformation. Multivalent metal ions improve the FC of proteins. Kinsella [21] has reported that NaCl added to soy protein suspensions increased FC, however decreased FS. The FC increase was due to higher protein solubility at the air/water interface during foam formation. However, added NaCl retarded the partial denaturation of polypeptides of proteins that is necessary for protein-protein interaction and stability. The improvement of foaming properties due to salt is due to the reduction in the rate of protein denaturation. Addition of NaCl can change protein conformation and solubility depending on the ions and proteins involved.

The additive most widely utilized in the production of foams in the U.S. is sodium dodecyl sulphate. This additive is accepted to standardize the foaming power of dried egg albumen at a level of 0.1% on a dry basis [23]. The foaming properties of yolk-contaminated egg white have been restored by addition of anionic sodium dodecyl sarcosinate or cationic dodecyl ammonium chloride.

Foam Inhibition

Foam inhibitors are water-insoluble substances, and they can cause rupture of a protein film at air bubbles. Among foam inhibitors are some of the lipids with high surface activity. Active antifoaming agents are yolk components, and they exhibit a deleterious effect on the FC of egg white. Foam inhibitors could cause local ruptures in the film and lead to the collapse of bubbles. Yolk contamination in fresh albumen at a level 0.03% can completely inhibit the FC of egg white [23]. Contamination of the egg white by yolk is caused by diffusion through the membrane during prolonged storage at a low relative humidity, and by poor quality egg breaking and separation of white and yolk. Yolk phospholipid lecithin exhibits strong antifoaming capacity. Positively charged lecithin can interact with negatively charged proteins and prevent protein association at the air/water interface. Effective foam inhibitors are soy proteins with a high content of phosphatides. The effect of lipids is important for the foaming properties of protein ingredients including low levels of lipids involved in the foaming process, i.e. 0.1% of soy phosphatides.

Low levels of contaminating lipids (less than 0.1%) impair the foaming properties of proteins. A lipid component in protein is detrimental to FS because it destabilizes protein films. High FC was established for egg white proteins free from yolk lipids, and defatted whey and soy proteins free from phosphatides. Extraction of either neutral or polar lipids from soybean or pea proteins significantly enhanced their foaming properties. Surface active, polar lipids interfere with the most desirable conformation of adsorbed protein films by dislocating at the air/water interface. Lipids can weaken protein-protein interactions by interfering with hydrophobic surfaces. Lipids disrupt protein-protein interactions at the air/water interface and inhibit foaming as a result of displacement of proteins from the interface. Highly polar lipids (lecithin) are more effective in their antifoaming action than neutral triglycerides because they possess a higher surface activity and can form complexes with acidic proteins. Lipoproteins impaired the foaming properties of proteins by weakening the cohesive forces between protein layers surrounding the air bubbles.

Protein foaming agents with lipid tolerance should be developed. Poole et al. [24] prepared a foaming agent with the combination of basic and acidic proteins. The basic proteins with pI greater than 7.0 such as clupeine (pI = 12) and lysozyme (pI = 10.7) do not form stable foams when used alone. However, in neutral solution, basic proteins added at level 20% increased the foaming power of acidic proteins such as whey or serum albumin (pI ~ 5). Foams with mixed protein foaming agents are tolerant to glycerides at a 25 vol% level and to lecithin. The foaming property of fat-containing protein systems can be enhanced by the incorporation of a small amount of basic proteins and as a result of preparing mixtures of acid and basic proteins.

Some of the proteins such as BSA can enhance FS in the systems containing lipids. Foam can be stabilized and stable film formed even when a lipid is associated with it [24]. Basic protein lysozyme is from egg white and clupeine from herring roes and the water-insoluble histones from cell nuclei. Basic proteins such as clupeine could be used to upgrade the foaming capacity of lipid-contaminated proteins. Clupeine significantly improved foaming properties of egg albumen contaminated with yolk [24]. Foaming capacity was restored even at high yolk concentrations (5%). Clupeine enhanced the foaming capacity of two whey proteins with foaming properties suppressed by fat. Charge of the basic protein is an important factor influencing foaming properties. Clupeine has better foaming properties than lysozyme which has a lower pI [9]. The antifoaming properties of fats can be minimized by foaming without the fat component and then mixing in the remaining ingredients containing the fat.

5.2.2 Foam Stability

Foam stability (FS) requires the specific properties of protein films as formation of cohesive, viscous, elastic, continuous, air-impermeable film around each gas bubble. Foams with high stability are prepared with high-molecular-weight globular proteins that produce thick adsorbed films. Foam stability is affected by the „self-healing“ ability of proteins, i.e their ability to move from a region of low interfacial tension to a region of high interfacial tension (Marangoni effect). Drainage of water from foam can be reduced if polar side chains of protein polypeptides interact with molecules of water within the lamella. Foam stability is determined by the physical properties of the film once formed. The foam stability is influenced by film thickness, mechanical strength, protein-protein interactions and environmental factors such as pH and temperature [25]. Higher stability was found for thicker films because of mechanical strength and better textural properties. The resistance of protein foam to coalescence and to the collapse of air bubbles is determined by ability of the protein to form a multimolecular matrix.

Lamellar liquids in the foam tend to drain and when the space between films decreases to ca. 50 Å, there is the gradual collapse of the foam [1]. The viscosity of the surface film and bulk liquid is a major factor affecting foam stability. Films with high surface viscosity form strong foams as a result of significant cohesive forces between protein molecules. Viscoelastic properties of the film result in the film being responsive to mechanical stresses by compression and expansion. In the stable foams, the interfacial film should be structurally stable and impermeable to the entrapped air. In stable foam, protein polypeptide chains are associated to form continuous intermolecular polymers enveloping the air bubbles.

The stability of the foam is influenced by the properties of the protein used, its concentration and solubility, and ionic strength. Generally, FS is directly affected by protein concentration in solution by influencing the thickness, mechanical strength and cohesiveness of the film. Foam stability rapidly increases with protein concentrations up to 0.1% depending on the protein surface properties [26]. A protein concentration of 0.1% before foaming would give a surface concentration of 1 mg/m^2 with formation of a coherent film of high surface rigidity.

Holding and ageing of protein solutions before foaming improves FS as a result of enhanced protein-protein interactions that lead to thicker adsorbed films. During foam ageing gravitational forces will cause water to drain, and air cells will come closer together. Separation of liquid from foam is either by leakage from the lamellae or by rupture of the bubbles. The reduction of water drainage in foam is obtained if proteins at the air-water interface bind water tightly and cause an increase in surface viscosity. Proteins with high rigidity and surface viscosity have low foaming power, however, they form stable foams [27]. The stability of the foam is at a maximum at the isoelectric point, when the surface elasticity is also at a maximum. Glutenin was a more effective foaming agent than either of its components, gliadin or gluten, but FC was lost if the -S-S- bonds were disrupted. Stable foams are produced by globular proteins that are soluble at the pI, when film of the greatest strength is produced. Strong protein films with high stability are formed due to the role of disulfide bonds at the interface. Foam stability was significantly reduced when disulfide bonds were broken.

Foam stability is influenced by the rheological properties of the protein film. An increase in the elasticity of the protein film reduces the drainage of water from the film which increases the FS. The maximum surface yield stress occurred in the pH range 5 to 6 close to the isoelectric region of BSA [16] and decreased rapidly at a pH above 6 and below pH 5. The surface yield stress is the characteristic of viscoelasticity of the film which depends on protein-protein interactions. Elasticity of the film in foam was measured using tensiolaminometry the and maximum was found at pH 4.5–6 [16]. At this pH range, higher inter- and intramolecular interactions were observed resulting in a more cohesive film. The foam stability of BSA should be at a maximum near the isoelectric point of proteins because of a reduction in the surface tension and an increase in film elasticity. The stability of foams of BSA to drainage was maximal in the isoelectric pH range because of the maximum surface pressure, surface viscosity, yield stress and elasticity. Electrostatic repulsions between the interfacial films of foam bubbles influence foam stability. The coalescence will be minimal if neighboring bubbles are repelled at a critical distance by electrostatic repulsions between proteins in the interfacial film [1].

Fluid drainage from the lamellae due to gravity and rupture of the film as a result of pressure causes foam instability. Drainage results from thinning of the lamellae and/or reduction in viscosity of the liquid phase. The large difference in density between the air and aqueous phases causes gravitationally-induced drainage. The rate of drainage usually falls with time and depends on the viscosity of the bulk liquid phase. For this reason, sucrose increased the stability of wheat gluten foams; different thickeners are also affected . Drainage of liquid from foam stabilized with two soy isolates, sodium caseinate, bovine plasma proteins and egg white as a function of incubation time is presented in Fig. 5.5 [28]. The incubation time necessary for curves to level off represented the maximum amount of drained liquid V. This time depended on the protein and varied from 20 min for WPC to about 350 min for egg white. Elizalde et al. [28] proposed the following empirical equation to describe the experimental drainage curves for different protein stabilized foams: $v(t) = V\, t/(B + t)$ [1] where: $v(t)$ is the volume of drained liquid at time t; V is the maximum volume of drained liquid; and B is the time needed to drain V/2. Foams with egg white, gelatin and caseinate showed the highest B values and they were most stable. The stabilizing activity of various proteins can be compared by using the foam stability B index.

The stability of a gluten foam increased when salt was present, and this was due to the adsorption of more protein in the interfacial layers. With increased whipping time, the size of air bubbles decreases and the more unstable they become. The interfacial area of the foam increases with decreased bubble size, and there is more air/water interface available for liquid drainage. Excessive denaturation causing coagulation and aggregation of proteins possesses destructive effect on protein foam stability. However, partial denaturation enhanced viscosity and rigidity of protein layers and foam stability. Food foams often contain solid particles trapped in the aqueous lamellae which enhances foam stability. Addition of solid particles to the foam, casein precipitates for example, reduces bubble disproportionation, i.e. disruption by rapid transfer from the smallest bubbles to the largest.

5.3 Milk Proteins

Milk proteins are major components for the unique functional role of milk and milk products in human diet. Milk and particularly whey proteins, whey protein concentrate (WPC) and whey protein isolate (WPI) are of considerable interest to the food industry because of their nutritional value and functional properties. The limiting factor in WPC and other milk proteins utilization is the variation in com-

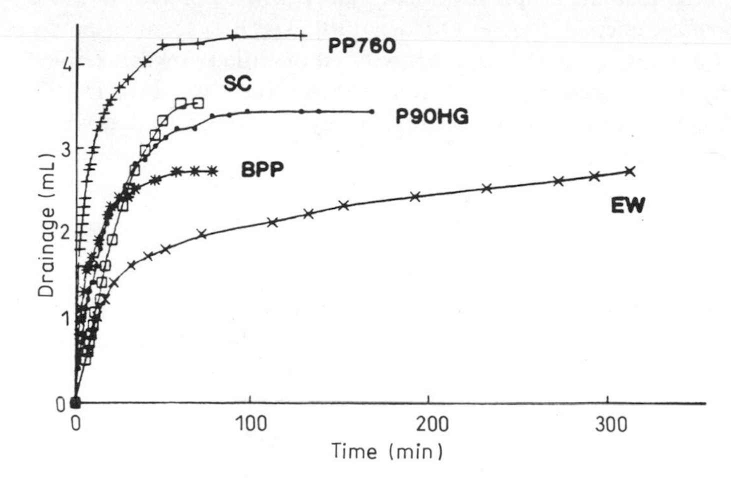

Fig. 5.5. Drainage of liquid as a function of time for different protein foams. *PP760*=Purina soy protein isolate; *SC*=sodium caseinate; *P90HG*=soy protein isolate; *BPP*=bovine plasma proteins; *EW*=egg white. From Ref. 28

positional and functional properties due to different sources and processing conditions. It is generally accepted that rather poor functional properties of whey proteins in food products are largely responsible for their limited application in foods. However, these functional properties of whey proteins might be improved by decreasing the minerals and fat content, and by increasing the protein content to improve foaming and other functional properties.

Factors influencing the FC of milk caseins are concentration of protein and of total solids, pH, ionic strength, degree of denaturation and extent of breakdown of the caseins to form protease peptones. Caseins have a high FC especially in whipped toppings. High quality coffee whiteners and whipped toppings may be produced using a modified skim milk powder and a vegetable oil. Because of higher prices, it is difficult to compete with less expensive vegetable proteins, except when functionality, such as foaming, emulsifying capacity and flavor are important.

The foaming capacity of milk protein isolate (MPI) was significantly higher than those for WPI, sodium caseinate and egg white [29]. The foaming properties of milk protein isolate reflect higher protein-protein interactions that were not observed in the casein or whey dispersions alone. The foam stability decreased with increasing concentration of protein co-precipitate prepared from NFDM probably due to stronger protein-protein interactions at high concentrations. Foam stability was affected by the amount of Ca present in the co-precipitate. An increase in NFDM utilization can be obtained by recovery of the milk proteins as co-precipitates and using them as food components. The stability of foam prepared with NFDM co-precipitate decreased with higher temperatures of precipitation or prolonged heating. The foaming properties of calcium NFDM co-precipitate were compared with some commercial protein products [30]. The best foaming power was exhibited by whipping agent D-100, a spray dried modified powder derived from soybeans. The foam expansion of Ca co-precipitate was lower than D-100, however it was higher than for sodium caseinate, whey powder and dried whole egg powder.

Whey Proteins

The annual production of whey solids in the U.S. is estimated to be 1|044|400 t which contains 3134|280 t of protein [31]. An estimated 40% of 16|337|400 t is disposed of as waste, resulting in a loss of a potential food resource. The composition of WPC is affected by the fractionation or isolation method used in manufacture. Whey protein (about 0.6 to 0.8% by weight) has the unusual property of being acid soluble. WPC has excellent ratio of the essential amino acids, making it a satisfactory source of protein for food fortification. However, the high content of mineral salts (0.3 to 0.7%) in whey proteins impart undesirable sensory properties to the protein product. Whey proteins can be utilized as a substitute for egg white proteins because they possess similar properties, especially high contents of sulfhydryl and disulfide groups, heat coagulability and foaming properties. Processing conditions of WPC influence its functionality in foods. An attempt was made to prepare an egg white substitute from WPC by its modification in order to obtain good foaming properties and gel strength [32]. The overrun of the modified WPC was improved over the value for dried egg white. The largest overrun score (183%) was obtained with addition of cysteine (1% of protein) and pepsin (2.5% of protein) to a 10% solution of WPC. The tested WPC, cysteine and pepsin solution was incubated for 52 min at 35 °C and pH 2.0 and held at 41 °C for 5 min. The overrun score 176% was obtained for spray dried egg white protein.

Whey proteins possess excellent foaming properties because they are molecularly soluble and stabilize foam by increased viscosity. A large number of WPCs have become available with a variety of functional properties. Foaming capacity increased with increase in WPC solubility. To serve as an effective foaming agent, whey protein must be soluble in aqueous solution and to be able to unfold at the interface. WPC with a high protein concentration (~25%), a minimum degree of protein denaturation and a low lipid content exhibits high FC [33]. Foam stability of WPC is better at high pH values obtained by the addition of calcium hydroxide due to facilitated unfolding of protein polypeptide chains. Commercial WPCs vary markedly in foaming properties due to the different extent of denaturation of the proteins, high ash content and the presence of lipids. WPCs good foaming properties are impaired by other added components.

The functionality of various WPCs and WPIs was tested and three WPIs gave superior FC with maximum foam expansion 800% [34]. The WPIs performed better at higher pH values and lower temperature of pretreatment (25 °C better than 55 °C). Maximum foam expansion for WPCs ranged from 339 to 1300% and minimum foam expansion was observed at pH 7.0. Milk protein isolate (MPI) with a large amount of caseinate and caseinate-denatured whey protein complex did not form foam at pH 4.5, but exhibited good foaming at pH 7.0 and 9.5. The WPIs generally gave more stable foams than WPCs and MPI. The temperature of pretreatment and pH had little effect on WPI foam stability.

Partial denaturation of β-lactoglobulin during homogenization may improve the foaming property of β-lactoglobulin. Stability of β-lactoglobulin and α-lactalbumin against denaturation might be improved by addition of sugars resulting in longer foaming times required for whey proteins compared to egg white.

Whey protein isolate with 95% protein produced with ion exchange cellulose exhibited good foaming properties [35]. The WPI solution (11% protein) gave foam with the same specific volume as egg white foam (10 ml/g), whereas WPC produced foam with 2.5 ml/g. The WPI foam stability was 8 ml drainage after 30 min which was higher than with egg white (12 ml) and WPC (complete collapse after 30 min). The poor FC and FS of WPC prepared by ultrafiltration was due to protein content (5%) in WPC. Foaming properties of membrane-processed acid WPC were correlated with protein hydrophobicity, solubility, and β-lactoglobulin content. Liao and Mangino [36] demonstrated that the foaming properties of WPC were determined by the content of β-lactglobulin and by the extent of whey protein denaturation.

The maximum FS of BSA (pI = 4.9) has been reported in the pH range 5–6, showing the effect of electrostatic repulsive forces on the extent of molecular adsorption and packing in the film and influencing film elasticity [16]. The mechanical strength of BSA films in the pI range results in more stable foams. Foam volume increases with whey protein concentration to a maximum value after which it decreases again [37]. Maximum overruns for WPC foams have been observed at 8 and 12% (m/v) whey protein. Maximum overruns and FS have been observed between pH 4 and 5 as a result of maximum interpeptide electrostatic attraction of proteins around the pI. Kinsella [1] suggested that the volume of foam produced is dependent on the total amount (and not concentration) of protein in the solution being bubbled.

A rapid adsorption of β-lactoglobulin that increased at pH 5.3 was reported [38]. Maximum surface viscosity was in the pH range 5–6 and decreased by 40% at pH 7.0. Under these conditions (pH range) foams prepared with β-lacto-globulin exhibited maximum strength. BSA utilized for foam preparation showed maximum surface viscosity, film elasticity and surface yield stress in the pH range 5–6. BSA exhibited maximum FS in the pH range 5–6 demonstrating the relationship between film properties and foaming properties. Native β-lacto-globulin, α-lactalbumin, and BSA showed similar adsorption at the air/water interface. Heat treatment of α-lactalbumin and/or β-lactoglobulin accelerated the surface pressure development. The foaming properties of whey-potato protein composites prepared by wet-blending were improved as pH increased and the proportion of potato protein decreased [39]. The FC of whey-potato protein dispersions was affected by pH. A considerably lower FC was obtained for 100% potato protein compared with other whey-potato protein composites. Protein solubility influenced the FC of composites. In general, whey protein exhibited better functionality than potato protein and protein blends displayed intermediate properties.

Caseins

Proteins like β-casein exhibit good foaming properties because they rapidly reduce surface tension at the air/water interface; they form large foam volume with large air cells [27]. Foaming expansion values for sodium caseinate, WPI and ovalbumin as a function of protein concentration are presented in Fig. 5.6 [40]. Expansion of foam volume ranged from 100 to 300% compared to 300–2000% published by Halling [26]. Foam volume increased with protein concentration up to a limit value over which it decreased. Reduction of foam expansion at high concentration of proteins is related to decreased solubility of protein solutions, from 94% at 4% to 60% at 8% concentration. Insoluble proteins probably interfered with foam formation. Maximum foam expansion was recorded for sodium caseinate at 4%, and 8% WPI and ovalbumin. Tornberg [41] reported that caseins were better surface tension depressors than whey proteins in their native state.

Initial drainage rates of sodium caseinate, WPI and ovalbumin foams plotted as a function of protein concentration are presented in Fig. 5.7 [40]. The highest drainage rates were recorded for foam prepared with ovalbumin, followed by WPI and sodium caseinate. The rate of drainage decreased at higher protein concentrations and allowed more water to remain in the foam matrix. The effect of protein concentration on serum drainage was affected by the type of protein used as a foaming agent. Maximum apparent viscosity for milk protein-stabilized foams was recorded at a concentration $< 0.5\%$. Increased protein concentration in foams caused the formation of thicker and more rigid interfacial film. The foaming capacity of sodium caseinate and αs-/K-casein-enriched caseinate was higher at 0.35% protein compared to 0.25% protein but lower for the β-casein-enriched caseinate [42]. Foams formed from 0.35% solutions of sodium caseinate and the β-casein-enriched caseinate were more stable than those prepared from 0.25% solutions whereas the αs-/K-casein-enriched caseinate foam was less stable at the higher concentration. The β-casein-enriched foams were most stable at 0.25 and 0.35% concentrations.

Lysozyme, the highly structured protein, forms small foam volume with small air cells, and lysozyme-stabilized foams are creamier than β-casein foams [27]. The foams stabilized with β-casein have low resistance to shear and low stability. Stability to collapse of foams prepared by shaking decreased in the order lysozyme > BSA > β-casein [18]. This was probably related to protein structure in lysozyme films, while β-casein exhibited a flexible, random structure even in solution. The foaming capacity of sodium caseinate solutions exhibited low dependence on pH and temperature [43].

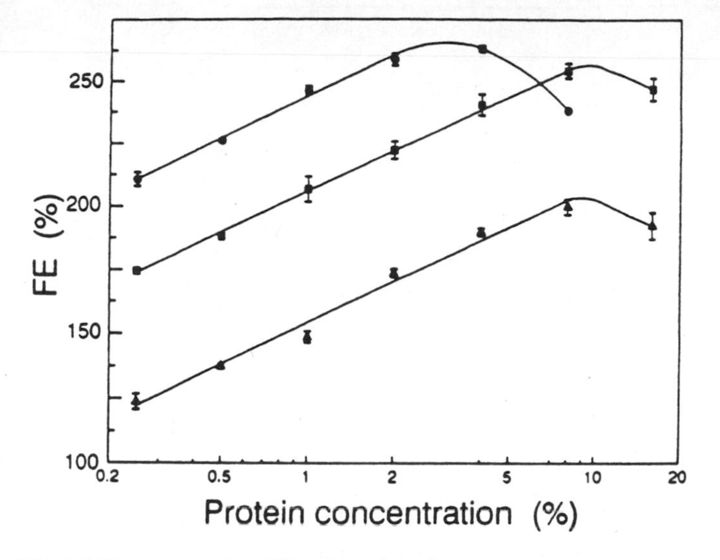

Fig. 5.6. Foam expansion (*FE*) of protein solutions. ●, sodium caseinate (CAS); ■, whey protein isolate (WPI); ◆, ovalbumin (OVA). From Ref. 40

The Effect of Protein Structure

Various functional properties of milk proteins are mainly determined by unique molecular structures of caseinates and whey proteins. Understanding of the relationship between structure, physicochemical properties and functionality of milk proteins is needed to extend their utilization in food products. The behavior of proteins at the interface is affected by physical interactions that are influenced by the composition and conformation of the protein in solution and at the air/water interface [44]. There are limited studies on the foaming properties of pure individual proteins or their mixtures to elucidate their behavior at the interface. It has been generally suggested that randomly structured proteins show higher surface activity and spread more fully at the interface than more highly structured proteins.

Structural properties of sodium caseinate and WPI, i.e. active sulfhydryl groups, flexibility and hydrophobicity influence its foaming properties. Sodium caseinate and WPI showed similar trends in hydrophobicity values as a function of pH and temperature [43]. The surface hydrophobicity values were not consistent with differences between hydrophobic amino acid concentrations in caseins and whey whey proteins [45]. It was suggested that the protein molecules reorient in solution to exhibit the minimum concentration of hydrophobic amino acid residues to the polar phase of the solution at various pH and temperature. Caseins are

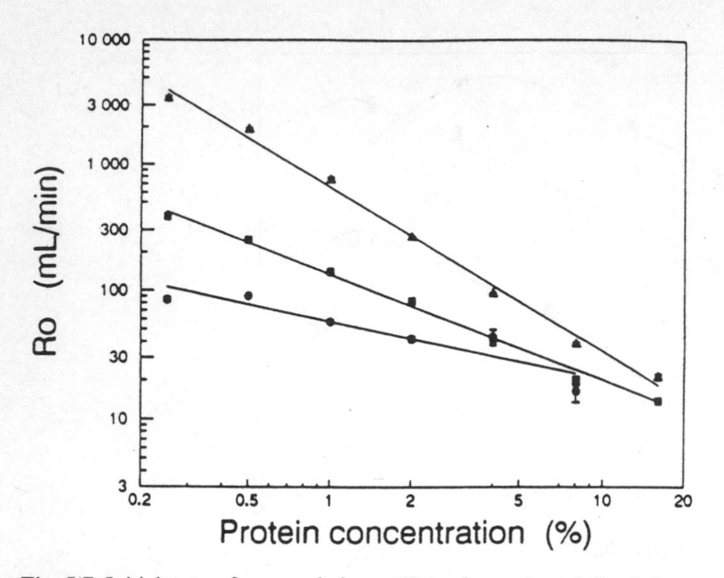

Fig. 5.7. Initial rate of serum drainage (R_o) of protein-stabilized foams. ●, sodium caseinate (CAS); ■, whey protein isolate (WPI); Δ, ovalbumin (OVA). From Ref. 40

highly susceptible to reversible intermolecular interactions with other proteins via hydrophobic ionic bonding and disulfide bonding mechanism.

The stability of whey protein foams may be increased by the opening up the globular structures of α-lactalbumin and β-lactoglobulin and exposing the -SH groups in these proteins. However, foam overrun was not improved. Hydrophobicity of proteins is considered as being responsible for the functionality of proteins as foaming agents, emulsifiers and gelling agents. Mangino et al. [31] compared hydrophobicities of whey proteins at various stages of processing and in finished WPCs. The surface hydrophobicity (S_o) was not significantly changed by ultrafiltration, when spray drying increased the S_o markedly.

The protein film-forming capacity is affected by the hydrophilic-hydrophobic balance, net charge, and osmotic and steric effects. The film-forming ability of proteins facilitates foam formation. Kim and Kinsella [16] showed a relationship between the film-forming capacity of β-casein and BSA and their foaming properties. β-Casein forms foam rapidly, however, because of limited protein-protein interactions, the foam is unstable. BSA, a globular protein with considerable tertiary structure at the interface, forms stable foams because of the formation of an extensive intermolecular network. The globular proteins form foams with small air cells and stable films resistant to deformation [27]. Less stable foams with larger cells are formed with β-casein, though more surface-active than lysozyme and albumin. During protein adsorption at the interface there should be numerous protein-protein interactions (hydrogen bonding, ionic, hydrophobic, van der Waals) to form a three-dimensional system. Casein easily absorbs at the interface and it can displace other proteins from air-water interfaces.

Correlation between the foaming properties of β-casein, lysozyme and BSA and their behavior at an air/water interface was established [18].

The higher FC of sodium caseinate was related to its ability to diffuse to the air-liquid interface, with proper orientation, decreasing interfacial tension and air entrapment. The rapid decrease in interfacial tension at very low concentrations was found for Na caseinate and WPC at the air-water interface using the drop volume technique [46]. This was observed, particularly at higher protein concentrations (> 0.1%). WPCs are weaker foaming agents than caseinates. This difference in functionality between whey proteins and caseinates is due to conformational differences in the two groups of proteins. Because of the lack of amphiphilic conformation whey proteins do not orient completely at air/water interfaces.

5.3.1 Factors Affecting the Foaming Properties of Milk Proteins

The Effect of Lipids

Lipids as components of undefatted or partially defatted proteins impair the foaming properties of milk proteins. Lipids incorporated in foam rupture the foam lamellae as a result of their higher surface activity. The surface hydrophobicity of proteins is negatively related to the lipid content because lipids and *cis*-paranaric acid compete for binding sites on the proteins . If hydrophobic sites are reserved by lipids they are blocked and do not participate in other hydrophobic interactions. A high level of WPC defatting will increase surface hydrophobicity and improve functionality, especially foaming and emulsifying properties. Mangino et al. [31] reported that removal of residual lipid from the whey markedly increased overrun of WPC. The FC of WPC was negatively correlated with total and free fat content [47]. The FC of WPCs prepared from fat-modified Cheddar cheese wheys was in the range from 330 to 648%. Very poor foaming capacity was found for WPC prepared from whole milk whey. No measurable foam was observed up to 15 min of whipping due to higher fat content. Whipping properties of protein systems in whipped toppings were affected by the type of lipid [48]. Poor whipping was obtained with non-hydroge-nated fats of low mel-ting point. Good whipping power was found for blends of cocoa butter and hydro genated oil. In a model system both low and high pH supressed the overrun of the product. The best foaming capacity was reported at pH 6.0.

Milk fat impaired the foaming properties of WPC and their fractions. In the simple foaming tests foaming properties, FC and FS of dilute aqueous solutions of WPC were inhibited by residual lipids, both total and phospholipids. Foam stability determined 15 min after foam formation can be predicted by the concentration of lipids, disulfides and ash (Fig. 5.8) [49]. These three variables gave an r^2 value of 0.829. However, in a complex colloidal system such as a high-fat whipped topping, fat and phospholipid content was positively correlated with the foam overrun. These data corresponded with Schmidt's and Van Hooydonk's

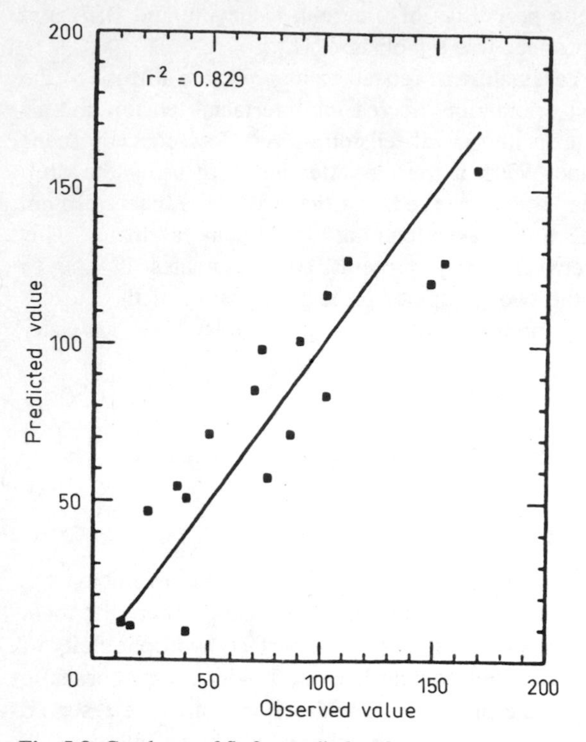

Fig. 5.8. Goodness of fit for predicting foam stability (FS) of 1% aqueous solutions of WPC and their soluble and insoluble fractions. From Ref. 49

[50] observations which reported a positive effect of phospholipids on foaming of cream. Whipping time decreased and overrun increased when concentration of phospholipid increased. However, Poole et al. [24] showed that foaming properties are affected by the type of acidic protein and lipid; more effective impairment of foaming properties was observed for lecithin than other lipids (corn oil, butterfat) due to effective disruption of the protein film. Unstable foams have been prepared with WPC in the presence of small amount of fat, particularly if monoglycerides and polar lipids are present in the lipids. The negative effect of lipids on foaming properties of WPC has been reduced by heating WPC solutions due to increase interaction between lipids and proteins. Several other methods have been developed to reduce the fat content of WPCs including centrifugal separation and filtration [51]. The use of ion exchange celluloses for protein recovery has it made possible to isolate WPI of high purity.

The Effect of pH

The foaming properties of whey proteins are affected by pH. The electrostatic interactions in foams have not been fully studied, however, it is recognized that the net negative charge on the outer film lamella stabilizes foams via net repulsion of

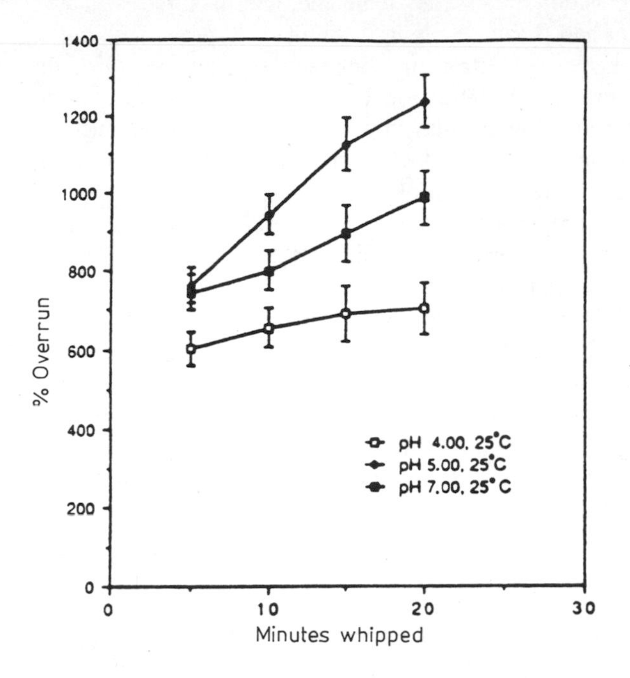

Fig. 5.9. Effects of pH on the overrun of WPI (5% concentration). From Ref. 53

adjacent films. Phillips et al. [52] showed that the addition of 0.5% lysozyme to WPI increased the foam volume by 981%, foam stability by 114%, and foam heat stability by 12%. The enhancement of foam volume and stability was due to electrostatic interactions between basic and acidic proteins. The effects of pH and prior heat treatment on foaming properties, i.e. overrun and foam stability of WPI were demonstrated by Phillips et al. [53]. The WPI was predominantly β-lactoglobulin (92% of WPI). The pH change 4.0, 5.0, and 7.0 had little effect on the surface pressure. The overrun of the WPI solution was affected by pH with the highest percentage overrun at pH 5.0 (Fig. 5.9) [53]. The FC at pH 5.0 increased with foaming time from 5 to 20 min. The high resistance of β-lactoglobulin to heating requires longer heating time for unfolding and formation of a stable film.

Foam overrun values ranged from 753% for WPI at pH 7 and 25 °C to 1536% for sodium caseinate at pH 6 and 65 °C [43]. The foaming capacity of WPI showed a direct dependence on pH and was not well related to temperature at pH 6 and 7. Le Meste [54] reported that WPC exhibited maximum FC and FS between pH 4 and 5. However Lee et al. [43] found higher FC and FS values at pH 8 of WPI solutions heated at 55 and 65 °C. Foams with poor stability have been prepared with sodium caseinate at pH 6 and 8, presumably due to the inability to form intermolecular disulfide bonds at those pH values. Stable foams have been prepared with WPI. The wide variations of the pH effect on FC of WPC have been noted depending on the composition of the whey foam and other factors.

The mixture of acidic and basic proteins exhibited improved foaming properties because of enhanced electrostatic interactions between protein molecules at the water-air interface [55]. The property of lysozyme and other basic proteins with pI > 9.0 to enhance the foaming power of acidic proteins such as BSA and β-lactoglobulin is related to the molecular size and conformation of the complex and to the net charge of the protein molecules [55]. Poole et al. [9] found positive effects on the foaming properties of several acidic milk proteins (pI, 4.7–6.0) when a small amount of basic proteins, such as lysozyme (pI, 10.7) and clupeine (pI, 12) were added. Improvement in foaming properties is due to complex formation between positively and negatively charged proteins. The FC of acidic proteins, i.e. WPC, WPI, BSA was enhanced by the addition of basic proteins (clupeine) due to increased tolerance to lipids in protein foaming system [24].

Clupeine was more tightly associated with β-lactoglobulin than lysozyme. Utilization of basic proteins (clupeine) in conjunction with acidic proteins (WPI) may improve foam stability.

The Effect of Additives

Electrolytes influence the foaming properties of milk proteins. In the presence of electrolytes increase in whey protein diffusion into the air/water interface was observed [46]. These data explain the positive correlation between ash concentration and foaming properties. Different ions may exhibit various foaming properties because of their effect on the structure of water and on protein conformation [1]. Halling [26] reported that addition of sodium chloride improved foaming properties by modifying the net charge, increasing adsorption and possibly decreasing the level of protein denaturation during foam formation. The foaming capacity of WPC was positively correlated with Ca and ash contents [47]. Improvement of FC with increased Ca is probably due to the formation of calcium bridges between carboxyl groups of protein molecules at the air/water interface. As a result a cohesive and rigid protein film is formed around air cells.

The FC was decreased by adding sucrose or soluble starch to WPC solutions due to the increase in viscosity of the solution, and probably due to minimizing protein unfolding at the interface because of increased protein stability. Sucrose had a synergistic effect on the lysozyme foaming properties. Sucrose and lysozyme made the foaming properties of β-lactoglobulin and WPI superior to egg white (Table 5.1) [52]. Similar results were obtained by the addition of clupeine and sucrose to WPI. The self-association of β-lactoglobulin in solution can be strongly enhanced by sugars. The decrease in aggregation allowed more protein to be involved in film formation and stabilization. The addition of sucrose may cause formation of a better network in the film. For certain uses, foaming properties of WPC may be improved by blending with other proteins.

Table 5.1. Foaming properties of whey protein isolate, beta-lactoglobulin and egg white[a]

Protein[c]	Overrun (%)	Foam stability (min)	Foam heat stability (%)	Remaining overrun[b] (%)
ß-Lg	1346	31	19	251
ß-Lg, lysozyme & sucrose	1706	61	99	1689
WPI	640	26	76	488
WPI, lysozyme & sucrose	1192	67	99	1180
EWP	1162	42	57	667

[a] Each was whipped for 15 min using 5%, pH 8.00 solutions for whey protein and egg white and 2.5%, pH 8.00 solutions for beta-lactoglobulin.
[b] A remaining overrun = % overrun x % foam heat stability.
[c] ß-Lg = ß-lactoglobulin; *WPI* = whey protein isolate; *EWP* = egg white protein. From Ref. 52

The Effect of Heating

The functionality of whey proteins in food products might be improved by processing conditions which minimize denaturation. Processing factors that minimize protein aggregation result in more acceptable whipping and foaming properties. Morr [51] reported that whey pasteurization had little effect upon the composition and solubility of WPCs. WPI consists of four major proteins, i.e. β-lactoglobulin, α-lactalbumin, BSA and the immunoglobulins that exist in their native state as compact, globular proteins highly susceptible to heat-induced denaturation and polymerization via intermolecular disulfide bond formation. An increase in FC of WPC-stabilized foams was obtained by mild heating due to the partial unfolding of whey proteins. Mild heating increased overrun and enhanced the stability of WPC foams. However, the heating effect is influenced by the pH of the protein solution [24].

In food processing, the functionality of milk proteins is modified by heating to accelerate milk protein functionality at the air/water interface. Significant correlation was found between FC and solubility of WPCs. Increase in overrun and FS of whey protein foams was achieved by prior mild heat treatment, ~40 °C [37]. This positive effect of heating on of FC and FS may be related to the formation of lipid-protein complexes. During heating, lipids may form lipoprotein complexes which precipitate out of the foaming mix and facilitate foam formation. The foaming properties of WPCs were improved by heat treatment that caused reduced whipping time, increased overrun and more stable foams. Spray dried WPCs and α-lactalbumin enriched samples derived from acid whey formed relatively stable foams with average overruns of 784% [33]. These foams were comparable to samples with freeze dried egg whites with average overruns of

633%. However, freeze dried WPCs exhibited low foaming ability when compared to spray dried WPCs and freeze dried eggs.

Heating WPI at 80 °C significantly reduced the FC at all pH's tested. The largest reduction in FC was observed after heating of WPI at pH 5.0 (63% after 20 min heating at 80 °C) [53]. At high heating temperatures, excessive interactions can cause protein aggregation and reduction of film formation. Heating of WPI at pH 7.0 and at 5580 °C increased the overrun of WPI by 22%. The highest overrun values of WPI were found at pH 5.0 with no heat treatment and at pH 7.0 after heating at 55 °C. The mild heating (55 °C) caused partial unfolding of β-lactoglobulin and increased protein-protein interactions. To enhance the FC of WPI, mild heat treatment (55 °C) and pH below 5.0 are recommended to minimize protein-protein interactions. Effect of heat treatment of WPI on FS at pH 4.0 or 7.0 was insignificant but at pH 5.0 and 80 °C, FS increased by 65% [53]. The conditions of foaming, especially temperature and pH, should be controlled to obtain foams with acceptable overrun and stability.

Significant effects of water activity, storage temperature and storage time on the FC of WPC has been reported [56]. The values of overrun of all samples ranged from 650 to 720%. The FS as a function of storage time and water activity at 20 °C was similar in air and nitrogen activity atmospheres. Foaming properties of WPCs prepared with HTST heating were affected by pH and pretreatment temperature [57]. Maximum foam expansion values ranged from 460 to 1050% for pH 4.5 and 25 °C pretreated WPC and 780–1275% for pH 4.5 and 55 °C pretreated WPC. Poor FC was found for WPCs with pH 9 and 25 °C pretreatment. Higher temperature treatment may have modified the proteins conformational structure improving their capacity to retain air and FC. HTST pasteurization of milk increased FC of WPCs, however, no effect was found for heating whey and ultrafiltration retentates on WPCs foaming capacity. Morr [57] suggested that minor changes in protein composition and conformation are necessary to improve functional properties.

The Effect of Protein Modification

The term lipophilization has been used in a broad sense to describe the increase in hydrophobicity of proteins by modification. The attachment of the fatty acid was between the carboxyl group of the fatty acid and the amino groups of the protein. This resulted in decrease in available amino groups. Fatty acid attachment was selected to modify the hydrophobicity of β-lactoglobulin because fatty acids are naturally occurring in the body. The stearic acid was incorporated into β-lactoglobulin and functionality was reported [58]. The foaming properties of modified β-lactoglobulin were measured using the bubbling technique. Nonsignificant increase in FC was observed at low and medium levels of fatty acid incorporation, however high level (13.1 mol) gave a very poor foam capacity. The FS increased at medium levels of fatty acid incorporation. The maximum FS was observed when 1–4 mol of stearic acid were attached per mole of the protein. FS decreased at higher levels of incorporation. The foam with 13.1 mol attached was found to be extremely unstable and collapsed almost immediately after formation. However, Haque and Kito [59] demonstrated increase in FS with an increase in fatty acid incorporation. Palmitoyl residues were covalently incorporated into αs_1-casein. FC increased until 6 mol of palmitoyl residues were attached per mole of protein, then it started to decrease with further incorporation. Increase in FC was related to improved amphiphilicity of the proteins. The enzymatic hydrolysis of whey proteins caused an increase in foam specific volume at the first step of beating then decreased with time. A slight increase (25%) in specific foam volume was observed when limited hydrolysis of whey protein was tested [54]. The flexibility of casein was altered by: 1) pH change, or 2) chemical modification (covalent binding of monosaccharide on the lysyl residues). A high level of glycosylation (80% galactosylated samples) increased protein flexibility, FC and FS of foams obtained with casein.

5.4 Egg Proteins

Egg White Proteins

Egg white is pure protein solution because proteins in egg white are mostly present in soluble form and can be easily isolated and studied. Egg white consists mainly of proteins (10.5%) and water (85%). Egg white contains as many as 40 different proteins. An important functional property of egg white is its high viscosity which decreases with aging as the ovomucin-lysozyme complex breaks down. This complex decomposition occurs at a higher pH which increases during storage from an initial value of 6.8 to over 9.0. Egg white during heating and coagulation is forming a continuous protein-network, trapped-water system. Egg white is the most widely used foaming agent, and due to unique combinations of proteins, the egg white produces large volume foams stable to heating. The high FC of egg white reflects the unique combination of proteins with different physical properties.

Egg albumen is a heterogeneous protein system, which means that protein-protein interactions occur during foaming. The mechanism of egg foam formation includes surface denaturation of albumen proteins at the liquid-air interface, and foam stability is related to albumen insolubility. The basic functionality of albumen in foam is the prolonged retention of a large volume of gas. However, the FC of albumen is not unique because most of the proteins have foaming properties. Native albumen contains water-soluble, surface-active proteins that can migrate to the air/water interface [26]. The unfolding of egg proteins at the air/water surface is caused by surface denaturation. The rate of surface denaturation is faster for highly hydrophobic secondary-structured proteins than for proteins with tertiary and quaternary structures. Foaming properties of albumen are related to surface denaturation of the globulin fraction of albumen followed by protein-protein interactions [10]. Egg albumen is easily denatured by most of the denaturing treatments. Aggregated protein films are produced from denatured proteins of albumen through a variety of physical and chemical bonds. These films accelerate air incorporation during beating egg foam. Aggregated particles of proteins play an important role in the formation of stable egg foam by holding water in the lamellae and providing rigid and elastic properties to the protein film [26]. The FC of six proteins of egg albumen have been reported. The FC of solutions was related to the performance of proteins in angel food cakes [60]. There was an inverse relationship between the foaming index of the protein solutions and the volume of the cake. The foaming index was measured as the specific volume (cm^3 g^{-1}) of a whipped protein solution divided by whipping time.

The major protein in albumen is ovalbumin, ~54% of the total egg albumen solid. The isoelectric region for ovalbumin is pH 4.64.8. Ovalbumin, highly structured globular protein is a phosphoglycoprotein in which phosphate and carbohydrate are attached to the polypeptide chain. Carbohydrate increases film viscosity and foam stability. Ovalbumin, a hydrophobic protein, denatures and coagulates in solution relatively easily when exposed to new surfaces. Heating of ovalbumin enhances rapid film formation, but decreases foam stability. Ovalbumin molecules are absorbed at the air-water interface with orientation of their hydrophobic portion toward the air thus exposing cysteinyl residues for reaction [61]. The foaming properties of ovalbumin related to protein concentration reached limiting values at pH ~4.0, i.e. below the pI. Extremely high sensitivity to surface denaturation was exhibited by ovalbumin, and it was found in a precipitated form in egg white foams. Surface coagulation of ovalbumin is presumably an important factor in the overbeating of egg white. The next most abundant protein in egg albumen is conalbumin or ovotransferrin which makes up about 12% of the egg albumen solids. Conalbumin can bind with metals and is more heat sensitive but less susceptible to surface denaturation than ovalbumin. Thermal coagulation of conalbumin in albumen is effectively inhibited by the addition of 2% sodium hexametaphosphate.

Ovomucoid, which makes up about 11% of the total egg albumen solids, is a glycoprotein and contains about 20-25% carbohydrate. Ovomucoid is an effective foaming agent resistant to heat in acid and alkaline media. Ovomucoid coagulates at temperatures above 60 °Cif lysozyme is present in an alkaline solution of protein. Ovomucin is a glycoprotein not soluble in water, but soluble in a dilute salt solution at pH 7 or above. Ovomucin stabilizes foams formed during short whipping time. Ovomucin and the interacting variables of ovomucin with globulins, ovomucoid, conalbumin, and ovalbumin correlated positively with foaming index [62]. Improvement of FC in ovomucin solution is mostly due to increased viscosity. Ovomucin increased the foaming index of solutions that contained no lysozyme and various levels of globulins. The greatest foaming index was produced by globulins while ovomucin and ovomucoid did not show a very high foaming property [62]. Ovomucin and ovomucoid exhibited high resistance to denaturation possibly due to the large number of disulfide linkages stabilizing the protein structure. Inter- and intramolecular sulfhydryl-disulfide reactions were involved in foam formation. Johnson and Zabik [60] observed a network of parallel filaments that included ovomucin, lysozyme, and globulins. Protein solutions with high ovomucin content and without lysozyme showed good FC but produced smaller cakes [62]. This effect of ovomucin has been related to its insolubilization at the air-albumen interface. This facilitated obtaining stable foams with lower expansion. Ovomucin increased film viscosity and facilitated film formation.

Lysozyme has the highest isoelectric point (pH ~10.7) of all the other albumen proteins and the lowest molecular weight. Lysozyme forms viscous films and enhances foam stability. Lysozyme influenced the foaming properties of other egg proteins. Lysozyme, especially at higher concentrations, depressed the FC of globulins, conalbumin, and ovalbumin. The negative effect of lysozyme on foamability may be related to the increase in the rate of formation of an ovomucin-lysozyme complex. Lysozyme exhibited improved foaming power in the absence of ovomucin. The foaming index of mixtures of egg proteins was reduced when lysozyme was added at pH 5.3 and an ionic strength of 0.20. This was due to a specific interaction between egg proteins, ovomucin and lysozyme. Globulins effectively reduced the interfacial tension and increased the FC of protein solutions at any level of lysozyme and ovomucin but especially in the absence of lysozyme and with 2.5% ovomucin. Johnson and Zabik [62] reported that the most desirable foams and cake volumes were prepared with the protein levels: 0.2–1.0% ovomucin, 0.–0.8% lysozyme, and 12.2–14.8% globulins. Lysozyme interacts with other egg proteins with conalbumin, ovalbumin, and ovomucoid, and this interaction may influence the chemical and physical properties of egg white. Carbohydrate components of ovomucin participated in electrostatic associations with lysozyme. The formation of the ovomucin-lysozyme electrostatic complex may be sufficient to produce a change in foam stability. The formation of a damaged network of the ovomucin-lysozyme complex produces a „dry“ foam that can be damaged by mechanical action. Johnson and Zabik [63] observed some variability in membrane thickness; globulin and ovalbumin films looked very thin; conalbumin film seemed slightly thicker and less flexible while lysozyme film was considerably thicker.

The foaming properties of mixtures of lysozyme and bovine serum albumin (BSA) have been reported [64]. Addition of lysozyme to a 0.5% BSA increased the foaming properties of the system when the BSA:lysozyme ratio was 5:1. It was found that at pHs distant from pI of BSA the mixture of BSA and lysozyme had much enhanced foaming properties. Foam expansion and foam liquid stability of BSA (0.5%, pH 8.0) were affected by lysozyme addition [64]. Foaming properties increased linearly with a lysozyme level up to 0.1%. Increased foam expansion in a BSA system with lysozyme is due to the overcoming of the charge barrier which retards the absorption of BSA molecules at the interface. The foaming properties of BSA were impaired at higher concentrations, more than 0.1% of lysozyme added, probably because of the large amount of protein insolubilized. BSA and lysozyme interact electrostatically and form a complex and migrate to the air interface as a complex or separately. As a result, an increase in the strength of the film and FS was obtained. The foaming power of BSA alone (at pH 8.0) can be increased by the addition of NaCl, presumably because electrostatic repulsion reduced intermolecular interactions [64]. When electrostatic attractions promote association between the two proteins, foam expansion decreased.

The storage time of the eggs influenced foaming and other functional properties. Eggs stored for one week reached their maximum foam height more rapidly than eggs stored for one day. Older eggs gave foams with higher volume but lower stability. Egg whites beaten at room temperature reach a greater volume more quickly than egg whites beaten at refrigerator temperatures. Warmer egg whites probably have lower surface tensions than cooler egg whites. However, foams prepared with egg whites at room temperature were not as stable as egg whites beaten at refrigerator temperatures.

Control of egg white pH i.e. lowering the pH would bring ovalbumin closer to its isoelectric point. At a pH range close to the isoelectric point proteins coagulate more easily and form more stable foams. If the pH is lowered to ranges below pI, peptization will occur with decreased foam formation and stability [64].

Foam overwhipping can disrupt film formation and reduce foam volume. For example, in egg white, foam formation of an insoluble lysozyme/ovomucin complex was observed [62]. Foams stabilized by albumen are generally stiff due to the gel structure of albumen. Factors influencing FC and FS such as the time of beating, speed of agitation, nature of the beater, temperature, the type of beater have been standardized in the industry for optimal results and the norms are established. The FS is also affected by moisture content of the albumen, and addition of water - up to 33% increases the foam volume.

Yolk Proteins

Information related to the foaming properties of individual proteins in egg yolk is much more limited than related to egg white proteins. Individual yolk lipids, proteins, and yolk fractions containing free lipid have no foaming capacity. Lipovitellenin is the major ingredient of yolk that participates in foam formation and stabilization. Lipovitellin inhibits foam formation but participates in foam stabilization already formed by other proteins, including lipovitellenin and albumen. The negative effect of yolk in liquid albumen is due to the formation of the electrostatic complexes between lysozyme and yolk lipids. Addition of 1% yolk to egg albumen showed an adverse effect on the specific volume of angel food cake, indicating the negative effect of yolk contamination. LDL showed low foaming properties and these properties were considerably damaged by heating at temperatures above 77.5 °C [65].

5.4.1 The Effect of Processing on Foaming Properties of Egg Proteins

Heat Treatment

The preservation of the functionality of egg albumen and yolk during pasteurization and dehydration is of significant practical importance.

Pasteurization of liquid egg albumen damaged its physical and functional properties. The FC of albumen is dependent on the pasteurization temperature and decreased with increased heating temperature. Ovomucin and lysozyme are two proteins responsible for protein damage by heat and damage to foaming properties. The influence of heating on the FC of albumen is related to heat denaturation of the ovomucin-lysozyme complex. Denaturation of the ovomucin-lysozyme complex presumably causes the formation of weak lamellae which are easily destroyed during foam beating. Heating of egg white causes damage to the foaming properties and this damage is influenced by the level of lysozyme. Heat damage is reduced when lysozyme concentration is less than that normally found in egg white. The formation of an ovomucin-lysozyme complex explains the impairment effect of pasteurization on the whipping capacity of egg proteins. The formation of ovomucin-lysozyme complex is accelerated by heating and followed by denaturation and aggregation causing a longer whipping time and a loss of foaming properties of the egg white.

Dehydration and Freezing

Commercially produced egg powders are spray-dried. Egg whites are also dried in trays or as a film on a continuous belt. Coagulability of spray-dried eggs is lower than frozen eggs. The foaming properties of spray-dried albumen are impaired as a result of the shear forces during atomization. FC and FS values were compared for products fractionated from nonspray-dried and spray-dried egg yolk in distilled water and 0.1-0.3 M NaCl solutions [66]. LDL was found to be responsible for foaming capacity. This was supported by the apparently greater foam capacity of the spray-dried granule fraction compared to that of the nonspray-dried granule fraction.

With a proper freezing and thawing process there is no significant damage to the functional properties of egg white. However, in commercial conditions poor functionality, especially the foaming property, is observed in frozen and thawed egg whites. There changes have been found in total nitrogen and amino acid composition. Denaturation of proteins in frozen egg white is probably the reason for egg white deterioration. Denaturation is possible during freezing, frozen storage and thawing. The amount of water soluble nitrogen in egg white was used as an indicator of protein denaturation and no change in soluble nitrogen was found after various storage periods. The effect of the freezing rate on egg white textural and foaming properties, foam specific gravity and foam instability of the thawed samples has been reported [67]. A decrease in the freezing rate from 3.10 to 0.03 °C/min caused a decrease in viscosity from 4.42 to 3.71 cPs. This was accompanied by a decrease in foam instability. The specific gravity of foams remained relatively constant at all freezing rates. The effect of thawing conditions on egg white viscosity and foaming properties was nonsignificant. The egg white properties were affected by the time of frozen storage. With increased storage time, increased losses in enthalpy of denaturation (<6,172> H_D) were found. The effect of frozen storage on the percentage loss of enthalpy of denaturation was more significant for conalbumin than ovalbumin. Changes of <6,172> H_D of conalbumin and ovalbumin were accompanied by a decrease in viscosity with increased storage time. Storage time caused a decrease in foam instability and an increase in foam specific gravity.

The Effect of Irradiation

The overrun of egg white was not significantly changed by irradiation at low doses but decreased at 4 kGy [68]. Enhancement in whipping properties of albumen was observed in irradiated shell eggs.

The Effect of Additives

Sugar added at the early stages of the beating process increased FS as a result of increased viscosity of the foam [69]. Addition of salt decreased the stability of foams prepared with dry egg whites and fresh egg whites beaten for a short time. Salt does not influence the stability of foams prepared from fresh eggs beaten for an extended time [9]. Sodium lauryl sulfate improved the FC of egg proteins and reduced surface tension, beating time and increased the cake volume. The foaming capacity and other functional properties of commercial egg white can be improved by the addition of sodium dodecyl sulfate (SDS). SDS forms complexes with egg white proteins and reduces surface tension. Van Elswyk et al. [70] reported that volumes of sponge cake prepared with n-3 fatty acid-enriched eggs did not significantly differ from those prepared with control eggs.

5.5 Blood Proteins and Gelatin

The blood globin protein exhibited a minimum FC at pH 7-10 [71]. At pH 3.4, the foam was stable for the first hour of the quiescent period. However, after the 2nd hour, the foam volume dropped to one third of its original value. The FS of the globin product was comparable to that of egg white at pH 6.0 and equivalent protein concentration.

Phosphated blood plasma proteins exhibited better foaming properties than ultrafiltered plasma proteins [72]. A substantially greater foam volume was produced by phosphated plasma proteins than egg proteins. The foam volume increased with protein concentration and decreased with holding time. The foam coalescence and leakage of free liquid increased at the lower protein concentrations. The FC of ultrafiltered plasma protein was similar to egg albumen. Differences in foaming properties between phosphated and ultrafiltered proteins were probably due to different amino acid composition and electrophoretic properties. The phosphated plasma protein had reduced levels of methionine and tryptophan.

Protcinaceous modified surfactants produced from gelatin were used in snow jelly, ice cream, and mayonnaise to replace conventional surfactants [73]. An increase in foam expansion was observed with the addition of gelatin modified with leucine C_{12} alkyl ester incorporated for ice cream and leucine $C_{2\text{-}6}$ alkyl ester incorporated for snow jelly. Proteinaceous surfactants especially gelatin-leucine-OC_{12} gave larger values of overrun than controls.

5.6 The Foaming Properties of Plant Proteins

Soy proteins

Significant variability in the foaming capacity between soy proteins is related to the different level of protein denaturation during preparation. Little information is available on the foaming properties of isolated 7S and 11S globulins. Functionality of glycinin (11S protein) may be limited by its relatively stable oligomeric structure. Molecular flexibility is important for use of 11S protein as a foaming agent. Molecular flexibility and FC was increased by the reduction of component disulfide bonds [74]. The improvement of foaming properties of soy proteins was obtained by moderate heating at 70–80 °C and partial proteolysis [75]. Increased unfolding of the polypeptides at the interface and facilitation of the hydrophobic associations contributes to improved foaming properties. The moderate heating at 70–80 °C may increase unfolding of soy protein at the air-water interface during foam formation and enhance intermolecular interaction to produce a stable film.

A membrane isolated by ultrafiltration of soy and cottonseed proteins can be competitive functionally with the conventionally produced commercial soy isolate. The foaming properties were higher with soy protein isolate and cottonseed nonstorage protein. Soy isolates obtained by the membrane isolation process showed a considerably better FC than commercial soy isolate [76]. Foaming properties of freeze-dried protein powders of soy, cowpea, and peanut preparations prepared from unfermented and fermented extracts are presented in Fig. 5.10 [77]. The FC and FS of freeze-dried powders prepared from legumes were greater than a commercial freeze-dried cultured butter milk powder. The similar initial foam volume of thrcc tested legume proteins was found. All tested proteins performed differently during foam incubation. Unfermented freeze-dried cowpea milk protein gave the most stable foam, followed by unfermented milks extracted from soybeans and peanuts. A preliminary fermentation process of the protein milks decreased the stability of foams prepared with freeze-dried powders. The fermentation process probably caused the acid coagulation of proteins in the legume milks affecting the capacity of these proteins to retain their foam stability. Unfermented freeze-dried cowpea protein produced foam of the greatest stability and this is related to a very low residue fat content when compared to the peanut and soybean proteins.

The foaming power of soy proteins can be enhanced by protein modification, by partial hydrolysis of soy proteins. Partially hydrolyzed soy proteins form a more stable interfacial membrane, however, stability of foams with these proteins is lower than that of the nontreated proteins [26]. Foam stability is influenced by the retention of a high degree of tertiary structure in the protein. Foams of lower stability have been prepared with increasing alkaline or enzymic hydrolysis of soy proteins. The foam quality of high molecular weight partly digested protein was superior than with native proteins. The soy isolate, canola protein isolate and sunflower concentrate have been modified to improve foaming properties [78].

The proteins were solubilized by treatment with trypsin, potassium linoleate, and SDS. Treatment with trypsin and SDS increased the FC, while linoleate increased the FC only for soy protein. The foam stability was different for tested proteins and methods of modification. Foam stability increased for soy protein modified by three tested methods. Foam drainage decreased for sunflower protein modified with SDS and linoleate. Less stable foams were prepared for canola protein modified with trypsin and SDS treatments.

Soybean glycinin has been modified by attachment of palmitoyl residues to hydrophilic acidic glycinin to improve the foaming activity and FS [79]. The results suggest that even for relatively hydrophobic proteins the amphiphilic balance could be improved by attaching hydrophobic palmitoyl residues. The thiolation, i.e. introduction of new SH groups into the soy protein isolates, considerably (2-2.5 times) increased foam expansion, however showed a drastically decreased foam stability [80]. Succinylation caused changes in viscosity and hydrophobicity of oligomeric glycinin that showed changes of its conformation and dissociation. The unfolding of glycinin caused an increase in viscosity. The FS is affected by interfacial film elasticity, and in succinylatd glycinin, dynamic elasticity increased at 20 and 50% of succinylation and decreased upon further succinylation. The FS measured as the half life time of liquid in the foam increased at 25% of succinylation and then decreased [81]. The FS was also higher than that of native glycinin at high levels of succinylation. A high degree of glycinin succinylation caused complete dissociation, reduced protein-protein interactions and decreased the strength of the interfacial films.

The functionality of soy isolate is affectd by the process of phytate removal. The phytate, the hexaphosphate in soy isolate, is a highly reactive compound capable of forming complexes with negatively charged protein molecules through a Ca and Mg bonding mechanism. The process for phytate removal was developed, and phytate-reduced (77%) soy isolate was prepared [82]. Phytate-reduced soy isolate gave maximum foam expansion values (1336%) comparing with control soy isolate (984%). The FC of phytate-reduced soy isolate was independent of preheating temperature. The greatest FC was observed at pH 3 and 5 and 10% protein concentration.

Peanut Proteins

Highly defatted peanut flour, concentrate and isolate are considered as potential food ingredients with specific functionality. The foam capacity and stability of peanut protein was affected by adjusting the pH of the protein suspension [83].

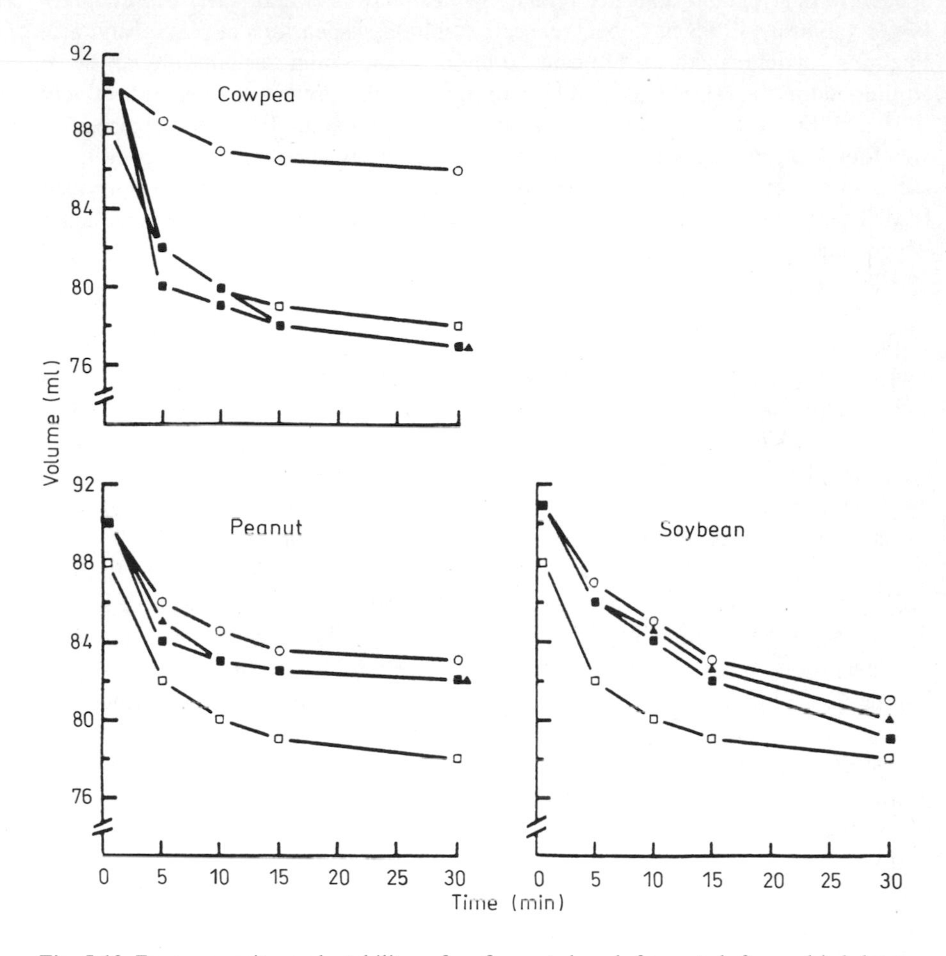

Fig. 5.10. Foam capacity and stability of unfermented and fermented freeze-dried legume powders and cultured buttermilk powder (CBMP). ❑, cultured buttermilk powder; ❍, unfermented seed extract, Δ, fermented with *L. bulgaricus*; and ■, fermented with *S. thermophilus*. From Ref. 77

The highest FC was obtained at pH 1.5. However, foam formation at this pH was depressed by increasing the concentration of NaCl in suspension to 1.0 M NaCl. There was a little difference in the FC at pH 4.0 between water and salt suspensions. A low FC was obtained at pH 6.7 and 8.2 for peanut protein meal suspensions. The foaming capacity increased as salt concentration increased at pH 6.7. The most stable and acceptable foams have been formed with peanut protein suspensions after two-step pH adjustment from 6.7 to 4.0 to 8.2, due to increased percentage of soluble protein (from 45.6 to 70.0%) [83]. The adjustment of pH could be an effective procedure for improving the foaming properties of peanut protein suspensions. The stability of the foam showed little variation among suspensions at pH 1.5 and 4.0, remaining between 35% and 45%. Significantly larger variations in FS have been observed among suspensions at pH 6.7 and 8.2. Foaming capacities and FS of peanut protein isolates from various cultivars were similar and in the range 6.3-7.5 ml/g and 5.8-7.5 ml/g protein isolate, respectively [84]. The FC was mostly inversely related to protein solubility, i.e. increased FC was found in proteins with decreased protein solubility. The foaming capacities of the fractions of peanut protein were modified by hydrolysis with α-chymotrypsin [85]. The foaming capacities were in the order total protein > arachin ≥ conarachin I > conarachin II.

Bean Proteins

The foaming capacity of navy and kidney bean flours and their concentrates was significantly higher than comparable soy products, 40.1%, 55.3%, and 20.3%, respectively. However, foams prepared with dry bean concentrate were less stable than that of the soy concentrate (Fig. 5.11) [86]. The higher FC and lower FS of dry bean protein products than soy products may be related to compositional and structural differences between navy, kidney beans and soy products.

Tepary beans contain 19.7 to 32.2% protein soluble in water and salt solutions [87]. Foaming capacities of both the tepary albumin and tepary flour were higher than those of tepary globulin and soy protein isolate. The higher FS was observed in tepary flour possibly due to other components of the tepary flour and their interactions with proteins. Because of the excellent FC and FS of tepary flour and tepary albumin, they can be used in bakery and other foods where good foaming is necessary. The black gram bean protein concentrate exhibited good foaming properties [88]. The highest increase in foam volume (126%] was at pH 2.0, while in 1% NaCl solution increase in foam volume was 86%. Faba bean and field pea protein products had excellent FC and FS compared to soy protein products [89]. They developed high initial foam volume stable during 2 h incubation. There was no difference in the FC between fine and coarse fractions of the pin-milled bean fraction [90]. The foaming properties suggest a potential application in beverages, and bakery products. The foam stability showed a trend similar to FC, however, changes of foam volume for the high-protein fractions were less dramatic than for the high-starch fractions. Roasting affected the FS of all fractions. Chick-pea fractions with high lipid content exhibited lowest foaming capacity and specific volume. An increase in the FC of winged bean protein flour with an increased degree of acylation has been reported [91]. The FC increased by 38% at 90% acetylation and 63% at 87% succinylation. However, acetylation resulted in poor FS in contrast to the succinylation that produced stable foams.

Guar Proteins

The foaming properties of guar protein were comparable to soy protein. The FC of guar protein meal increased with increased concentration up to 5% [92]. The FC of guar protein was affected by pH, and minimum FC was found at pH 4.5 in the iso-

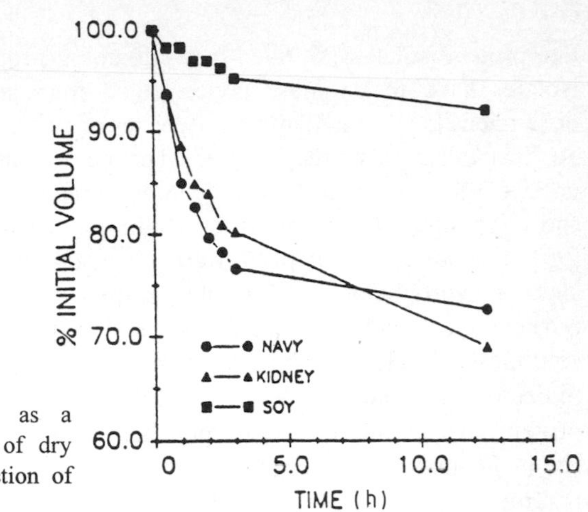

Fig. 5.11. Foam stability, expressed as a percentage of initial foam volume, of dry bean and soy concentrates as a function of time. From Ref. 86

electric pH range. The minimum FC at this pH was probably due to the low protein solubility. The FC of guar protein was higher than soy protein at different pH's, except at the minimum point, where soy protein had higher FC values. Soy isolate showed higher FC values than guar protein isolate only around pH 4.5 and no difference in the FC was found at pH's below 3 and above 5.5. Absence of the protein unfolding and spreading at the isoelectric point is a possible reason for low FC values. The FC values of guar proteins increased with NaCl concentration to a maximum (190–215%) at 0.3 M NaCl and then decreased. Guar protein meal contains saponins which also contribute to the foaming properties. Guar protein meal and guar protein isolate showed good foam stability. Guar protein isolates prepared by various methods exhibited a FC from 94 to 117% against 125% for the control defatted meal isolate [93]. Guar meal isolate extracted with aqueous 2-propanol showed highest the FC (150%) probably due to its higher protein content and nitrogen solubility at pH 7. The stability of foams prepared with guar protein isolates was comparable with the respective meals and equal to 47–62%.

The pH-FC profile (pH range 2–11) of guar protein meal extracted by different methods have a similar trend to that of the pH-nitrogen solubility profile [94]. The low FC values were found in the pH range 4.5–6.5. The defatted guar protein meal showed a higher FC than defatted soy protein meal. Nath and Narasinga Rao [92] reported different data that guar meal had a lower FC than defatted soy protein meal in the pH range 4.5–6.5. The lowest FC was found for acid extracted proteins. Isopropanol extracted protein showed the highest FC at all pH values, followed by methanol and ethanol extracted. The specific property of isopropanol-extracted protein was the highest FC (95 ml/g protein) at pH 4.7 compared to 20–80 ml/g protein for samples extracted by other procedures. A high FC was found for foam prepared with defatted guar protein meal (44%), followed by raw guar protein meal (33%), autoclaved (24%) and soy protein meal (30%). Extraction methods affected the FS of foams.

Pea Proteins

Pea protein isolates exhibited better foaming properties than soy isolates, but pea isolates have to be more concentrated than soy isolates to produce viscous dispersions [95]. The foaming properties of pea isolates were adversely affected by all heat treatments except 70 °C. The pea isolate/whey concentrate exhibited a greater FS than NFDM. The legumin preparations of pea protein concentrate showed a higher FC, whereas the vicilin preparations produced more stable foams [96]. Foams formed at pH 7 were made up of large unstable air bubbles. The surface hydrophobicity of soluble protein at pH 3 was greater than at pH 7. Surface hydrophobicity and FC were significantly increased by heating protein solutions with pH 7 at 90 °C for 5 min. Increased surface hydrophobicity was the indicator of protein molecules unfolding. The surface hydrophobicity of pea protein concentrate increased upon heating for 1 min at 80 °C, at pH 5.8. More stable foams were prepared with vicilin because vicilin contains fewer cysteine residues and fewer potential disulfide bridges therefore forming more flexible films. The sulfur amino acid content is higher in legumin and during heating there is a slow unfolding and a relatively low FS is obtained. Spray-dried pea sodium proteinate exhibited the highest FC (433%), followed by spray-dried isoelectric pea protein (412%) and soy isolate (250%) [97]. These proteins also showed good foam stability.

Sunflower Proteins

Sunflower protein flour and concentrate showed a higher FC and FS than soy flour and soy concentrate. The sunflower albumin fraction exhibited better foam expansion than sunflower meal in the pH range 2 to 10 [98]. There was no difference in foam volume at pH 10. The good foaming properties of the albumin fraction are probably related to its high solubility. The significant enhancement of foam expansion of the albumin fraction as compared to the sunflower meal was observed in the acid region, while the lower increase was found in the neutral and alkaline region. Sunflower meal formed less stable foams in the acidic pH region, but the foams were more stable than with the albumin fraction in the pH range 6–9. Sunflower protein exhibited a maximum FC near the isoelectric point with a constant volume of foam (FS) at pH 5.5 to 7 during 2 h incubation [99]. NaCl increased the FC of sunflower protein, and the FC was not affected by ionic strength in the NaCl range 0.25–2.0 M. However, sunflower foams prepared in NaCl solutions were unstable and collapsed after 2 h of incubation. Two sunflower concentrates showed significant foam expansion in the low range of concentrations, reaching a maximum at 2%. Foam stability was higher for soy (leakage < 20%) than sunflower proteins. Stability of foams increased at higher concentrations (4–12%) of sunflower flour, concentrate, and soy proteins. Foam stabilization at higher concentrations is related to the stabilizing effect of insoluble particles.

Heating influenced foaming properties of sunflower protein [100]. An initial decrease in foam expansion at 25–35 °C was observed, and an increase in foam volume in the range 35–55 °C due to protein denaturation and unfolding and a slow decrease at 60–90 °C due to denaturation and loss of protein solubility. The foaming capacity and foam drainage of sunflower proteins were slightly affected by whipping time. The maximum foam expansion was observed during the first minutes with a slight decrease at longer whipping times.

Fermentation increased the FC of sunflower protein meal, producing fine-textured foams [101]. The FC of incubated protein increased about 25% in the acid pH region and 40% in the neutral and alkaline pH range. However, foams prepared with fermented sunflower protein exhibited lower stability, especially in the acid pH range. The foam expansion with enzymatically modified sunflower isolate was markedly increased after digestion with pepsin for 1 h (5% hydrolysis) and enzyme inactivation by change of pH [102]. The foam expansion values were higher than for egg albumin. Reduction of foam expansion was observed after digestion longer than 1 h. Succinylation of sunflower concentrate increased the FC and FS. Modified sunflower concentrates showed more pH dependence of their foaming properties. Succinylated (50%) sunflower concentrate with highest FC showed maximum FS in the isoelectric pH range [103]. Succinylation and acetylation improved the foaming properties of sunflower isolate at pH range from 4 to 10. The FC of egg albumin was lower than acylated sunflower isolate at pH 7–10 and similar to nonmodified sunflower protein. The limiting factor of sunflower protein utilization as an ingredient in human foods is the presence of chlorogenic acid, the major phenolic constituent of the seeds. Effective methods of chlorogenic acid removal from sunflower protein meal have been developed [100].

Rice Protein

Full-fat and defatted rice bran protein concentrate showed a positive relationship between pH and FC [104]. The foaming capacity of a defatted sample was slightly higher than a full-fat protein, this was related to presence of lipids. Full-fat proteins contained more than 18% fat. Defatted rice concentrate showed a maximum FC, a 60% volume increase at pH 10.5. Low solubility of both protein concentrates caused their poor foaming properties. A low salt concentration (0.1 M NaCl) increased the FC of both protein concentrates, probably by enhancing protein solubility. However, at a higher NaCl level (1.0 M) the FC was reduced in the defatted sample and slightly increased in the full-fat protein concentrate.

Rapeseed Proteins

Rapeseed meal and rapeseed protein isolate showed a better FC than soy protein products [105]. The foaming properties of rapeseed proteins were concentration dependent. Kinsella [106] reported that salts and increased ionic strength destabilized protein foams because of the reduction of the electrostatic forces between polypeptide chains. Addition of sucrose slightly increased the FC and FS probably because of increased viscosity of the lamellar water. The content of phytic acid in rapeseed protein products did not affect foaming properties [105].

Rapeseed protein concentrate prepared according to Thompson et al. [107] as a 3% dispersion showed markedly higher foam volumes (628%) than rapeseed flour (554%) and soy isolate. When a 5% dispersion was used, a higher volume (851%) was obtained than for egg white (687%). The stability of foam with rapeseed concentrate was comparable to that of egg white. Rapeseed concentrate can be used as an additive to improve foaming properties of egg albumen in angel food cakes and as an egg white replacement.

Cottonseed Proteins

The foaming properties of cottonseed, soybean, and peanut flours and isolates were affected by the pH of aqueous suspensions [108]. Co-isolates of soy and peanut proteins containing cottonseed proteins exhibited a higher foam volume expansion than soybean-peanut co-isolates at all pH values tested. Cottonseed isolate showed a higher foam volume increase than soybean and peanut protein isolates. An increase in FC (~ 40%) of cottonseed flour treated with protease at levels of 30 and 70 mg protease/g protein was found [109]. The FC decreased slightly at the 100-mg protease level, but remained higher than the FC for control protein flour. Cottonseed protein hydrolysis did not affect foam stability.

Oat Proteins

Fractions of oat albumins exhibited a higher FC than globulins, prolamins and glutelins [110]. All these protein fractions except glutelin showed a high FS. The pH-foamability curves of these fractions were similar to the pH-solubility curves, with the lowest foamability at slightly acidic pH. Defatted oat protein concentrate and isolate exhibited FC comparable to that of soy isolate and gluten [111]. The foaming properties of oat protein isolates did not depend entirely on the quantity of soluble proteins. The FC and FS of defatted oat protein concentrate was higher than that of the undefatted one [112]. The oat protein isolate when heated in the presence of 1.5% SDS exhibited surface hydrophobicity values comparable with soy isolate and gluten [111].

Wheat Proteins

The foam properties of wheat proteins are important in baked products. Bread, during proofing and breadmaking expands as a foam-like structure as a result of the entrapment of enormous numbers of gas bubbles. Commercial samples of wheat flour solubles exhibited different foaming properties due to salt and protein content [113]. The FC of OM-25 was higher than the others, due to a higher protein content. The pH/FC profile of OM-25 exhibited the highest FC at pH 10 due to an increased amount of soluble proteins. The FS was highest at pH 2 and remained relatively constant between pH 4 and 12. Dense and stable foams were formed at pH 7.5, i.e in the isoelectric range of gluten. The high stability of gluten foams is due to the high level of hydrogen bonding between the large number of

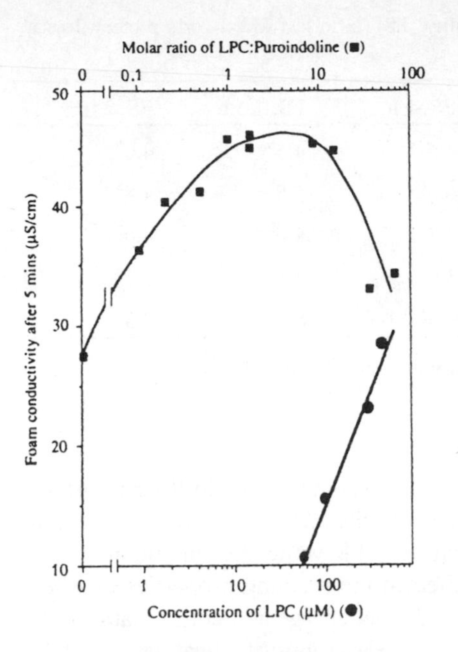

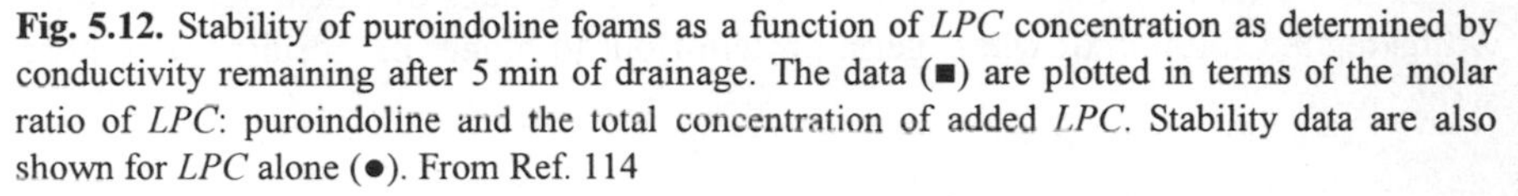
Fig. 5.12. Stability of puroindoline foams as a function of *LPC* concentration as determined by conductivity remaining after 5 min of drainage. The data (■) are plotted in terms of the molar ratio of *LPC*: puroindoline and the total concentration of added *LPC*. Stability data are also shown for *LPC* alone (●). From Ref. 114

glutamine residues. Glutenin showed a superior FC and FS than gliadin or gluten itself. Good foaming properties of glutenin are related to the large size of the molecules - polymers with disulfide links. Glutenin forms a more viscous film than gliadin because of a different degree of polymerization. Extensive disulfide bonds in the structure of glutenin influence its elastic properties. The gliadin, alcohol-soluble fraction of wheat gluten can be utilized in food foams. Non-nitrogenous substances in gliadin impaired foaming power, prolonged the whipping time required and reduced the FS.

The foaming properties of a low molecular weight lipid-binding protein, puroindoline isolated from wheat flour were enhanced in mixtures containing lysophosphatidylcholine (LPC) [114]. The maximum foam stability was obtained in the range of LPC to puroindoline molar ratios of 1-10 (Fig. 5.12). The enhancement of foam active properties of puroindoline by addition of LPC was related to the formation of the complex, which reduces puroindoline aggregation at the interface. The LPC and puroindoline may act synergistically. Puroindoline alone had excellent FC and FS properties. The good foaming properties of puroindoline may be related to its tryptophan-rich domain. Puroindoline could have technological applications in the food industry. The disruptive effect of lipid adsorption in protein-stabilized foams may be eliminated by interaction with puroindoline at the interface.

Table 5.2. Foam Expansion /FE), and foam stability (FS) (in ml) of wheat germ protein flour at different pHs*

pH	FE, 30 s	FS, 0.5 h	FS, 1 h	FS, 1.5 h	FS, 2 h
4	9.5[c]	5.5[b]	4.6[b,c]	4.5[b,c]	4.0[b,c]
5	10.8[b]	6.5[b]	6.0[b]	5.5[b]	4.8[b]
6	10.3[b,c]	5.6[b]	4.3[b,c]	4.0[b,c]	4.0[b,c]
7	8.0[d]	4.3[b]	3.6[c]	3.1[c]	3.1[c]
8	17.3[a]	11.8[a]	9.5[a]	8.3[a]	7.3[a]

[a,b,c,d] Means with the same letter are not significantly different ($P < 0.05$).
* Mean of three replications. From Ref. 19

The foam expansion and foam stability of wheat germ protein flour at pH 4-8 and incubation for 0.5-2.0 h is presented in Table 5.2 [19]. The highest foam expansion and foam stability was obtained at pH 8 while the minimum was at pH 7. The pH of the dispersing medium affected the foaming properties of wheat germ protein flour due to its direct effect on the net charge and conformation. The pH might influence protein solubility and active-SH groups of wheat germ protein. Adjusting the solution of wheat germ protein flour to higher pH resulted in a more stable foam.

Safflower Protein

Safflower protein isolate exhibited a greater FC and FS than soy concentrate or isolate and foaming properties were affected by the physical properties and composition of the protein [115]. Soy protein produced foams with an expansion of ~ 200%, whereas safflower protein isolate gave a volume increase of ~ 300%. Drying methods, by freeze- and spray-drying did not affect the FC of safflower protein isolate. The FS of foams prepared with safflower protein isolates was affected by pH; stable foams were obtained from pH 6.5 to 6.9. Safflower protein isolates obtained by the micellization procedure (dilution in cold water) showed a higher FC than those obtained by isoelectric precipitation [116]. Protein isolates obtained by micellization showed the highest FS at pH 4, whereas the highest stabilities for isolates obtained by isoelectric precipitation were found at the pH extremes.

Linseed Proteins

The foaming capacity of linseed protein flour was more or less constant in the pH range 2 to 6, i.e. in the range of minimum protein solubility, and significantly increased at pH above 8.0 [117]. A decrease in protein extractability is responsible for the lower FC of heated linseed protein flour. Soy protein flour showed a better FC than linseed protein at all pH values possibly due to a higher protein content and the different properties of the proteins. An increase in the FC of heated linseed protein was nonsignificant. In general, mild heating results in an improved FC due to limited surface denaturation and the exposure of the hydrophobic sites of the protein. The FS of foams with linseed protein at 0.4 M NaCl had a 70% foam volume during 20–120 min holding and decreased at 0.8–1.0 M NaCl levels. Heated linseed protein exhibited a low FS, probably because of the low amount of protein extracted near neutral pH or in the presence of NaCl. A negative effect of extensive heating of plant proteins on their FS has been reported [94].

Potato Proteins

Potato protein concentrate exhibited good foaming properties [118]. The volume increase in potato protein dispersions during whipping was 5 times that of the initial volume. Incubation of foams prepared with potato protein concentrate showed a 10% decrease in initial foam volume. Superior foaming properties were found for spray dried potato protein concentrate. Wojnowska et al. [119] reported that the ability to form stable foam depends upon saponin concentrations. The potato protein after ultrafiltration and concentration to 1:10 exhibited a 1000% volume increase, to 1:15 exhibited a 1100% volume increase, and dried protein preparations a 650% volume increase.

Jojoba Proteins

The maximum FC and FS of jojoba protein concentrate was obtained with a whipping time of between 9 and 15 min [120]. The FC of jojoba protein concentrates was highest at pH 7, and FS was highest at pH 4. The FC and FS of jojoba protein concentrates were similar to those of egg albumin and increased as the temperature of dispersions incubated for 30 min increased from 4 to 80 °C . The effect of incubation temperature on FC is probably due to increased protein solubility. Addition of NaCl at low concentration increased protein solubility and the FC of jojoba protein.

Alfalfa Leaf Proteins

Stable foams have been prepared from 2, 4, and 10% solutions of alfalfa leaf protein concentrate produced by spray-drying at an outlet temperature of 85 °C [121]. The increase in volume of the solution was about 10 times the same as that obtained with frozen-thawed egg white. The low foam volume was obtained at a pH above 6, and foam was most stable at a pH near 4.5. The foam was stable to the heating of baking. The pH-foam volume profile and FS were similar to the pH-protein solubility profile, i.e the minimum FC was found in the isoelectric pH range.

Spirulina Proteins

Protein flour (18% protein) and concentrate (27% protein) of Spirulina cells exhibited foaming properties comparable to soybean meal [122]. The pH-FC profile of Spirulina flour and concentrate, and soybean meal resembled the shape of their pH-nitrogen solubility profiles. The foaming property was affected by solubilized protein. The FC of Spirulina flour and concentrate was at its minimum at pH 3.0. The maximum FC for all tested products was at pH 10.0. Spirulina protein concentrate exhibited greater FC in the alkaline pH range than soy proteins.

Yeast Proteins

The foaming properties of phosphorylated yeast protein with 73% protein yeast nucleoprotein and BSA at various pH have been reported [123]. Yeast nucleoprotein produced weak foams at pH 7.0 and above. Phosphorylated yeast protein exhibited markedly better FC and FS at all pH values. Halling [26] reported that phosphorylation increased the solubility of yeast protein. Improvement of FS was related to the increase in electrostatic repulsions between proteins of neighboring surface films by modification of the ε-amino groups with negatively-charged phosphoryl groups. The stability of foams prepared with phosphorylated yeast protein increased from $t_{1/2}$ =1.5 to $t_{1/2}$ =13 min as pH increased from 5 to 8. The FS was improved with increased protein concentration up to 0.2%.

References

1. Kinsella, J. E. (1981). Functional properties of proteins: possible relationships between structure and function in foams, *Food Chem.*, *7*: 273.
2. Purdon, A. D. (1980). The temperature dependence of surface tension and critical micelle concentration of egg lysolecithin, *Colloid Polym. Sci.*, *258*: 1062.
3. Kitabatake, N. and Doi, E. (1982). Surface tension and foaming of protein solutions, *J. Food Sci.*, *47*: 1218.
4. Graham, D. and Phillips, M. (1975). *Foams* (R. J. Ackers, ed.), Academic Press, London, p. 237.
5. MacRitchie, F. (1978). Proteins at interfaces, *Adv. Protein Chem.*, *32*: 383.
6. Graham, D. E. and Phillips, M. C. (1980). Proteins at liquid interfaces. I. Kinetics of adsorption and surface denaturation, *J. Colloid Interface Sci.*, *70*: 403.
7. Graham, D. E. and Phillips, M. C. (1979). Proteins at liquid interfaces. III. Molecular structure of adsorbed films, *J. Colloid Interface Sci.*, *70*: 427.
8. Cherry, J. P. and McWatters, K. H. (1981). Whippability and aeration, In *Protein Functionality in Foods*, ACS Symposium Series 147 (J. P. Cherry, ed.), Amer. Chem. Soc., Washington, D. C.
9. Poole, S., West, S. I., and Walters, C. L. (1984). Protein-protein interactions: their importance in the foaming of heterogeneous protein systems, *J. Sci. Food Agric.*, *35*: 701.
10. Townsend, A. M. and Nakai, S. (1983). Relationships between hydrophobicity and foaming characteristics of food proteins, *J. Food Sci.*, *48*: 588.

11. Kim, S. H. and Kinsella, J. E. (1987). Surface active properties of food proteins: effects of reduction of disulfide bonds on film properties and foam stability of glycinin, *J. Food Sci.*, *52*: 128.
12. Hayakawa, S. and Nakai, S. (1985). Relationship of hydrophobicity and net charge to the solubility of milk and soy proteins, *J. Food Sci.*, *50*: 486.
13. Kato, A. and Nakai, S. (1980). Hydrophobicity determined by a fluorescence probe method and its correlation with surface properties of proteins, *Biochem. Biophys. Acta*, *624*: 13.
14. Kato, A., Tsutsui, N., Matsudomi, N., and Kobayashi, K. (1981). Effects of partial denaturation on surface properties of ovalbumin and lysozyme, *Agric. Biol. Chem.*, *45*: 2755.
15. Kato, A., Osako, Y., Matsudomi, N., and Kobayashi, K. (1983). Changes in the emulsifying and foaming properties of proteins during heat denaturation, *Agric. Biol. Chem.*, *47*: 33.
16. Kim, S. H. and Kinsella, J. E. (1985). Surface activity of food proteins: Relationships between surface pressure development, viscoelasticity of interfacial films and foam stability of bovine serum albumin, *J. Food Sci.*, *50*: 1526.
17. Graham, D. E., Levy, S., and Phillips, M. S. (1976). The conformation of proteins at interfaces and their role in stabilizing emulsions, In *Theory and Practice of Emulsion Technology* (A. L. Smith, ed.), Academic Press, London, p. 57.
18. Graham, D. E. and Phillips, M. C. (1976). The conformation of proteins at the air-water interface and their role in stabilizing foams, In *Foams* (R. J. Akers, ed.), Academic Press, New York, p. 131.
19. Bolnedi, V., and Zayas, J. F. (1993). Foaming properties of wheat germ protein flour in comparison to plant and animal proteins in model system. *Annual IFT Meeting, 1993. Book of Abstracts*, p. 160.
20. Baldwin, R. E. (1986). In *Egg Science and Technology*, 3rd edit. (W. J. Stadelman and O. J. Cotterill, eds.), AVI Publi. Co., Inc., Westport, CT, p. 345.
21. Kinsella, J. E. (1976). Functional properties of proteins in foods: a survey, *Crit. Rev. Food Sci. Nutr.*, *7*: 219.
22. Lee, J. C. and Timasheff, S. N. (1981). The stabilization of proteins by sucrose, *J. Biol. Chem.*, *256*: 7193.
23. Bergquist, D. H. (1986). In *Egg Science and Technology*, 3rd edit. (W. J. Stadelman and O. J. Cotterill, eds.), AVI Publi. Co., Inc., Westport, CT, p. 285.
24. Poole, S., West, S. I., and Fry, J. C. (1986). Lipid tolerant protein foaming systems, *Food Hydrocolloids*, *1*: 45.
25. Kinsella, J. E. and Whitehead, D. M. (1989). Proteins in whey: Chemical, physical, and functional properties, *Adv. Food Nutr. Res.*, *33*: 343.
26. Halling, P. J. (1981). Protein-stabilized foams and emulsions, *CRC Crit. Rev. Food Sci. Nutr.*, *13*: 155.
27. Phillips, M. C. (1981). Protein conformation at liquid interfaces and its role in stabilizing emulsions and foams, *Food Technol.*, *35*: 50.
28. Elizalde, B. E., Giaccaglia, D., Pilosof, A. M. R., and Bartholomai, G. B. (1991). Kinetics of liquid drainage from protein-stabilized foams, *J. Food Sci.*, *56*: 24.
29. Phillips, L. G., Haque, Z., and Kinsella, J. E. (1987). A method for the measurement of foam formation and stability, *J. Food Sci.*, *52*: 1074.
30. Kosaric, N. and Ng, D. C. M. (1983). Some functional properties of milk protein calcium co-precipitates, *Can. Inst. Food Sci. Technol. J.*, *16 (2)*: 141.
31. Mangino, M. E., Huffman, L. M., and Regester, G. O. (1988). Changes in the hydrophobicity and functionality of whey during the processing of whey protein concentrate, *J. Food Sci.*, *53*: 1684.
32. To, B., Helbio, N. B., Nakai, S., and Ma, C. Y. (1985). *Can. Inst. Food Sci. Technol. J.*, *18 (2)*: 150.

33. Slack, A. W., Amundson, C. H., and Hill, Jr., C. G. (1986). Foaming and emulsifying characteristics of fractionated whey protein, *J. Food Proc. Pres.*, *10*: 81.
34. Morr, C. V. and Foegeding, E. A. (1990). Composition and functionality of commercial whey and milk protein concentrates and isolates: A status report, *Food Technol.*, *4*: 100.
35. Burgess, K. J. and Kelly, J. (1979). Technical note: selected functional properties of a whey protein isolate, *J. Food Technol.*, *14*: 325.
36. Liao, S. and Mangino, M. (1987). Characterization of the compositional, physiochemical and functional properties of acid whey protein concentrates, *J. Food Sci.*, *52*: 1033.
37. De Wit, J. N. (1989). Functional properties of whey proteins, In *Developments in Dairy Chemistry*, v. 4, Functional milk proteins, Elsevier Appl. Sci., London, New York, p. 285.
38. Waniska, R. D. and Kinsella, J. E. (1985). Surface properties of β-lactoglobulin: Adsorption and rearrangement during film formation, *J. Agric. Food Chem.*, *33*: 1143.
39. Jackman, R. L. and Yada, R. Y. (1988). Functional properties of whey-potato protein composite blends in a model system, *J. Food Sci.*, *53*: 1427.
40. Britten, M. and Lavoie, L. (1992). Foaming properties of proteins as affected by concentration, *J. Food Sci.*, *57*: 1219.
41. Tornberg, E., Granfeldt, Y., and Hakansson, C. (1982). A comparison of the interfacial behavior of three food proteins adsorbed at air-water and oil-water interfaces, *J. Sci. Food Agric.*, *33*: 904.
42. Murphy, J. M. and Fox, P. F. (1990). Functional properties of αs-/K-caseinate- or β-rich casein fractions, *Food Chem.*, *39*: 211.
43. Lee, S. Y., Morr, C. V., and Ha, E. Y. W. (1992). Structural and functional properties of caseinate and whey protein isolate as affected by temperature and pH. *J. Food Sci.*, *57*: 1210.
44. Phillips, L. G., Davis, M. J., and Kinsella, J. E. (1989). The effects of various milk proteins on the foaming properties of egg white, *Food Hydrocolloids*, *3*: 163.
45. Fox, P. F. (1989). The milk protein system, In *Developments in Dairy Chemistry – 4* (P. F. Fox, ed.), Elsevier Applied Science, New York.
46. Tornberg, E. (1978). The interfacial behavior of three food proteins studied by the drop volume technique, *J. Sci. Food Agric.*, *29*: 762.
47. Patel, M. T. and Kilara, A. (1990). Studies on whey protein concentrates. 2. Foaming and emulsifying properties and their relationships with physicochemical properties, *J. Dairy Sci.*, *73 (10)*: 2731.
48. Harper, W. J. (1984). Model food system approaches for evaluating whey protein functionality, *J. Dairy Sci.*, *67*: 2745.
49. Peltonen-Shalaby, R. and Mangino, M. E. (1986). Compositional factors that affect the emulsifying and foaming properties of whey protein concentrates, *J. Food Sci.*, *51*: 91.
50. Schmidt, D. G. and Van Hooydonk, A. C. M. (1980). A scanning electron microscopical investigation of the whipping of cream, *Scanning Electron Microsc.*, *111*: 653.
51. Morr, C. V. (1985). Composition, physicochemical and functional properties of reference whey protein concentrate, *J. Food Sci.*, *50*: 1406.
52. Phillips, L. G., Yang, S. T., Schulman, W., and Kinsella, J. E. (1989). Effects of lysozyme, clupeine, and sucrose on the foaming properties of whey protein isolate and β-lactoglobulin, *J. Food Sci.*, *54*: 743.
53. Phillips, L. G., Schulman, W., and Kinsella, J. E. (1990). pH and heat treatment effects on foaming of whey protein isolate, *J. Food Sci.*, *55*: 1116.
54. Le Meste, M. L., Colas, B., Simatos, D., Closs, B., Courthaudon, J. L., and Lorient, D. (1990). Contribution of protein flexibility to the foaming properties of casein, *J. Food Sci.*, *55*: 1445.
55. Poole, S., West, S. I., and Fry, J. C. (1987). Charge structural requirements of basic proteins for foam enhancement, *Food Hydrocolloids*, *1*: 227.
56. Hsu, K. H. and Fennema, O. (1989). Changes in the functionality of dry whey protein concentrate during storage, *J. Dairy Sci.*, *72*: 829.

57. Morr, C. V. (1987). Effect of HTST pasteurization of milk, cheese whey and cheese whey UF retentate upon the composition, physicochemcial and functional properties of whey protein concentrates, *J. Food Sci.*, *52*: 312.
58. Akita, E. M. and Nakai, S. (1990). Lipophilization of β-lactoglobulin: Effect on hydrophobicity, conformation and surface functional properties, *J. Food Sci.*, *55*: 711.
59. Haque, Z. and Kito, M. (1983). Lipophilization of αs_1-casein. 2. Conformational and functional effects, *J. Agric. Food chem.*, *31*: 1231.
60. Johnson, T. M. and Zabik, M. E. (1981). Response surface methodology for analysis of protein interactions in angel food cakes, *J. Food Sci.*, *46*: 1226.
61. Kitabatake, N. and Doi, E. (1987). Conformational change of hen egg ovalbumin during foam formation detected by 5,5'-dithiobis (2-nitrobenzoic acid), *J. Agric. Food Chem.*, *35*: 953.
62. Johnson, T. M. and Zabik, M. E. (1981). Egg albumen protein interactions in an angel food cake system, *J. Food Sci.*, *46*: 1231.
63. Johnson, T. M. and Zabik, M. E. (1981). Ultrastructural examination of egg albumen protein foams, *J. Food Sci.*, *46*: 1237.
64. Poole, S., West, S. I., and Walters, C. L. (1984). Protein-protein interactions: Their importance in the foaming of heterogeneous protein systems, *J. Sci. Food Agric.*, *35*: 701.
65. Tsutsui, T. (1988). Functional properties of heat-treated egg yolk low density lipoprotein, *J. Food Sci.*, *53*: 1103.
66. Dyer-Hyrdon, J. N., and Nnanna, I. A. (1993). Cholesterol content and functionality of plasma and granules fractionated from egg yolk. *J. Food Sci.*, *58*: 1277.
67. Wootton, M., Hong, N. T., and Pham Thi, H. L. (1981). A study of the denaturation of egg white proteins during freezing using differential scanning calorimetry, *J. Food Sci.*, *46*: 1336.
68. Ma, C. Y., Harwalkar, V. R., Poste, L. M., and Sahasrabudhe, M. R. (1993). Effect of gamma irradiation on the physicochemical and functional properties of frozen liquid egg products. *Food Research Internat.*, *26*: 247.
69. Freeland-Graves, J. H. and Peckham, G. C. (1979). *Foundations of Food Preparation*, *30*: 264.
70. Van Elswyk, M. E., Sams, A. R., and Hargis, P. S. (1992). Composition, functionality, and sensory evaluation of eggs from hens fed dietary menhaden oil, *J. Food Sci*, , *57*: 342.
71. Shahidi, F., Naczk, M., Rubin, L. J., and Diosady, L. L. (1984). Functional properties of blood globin, *J. Food Sci.*, *49*: 370.
72. Etheridge, P. A., Hickson, D. W., Young, C. R., Landmann, W. A., and Dill, C. W. (1981). Functional and chemical characteristics of bovine plasma proteins isolated as a metaphosphate complex, *J. Food Sci.*, *46*: 1782.
73. Watanabe, M., Shimada, A., Yazawa, E., Kato, T., and Arai, S., (1981). Proteinaceous surfactants produced from gelatin by enzymatic modification: application to preparation of food items, *J. Food Sci.*, *46*: 1738.
74. Kim, S. H. and Kinsella, J. E. (1986). Effects of reduction with dithiothreitol on the molecular properties of soy glycinin, *J. Agric. Food Chem.*, *34*: 623.
75. Horiuchi, T., Fukushima, D., Sugimato, M., and Hattori, T. (1978). Studies on enzyme-modified proteins as foaming agents: effect of structure on foam stability. *Food Chem.*, *3*: 35.
76. Manak, L. J., Lawhon, J. T., and Lusas, E. W. (1980). Functioning potential of soy, cottonseed, and peanut protein isolates produced by industrial membrane systems, *J. Food Sci.*, *45*: 236.
77. Schaffner, D. W. and Beuchat, L. R. (1986). Functional properties of freeze-dried powders of unfermented and fermented aqueous extracts of legume seeds, *J. Food Sci.*, *51*: 629.

78. Paulson, A. T., Tung, M. A., Garland, M. R., and Nakai, S. (1984). Functionality of modified plant proteins in model food systems, *Can. Inst. Food Sci. Technol. J.*, *17 (4)*: 202.
79. Haque, Z., Matoba, T., and Kito, M. (1982). Incorporation of fatty acid into food protein: palmitoyl soybean glycinin, *J. Agric. Food Chem.*, *30*: 481.
80. Sung, H. Y., Chen, H. J., Liu, T. Y., and Su, J. C. (1983). Improvement of the functionality of soy protein by introduction of new thiol groups through a papain-catalyzed acylation, *J. Food Sci.*, *48*: 708.
81. Kim, S. H. and Kinsella, J. E. (1987). Surface active properties of proteins: Effects of progressive succinylation on film properties and foam stability of glycinin, *J. Food Sci.*, *52*: 1341.
82. Chen, B. H. Y. and Morr, C. V. (1985). Solubility and foaming properties of phytate-reduced soy protein isolate, *J. Food Sci.*, *50*: 1139.
83. McWatters, K. H., Cherry, J. P., and Holmes, M. R. (1976). Influence of suspension medium and pH on functional and protein properties of defatted peanut meal, *J. Agric. Food Chem.*, *24 (3)*: 517.
84. Kim, N. S., Kim Y. J., and Nam, Y. J. (1992). Characteristics and functional properties of protein isolates from various peanut (*Arachis hypogaea L.*) cultivars, *J. Food Sci.*, *57*: 406.
85. MOnteiro, P. V., and Prakash, V. (1994). Functional properties of homogeneous protein fractions from peanut(*Arachis hypogaea L.*). *J. Agric. Food Chem.*, *42*: 274.
86. Kohnhorst, A. L., Smith, D. M., Uebersax, M. A., and Bennink, M. R. (1990). Compositional, nutritional and functional properties of meals, flours and concentrates from navy and kidney beans (*Phaseolus vulgaris*), *J. Food Quality*, *13*: 435.
87. Idouraine, A., Yensen, S. B., and Weber, C. W. (1991). Tepary bean flour, albumin and globulin fractions functional properties compared with soy protein isolate, *J. Food Sci.*, *56*: 1316.
88. Sathe, S. K., Deshpande, S. S., and Salunkhe, D. K. (1983). Functional properties of black gram (*Phaseolus Mungo L.*) proteins, *Lebensm.-Wiss. u.-Technol.*, *16*: 69.
89. Sosulski, F. W. and McCurdy, A. R. (1987). Functionality of flours, protein fractions and isolates from field peas and faba bean, *J. Food Sci.*, *52*: 1010.
90. Han, J. Y. and Khan, K. (1990). Functional properties of pin-milled and air-classified dry edible bean fractions, *Cereal Chem.*, *67 (4)*: 390.
91. Narayana, K. and Narasinga Rao, M. S. (1984). Effect of acetylation and succinylation on the functional properties of winged bean (*Psophocarpus tetragonolobus*) flour, *J. Food Sci.*, *49*: 547.
92. Nath, J. P. and Narasinga Rao, M. S. (1981). Functional properties of guar proteins, *J. Food Sci.*, *46*: 1255.
93. Tasneem, R. and Subramanian, N. (1986). Functional properties of guar (*Cyamopsis tetragonoloba*) meal protein isolates, *J. Agric. Food Chem.*, *34*: 850.
94. Tasneem, R., Ramamani, S., and Subramanian, N. (1982). Functional properties of guar seed (*Cyamopsis tetragonoloba*) meal detoxified by different methods, *J. Food Sci.*, *47*: 1323.
95. Swanson, B. G. (1990). Pea and lentil protein extraction and functionality, *J.A.O.C.S.*, *67 (50)*: 276.
96. Koyoro, H. and Powers, J. R. (1987). Functional properties of pea globulin fractions, *Cereal Chem*, *64 (2)*: 97.
97. Sumner, A. K., Nielsen, M. A., and Youngs, C. G. (1981). Production and evaluation of pea protein isolate, *J. Food Sci.*, *46*: 346.
98. Canella, M., Castriotta, G., Bernardi, A., and Boni, R. (1985). Functional properties of individual sunflower albumin and globulin, *Lebensm.-Wiss. u.-Technol.*, *18*: 288.
99. Rossi, M. and Germondari, I. (1982). Production of a food-grade protein meal from defatted sunflower. II. Functional properties evaluation, *Lebensm.-Wiss. u.-Technol.*, *15*: 313.

100. Canella, M. (1978). Whipping properties of sunflower protein dispersions, *Lebensm.-Wiss. u.-Technol.*, *11*: 259.
101. Canella, M., Bernardi, A., Castriotta, G., and Russomanno, G. (1984). Functional properties of fermented sunflower meal, *Lebensm.-Wiss. u.-Technol.*, *17*: 146.
102. Kabirullah, M. and Wills, R. B. H. (1981). Functional properties of sunflower protein following partial hydrolysis with proteases, *Lebensm.-Wiss. u.-Technol.*, *14*: 232.
103. Kabirullah, M. and Wills, R. B. H. (1982). Functional properties of acetylated and succinylated sunflower protein isolate, *J. Food Technol.*, *17*: 235.
104. Bera, M. and Mukherjee, R. K. (1989). Solubility, emulsifying and foaming properties of rice bran protein concentrte, *J. Food Sci.*, *54*: 142.
105. Dev, D. K. and Mukherjee, R. K. (1986). Functional properties of rapeseed protein products with varying phytic acid contents, *J. Agric. Food Chem.*, *34*: 775.
106. Kinsella, J. E. (1981). Functional properties of proteins: possible relationships between structure and function in foams, *Food Chem.*, *7*: 273.
107. Thompson, L. U., Liu, R. F. K., and Jones, J. D. (1982). Functional properties and food applications of rapeseed protein concentrate, *J. Food Sci.*, *47*: 1175.
108. Berardi, L. C. and Cherry, J. P. (1981). Functional properties of co-precipitated protein isolates from cottonseed, soybean and peanut flours, *Can. Inst. Food Sci. Technol J.*, *14 (4)*: 283.
109. Rahma, E. H. and Narasinga Rao, M. S. (1983). Effect of limited proteolysis on the functional properties of cottonseed flour, *J. Agric. Food Chem.*, *31*: 356.
110. Ma, C. Y. and Harwalkar, V. R. (1984). Chemical characterization and functionality assessment of oat protein fraction, *J. Agric. Food Chem.*, *32 (1)*: 144.
111. Ma, C. Y. (1983) Preparation, composition and functional properties of oat protein isolates, *Can. Inst. Food Sci. Technol. J.*, *16 (3)*: 201.
112. Ma, C. Y. (1983). Chemical characterization and functionality assessment of protein concentrates from oats, *Cereal Chem.*, *60 (1)*: 36.
113. Oomah, B. D. and Mathieu, J. J. (1987). Functional properties of commercially produced wheat flour solubles, *Can. Inst. Food Sci. Technol. J.*, *20 (2)*: 81.
114. Wilde, P. J., Clark, D. C., and Marion, D. (1983). Influence of competitive adsorption of a lysopalmitoylphosphatidylcholine on the functional properties of puroindoline, a lipid-binding protein isolated from wheat flour. *J. Agric Food Chem.*, *41*:1570.
115. Betschart, A. A., Fong, R. Y., and Hanamoto, M. M. (1979). Safflower protein isolates: functional properties in simple systems and breads, *J. Food Sci.*, *44*: 1022.
116. Paredes-Lopez, O. and Ordorica-Falomir, C. (1986). Functional properties of safflower protein isolates: water absorption, whipping and emulsifying characteristics, *J. Sci. Food Agric.*, *37*: 1104.
117. Madhusudhan, K. T. and Singh, N. (1985). Effect of heat treatment on the functional properties of linseed meal, *J. Agric. Food Chem.*, *33*: 1222.
118. Knorr, D. (1980). Functional properties of potato protein concentrates, *Lebensm.-Wiss. u.-Technol.*, *13*: 297.
119. Wojnowska, I., Poznanski, S., and Bednarski. W. (1981). Processing of potato protein concentrates and their properties, *J. Food Sci.*, *47*: 167.
120. Wiseman, M. O. and Price, R. L. (1987). Functional properties of protein concentrates from pressed jojoba meal, *Cereal Chem.*, *64 (2)*: 94.
121. Knuckles, B. E. and Kohler, G. O. (1982). Functional properties of edible protein concentrates from alfalfa, *J. Agric. Food Chem.*, *30 (4)*: 748.
122. Devi, M. A. and Venkataraman, L. V. (1984). Functional properties of protein products of mass cultivated blue-green algae Spirulina platensis, *J. Food Sci.*, *49*: 24.
123. Huang, Y. T. and Kinsella, J. E. (1987). Effects of phosphorylation on emulsifying and foaming properties and digestibility of yeast protein, *J. Food Sci.*, *52*: 1984.

Chapter 6

Gelling Properties of Proteins

6.1 Introduction

The gelling capacity of food proteins is an important functional attribute for food manufacturing. Large numbers of important foods are gels in which the gelling ingredients are proteins. Together with pectins, starches and gums they form strong gels. The food industry uses different proteins to produce gels or gel-containing products which exhibit various rheological properties, appearance and gel point. Gelation is a basic process in the processing of various foods, milk gels, comminuted meat and fish products, other meat products, fruit jellies, bread doughs, pie and cake fillings, coagulated egg white and others.

Gelling capacity is the criterion that is frequently used to evaluate food proteins. The quality characteristics of many foods, especially textural properties and juiciness are determined by the gelling capacity of proteins. Modern methodology of the physical chemistry of polymers has been applied to study protein gelation. Gels can vary markedly in their rheological properties such as cohesiveness, hardness, stickiness, and adhesiveness. A unique property of protein gels is to behave as a solid-like material, but, at the same time, they possess many characteristics of a fluid. The gelation phenomenon is responsible for the solid-like, viscoelastic properties of foods, increased viscosity, adhesiveness and improved water retention. These specific properties of gels are due to the presence of a three-dimensional network of proteins.

It is of practical interest to control the gelation process to predict the properties of the finished products, mainly textural characteristics. The formulations of most foods include salt, sugar, and fat which modify the properties of the gels. Utilization of plant protein concentrates as food ingredients is largely determined by their functionality, including their gelling properties. A knowledge of protein gelation properties is required for the substitution of one protein for another in food systems. Protein gels may be utilized to simulate the textural properties and mouthfeel of lipids.

The gel forming ability of proteins affects other functionalities such as water retention, and fat binding. The gelation phenomenon plays a major role in stabilizing emulsions and foams. The property of proteins to form a gel and retain significant amount of sugars, flavor and other food ingredients in a three-dimensional matrix is widely utilized in food processing and in the development of new food products [1]. Some proteins do not exhibit gelling properties. Bovine serum albumin (BSA) is the gel-forming protein in cheese whey while α-lactalbumin (α-La) is a non-gelling protein. However, the gelling capacity of BSA is increased when α-La is added [2].

6.2 The Mechanism of Protein Gel Formation

Gels are formed when partially unfolded proteins develop uncoiled polypeptide segments that interact at specific points to form a three dimensional cross-linked network. Partial unfolding of proteins with slight changes in secondary structure is required for gelation [3]. Partial unfolding of the native structure can be related to the action of various factors, heating, treatment with acids, alkali and urea.

The gelation process is dependent on the formation of a three-dimensional protein network as a result of protein-protein and protein-solvent (water) interactions. These interactions and gelation are accelerated at high protein concentrations because of more intense intermolecular contacts. Protein gel formation is a result of intermolecular interactions resulting in the three-dimensional network of protein fibers which develop high structural rigidity. Kinsella [1] defined gelation as a hydration, structural, textural, and rheological property of proteins. Schmidt [4] defined gelation as a protein aggregation phenomenon in which polymer-polymer and polymer-solvent interactions are so balanced that a tertiary network or matrix is formed. Foegeding [5] described gels as consisting of interconnected unit structures with the liquid phase throughout the three-dimensional matrix. In a gel, the liquid prevents the three-dimensional matrix from collapsing into a compact mass and the matrix prevents the liquid from flowing away.

A gel matrix has been described as including interconnected cage-like unit structures saturated with a continuous solvent phase [5]. The textural properties of gels are determined by protein-protein, protein-solvent interactions, the strength of junctions, and the flexibility of the polypeptide chains. If protein-protein interactions are extremely strong, the three dimensional network will collapse, and water will be released from the structure. Electron microscope studies showed that linear aggregation of protein molecules was more important than their random aggregation [6]. In this case, a more stable network composed of strands of proteins was formed. Proteins with good gel-forming properties normally possess a high degree of asymmetry of structure. The mechanism of gel formation is different for various proteins, predominantly in the type of protein-protein and protein-water interactions that stabilize the gels [7].

Gels with high strength and stability can be formed as a result of cross-linking which gives the fluidity, elasticity, and flow behavior of gels. Gel formation is a result of hydrogen bonding, ionic and hydrophobic interactions, Van der Waals forces, and covalent disulfide bonding. Cross-links are the determining factor in the formation of the rubbery nature of protein gels [8]. Decreasing the number of cross-links should decrease gel hardness. Rheological properties are a useful source of information related to the cross-linking characteristics of protein gels. The simplest way to identify the cross-linking mechanism in gel systems is to add reagents such as an inhibitor of disulfide bonds to the gels prior to heating and to determine their effect on the gelation process. The effect of disulfide bonds on the formation of the gel network was demonstrated by the addition of dithiothreitol, which prevents the formation of disulfide bonds and causes a decrease in gel strength [9].

High molecular weight proteins and a high percentage of amino acids with hydrophobic groups tend to establish the strong network of the gel system. The presence of hydrophobic amino acids also influences changes of the proteins during heating. An increase in the number of -SH and -SS-groups during denaturation strengthens the intermolecular network. A reaction important for gel formation is disulfide bridge formation. During heating there is a cleavage of disulfide bonds and an „activation" of buried sulfhydryl groups as a result of the unfolding of the protein polypeptide chains. Reactive SH groups can form new intermolecular disulfide bonds. Heat-irreversible gels are formed if covalent disulfide cross-links prevail in the protein structure as in β-lactoglobulin (β-Lg) and ovalbumin gels.

Models for predicting the gelling ability of proteins have been developed. The gelation was affected by surface hydrophobicity measured after the unfolding of the protein molecules, and sulfhydryl groups (free and buried SH), total sulfhydryl groups (SH + reduced SS) and disulfide groups of the proteins [10]. Nakai et al. [11] found a strong correlation between the heat-induced gel strength and the hydrophobicity of the proteins. Gel strength and hardness is related to the size and shape of the polypeptides in the gel network, rather than to their chemical composition, i.e. amino acid composition [12]. A correlation coefficient of 0.913 was reported between average molecular weight of proteins and the square root of the hardness of the gel.The critical minimum molecular weight for gel formation was 23|000. A sufficient number of protein molecules must be available to form a continuous matrix and the aggregation must proceed at a slower rate than the unfolding step. Slight changes in the secondary structure of proteins are needed for the gelation of some of the proteins [3]. Globular proteins in the native state have a characteristic secondary structure, i.e. a specific content of α-helix, β-sheet and disordered peptide chain conformations. The circular dichroism technique was applied to study the secondary structure in protein systems during gelation [3]. Aggregation of proteins was obtained by heating BSA solutions (pH 8) at 75 °C for 2 h, then cooling to 25 °C. Linear aggregates are formed as small polymers, possibly dimers and trimers.

6.2.1 Heat-Induced Gelation

There are two basic types of heat-induced gel structures depending on the conditions involved: 1) Thermo-set (or „set“) or reversible and 2) Thermoplastic or irreversible gels. In thermo-set gelation, the sol or progel is formed upon heating which is usually accompanied by increased viscosity. Thermo-set gels can be melted by subsequent heating and form a gel upon cooling. In thermoplastic gels melting or reversion to the progel does not occur under normal conditions.

The heat-induced gel formation was suggested as being a two stage process [13, 14]. The first step includes denaturation of the native protein followed by protein-protein and protein-solvent interactions resulting in a three-dimensional protein network which forms the gel. The following mechanism of gel formation was proposed by Ferry [13]:

♣ heat heat and/or cooling $x\,P_N$ ----> $x\,P_D$ -------------------> $(P_D)\,x$♣

where x is the number of protein molecules, P_N is the native protein and P_D is the denatured protein. According to the Ferry theory, the final gel state corresponds to aggregates of partly denatured protein.

In gel formation, the transition from a native to a denatured state is an important precursor to the protein-protein interactions. Denaturation cannot be considered only as the physical process of unfolding polypeptide chains. Samejima et al. [15] reported that the denaturation process was evaluated using fluorescence and different spectra indicated local irreversible changes involving aromatic amino acids. Some amino acids, especially tryptophan, are transferred from a moderately hydrophobic environment to a more polar environment. The degree of protein denaturation necessary for gel formation is a matter of some dispute. The gel network can be formed after partial protein denaturation and the protein molecules are „fixed“ in a partially denatured state.

The modern view on heat gelation of globular proteins is that only a relatively small amount of unfolding of polypeptide chains is necessary for the formation of continuous network [3]. The mechanism of this process could be explained by the fact that, at high subphase concentrations, the degree of uncoiling in the film decreases below the level at which maximum cross-linking occurs. Temperature dependence of the degree of cross-linking in gels is not significant and the increase in gel strength during cooling is caused by the increased stability and growth of existing junctions. The temperature used for gel formation should balance the rate of protein unfolding with that of aggregation. The formation of heat-induced protein gels is governed by a balance between attractive and repulsive forces. The attractive forces are considered to be induced by the various functional groups exposed by the thermal unfolding of protein and repulsive forces are considered to be created by the surface charge. In the globular protein solutions denaturation precedes aggregation and is the basic reason for protein-protein interactions. Oilseed protein isolates may form gels without heating because denaturation has occurred during processing, heating, extraction by solvent and treatment with alkali.

6.2.2 Protein-Water Interactions in Gels

Protein-water interaction is the most important phenomenon in the formation of heat set protein gels and at a low degree of interaction, aggregation or precipitation of the protein is obtained. Protein-water interactions in gels are of great practical interest and have been studied intensely. The water binding capacity of proteins in gels is a limiting factor in the production of textured foods in which water becomes a structurally integral component. The interaction between proteins and water plays an important role in gel formation, especially during the sol→gel phase transition.

In the three-dimensional network of protein gels, a large amount of water is physically entrapped, in some gels 10 g water per gram protein. Comparatively strong gels are prepared with 98% water and 2% gelling agent. In some cases the water content in gels is up to 99%. Gel properties are determined by protein-water interactions and water may be bonded by hydrogen bonds to polar groups, by dipolar interactions with ionic groups or retained as structured water. Hydrogen bonds are necessarily very weak because water is itself a hydrogen bond donor and acceptor. Some water molecules are so tightly bound that they fail to freeze even at temperatures as low as -60 °C.

The water, which is held in the three-dimensional structure of a gel can be classified into bound and free water. In most protein gels, water is so strongly immobilized that it cannot be „squeezed out“. The intensity of water imbibition varies with the source of proteins, presence of carbohydrates, lipids, minerals, size of protein particles, pH, temperature, ionic strength, and diameter of pores in the gel microstructure.

Water can be held back in the pores of the protein network through capillary mechanisms. The diameter of the pores in the microstructure is the determining factor for moisture retention. In gels with pore sizes in the range of 1–100 nm, water is strongly held by capillary forces and thus, water is separated during gel shrinkage, when the gel is heated at temperatures above gelation [16]. However, larger pore sizes (100–2000 nm) are also important for water binding. Changes in the pore size 100–2000 nm cause changes in moisture loss from 10 to 50%.

Interactions of amino acids with water have been well described. Studies of hydration capacity of amino acids showed the range from one H_2O molecule per amino acid (for amino acids with nonpolar side groups) to seven H_2O molecules for basic ionic molecules such as tyrosine in solution [17].

Shrinkage of gels will occur if protein-protein interactions are uniform in all the gel sample. As a result of shrinkage, water will be removed from the gel. The temperature of heating and the level of protein denaturation influence the water holding capacity of gels and the degree of phase separation. Increasing the heating temperature above the gelation point will cause a loss of moisture [18]. For whey protein and blood plasma protein such as BSA, an increase in the heating temperature during gel production caused a large moisture loss. Heating at 95 °C caused a coarser structure with the formation of larger complexes and pores than heating to 75 °C. Soy glycinin has a complex quaternary structure and dissociation occurred at a certain heating temperature and reassociation of subunits into strands under the proper conditions. Microstructural analyses of gels by observing physical changes of gel structures helped to explain why water is retained within a rigid matrix of molecules.

6.2.3 Factors Affecting the Properties of Gels

The development of the three-dimensional network during gel formation is influenced by the method of protein preparation, processing conditions, including the length and rate of heating and cooling, and environmental factors (pH and ionic strength). The gelling ability of proteins is influenced by protein concentration, amino acid composition, molecular weight, and hydrophobicity of the proteins [19].

Gelling properties of various proteins have been studied in the range of protein level from 0.5 to 25%. The strength of the protein gels increased with protein or solids concentration. Protein concentration required for gel formation varies depending on the protein properties. Gelatin will form a gel at relatively low concentration. Globular proteins require considerably higher initial concentration for gel formation. Gelation of proteins will not occur below a certain concentration that is dependent on protein origin and properties. At low protein concentration, the interactions between proteins tend to occur within molecules rather than between molecules and the gel network is not formed. The intermolecular interactions are possible at increased protein concentration, resulting in firmer gels as more water is tightly bound to protein molecules. In the gels, as the protein concentration increased, the net matrix area occupied by the protein network and interlinks increased.

The critical concentration of protein sufficient for gel formation can be used for comparison of various gelling agents. At the above-mentioned critical concentration of protein, the gel matrix is formed as a result of association of denatured protein molecules with the formation of aggregates. In various protein gels, increasing the protein concentration above the critical level shortens the time necessary to form the gel.

The Effect of pH and Ionic Strength

Protein denaturation, protein-protein, and protein-water interactions are affected by pH and the proper pH can prevent the collapse of the gel network from charge repulsion. Charge distribution among the amino acid side chains is altered by pH and the ionic strength of the protein environment. Gels formed at the isoelectric pH of proteins are less hydrated and less firm because of the lack of repulsive forces. The type and stability of the gel structure is significantly affected by the net charge of the protein [18]. The structure of gels will change from an aggregated to an ordered strand structure as the result of increased repulsive forces between molecules and suppression of the random aggregation.

Gelation pH ranges are dependent on protein concentration for proteins with large molar percentages of hydrophobic amino acids such as egg albumin. They are not dependent on protein concentration for proteins with small molar percentages of hydrophobic amino acids as in ovomucoid, gelatin, and soy proteins. Considerable changes of globin viscosity on heating occurred at pH 5 to 6 [20]. The maximum viscosity was recorded at a heating temperature of 95 °C at pH 5.3.

The effects of sodium chloride and other electrolytes on gels properties have practical significance in food processing.

Addition of NaCl increased the degree of random aggregation for soy protein, whey protein and BSA gels. Moisture loss increased with increasing salt concentration in the gels of whey protein and BSA. Higher levels of random aggregation and coarser structure are obtained with NaCl addition. Hermansson et al. [21] reported that gels formed at low ionic strength (0.25 M KCl) showed a fine microstructure, and gels formed at high ionic strength (0.6 M KCl) showed coarsely aggregated microstructure. Increasing the ionic strength (0.31 and 0.61 M) in beef myofibrillar protein gels produced more elastic gels with increased storage modulus, G' a measure of the total force or energy required to deform a sample [22].

6.3 Gelling Properties of Meat Proteins

Introduction

The capacity of meat proteins to form gels during production of meat products is an important functional property contributing to textural properties and binding of structure. The interactions between protein-protein, protein-water and protein-lipid are responsible for the formation of a stable matrix and structure of processed meats. Textural properties and the yield of processed meats are influenced mostly by gelling, emulsifying and the binding capacity of meat proteins, especially myofibrillar proteins. Gel formation of myofibrillar and sarcoplasmic proteins is responsible for retaining and stabilizing water and fat in comminuted meats and for binding meat pieces in restructured meat products. A different gelling capacity was found for salt soluble, insoluble myofibrillar and connective tissue proteins [23]. Salt soluble muscular proteins contribute mostly to the formation of gels. These protein fractions should be utilized by manufacturers to improve existing or to develop new meat products.

In batters of comminuted meats, gelation of myofibrillar proteins increases the heat stability of meat emulsions and as a result decreases heating losses. Gel formation in processed meats, particularly comminuted meats is influenced by salt concentration, pH of meat and processing conditions. Acton [24] studied protein-protein interactions during gelation of myosin as affected by heat treatment. Thermal gelation of myofibrillar proteins from skeletal muscles is affecting textural properties of processed meats. Textural properties of processed meats are mainly influenced by the gelation of myosin, actomyosin and actin during heating. Improving the quality of processed meats and development of new, particularly restructured meat products requires understanding of the physicochemical and biophysical interactions of meat proteins. Temperature and time of heating during production of comminuted meats are not sufficient for significant collagen solubilization. However, stroma proteins may strengthen the gel structure formed by myosin, actin and other meat proteins. The gelation mechanism and gelling properties of processed meats have been mainly studied in model conditions using isolated myofibrillar proteins.

6.3.1 Myofibrillar Proteins

Myosin and Actin

The functionalities of myofibrillar proteins in processed meats are the determining factors of their quality. Myofibrillar proteins are considered as the most important in binding meat structure. From individual isolated myofibrillar proteins myosin and actomyosin produced the gels with the greatest gel strengths. The binding properties of processed meat and fish products are mainly determined by the gelling properties of myosin [25]. The main gel-forming protein of muscles is myosin and actin affectes myosin gelation. The binding mechanism involves the interaction of contractile proteins from myofibrils and gelation of these proteins after thermal treatment. Solubilized myofibrillar proteins exhibited high low-temperature gelation ability which together with swelling of muscle fibers increased viscosity of the protein matrix and stabilized fat in the protein matrix [9, 26].

Gel formation in sausages and restructured meats is related mainly to the gel-forming properties of myosin and actomyosin. Samejima et al. [15] suggested that heat-induced gelation of myosin involves two basic reactions: 1). formation of aggregates of the globular myosin heads with involvement of disulfide bonds, unfolding of the tail portion, and 2). formation of a myosin network as a result of noncovalent interactions between tail portions of myosin and unfolding of the helical tail segment. The different gel-forming capacity of myosin is related to different filament-forming ability and gel-forming capacity of head and tail segments of myosin [27]. The capacity of the myosin tail to form cross-links within the myofibrils is observed in the native structure of myosin thick filaments. The aggregation of unheated myosin often occurs in a very irregular way, giving gels with very long aggregates. Gels of much higher rigidity have been formed from fine strand myosin than the aggregated myosin [21].

Heat induced gelation of myosin developed due to aggregation and formation of the three-dimensional network was found to be imparted by the subfragments obtained from the globular heads of myosin and the myosin rod obtained from the helical tail [15]. Samejima et al. [15] demonstrated the difference in the strength of the gels formed by myosin fragments, myosin rods, S-1 fragments and intact myosin molecules.

During the heating process and gelation, myosin loses its noncovalently stabilized α-helix structure, followed by intermolecular association and aggregation with stabilization of covalent (disulfides), and noncovalent interactions and development of the rigid structure. Extended three-dimensional network systems are produced from myosin and myosin rods due to conformational changes arising from a partially irreversible helix-coil transition during heating [15]. Egelandsdal et al. [28] suggested that under conditions favoring myosin filament formation, storage modulus transitions involved two types of reactions, i.e. reversible and irreversible associations of myosin filaments. At equivalent protein concentrations, the gel strength of different forms of myosin or its subfragments is in the order: myosin filaments > myosin heavy chain ≈ myosin monomers > total rod > LMM > S-2 fragments= S-1 fragments [29]. Yamamoto et al. [30] reported that myosin gel properties are determined by the length of native myosin filaments. Longer filaments of myosin formed finer, more rigid gel structures. The gel strength is affected by the structure of myosin. Filamentous myosin forms more rigid gels than monomeric myosin.

The effects of temperature, pH, ionic strength, and protein concentration on gel rigidity indicated that heat-induced gelation of the head portion of myosin differed from the heat-induced gelation of myosin. No difference was found between the tail portion of myosin and intact myosin. Samejima et al. [15] suggested that myosin produced gels upon heating by head to head binding and then self-association through cross-linking of the tail portions into a gel.

The gelling properties of heavy meromyosin (HMM) and light meromyosin (LMM) have been demonstrated as a heat-induced network forming ability in 0.6 M KCl at a pH 6.0 [25]. The reversible helix-coil transition of LMM has been recorded in the 40–70 °C range, with no evident aggregation as determined by measuring the turbidity of LMM dispersions. The irreversible association of HMM at pH 5.0 and 0.1 M NaCl produced a gel with increased rigidity. There was no change in SH group content in LMM in the temperature range 20 to 70 °C. In heated HMM, a decrease of 8.6 mol of SH groups per mol of HMM was found [25]. Oxidation of SH groups in the head region of HMM is involved in the heat-induced gelation. Samejima et al. [31] reported that the myosin heavy chains are responsible for thermal gelation of myosin and no difference was found between intact myosin and heavy chain myosin in gel strength level. At the same time, HMM showed poor gel-forming capacity [32]. Because the head portion of myosin is globular and about 10 nm in diameter, while the tail portion is rod-like, approximately 150 nm long, the head portion might be less important in gel formation in the relatively low temperature range. HMM did not form a gel during heating and was transferred into a curdy matter. The aggregation of HMM corresponded to that of myosin alone in the temperature range above 50 °C.

Light meromyosin formed a gel on heating. It was concluded that the first development of gel elasticity of myosin, which occurs in the 30–44 °C range, is attri-

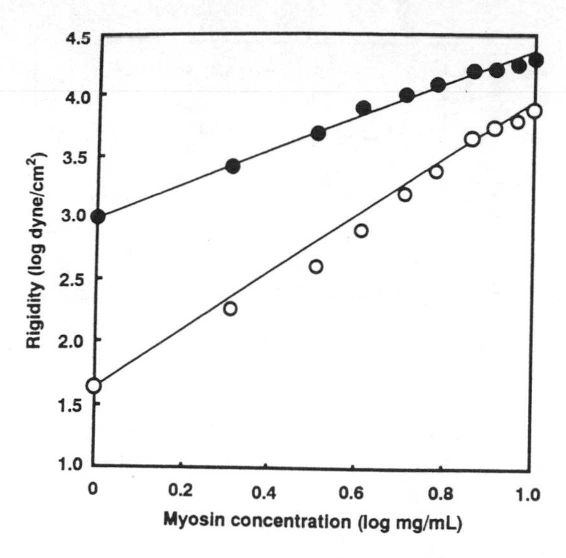

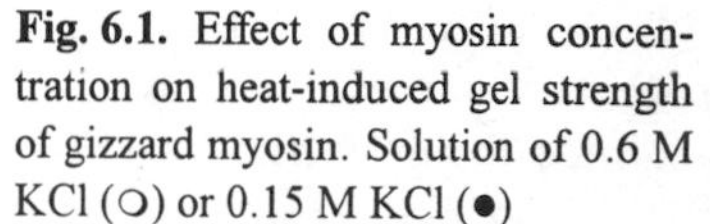
Fig. 6.1. Effect of myosin concentration on heat-induced gel strength of gizzard myosin. Solution of 0.6 M KCl (○) or 0.15 M KCl (●)

butable to the tail portions of myosin and the second step of gel elasticity formation in the 51–80 °C range is related to the head portion of myosin. Ishioroshi et al. [33] reported the maximum gelation of myosin and gel rigidity at 65 °C at pH 5.5 and 6.0 and ionic strength 0.2 and 0.6 M KCl, respectively.

The effects of gizzard myosin concentrations from 1 to 10 mg/ml were tested and gel rigidity was proportional to myosin concentration $(mg/ml)^{2.5}$ and $(mg/ml)^{1.4}$ in 0.6 M and 0.15 M KCL, respectively (Fig. 6.1) [34]. Rigidity of heat-induced gels of skeletal muscle myosin increased proportionally to $(mg/ml)^{1.8}$ at 0.6 M KCL. The effect of muscle protein concentration on the shape of the storage modulus (G') thermogram is presented in Fig. 6.2 [22].

The apparent onset of gelation at 49–52 °C was not affected by myofibrillar protein concentration, however the magnitude of storage modulus is concentration-dependent.

Actin has a different sensitivity to heating than myosin. It forms a gel by polymerization of native molecules and does not gel when heated [35]. Actin without myosin is able to form a gel with the formation of a three-dimensional network of actin linear polymers. Actin added to myosin gel increased the rigidity when heated for 20 min at 65 °C and pH 6. Actin-myosin interaction in the gel of two mixed proteins accelerated the association process with the formation of more rigid gels. Actin showed an enhancing effect on the gel strength of myosin, and the actin effect was more apparent in red muscle myosin than in white muscle myosin of chicken and was dependent on pH [36].

The gelling capacity of myosin depends on the myosin/actin weight ratio and state of the myosin before heating [29]. The maximum increase in gel rigidity was observed at a myosin to actin weight ratio of 15 [35]. Gels with a maximum strength in 0.6 M KCl, pH 6.0 have been formed at a free myosin-to-F-actin mole

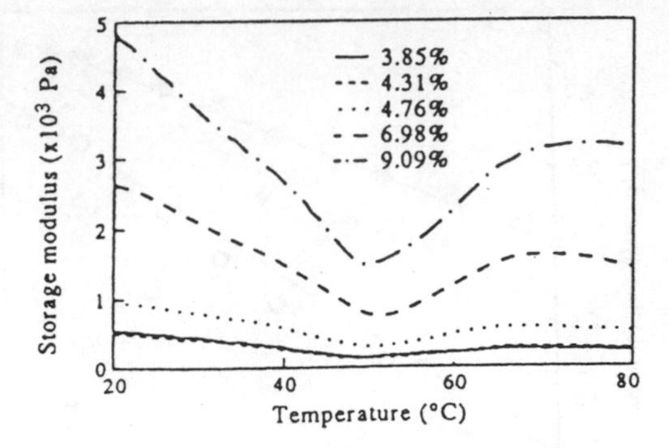

Fig. 6.2. Mechanical thermogram showing changes in storage modulus (G') for a myofibrillar protein extract of varying concentration as a function of increasing temperature

ratio of 27:1 equal to a weight ratio of 15:1 [37]. Actin exhibited a synergistic property, markedly increasing the binding capacity of myosin. Ishioroshi et al. [38] reported that in the 0.2 M KCl system, actin interfered with cross-linking between the myosin heads and inhibited gelation capacity of myosin filaments.

Actomyosin

The maximal gelling ability, i.e. gel strength of actomyosin was reported at pH 5.0 (natural actomyosin) and 5.5 pH (crude actomyosin) with a decrease in strength on both sides of this pH range [39]. Authors concluded that additional proteins associated with the actin-myosin complex, i.e. actinins and troponin, contribute to gel network formation and gel strength. The gel strength of actomyosin gels increased exponentially with concentration from 3.75 to 10.0 mg/ml. Natural actomyosin solution formed a gel at 30 °C when held for 30 min and gel strength continued to increase at 30–60 °C and higher. The increase in actomyosin gel strength from 60 to 80 °C may be related to actin interaction within the gel structure. The denaturation temperature of actin is in a higher range than that observed for myosin. This study demonstrated that the mechanism of gel structure formation at higher actomyosin concentration is the same as that occurring in a diluted solution of actomyosin [39].

The actomyosin complex formed a more rigid gel than myosin alone. A significant increase in the rigidity of actomyosin gels was obtained at a myosin:actin mole ratio of 1.5–2.0 and it was dependent on pH and KCl concentration [35]. Data obtained by Yasui et al. [35] corresponded with findings by Sano et al. [40] that the highest rigidity modulus was obtained at a F-actin/myosin ratio of 0.067. The F-actin/myosin sample was obtained using rabbit skeletal muscle. Elasticity of the gels decreased with an increase in the F-actin/myosin ratio. The gels of highest elasticity and gel strength were obtained with myosin alone. Conse-quently, actin did not affect the elasticity development which occurred in the 36–46 °C range. However, F-actin provided the viscosity of the cooked gels. The thermal gelation mechanism of muscle proteins could be understood by studying the role of F-actin during the thermal gelation. However, the synergistic effect of actin on myosin gel-

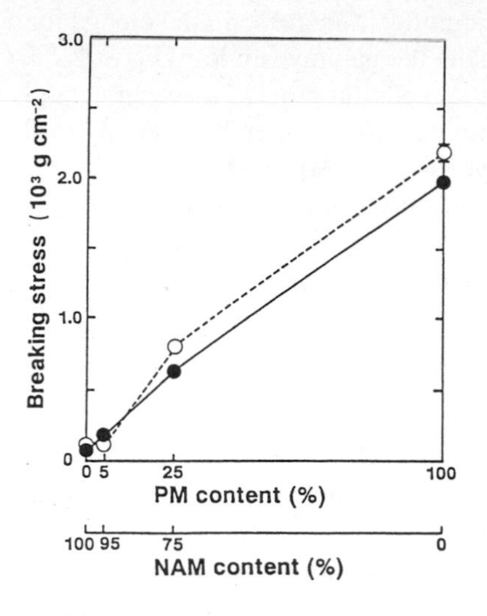

Fig. 6.3. Effect of ATP on the changes in breaking stress of the paramyosin *(PM)*-natural actomyosin (*NAM*) system gels. (-●-), with ATP; (---○---), without ATP

ling properties was ionic strength dependent. Ishioroshi et al. [33] found a progressive decrease in the heat-induced gel strength of myosin with the addition of 0.5 to 3.0 mg/ml of actin to myosin in 0.2 M KCl solution. There were no significant differences in the gel ultrastructure between myosin and actomyosin [41]. Yasui [42] suggested a specific myosin to actin ratio to obtain gels with higher rigidity than with myosin alone. Gels with maximum rigidity were obtained at a myosin to actin mole ratio of 2.7, i.e. the weight ratio of myosin to actin ≈15. Gels of lower rigidity were obtained with an increased proportion of myosin. The effect of actin on gel rigidity was excluded by the addition of pyrophosphates or ATP which dissociate the actomyosin complex to actin and myosin [35]. There is a variation among myosin isoforms in gelling properties. In sausage batters, heat-induced gel formation is possible by cross-linking free and bound myosin molecules with actin [42].

Gelation of HMM and S-1 was not affected by actomyosin but an increase in gel rigidity was observed when actomyosin was combined with rod or LMM [42]. Addition of troponin and tropomyosin did not affect the gelation of actomyosin. Acton et al. [39] reported an acceleration effect of these proteins on the actomyosin gelling capacity. It is generally accepted that native tropomyosin does not contribute significantly to the microstructural characteristics of myosin gels [43]. Foegeding [44] showed that suspensions of myosin/actomyosin (79–86% myosin) could form self-supporting gels at 10–20 mg/ml, while salt soluble proteins (52–54% myosin) required 25–40 mg/ml. The consistency of sols of natural actomyosin and the mixture of paramyosin and natural actomyosin was markedly decreased by the addition of the ATP solution. The breaking stress of the paramyosin-natural actomyosin gels prepared without ATP increased linearly on increasing the paramyosin content (Fig. 6.3) [40]. The same trend in increasing breaking stress was observed in the gel system with ATP. However, the breaking strength values were smaller than those without ATP, and the difference increased at a paramyosin content of 25% and more. A significant increase in the elongation was found for the gels without ATP on increasing the paramyosin level up to 25%. Significant differences were found in the thermal gelation process between natural actomyosin and myosin. Using a laser scanning microscope, Kim et al. [45] demonstrated the acceleration of beef actomyosin polymerization induced by the addition of transglutaminase. The polymerized actomyosin was increased while the myosin monomer was decreased. The polymerization resulting from cross-linking of actomyosin was induced by the transglutaminase.

6.3.2 Sarcoplasmic Proteins

Sarcoplasmic proteins after heat coagulation contribute to gel formation and the binding structural elements of meat. Sarcoplasmic proteins make a significant contribution to the binding ability of the system when the ionic strength is low (below 0.4 M). With an increase in the ionic strength (more than 0.4 M) they do not contribute to binding of the system [46]. The synergistic effect appeared between myosin and sarcoplasmic proteins, however binding obtained by sarcoplasmic proteins was too small to be measured. Considerably more rigid gels have been obtained with myofibrillar proteins than with sarcoplasmic ones of broiler breast muscles. Gel formation is influenced by sarcoplasmic proteins to a small extent because these proteins coagulate and precipitate when heated above 40 °C.

6.3.3 Gelation of Red and White Muscle Proteins

Actomyosin from different species, types of muscle (red and white), gives rise to gels with different properties. The gelation patterns in actomyosin of pork and chicken were similar and in hake different [47]. Hake gels were more sensitive to heat than chicken or pork gels.

A significant difference in gelation properties between white and red myofibrillar proteins in poultry meat was reported [48]. Under similar processing conditions, white muscle proteins showed better gelling properties, i.e. formed stronger gels, than red muscle proteins. Breast and leg salt-soluble proteins showed 1–3 transitions in protein-protein interactions within pH 5.5–6.5. The combined breast/leg myofibrils formed stronger gels than breast myofibrils alone at pH < 6.0. Puncture force of gels from chicken breast (20 mg/ml, pH 6.5) was higher than similar chicken leg myofibril gels [49].

Myosin isolated from white broiler muscle exhibited greater gel strength than from red muscle under similar pH, ionic strength and protein (myosin) concentration [29]. The gel strength of myosin from white muscle was about three times greater than that from red muscle at 5 mg/ml protein content in the model system. Gel forming and the water-binding capacity of myosins isolated from two bovine muscles, M. cutaneous trunci and M. masseter were different [50]. Textural studies showed that white myosin formed the superior gels. Gel strength expressed as storage modulus, was consistently higher at all temperatures above the gel-inducing minimum at NaCl concentration 0.2 M and 0.6 M. The storage moduli had practically the same concentration dependence with white myosin and with red myosin. Both myosin gels formed in 0.2 M NaCl showed the same liquid loss after centrifugation, regardless of pH, i.e. about 15% was lost. When the two myosins were mixed at pH 5.65 the gel-strengthening effect was observed. Different gelling capacity of white and red broiler muscles should be considered when formulations of restructured meat products are developed.

Chicken breast myosin formed more rigid gels than chicken leg myosin, however at the same pH and ionic strength conditions these gels have similar rigidity [36]. Myosins from chicken leg and breast showed optimum rigidity at pH 5.1 and 5.4, respectively. Myosins from chicken and rabbit exhibited similar gelling properties but myosin from tilapia (fish) showed a different gelling ability [51]. Different gelation capacities which are probably caused by pH level could affect the palatability of processed products from turkey meat. The relationships between the functionality of protein of myofibrils from turkey breast and thigh have been reported [52]. Maximum shear stress (increase 150%) was found for gels prepared after removal of water-soluble proteins and other components. Amato et al. [53] reported that the thigh gels of turkey exhibited higher shear stress and shear strain values than breast gels. However, differences in pH were not measured. Gels prepared from comminuted breast and thigh contained different amounts of connective tissue and lipids influencing gel properties. Dark meat with pH $\approx$ 6.2 that is higher than in white meat (pH $\approx$ 5.6) exhibited more favorable molecular interactions which caused gelation [53]. A significant amount of smooth muscle is produced as byproducts by the meat industry. The gelation properties of myosin from smooth muscles are important for exploring effective uses of animal products.

6.3.4 Factors Affecting the Gelling Properties of Meat Proteins

Effect of the Pre-Rigor and Post-Rigor State

Differences in gelling and other functional properties between pre- and post-rigor meat muscles are related mainly to protein extractability. Gel strength increased with heating temperature, indicating a continuous denaturation of proteins to facilitate gel matrix formation (Fig. 6.4) [49]. Rigor mortis affected the gelling properties of myofibrillar suspensions. The stronger gels have been prepared from post-rigor breast myofibrils than from pre-rigor myofibrils. This difference between pre- and post-rigor myofibrils was not observed for leg samples. Differences were found between the onset temperatures of gel formation for breast and leg and for pre-rigor vs. post-rigor myofibrils (Fig. 6.4). Post-rigor chicken breast meat is more functional than pre-rigor breast meat for making gel-type products. Similar data were reported for beef myofibrils by Samejima et al. [54].

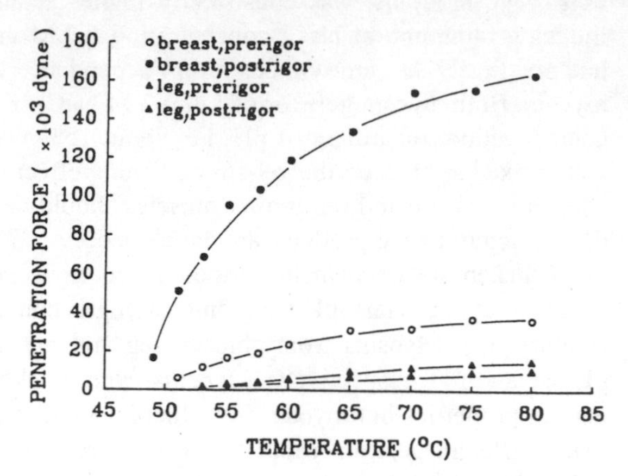

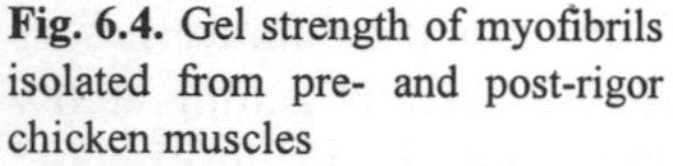

Fig. 6.4. Gel strength of myofibrils isolated from pre- and post-rigor chicken muscles

The gelling ability of intact myosin molecules is higher than ones disintegrated by comminution and myosin after disintegration failed to form a gel of sufficient strength [43]. According to other data [55] the gelling properties of unaged meat were better than aged meat when used in sausage production. Salt-soluble proteins in pre-rigor red meat exhibited higher extractability than salt-soluble proteins in post-rigor red meat. The expressible moisture of salt-soluble protein gels decreased as protein concentration increased from 4 to 8% suggesting that salt-soluble proteins were responsible for the gel properties [56]. The gel rigidity from fresh muscle was 3200 dyne/cm^2 and during 7 days aging period the gel rigidity decreased to 2600 dyne/cm^2. Gelling properties of chicken breast muscle samples and their salt-extractable proteins were generally poorer after 2 days of storage, compared to no storage [57]. However, the gelation capacity after 7 days of storage was restored and was higher than 2 day-stored samples and in most cases higher than nonstored samples. Pre-rigor chilling or freezing was not detrimental to protein gelation at pH 6.0 and 0.6 M NaCl [58]. Cold-shortened muscle produced a stronger dynamic salt-soluble protein gel network than muscle stored at -29 °C or 15 °C. Smith [59] reported the effect of frozen storage and the degree of protein denaturation on the gelling properties of protein from chicken muscle.

The Effect of Heat Treatment

The mechanism and factors of heat-induced gel formation of myofibrillar proteins were investigated extensively. Gelation of myofibrillar proteins is generally heat-induced and the sol to gel transformation results in the formation of a three-dimensional network. The formation of the three-dimensional myofibrillar protein network physically and chemically stabilizes water and holds dispersed fat particles structurally in comminuted meats.

The aggregation of meat proteins during heating is a prerequisite for the development of the protein network and gel formation. Physico-chemical changes of myosin and actomyosin during heating play an important role in determining the binding of basic ingredients in processed meats. Myosin which is a basic protein in the formulation of comminuted meats with relatively low gelling temperature is a limiting factor in the normal temperatures of heat treatment. The mechanism of myosin gelation involves the formation of fairly stable bonds as a result of heating and irreversible changes in its quaternary structure.

The thermal transition of beef proteins was observed at three temperature ranges: T_1 (45–60 °C) attributed to myosin, T_2 (60–76 °C) attributed to sarcoplasmic proteins, actomyosin and proteins of connective tissue and T_3 (76–92 °C) attributed to actin [60]. The transition temperature for the different molecular interactions and conformational changes are affected by the method of protein preparation and the heating rate. The transition temperatures of myosin coincide with those for thermal unfolding of the helical structure of the myosin rod. This suggests that the unfolding of the myosin tail portion may mostly affect the heat-induced gelation of myosin. Changes in dilute (0.5 mg/ml) and concentrated (20 mg/ml) myofibrillar protein suspensions heated at temperatures 20 to 70 °C are presented in Fig. 6.5 [61]. In dilute suspensions, protein-protein interactions began at 43 °C with the formation of aggregates influencing optical density and the storage modulus (G'). The initial increase in G' was detected at 41 °C and was related to crosslinking of myosin filaments due to denaturation of heavy meromyosin. As the temperature of protein suspension increased, protein molecules continuously aggregated with increasing turbidity. However, the magnitude of G', after reaching a maximum at 50 °C, decreased sharply to a minimum at 55 °C (Fig. 6.5). The G' decrease at >50 °C was related to denaturation of light meromyosin, and G' increase at >60 °C probably resulted from formation of irreversible myosin filaments or complexes.

The heating rate influences the gelation of meat proteins, which is a limiting factor in comminuted meats. Gelling properties of myosin gels with high protein concentration (>10 mg/ml) have been reported by Camou et al. [62], especially at increasing heating rates. The gel strength decreased when heating rate was increased from 17 to 38 and 85 °C/h.

The heating temperature and pH range applied for muscle foods would suggest minimal modification of amino acids. For this reason, the major cause of irreversibility of protein denaturation is protein association and slow refolding. Gelation of individual proteins, myosin, albumin, and fibrinogen was affected by the heating rate [63]. When rapid heating was utilized at 70 °C for 20 min a less stable gel was formed.

At low temperatures (40–50 °C), hake gels were stiffer than chicken or pork gels. However, at higher temperature (60 °C) actomyosin from hake, chicken, and pork formed gels with similar textural properties [47]. Manufacturers can produce products with various textural properties by varying the rate and tempe rature of heating to control protein-protein interactions. Thermally induced changes in gel viscoelastic moduli for myofibrillar proteins were established [58]. Considerable differences in viscoelastic properties of myofibrils and salt-soluble proteins of the three tempered muscle samples were found. The complex modulus G^* increased

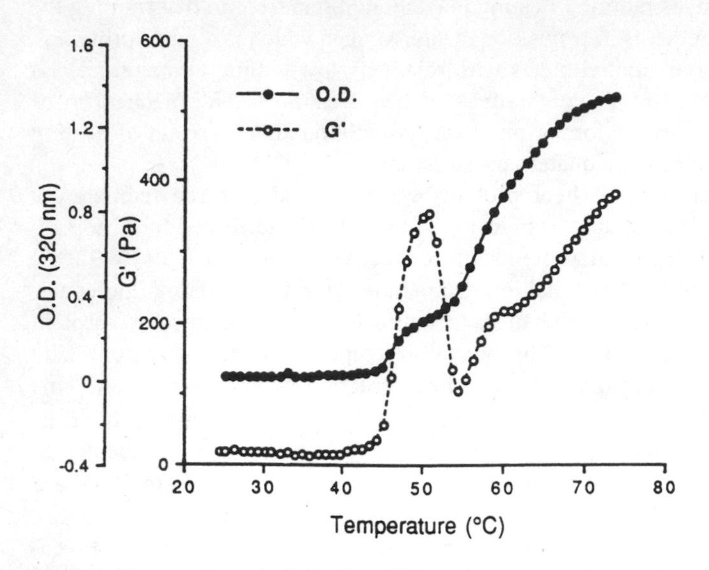

Fig. 6.5. Changes in optical density (*OD*) and storage modulus (*G'*) during linear heating of myofibrillar protein suspensions

rapidly between 40 and 55 °C for myofibrils of all three muscles (Fig. 6.6), suggesting the formation of a viscoelastic gel matrix [58]. A drastic decrease in G^* was found at above 55 °C with a minimum at 60 °C and then G^* recovered at higher temperatures (65–75 °C). Salt-soluble proteins exhibited similar changes in G^* to those of the myofibrillar samples. However, G^* of salt-soluble proteins was about twice that of myofibrils.

The Effect of pH and Ionic Strength

In the production of comminuted meats the pH must be carefully controlled in order to increase extractability of myofibrillar proteins. pH significantly influences gel forming properties of individual proteins of myofibrils such as myosin and protein complexes such as actomyosin [39]. Beef actomyosin gels with maximum gel strength were prepared in the pH range 5.5–6.0. The effect of pH and ionic strength on the heat-induced gel characteristics of myosin may be related to the morphology of myosin filaments or aggregates. The effect of pH and ionic strength on the gelation of various myosin subfragments (gel rigidity) will vary with myosin subfragments [25]. The maximum gel strength of myosin gels containing 4.5 mg/ml protein in 0.6 M KCl was obtained at pH 6.

The myosin gels prepared at higher pH's were relatively translucent and, at pH's lower than 6.0, gels exhibited symptoms of syneresis.
The gelation of protein from breast myofibril was more pH dependent than those of leg myofibril [49]. Gels with high strength were prepared from chicken myofibrils at pH 6.0. The shear stress and shear strain of the gels from turkey breast

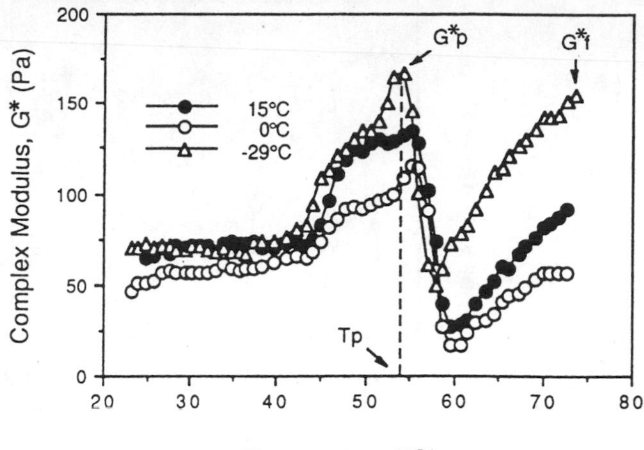

Fig. 6.6. Typical viscoelastic curves showing dynamic changes in the complex moduli of myofibril suspensions of cold-shortened (0 °C), thaw-rigor (–29 °C), and control (15 °C) bovine sternomandibularis muscles during thermal gelation. *G*p* is the complex modulus at transition peak at temperature *Tp*; *G*f* the complex modulus at the end of heating

breast and thigh increased between pH 6.0 and 6.4 [64]. These properties could be related to water holding capacity.

Actomyosin gels formed at pH range 5.0–5.5 were described as „spongy", with obvious syneresis and at pH 6.0 and above were characterized as uniform and opaque. Yasui [35] reported that optimal pH for gelation is dependent on the concentration of myosin and actin in solution. The effect of pH on the relative gel strength of myosin gels was reported by Fretheim et al. [9]. The pronounced maximum of pH for gel strength was found at about pH 4.5. A significant decrease in relative gel strength was observed at pH lower than pH 4.0 and higher than pH 5.0. Relatively weak gels were obtained at pH≥5.00, because higher pH values imply that higher amounts of myosin were dissolved in the liquid phase of the gel. In this pH range a lower amount of protein is in the gel network.

Changes in the state of myosin molecules are important for gelling capacity since gel formation results from protein-protein and protein-solvent interactions.

State of the myosin molecules is influenced by the ionic strength. Myosin molecules associate and form filaments at low ionic strength (<0.3 M). However, dispersion of filaments to the monomers has been observed at ionic strengths above 0.3 M. Sano et al. [65] suggested that the interaction between myosin molecules occurred more easily on increasing the ionic strength. In 0.1–0.2 M KCl interfilamental aggregation of myosin heads on filament surfaces is responsible for gel formation, and gel strength is positively correlated with salt concentration [15]. Acceleration of gelation was observed at low ionic strength (0.2 M KCl) with the formation of a fine-stranded gel microstructure, with greater rigidity than at high ionic strength (0.6 M KCl) [21].

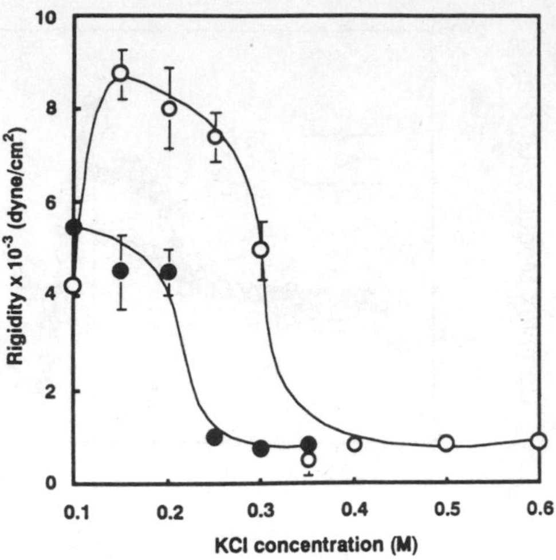

Fig. 6.7. Effect of ionic strength on heat-induced gel strength of gizzard myosin. Solution of 0.4 mg/ml myosin, 40 mM potassium phosphate buffer (pH 5.9), 0 mM (○) or 10 mM (●) ATP

The effect of the ionic strength on the rigidity of gizzard myosin gel showed rigidities higher at low ionic strength than at high ionic strength (Fig. 6.7) [34]. Gel rigidities at 0.15–0.25 M KCl were much higher than those at 0.35–0.6 M those at 0.35–0.6 M KCl. The higher gel strength at low ionic strength was due to the formation of filaments. The rigidity of gizzard myosin gels in 0.6 M KCl increased with temperature above 50 °C and after 65 °C rigidity remained constant.

Myosin solutions formed finer three-dimensional network at 0.2 M KCl than at 0.6 M KCl. At high ionic strength myosin molecules exist as monomers that upon heating form aggregates. Actomyosin gels exhibited two optima of rigidity, the first at ≈0.2 M KCl, and the other at ≈0.6 M KCl [42].

The properties of gels produced from the high ionic strength (HIS) soluble protein fraction and in combination with the low ionic strength (LIS) soluble or insoluble protein fraction have been reported [65]. Apparent stress at failure was greatest in skeletal HIS 6% (w/w) protein gels, followed by heart, lung, and spleen (Fig. 6.8) [66]. The spleen contains a higher ratio of actin to myosin which may explain the lower gel strength of spleen HIS proteins in comparison to skeletal muscle proteins. The percentage of myosin in the HIS protein fraction may influence the different properties of gels. Numerous studies have established a positive correlation between protein extractability and the binding properties of myosin for various heating conditions [67]. The effect of salt on the binding capacity of myosin is related to its ability to dissolve myosin and the formation of a firmer gel. The positive effects of KCl and NaCl on gel strength and cohesiveness are connected to accelerated hydrophobic interactions within the actomyosin gel structure. Hydrophobic interactions and hydrogen bonding are the main molecular forces involved in the crosslinking of polypeptide chains in heat-induced actomyosin gels. Addition of NaCl (0.3–0.6 M) and pyrophosphate increases protein extractability and the storage moduli of non-heated myofibrils [54].

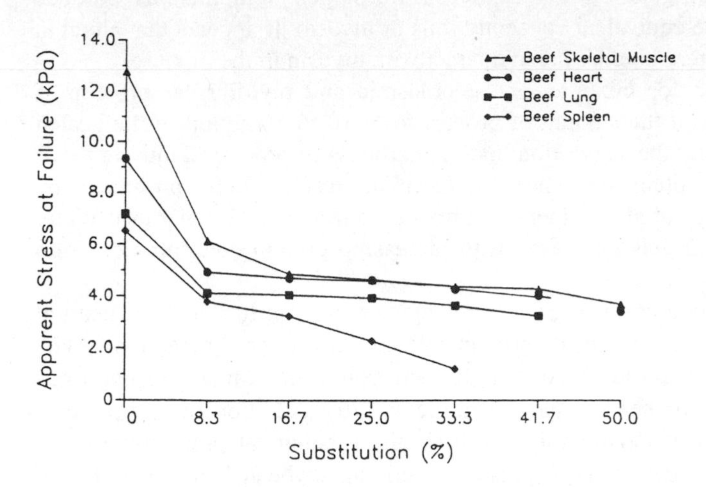

Fig. 6.8. Apparent stress at failure of heat-induced gels prepared with high (HIS) and low ionic strength (LIS) soluble protein fractions from beef skeletal, heart, lung and spleen tissues

Extractability of proteins from non-heated myofibrils increases in the 0.1–0.6 M NaCl concentration range.

The presence of neutral salts affects the rheological properties of heat-induced actomyosin gels and the relative effects of the anions on the compressive strength and cohesiveness of actomyosin gels is in the order [68]: ♣F>Cl>Br>SCN>I♣

The relative effects of the cations on the heat-induced gelling properties of actomyosin are in the order: ♣$K^+ \approx Na^+ > NH_4^+ > Mg^{2+} > Ca^{2+}$♣

The increased binding properties of myofibrillar proteins by adding polyphosphates is due to increased solubility of myosin [41]. The binding capacity of these proteins is dependent on the presence of myosin in the dissociated state. Myosin forms a three-dimensional network of overlapping fibers in the presence of salt and phosphate and a sponge-like network in their absence. The increase in binding strength of myosin caused by polyphosphate may be attributed mainly to specific protein-polyphosphate interactions that increase myosin solubility by dissociation of actomyosin.

6.3.5 Myosin Blends with Other Proteins and Lipids

The gelling properties of myosin-albumin and myosin-fibrinogen blends are affected by the heating rate at pH 6 [69]. In a myosin-fibrinogen blend there was an interaction between these two proteins and the gel strength of the blend was higher than sum of the gel strengths for the two individual proteins. The strength of the myosin-fibrinogen gels was affected by the heating rate at temperatures between 55 °C and 70 °C. Myosin and fibrinogen form a gel matrix when heated at temperatures encountered in meat processing (60–70 °C). These gels are stronger than myosin-albumin gels. At 50 °C myosin-fibrinogen combinations produce stronger gels than the equivalent concentrations of myosin. It appears that albumin must be thermally altered before it can interact with myosin in the mixture.

The gel strength for mixtures of sarcoplasmic and myofibrillar proteins is directly proportional to the amount of protein from 10 to 50 mg/ml in the heated solution. This supports the suggestion that the hardness of processed muscle foods is influenced by protein concentration. Similar results were presented by Foegeding et al. [63] but at considerably lower protein levels (3–6 mg/ml). Water loss from compressed gels decreased with increasing protein concentration from 10 to 50 mg/ml.

Soy proteins incorporated in comminuted meats influence the textural properties as a result of meat protein-soy protein-water interactions. The rheological properties and the °C of gels formed from ground meat are improved by the incorporation of heat-treated soy proteins [70]. Addition of soy proteins with various degrees of denaturation modifies the rheological properties of raw sausage batters [71]. Peng et al. [72] also showed that soybean 11S protein has to be thermally altered before it can interact with myosin. β-conglycinin and the 7S protein fraction interacts with myosin at temperatures above 85 °C. The WHC of soy protein gel is improved after preheating during preparation of soy protein.

The gelling of meat proteins with the formation of gels of maximum stability takes place at temperatures between 60 °C and 70 °C. In this temperature range, soy proteins form weak and unstable gels [73]. Stable and firm gels are formed at temperatures of 90–100 °C. Heat treatment at 70 °C in sausage production does not promote the formation of soy protein gels. Modification of soy proteins is recommended to give firm and stable gels at lower temperatures (60–70 °C). The gelling ability of soy isolate is considered to be responsible for its superior performance in sausage emulsions. The poor emulsion-stabilizing effect of soy concentrate is related to its limited gelling capacity due to its fiber content. The effects of interactions between myosin and lipids on gel properties have been reported [74]. Fatty acids bind to the myosin causing increased electrostatic repulsion between the myosin chains and, as a result, increase gel strength. Fatty acid salts influence the properties of myosin gels when incubated in the temperature range 40–75 °C. Potassium decanoate is a powerful enhancer of myosin gel strength. All tested fatty acid salts exhibited a denaturing effect on myosin. In the case of potassium decanoate myosin mixtures, the storage modulus was higher than that of the pure myosin system and increased with the amount of fatty acid salts added.

Oxidized lipid-protein interactions in processed meats accelerate the insolubilization of myofibrillar proteins during frozen storage and decreased myofibrillar gel strength. In a system including myofibrillar proteins and metal oxidizing agents a decrease in myofibrillar gel strength and water holding capacity was observed [75]. Crude myofibrils isolated from heart by repeated washing in the presence of antioxidants (propyl gallate, ascorbate and tripolyphosphate) formed stronger gels with lower TBA values in the pH range 5.8–7.0 and in 0.6 M NaCl than the control myofibrils [76]. The improved gel strength resulted from the inhibition of lipid oxidation and possible physical and chemical changes of proteins during the washing and heating process.

The fastest heating rate resulted in the weakest gel strength. The gel strength for a mixture of sarcoplasmic and myofibrillar proteins was directly proportional to the amount of protein from 10 to 50 mg/ml in the heated solution. This supports the suggestion that the hardness of processed muscle foods is influenced by protein concentration. Similar results were presented by Foegeding et al. [75] but at considerably lower protein levels (3–6 mg/ml). Water loss from compressed gels decreased with increasing protein concentration from 10 to 50 mg/ml.

6.3.6 Fish Proteins

The gelling capacity of fish proteins in comminuted fish products is one of their most important functional properties. The gelling properties of proteins in surimi products have already being utilized commercially in the production of imitation shellfish meats, such as crab meat, scallops and shrimp. The gelling capacity of myofibrillar fish proteins influences the textural properties of the minced products manufactured with fish meat. The myofibrillar proteins of fish form firm gels upon heating and cooling, while the albumins mostly undergo thermal coagulation. The gel-forming capacity of the myofibrillar fraction is significantly higher than a mixture of muscle proteins such as that existing in the meat.

The gelling ability of fish protein mixtures after heating is significantly affected by the solubility of the protein components. Fish muscle proteins differ in their critical gelling temperatures compared to the proteins of egg white, whey, pork or beef muscle. More elastic and stronger gels can be prepared by preheating minced fish paste at 40 °C. The optimum solubility and gelling properties of myofibrillar proteins are obtained when they are in the undenatured state [77]. The main gel-forming protein in fish is myosin and the role of actin has not been studied.

The gel forming ability differed little between myosin and actomyosin and was not altered by the presence of F-actin. Myosin is an essential protein for the elasticity development of fish gels. Sano et al. [78] reported the role of F-actin in gel formation when F-actin concentration was increased by adding natural actomyosin to myosin. It is difficult to prepare a sufficient amount of actin from fish muscle. In Fig. 6.9, the storage modulus of rigidity is plotted as a function of the F-actin/myosin ratio. Storage modulus and loss modulus represent the elastic and the viscous components, respectively. The storage modulus value was lowest for the F-actin/myosin ratio 0 and increased on increasing this ratio to 0.13. If the ratio F-actin/myosin increased above 0.13, the modulus of rigidity values decreased decreased slightly. The storage modulus values at 80 °C were highest at the F-actin/myosin ratio 0.061. The storage modulus values decreased on further increase in the F-actin/myosin ratio.

The role of hydrophobicity in gelation of fish muscle paste and sol was investigated by Niwa et al. [79] and they were able to relate the gelling capacity of

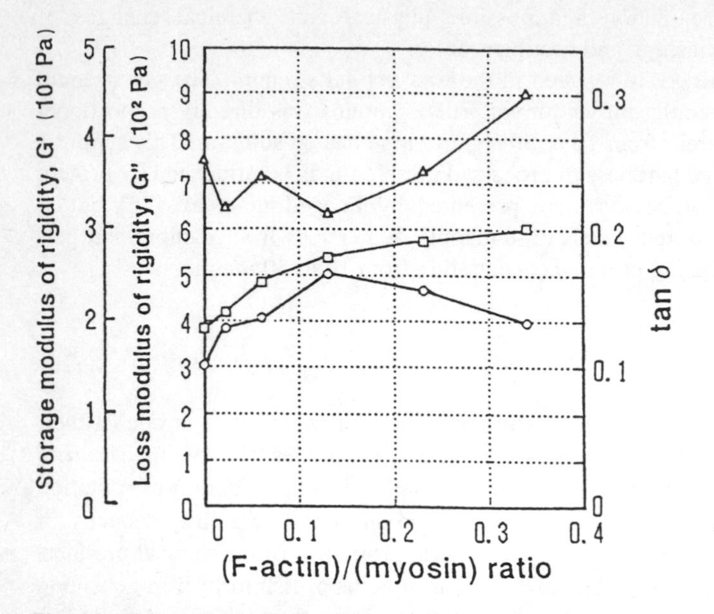

Fig. 6.9. Changes in storage modulus (-O-), loss modulus (-□-), and tan δ (-<6,136>-) values before the rise in temperature (at 5 °C)

the muscles from different species to the abundance of hydrophobic groups of the myosin molecule. Differences in the gelling ability of various species of fish are attributable to differences in cross-linking of heavy myosin chains [80]. Species differences were related to the surface hydrophobicity of the unfolded domains of heavy myosin chains and heating temperature. Chan et al. [81] reported that the thermal aggregation ability of myosin subfragments increased linearly with increased surface hydrophobicity (So) of myosins from cod and herring. This was not observed in heavy meromyosin from cod where the increase in So between 45–55 °C did not correspond to an increase in cross-linking ability. Studies of the viscoelastic properties and the turbidity of isolated carp heavy (HMM) and light meromyosins (LMM) suggested that the initial stage in gel formation was associated with changes of LMM at 30–44 °C [32]. Protein-protein interaction studies [82] showed that in the myosin fragment solutions, the thermal gelation of fish myosin consisted of two reactions. The first at 30–40 °C mainly involved the myosin head portion and the second was associated with the LMM aggregation at above 50 °C. Lanier et al. [83] reported the „setting“ of fish muscle proteins at below 40 °C when partially denatured proteins interacted noncovalently and formed an elastic network. Translucent elastic fish protein gel is formed after hea ting fish muscle sol at 40 °C or holding at 0 °C for 12–24 h. Gels of greater firmness can be prepared by heating fish muscle sol at higher temperatures. Both methods of fish gel production were applied in commercial conditions and each produced gels with unique textural properties.

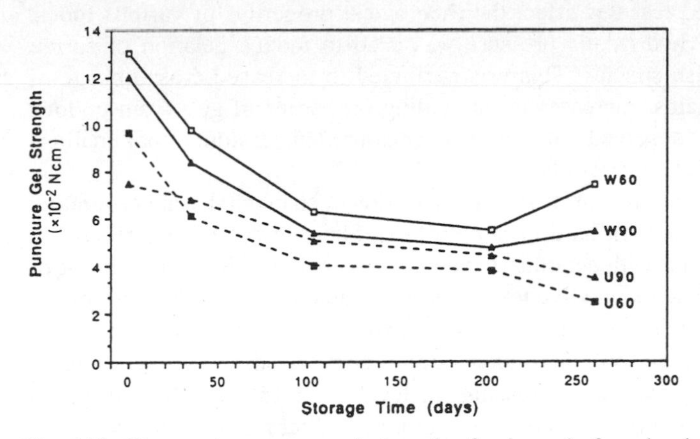

Fig. 6.10. Changes in puncture gel strength of gels made from headed and eviscerated hoki stored frozen at –29 °C

The five factors affecting the gelling properties of fish actomyosin were tested: concentration of protein, pH, ionic strength, time and temperature of heating [84]. A negative linear effect of the pH on the logarithm of gel strength values was established independently of the rest of the variables. The lowering of the pH and increasing protein concentration increased the gel strength of the heat-induced cod actomyosin gels. The effect of pH was strong at high actomyosin concentrations (25 mg/ml) and weak at low actomyosin concentration (5 mg/ml). Sodium chloride (3%) added to surimi caused a shift in the T_{max} values (denaturation temperature) to lower temperatures showing a decrease in stability to thermal denaturation [85]. Fish proteins with good binding and texturizing properties in fabricated foods are manufactured by a washing process that converts colored and strong flavored fish into bland and light-colored muscle protein [83].

The gel strength of the washed gels cooked at 90 °C decreased by 48% after 35 days of storage. The gel strength of gels heated at 60 °C and made from washed minces was greater than of those made from unwashed minces (Fig. 6.10) [86]. There was no difference between the samples after 50 days storage. The greatest decline of gel strength was detected after 1 month of frozen storage with a further decline until 100 days of storage after which there was a levelling off.

Freezing and frozen storage decreased the gelling ability of fish proteins. The loss of the gel-forming capacity in frozen fish has been related to denaturation and aggregation of the myofibrillar proteins. The gelling properties of minced fish declined with increasing frozen storage at -29 °C for 260 days. The temperature fluctuation during frozen storage caused reduction in gel-forming capacity of fish proteins. French et al. [87] reported protein hydrolysis in partially frozen salmon samples stored at 0 °C, -1 °C, -2 °C, -3 °C, and -20 °C for 25 days. The highest rates of hydrolysis were in fish held at -2 °C and -3 °C. Study of the gel-forming properties of surimi prepared from two fish species after various numbers of freeze-thaw cycles showed a reduction in the gel strength and deformability [88].

High hydrostatic pressures affect the rheological properties of various foods. Treatment with high hydrostatic pressure was used to induce gelation of surimi paste from several fish species. This was attributed to increased cross-linkage of the myosin heavy chains. Increases in the gelling properties of gels from pollock and whiting were observed at various pressure/temperature combinations compared with heat-set controls [89].

The physical properties of fish gels have been changed by incorporating shortening such as a medium hard plastic fat [90]. Shear forces of gels reached a maximum when 15% fat with a hardness index of 1.8 cm was added. As a result of hard fat incorporation a sponge-like texture was prevented and the plasticity of the gels increased. Weakening of fish gel texture is related to protein degradation by alkalic protease with an optimum temperature corresponding to the cooking temperature. However, with an increasing fat level from 15% to 30%, fish gels were less sensitive to the cooking temperature effect. This indicated that incorporation of hard fat minimized the textural variations in fish gels due to cooking conditions.

6.3.7 Collagen Gelation

Collagen is the most abundant protein in an animal carcass and an important component in terms of the physical properties of muscle. Collagen is the major protein component of skin, bone, tendon and other forms of connective tissue. Collagen has received considerable attention in the food industry because of its unique physical, chemical properties and specific functionality.

Collagen is a protein with unique gelling properties. It can form thermally reversible aqueous gels without additives over a wide pH range. Collagen and other connective tissue proteins play a negligible role in gel formation in sausage batters, mainly because of the heating temperature being insufficient for collagen gelatinization. However, functionality of connective tissue proteins in meats is important because meat products formulated without these proteins are soft, rubbery, with poor cohesiveness. The natural structure of edible bone collagen makes it a good carrier for other proteins and protein hydrolysates which are not textured.

The age of the animal influences the quantity and quality of connective tissue and its basic protein collagen can significantly affect the functional properties. Modification of collagen with age via cross-linking increases the toughness of meat. The number of cross-links in collagen increased with increasing age of the animal and this may explain why meat from older animals is tougher than that from younger animals. Generally, muscles from younger animals contain more collagen, however this collagen does not contribute to meat toughness as in the tissue of older animals. Collagenous tissue from older animals with more cross-linkages would be expected to be more resistant to swelling and have a lower water holding capacity. There is marked reduction in imbibition and swelling of collagenous tissue with an increase in biological age. This was related to the effect of the increase in cross-linking and molecular ordering of the collagen. Collagen from young animals is more easily solubilized but produces structures with low tensile strength. In contrast, collagen from old animals is difficult to solubilize and produces a structure with high tensile strength [91].

In muscle tissue, the amount of connective tissue is low. However, because of their high strength, connective tissue proteins are an important factor affecting the physical properties of muscular tissue. Sims and Bailey [92] reported that intramuscular collagen in beef is capable of exhibiting greater tension than tendon. The process of postmortem aging has an influence on the textural properties of connective tissue as a result of some enzymatic weakening of the structure. However, during postmortem aging there is little change in collagen properties and solubility.

The specific ability of collagen disintegrated from its native state to repolymerize and reform into a fibrous structure was utilized in the production of regenerated edible sausage casing. During heating, collagen fibrils shrink to less than one-third of their original length at the shrinkage temperature. Conversion of collagen to gelatin during heating plays an important role in the tenderness of meat. This role of collagen conversion is especially important for meat with a high connective tissue content.

The dislocation and concentration of the proline and hydroxyproline in the collagen chains influences the rate of gel formation and the textural properties, especially the strength of the gel. For example, the low gelling capacity of fish gelatins was due to their low pyrolidine content. The reduced gel-forming capacity of fish collagen chains is related to their lower content of pyrolidine residues.

In gel formation, partial renaturation of the collagen takes place. Renaturation of collagen involves the local establishment of a *trans-trans* configuration at each proline-hydroxyproline peptide bond. An additional factor is the formation of the interchain hydrogen bonds between the glycine residues. Hydrogen-bonded aggregation initiates helix formation. The total quantity of helix formation in the gelatin will be affected by the rate of cooling. A more ordered structure will be formed with slow cooling.

Gelatins are water-soluble at temperatures above 30 °C. Proteinaceous substances of gelatin are prepared by various procedures (thermal and chemical treatments) involving the destruction of the quaternary, tertiary, secondary structure, and in some cases some aspects of the amino acid sequence of the native collagen. The amino acid sequence of gelatin is virtually identical to its parent collagen. The degree and type of collagen degradation achieved in the manufacture of gelatins varies with the processing conditions. For this reason, commercially produced gelatins are heterogeneous mixtures of polypeptides.

Properties of gelatin gels are influenced by temperature, concentration, pH, and ionic strength. The critical concentration for the formation of a three-dimensional network in gelatin gels occurs between 0.5 and 1% gelatin by weight. Two types of gelatin are known: type A is obtained from acid-treated collagen with a high and broad isoelectric region (pH 7 to 9), and type B is obtained from alkaline-treated collagen with an isoelectric point at pH 4.8 to 5.0. Both types of gelatin can become partially insoluble in water and this can prevent gelling of the dessert, because of the lower level of soluble gelatin in the aqueous phase.

Two gelatins with the same average molecular weight from similar sources and identical maturation may have a different gel strength. The two main reasons of that are the heterogeneity, with regard to size and shape of the molecules and also the difference in the amount and distribution of the amino acids within the polypeptide chains [93]. The method of gelatin preparation also determines the strength of gelatin, especially the mechanism and conditions of hydrolysis. Gelatins made by denaturation of solubilized collagen under mild processing conditions had high gel strength. Partial insolubilization of gelatin is caused by prolonged heating. Gelatin insolubilization is caused by polymerization as a result of cross-linking and hydrogen bonding.

Strong gels can be formed by the interaction of proteins with polysaccharide gelling agents. Gels with high melting points (80 °C) were prepared as a result of nonspecific ionic interactions between positively charged gelatin and negatively charged alginates or pectates. Gelatins have many advantages over polysaccharide gelling agents; they are water soluble, but they need improved thermal stability. Use of gelatins is limited, because their peptide bonds in the presence of water are relatively labile and as a result of heating gelatins lose their gelling properties. Watanabe et al. [94] modified gelatin and improved the surface active properties of a hydrophilic gelatin by incorporation a proper amount of a lipophilic L-leucine by incubation with papain. Modified gelatins can be utilized in food formulations.

6.3.8 Blood Proteins

Proteins from animal blood are used as food ingredients. Blood plasma has been widely used in European countries as an ingredient in sausage production. Blood plasma contains several proteins that may contribute to the gel structure and properties. The most abundant blood plasma proteins are serum albumin, immunoglobulins, and fibrinogen. Edible blood plasma is an effective protein additive to meat products which are to be cooked or canned. Usage of plasma proteins in cooked sausages is advantageous because of a high WHC and gelling ability of these proteins, and products with good slicing properties are produced. Bovine plasma proteins can be spun into fibers which may be used in meat analogues.

The basic function of blood plasma as an ingredient in meat products is to improve gelling properties induced by heat treatment. Heating of the blood globin in water, pH 5.4–5.8 at 95 °C for 30 min markedly increased viscosity of the 5% globin solution. The function of plasma gel is to hold water and fat as a result of matrix formation. The gelling capacity of blood proteins determined by gel rigidity, elasticity, viscosity is a function of the types of protein, the temperature and time of heating, protein concentration, pH and ionic strength [95].

The bovin plasma proteins form a stronger gel structure than egg albumen proteins at the same concentration and under the same treatment conditions. The gels obtained from bovine plasma proteins showed a higher apparent elasticity modulus and these gels become increasingly rubbery in texture with increased heating time [96]. Bovine plasma proteins exhibited optimum gel properties at pH 7. Bovine and porcine plasma can form gels at a concentration of 3.5% protein after heating to 82 °C. With an increase in protein concentration to 5% in suspension, gelation was observed at 72 °C and pH 7 or 9 [18]. The change in the gel structure influences the capacity to bind water and gel textural properties. The water retention in plasma gels increased with pH from pH 6.5 to 10.5 due to the increase in the net negative charge and the lower tendency for random aggregation with increased pH. The highest water binding properties were found in the range 75–77 °C and the moisture loss increased at above 80 °C. Water holding properties and elasticity of gels made at pH 9 were slightly affected by protein concentration and no effect was determined at pH 7 [18]. Water binding decreased with an increase in degree of random aggregation of the protein gel network. The degree of aggregation increased with increasing protein and salt concentration and decreasing pH from 9 to 6. Both aggregation and denaturation started at 55 °C.

Increases in the gel strength of the gels with porcine and bovine plasma and porcine serum were observed as time and temperature of heating were raised [97]. The penetration force increased with increasing heating temperature of gels from 72 °C to 92 °C. The increase in firmness with the heating temperature is probably due to an increase in density and the formation of larger aggregates. The increase in firmness with protein concentration is due to the increased density of the protein phase. The addition of NaCl causes an increase in firmness and a decrease in elasticity of gels at pH 9.0, possibly due to the increase in the degree of aggregation. Gelling properties were similar for freeze- and commercially dried plasma proteins. The water holding capacity of freeze-dried globin gels increased significantly.

The interaction between blood plasma and egg proteins was studied using interaction index [98]. The interaction index was calculated using either the gel strength or breaking strength values. The heat treatment resembled the baking process: 95 °C for 20 min. For porcine plasma and the egg albumen, interaction was highest at 80 and 85 °C for 15–30 min. However, the interaction was higher at 90 and 95 °C. Whey proteins exhibited increased gel strength after interaction with plasma proteins. Weak gels resulted from interaction of soya isolate and sodium caseinate with plasma proteins.

6.4 Gelling Properties of Milk Proteins

Introduction

The gel forming capacity of milk proteins and curd formation are important in the processing of milk protein products. Gelation of milk proteins is directly responsible for the structure of cheese and semisolid dairy products and influences the texture of other dairy products such as yogurt. Structural characteristics of traditional milk products especially cheese, and yogurt depend on the formation of protein gels. Gelling functionality is important because gels produce the required textural characteristics of milk products and retain water, fat and other components. Technologically important properties of milk gels are fat- and water-holding capacities.

Whey proteins can be utilized as ingredients in various foods. The gelling capacity of whey proteins is an important functional property for their utilization in meat products. The utilization of milk proteins in processed meats depends upon their compatibility with meat proteins, particularly in their gel forming capacity.The heat-induced gelling properties of whey protein concentrate (°C) can improve textural properties of comminuted meats. Incorporation of WPC into the formulations of processed meats is limited by the difference in thermal properties between whey and myofibrillar proteins. WPC requires 80 °C for gelation while myosin gels at 60 °C. At production temperatures in comminuted meats (68–70 °C) there is no effective gelation of WPC.

The properties of gels obtained from whey proteins vary in the range from viscous fluids, curds to rubbery gels. WPC gels vary in cohesiveness, stickiness, hardness and color. Weak translucent WPC gels are obtained at low protein concentration and low ionic strength. Whey proteins form gels according to the mechanism of gelation similar to that of other globular proteins. The mechanisms responsible for the formation of the three-dimensional network of whey protein gels are not fully understood. Matsudomi et al. [99] determined the mechanism of the interactions between α-lactalbumin (α-La) and bovine serum albumin (BSA) in milk gels.

The gels formed by milk proteins are mostly irreversible and result from enzymatic reactions, from heat-induced reactions, from cation (calcium) interactions or from combined reaction mechanisms. In the manufacture of cheese the first step is obtaining of casein gel by the combined action of proteolytic enzymes, i.e. chymosin, rennet (animal and microbial origin), and calcium. During storage of milk products, as sterilized milk concentrate and UHT-treated milk age gelation, i.e. increase in gel forming capacity is a major factor limiting the storage stability. The new application of heat-induced WPC gelation is the manufacture of simplesse, which includes heat-induced whey protein microparticles in the size range of 0.5 to 2.5 μm [100]. This chapter reviews the mechanism of milk proteins gelation and the influence of intrinsic and environmental factors on gelation of milk proteins.

6.4.1 Gelling Properties of Whey Protein Concentrate, Isolate, and Individual Whey Proteins

The principal protein component of whey is β-lactoglobulin (β-Lg) and its gelling properties influence the gelling capacity of whey protein preparations. β-lactoglobulin comprises up to 50% (or 2–4 g per liter) of the total whey protein and is

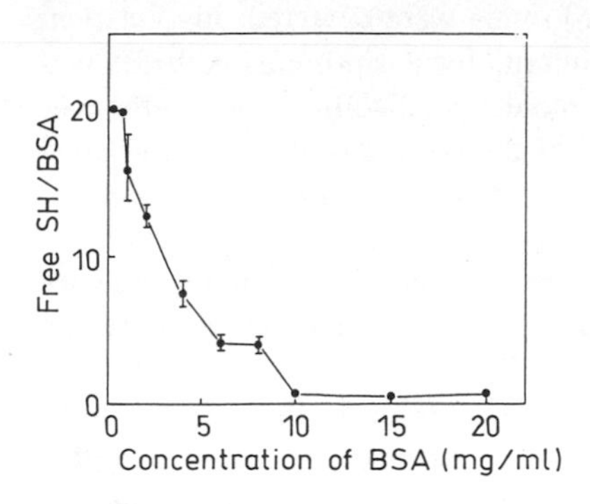

Fig. 6.11. Number of free sulf-hydryls in BSA

the major whey protein participating in gelling. The β-lactoglobulin content of WPC is highly correlated with gel strength and other functional properties of WPC and no significant relationship has been found between α-lactoglobulin (α-La) content and gel strength [101].

An essential property of whey proteins is their abundance of sulfhydryl amino acid residues that allows them to form intermolecular covalent bonds during heating. The stability of WPC and WPI gels is related to disulfide bonds and sulfhydryl groups. Intermolecular disulfides appear to be important to the elasticity of WPI gels. Gels prepared by heating whey proteins containing higher levels of SH groups are more opaque. β-Lg contains two disulfide groups and one free-SH group. The gelling properties of β-Lg were expressed by the water holding capacity of the gel matrix during centrifugation [102]. β-Lg and α-La showed a high level of sulfur-containing amino acids responsible for the gelling properties through -S-S- bond formation.

The relationship between total sulfhydryl content in WPC and gel time was established. However, this relationship was not well correlated experimentally. Shimada and Cheftel [103] demonstrated the significance of sulfhydryl/disulfide bonding in WPC/whey protein isolate (WPI) gel formation and an increase in gel firmness and elasticity. Cross-linking in the gels provides the elastic, fluid, and rigid properties. Gel properties can be changed by controlling of the extent of cross-linking. The heat-induced mechanism of gelation suggests that protein polymerization reactions lead to the formation of covalent bonding (disulfide linkages). These protein-protein interactions include β-Lg and k-casein interaction accelerated by heat.

The formation of a protein network represent an equilibrium between inter- and intra-particle disulfide bonds. The effect of disulfide bonds on gelation was tested by incubating bovine serum albumin (BSA) at different concentrations with >70 mM reduced glutathione (GSH) at 37 °C for 16 h, and determining the number of free SH groups [104]. The number of free SH/protein molecules decreased with increasing concentrations of protein (Fig. 6.11) [104]. At protein concentrations higher than 10 mg/ml, few SH-groups were detected. Involvement of cleared disulfide bonds in the formation of intermolecular protein disulfides was demonstrated by electrophoretic studies. Intermolecular disulfide bonds probably play the most important role in the formation of three-dimensional gel networks. The BSA gel hardness increases with increasing concentrations of BSA under the two buffer conditions.

Non-covalent interactions may also be responsible for protein network formation [105]. Beveridge [106] suggested that non-covalent cross-links can further stabilize the covalently bonded network formed in gels. Modification of proteins via disulfide reduction, thiol oxidation, and thiol-disulfide interchange reactions affect intermolecular cross-linking to improve viscosity and the formation of the protein network. Thiolation of β-Lg results in intermolecular disulfide cross-linking and the formation of high-molecular-weight polymers with increased viscosity and gelling properties [107]. Modified β-Lg in the presence of calcium ions formed strong, transparent gels. The fatty *N*-acylamino acids and their esters affected the gelling properties of spray-dried whey protein isolate. Both lauroyl (C_{12}) and myristoyl (C_{14}) derivatives enhanced the gelling properties of milk proteins and provided the highest increase in gel hardness and storage modulus. The increase in gel strength was due to binding of these amphiphiles to protein, causing repulsion between protein chains.

Nonsignificant conformational changes may increase protein-protein interaction and association to form long chains. There is limited information about the extent of protein unfolding prior to gelation. Clark et al. [109] demonstrated an increase in β-sheet and unordered structure with a decrease in the α-helical content of BSA.

The properties of BSA gels were affected by ionic strength. The clarity of BSA gels was decreased by adding salts or decreasing pH, i.e. minimizing of net charge. Using dynamic rheological technique Mitchell [110] measured the gelling properties of whey proteins. α-La exhibited poor gelling properties, β-Lg intermediate and BSA good gelling properties at pH 6.6, 1% NaCl, and heating at 30–95 °C. The optimum conditions for obtaining gels with maximum gelling properties of mixtures of α-La and β-Lg were established with heating at 80 °C [111]. Both glutathione and NaCl contributed to gel formation. Maximum gel strengths were obtained with glutathione and NaCl concentrations of 43 mM and 116 mM, and 45 mM and 83 mM, respectively. Enhancement of the gel formation resulted from conformational changes due to the interaction of glutathione with disulfide bonds in the α-La and β-Lg.

6.4.2 The Effect of Heating and Protein Concentration

The effect of heat-induced β-Lg-k-casein complex formation by preheating of skim milk on the syneresis of curd was reported by Pearse et al. [112]. Heat treatment at

80 °C for 10 min caused a reduction in syneresis. Samples of reconstituted skim milk heated at 60 to 80 °C for 10 min showed inhibition in syneresis. This inhibition of syneresis was dependent on the presence of β-Lg and to a lesser extent α-La. The evidence of the effect of β-Lg/k-casein complex formation was obtained by adding exogenous k-casein and β-Lg before heat treatment. A significant reduction of syneresis was observed when β-Lg and k-casein were added. The syneresis phenomenon during gel storage may be related to the formation of additional interprotein bonds, which decrease the number of active groups for binding water and reduce the intermolecular space for binding water through capillary forces.

The first step in WPC heat-induced gelation is weakening and breaking of hydrogen and disulfide bonds and disruption of native protein conformational structures. The second step includes polymerization of the dissociated protein molecules and formation of a three-dimensional protein structure with immobilizing a significant amount of solvent through intermolecular disulfide, hydrophobic and ionic bonds [113]. In protein gels, a substantial portion of the water is bound by the proteins, but most of the water is physically entrapped within the three-dimensional gel network [114].

Protein dispersions from WPC form gels with heating as a result of protein denaturation and aggregatoin reactions. Translucent and weak gels were prepared at low temperatures (55–70 °C) and lower protein concentration (5%). Opaque and firmer gels were obtained at higher heating temperatures (about 90 °C) and higher protein concentrations [4]. The critical concentration for protein gelation is usually highly pH dependent, with a minimum at the protein isoelectric point [115]. Commercial samples of WPC exhibited different gelling time, i.e. from 1 to 17 min at 100 °C under identical conditions. This was related to variation in whey composition, ion concentration and pH. The heating temperature of skim milk affected properties of prepared gels. Skim milks were acid coagulated at 60 and 80 °C for 1 h with and without glucono-delta-lactone [116]. Gels preheated at 70 °C were weak with low water-holding capacity. Preheating at 80 °C caused formation of the solid gels with high water holding capacity.

The effects of degree of insolubilization of WPC on gel properties have been reported [56]. The WPC with 98.1% solubility did not form a gel at protein concentrations 8–20% when heated to 65 °C because the denaturation temperature of β-Lg is above 65 °C. The gel hardness increased as the concentration of 98.1% soluble WPC was increased suggesting that WPC was contributing to gel hardness [56]. The heating temperature (70–90 °C) did not affect the expressible moisture levels of gels prepared from 98.1%, 41.0% or 27.5% soluble WPC. Expressible moisture was highest for 41.0% soluble WPC at all temperatures evaluated due to the formation of weak gels. The apparent stress of 98.1% soluble WPC gels in 0.1 M NaCl increased 11-fold when the heating temperature was increased from 70 to 80 °C [117]. Zirbel and Kinsella [118] reported that the strength of WPI gels increased at a heating temperature from 70 to 80 °C, but did not increase on heating to 90 °C. Increasing the temperature from 75 to 125 °C had no effect on water retention and the elasticity of WPI gels [119]. Hardness and deformability of gels decreased when gels were prepared by heating at 90 °C for 15 min with increased protein insolubilization. Significant variations in the textural and water retention properties of gels prepared with WPCs containing 27.5 to 98.1% soluble protein have been reported [117]. Those data showed that gelation properties of WPC can be enhanced with an increase in the insolubility of protein in WPC. Beuschel et al. [117] reported that WPC gels deformability decreased by 40% as WPC solubility decreased from 98.1 to 27.5% at both 80 and 90 °C. Gels prepared at 80 °C with 98.1% soluble WPC were most deformable. Commercial WPC may contain 20 to 50% denatured protein [120]. Insolubilized proteins may function in gels as rigid fillers to fill interstitial spaces of another protein gel matrix with increase gel hardness. However, an increase in WPC gel strength and the formation of harder gels was observed at temperatures from 80 to 110 °C – possibly because of an increase in intermolecular bonding. The transition of WPI dispersion from a viscous to a viscoelastic structure during heating is time/temperature and protein concentration dependent [2]. The gelling temperature of WPC is affected by the method of whey protein isolation.

The gelling properties of WPC and WPI preparations are determined by gelling capacity of individual proteins. The most significant changes in WPC gel strength would be expected near 80 °C, because the denaturation temperature of β-Lg is near 80 °C. The initiation temperature was concentration dependent and differed between proteins: BSA, 70–90 °C, and β-Lg, 75–80 °C [2]. The gelation onset temperature of β-Lg is in the range from 84 °C to 88 °C and is affected by pH[121]. Matsudomi et al. [122] reported that the hardness of BSA gels increased rapidly between 75 and 95 °C and β-Lg formed gels at lower temperatures (70–80 °C). The heating temperature above the minimum denaturation temperatures of the proteins is required for gel formation by BSA and β-Lg. The hardness of BSA gels increased with increased heating time (5–30 min). The transparent gels with high hardness were produced with the addition of 3% α-La to 6% BSA. BSA and α-La formed gels due to the interaction with formation soluble aggregates through thiol-disulfide interchange reaction.

Gelling porperties of WPI and β-Lg during storage were studied by an accelerated dry storage test at 80 °C [123]. Denaturation and protein-protein interactions were detected with a decrease in gelling capacity. As a result of a polymerization process, the relative concentrations of monomeric β-Lg in the WPI decreased from 60.6% to 33.3% after seven days of heat treatment. A similar trend in protein polymerization was found with α-La and BSA. Polymerization was apparent at 40 °C, the lowest temperature tested. The decrease in monomeric β-Lg is shown in Fig. 6.12 [123]. Monomers have been converted to high-molecular-weight material. The β-Lg monomer concentration correlated with the hardness of the corresponding gels.

The gel forming capacities of BSA, β-Lg and α-La have been studied by a continuous dynamic rheological method [2]. The minimum protein concentration required to obtain a gel was; for BSA 1%, β-Lg around 2%, α-La up to 20% (w/v). The probability of BSA gel formation should be lowest at the pI and increased with increasing positive or negative net charge. An increase in gelling capacity can

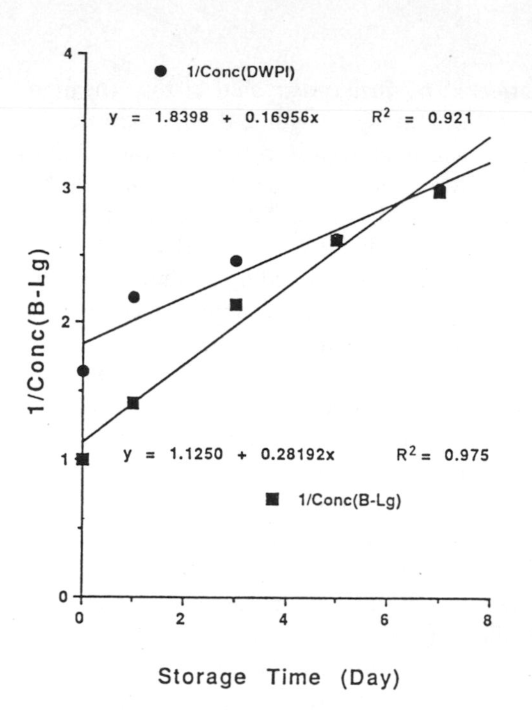

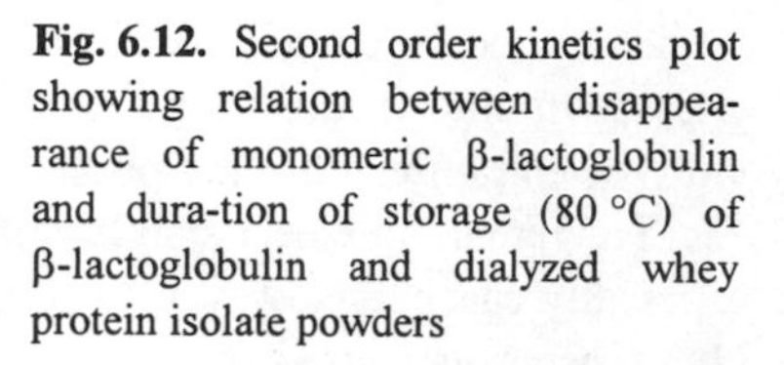
Fig. 6.12. Second order kinetics plot showing relation between disappearance of monomeric β-lactoglobulin and dura-tion of storage (80 °C) of β-lactoglobulin and dialyzed whey protein isolate powders

be obtained by mixing protein systems. The gelation process of the BSA and β-Lg mixture was not a result of an additive effect of the proteins. The gelation properties of BSA dominated the gelation process. However, the addition of 5% β-Lg slightly increased the viscosity of the BSA gels.

The firmness of milk gels is mainly affected by protein concentration. Excessive bridging between casein micelles is observed at higher protein concentrations presumably via strands of β-Lg. Commercial WPCs showed different gelling properties as a result of differences in protein/ion, protein/fat, or protein/lactose ratios. Schmidt [4] found that more than 7.5% (w/v) protein was required for the formation of strong gels from WPC heated at 100 °C for 10 min at pH >7.0 and the opacity of the gels increased with increasing protein concentration.

The gelling time of WPC gels is dependent on protein concentration and temperature. The gelling time of WPC decreased linearly from 4 to 1 min with increasing protein concentration from 8 to 12%. The lowest protein concentration required for gel formation might be used as a criterion of the gelling capacity of this protein. Weak reversible gels can be formed from whey proteins when lower concentrations of proteins are heated [124]. Under such conditions large polymers of covalently linked whey proteins are formed. Such protein solutions could form thermoplastic gels via H bonds and non-covalent interactions [125]. Both the type of protein and the environmental conditions determine the optimal concentration required for gelation.

6.4.3 Gelation of Casein

The properties of casein gels are affected by their processing history (heating conditions), the concentration of casein, pH, ionic strength and concentration of additives. The pH-induced gelation of casein is utilized in cottage cheese, cream, and acid casein. Micellar casein does not form gels unless the k-casein, held to be more concentrated at the micelle surface, is hydrolyzed. After that the micelles aggregate with increase in viscosity leading to formation of a firm gel or curd.

In cheese making, the gelation of whole milk is obtained by treatment with an enzymic coagulant. Normally, rennet, an extract from calf stomachs containing chymosin and bovine pepsin is utilized. As a result, the water and water-soluble components are mostly removed and the fat is retained in the curd. During coagulation, chymosin-altered k-casein micelles begin to aggregate steadily as small, chain-like aggregates. The characteristic features of rennet-treated casein gels are that they are less rigid and have a higher tendency to syneresis than heat-treated gel structures involving egg white and gelatin. As a result of syneresis the rennet-treated casein gel there is more curd packing and a cohesive structure of the cheese. In cheese, the formation of rigid, highly-cross linked network structures with the entrapment of fat globules was found by electron microscopy [126].

The remarkable heat stability of milk is due to the heat stability of its proteins, especially caseins. However, heating can cause milk coagulation or gelation. This phenomenon is caused by aggregation of casein micelles. Different theories have been put forward to explain the development of the gelation of milk with the formation of a custard-like, irreversible, three-dimensional gel matrix with minimal syneresis. These theories can be divided into related to heat-treatment effects and related to enzyme-induced aggregation. Heat coagulation of milk is accelerated by heat-induced interactions between β-Lg and k-casein and may also be involved in gel formation during storage. The enzyme-induced mechanism of milk gel formation requires the presence of proteases that are still active after heating, and are a result of microbial growth or are native to milk.

A typical milk gel is yoghurt and its important property is the ability to retain water in the gel structure. An increased level of incorporation of caseins and whey proteins increased the firmness of the gel and decreased syneresis of the resultant yoghurt [127]. Casein produced firmer yoghurt, while with whey proteins the yoghurt was smoother with a better appearance. The acid casein gels have a high strength and a tendency to separate whey (syneresis) which is important in the production of acid cheeses but is undesirable in yoghurt. The interaction between denatured whey proteins and casein is responsible for the formation of the characteristic texture of yoghurt. Protein-protein interactions are enhanced by lactic acid fermentation and lowering the pH.

6.4.4 Factors Affecting the Gelling Properties of Milk Proteins

The Effect of Lipids

Incorporation of milk lipids into milk protein gels improves the texture of gelled milk products (jellies, fresh cheeses). In dairy foods gel formation is not strictly confined to protein-protein and protein-water interactions, since protein-carbohydrate and protein-lipid interactions also influence the structure of the gels. The gelling properties of heat-induced WPC gels is influenced by protein hydrophobicity. Lipids in WPC samples may inhibit protein gelation by interfering with protein polymerization via formation of intermolecular hydrophobic bonding [4]. Lipids increase coagulation time by interfering with protein-protein polymerization. The amount of lipids which can be incorporated into whey protein gels without significant change in desired gel textural properties has been reported [128]. The firmness of heat-induced whey gels increases with the fineness of the emulsion and is influenced by the level of protein and fat in the formulation.

The textural properties of dialyzed WPI gels prepared with and without emulsified fat have been reported [129]. Incorporation of emulsified fat in the gel structure causes an increase in the gel strength. Small fat droplets were more effective in increasing the gel strength and the optimum mean droplet size was 1.85 μm. The viscosity and elastic moduli of WPI gels at pH 4.6 increased with fat content, whereas syneresis decreased with the addition of emulsified fat. Gels with incorporated fat have a tighter gel matrix with an increased amount of immobilized water. The microstructure of gels containing fat is more compact.

In different gelled milk products milk fat is included as a texture element. The gelling properties of oil/water emulsions are influenced by the dispersibility of the oil component [128]. Gelation and the firmness of heat-induced gels increases with increasing emulsion homogeneity. A high gelation capacity in emulsions has been obtained with a relatively narrow size distribution and for the mean diameter of the fat droplets was in the range 300–700 nm. Correlation between emulsion dispersivity/homogeneity and gelation capacity was established.Highly dispersed fat assumes a positive role in gel texture formation. Changes in gel rigidity were influenced by the mean diameter of the fat droplets, the mode and the percentage of droplets smaller than 1000 nm. Immobilization of lipid droplets with the protein matrix of the gel was enhanced by an increased amount of protein adsorbed at the lipid-water interface. Self-supporting gels were obtained at relatively low protein concentrations (4–5% w/vol) and with an oil concentration of more than 10%. To obtain gels of similar firmness without oil, the required protein concentration was at least 8%.

The Effect of Ionic Strength

Ionic strength influences the gelling ability of WPC and other milk proteins. The ions in the solution affect the attractive and repulsive forces between proteins. The calcium ions influence the gelling properties of proteins. A low level of Ca^{2+} may increase the hardness of WPC gels resulting from an increase in cross-linking in proteins [120]. A high Ca^{2+} concentration causes an excessive aggregation of proteins that can impair gelation. Addition of $CaCl_2$at 0.1 to 0.3 M results in increased gel strength [130]. Chen et al. [131] demonstrated that 7% hydrolyzed β-Lg forms gels with 20 mM $CaCl_2$ but not with 100 mM NaCl at 60 °C. Firmness of WPI gels was maximized at 10–20 mM calcium. The effects of salts on gelation of β-Lg were determined and it was demonstrated that solutions of β-Lg with 20 mM $CaCl_2$ have lower gel points and gel more rapidly than those that form in 100 mM NaCl [132].

The salt level affects the structure of whey gels with the formation of a coarse gel at 0.2 M NaCl. The amount of sodium and calcium ions influences the gelation of WPC and replacement of calcium with sodium ions increases hardness, springiness and chewiness of 10% WPC gels [133]. The gelling properties of β-Lg are influenced by NaCl and $CaCl_2$ levels [134]. Gels with maximum hardness are obtained at 200 mM NaCl and 10 mM $CaCl_2$. Increases in concentrations of NaCl and $CaCl_2$ produce a less ordered gel with lower hardness. At 400 mM NaCl, the gel continuous network collapses because of excessive protein-protein interaction. At high NaCl levels, unfolding of β-Lg is retarded thereby decreasing gel strength and releasing water. Determination of the rheological parameters showed that gels with higher gelling strength are obtained by adding $CaCl_2$ than by adding NaCl [134]. At a low level of $CaCl_2$ the primary functions of $CaCl_2$ are electrostatic interactions with Ca^{2+} acting to cross-link negatively-charged partially unfolded protein molecules to form the matrix [135]. The positive effect of Ca^{2+} compared to Na on the strength of β-Lg gels is of practical significance because of the dietary role of calcium. Rheological properties and the microstructure of Ca^{2+}-induced WPI gels and thermally induced WPI gels are different [136]. Ca^{2+}-induced gels formed by mixing protein in $CaCl_2$ solutions have a fine-stranded microstructure and lower gel penetration force values than thermally induced gels. Thermally induced gels have a particulate microstructure consisting of bead-like particles.

The hardness of both BSA and β-Lg gels is increased by the addition of $CaCl_2$ which is more effective than NaCl with the formation of more elastic and transparent gels [122]. The strengthen matrix of gels results from calcium bridging between negatively charged groups on unfolded protein molecules. However, when the calcium level exceeds the optimum, i.e. 2 and 5 mM for β-Lg and BSA, respectively, the matrix collapses yielding a coagulum. Calcium cations cause greater protein-protein interactions and three-dimensional network formation due to the reduction of electrostatic repulsion.

The effect of various electrolyte concentration on the level of cross-linking between denatured protein molecules has been reported [137]. There is a low level of cross-linking between protein molecules at low electrolyte concentration and gels with high gelling strength are formed. At high concentrations of electrolyte, interactions between proteins increased because the decrease in repulsive electrostatic forces favors a greater number of cross-links and leads to a collapse of the protein matrix.

The Effect of pH

Whey proteins in the native state and produced under mild conditions form irreversible gels at the appropriate pH. The gel properties, i.e. the type and number of protein cross-links during gelation are influenced by the pH. The pH of WPC solutions affects the sulfhydryl-disulfide interchange mechanism for protein polymerization and gelation. WPC prepared from acid (HCl) whey showed a poor gelling capacity at pH 3.0, good gelling capacity (gel firmness) at pH 5.1 (natural pH) and excellent gel properties (gel firmness) at pH >8 [138]. Demineralization of the whey gave the WPC with a natural pH 7.4 and good gelling capacity compared to the undermineralized whey at pH 3.0. The unfolding and aggregation of whey protein is pH and temperature dependent. Schmidt [4] reported that at low pH (pH < 6) gels are less elastic than those formed at pH 7–9. The strength of gels obtained by heating whey protein solutions at 100 °C for 15 min at pH 7, 9, 10, and 11 decreases from 100 to 57, 23, and 0 N/m^2, respectively.

Whey proteins exhibit a good thermal gelling ability above pH 7 that is affected by the solubility and aggregation properties of β-Lg [102]. These proteins below pH 7 possess the ability to coagulate and aggregate due to intense electrostatic interactions. Transparent gels of β-Lg are formed after heating β-Lg above pH 6.5. The increase in gelling properties at pH above 8 results from disulfide cross-linking and formation of a matrix via thiol-disulfide interchange [139].

The firmness of WPI gels decreases with an increase in pH from 6.5 to 9.5 [119] due to an increase in protein solubility and a decrease in protein-protein interactions. This effect of elevated pH can be reduced by addition of NaCl (0.1 M) which causes an increase in firmness and a decrease in solubility of WPI gels. Under these conditions (pH 7.5–9.5, 0.1 M NaCl) water retention of WPI gels increases with pH. However, water retention remains constant in the absence of NaCl. Gels of WPI prepared at pH 2.5–3.5 are inelastic and contain 75–80% soluble protein [119]. WPI gels formed at alkaline pHs are stabilized by hydrophobic and disulfide bonds, and gels formed at acid pHs are stabilized by hydrogen bonds.

Functionalities of 11 commercial WPCs and WPIs have been reported [139]. Nonpourable gels were formed at pH 4.5, near the pI of the major gel whey proteins, and higher WPC concentration was required for gel formation at pH 6.0 and 7.5. The pH effects on gelling the properties of β-Lg and BSA demonstrated that the net charge on protein molecules affects gel hardness [115]. The β-Lg and BSA gels exhibit maximum hardness at pH 6.5 and, as the pH increases above 6.5, the gel strength decreases [122]. Opaque gels are produced by heating β-Lg at pH 6.0, whereas they are completely transparent at pH < 3.5 or >7.0 [141]. The change in the rate and extent of whey proteins gelation at different pHs is due to the effect on protein conformation and net charge. The gelling properties of BSA and WPI with the addition of basic proteins, i.e. clupeine (pI=12), hydrolyzed clupeine, or modified β-Lg (pI=9.5) have been reported [142]. The basic protein clupeine may improve gelation by increasing the formation of cross-linkages of partially denatured acidic proteins via electrostatic attractive forces.

The mechanism of gel formation of BSA differs on one side of the isoelectric point to the mechanism on the other [125]. On the acid side of the isoelectric point, the intermolecular β-structure must be the main force in the formation of the gel network of BSA. On the alkaline side of the isoelectric point, S-S bonds and β-structure contribute to the gel formation of the BSA solution. The infrared spectra of BSA aggregates show the presence of a β-structure peak in this pH region. The attractive forces, particularly sulfhydryl groups are obtained from functional groups exposed by the thermal denaturation and unfolding of proteins. However, Hillier et al. [6] did not find any correlation between the content of SH groups of whey powders and the capacity of these powders to form heat-induced gels. The gelation of BSA was highly dependent on the concentration of reduced glutathione (GSH), a naturally occurring thiol. The effect of pH on the glutathio ne-dependent gelation of BSA is shown in Fig. 6.13 [104]. Gelation is observed at 37 °C in the presence of 70 mM GSH at pH 7.5 to 9.2. There is no gel formed in this range in the absence of GSH. The sharp peak of gel hardness is obtained at pH 8.2.

The Effect of Various Treatments

Ultrafiltration of whey proteins influences their gelling properties as a result of changing the lactose/protein and the salt/protein ratios. WPCs obtained from low-salt whey possess excellent gelling properties. An ion exchange process was applied for the removal of lipids and preparing WPC from cheese whey (63% protein) and the gelling properties were reported [143]. The gelation properties of Spherosil-QMA WPC were superior to those of commercial and ultrafiltration WPC's. The gelling ability of WPC can be improved as a result of diafiltration with removal of nonprotein components. The gel forming capacity of whey proteins has been influenced by the presence of minerals and lactose. Removal of minerals and lactose improved the gelation properties of WPC systems. Gels obtained from WPC's after dialysis had a higher hardness and cohesiveness. WPC after dialysis had lactose and salts removed. The effect of calcium removal and replacement by sodium has been reported [133]. Gelation time at both natural pH (4.6–5.2) and pH 6.7 increased as% calcium replacement increased. The gelling ability of protein components of WPC and WPI were tested after partial removal of lipids and low-molecular-weight solutes by centrifugal clarification and centrifugal gel filtration [144]. Purified protein fractions of WPC and WPI showed little effect on their gelation as compared with control when tested as a function of pH, protein concentration and mineral ion addition. The highest gel firmness values were obtained at pH 6 and 7.5, and at 0.1 M ion addition. Addition of polyphosphates to whey protein modified with pepsin improved its gelling ability [145].

The improved gelation properties of limited hydrolysates are of interest since such hydrolysates can be easily produced. Limited proteolysis of β-Lg by immobi-

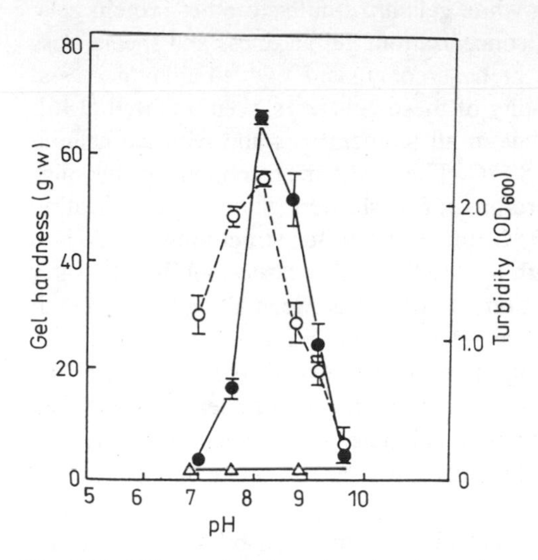

Fig. 6.13. Effects of pH on the glutathione-dependent gelation of BSA. BSA incubated at 35 mg/ml with 70 mM GSH (-●-, --○--) or without the thiol (-<6,136>-) in 0.1 M sodium phosphate buffer, at 37 °C, 16 h. Gel hardness (-●-, --<6,136>--) and turbidity (--○--)

lized trypsin improves the gelling properties of this protein [146]. The partially hydrolyzed β-Lg has a lower gel point and forms a gel more rapidly than native β-Lg. The gel formed by 15% hydrolyzed β-Lg was 61 times more rigid than that of native β-Lg gels. A gel time value of 7 min after reaching 80 °C was observed for hydrolyzed β-Lg, whereas the value was 38 min at 80 °C for native β-Lg. Since hydrolyzed β-Lg contains multiple protein fragments, the mixture may be considered a multiprotein system.

6.5 Gelling Properties of Egg Proteins

6.5.1 Gelation of Egg White

Egg white is widely used in the food industry because of its important functional properties. Egg white is a key ingredient in many processed food products because of its ability to coagulate upon heating. Egg white is a food ingredient with important functional properties that make it useful as a gelling agent, thickener and food binder. Proteins of egg white have unique physical and biochemical functional properties. Proteins in egg white are easy to separate due to the fact that egg white is a solution of many proteins with a low sugar and salt content. Thick egg white is a weak gel interpenetrated by a network of elastic fibers. Weakening of the gel of the thick white during storage is due to gradual depolymerization of the cross-linked, three-dimensional network.

The rheological properties of egg-white gels are similar to other protein gels; with an increase in egg white protein concentration, gel hardness and shear stress increases. Protein-water interactions in heat coagulated egg albumin gels and comparison of NMR studies with rigidity of these gels have been reported [146]. Gel strength increases with heating time at all temperatures and with the heating temperature in the range from 65 to 80 °C. The NMR measurement of the spin-lattice relaxation time of water protons (T_1) showed that, as the heating temperature increased from 60 to 90 °C, increased water structuring which was followed by an increase in gel strength resulted in a decreased mobility of water molecules in the network. The correlation coefficient between T_1 and gel strength at all heating times and temperatures was –0.835. Weakening, i.e. thinning of the thick albumen that occurs during storage is probably related to a decrease in the protein-protein interaction within the lysozyme-ovomucin complex. However, an increase in viscosity for thick , thin and mixed albumen during the initial 5 days of storage has been observed [147]. This increase was probably related to the observed changes in albumen pH.

The meat analogs prepared from vegetable proteins need binders to hold the texturized vegetable proteins together. Egg white is one of the best binders in the production of meat analogs. The most important advantage of egg white as a binder is that the protein will set on heating. Ovalbumin is a major protein in egg white, accounting for about 50% of the total solids. The unique functional properties of ovalbumin are related to the high content of -SH groups. After denaturation, all -SH groups are reactive.

Formation of the gel in egg white starts at 71–74 °C at a rate of heating 0.5 °C/min or 1 °C/min. The gel formation is measured as an increase in rigidity and a change of properties from viscous to viscoelastic [148]. Factors affecting egg white gelation include protein concentration, albumen composition, electrostatic forces, hydrophobic amino acid residue exposure, the formation of disulfide bonds, pH, ionic strength, and heating temperature [149]. Almost all of the protein aggregated when ovalbumin was heated to 80 °C. Aggregation of about 80% of the protein was found at 75 °C, regardless of albumin concentration. However, at 70 °C, a concentration of at least 1% is required for aggregation.

The gel matrix of ovalbumin is stabilized by disulfide and noncovalent bonds. Disulfide bonds were detected in ovalbumin aggregates formed at 90 °C [150]. The aggregation of egg ovalbumin resulted from intermolecular associations of β-sheets. During heating of egg albumin polymerization was observed by intermolecular sulfhydryl-disulfide exchange with network formation [151]. However, Hegg [102] did not find any correlation between disulfide or sulfhydryl group content and gel-forming capacity. Doi et al. [152] reported that the disulfide bonds in ovalbumin are not of primary importance in gel formation. However, the presence of sulfhydryl agents during heat treatment of ovalbumin influences its gelling ability. Mine et al. [153] reported that albumen aggregates were formed from the partially unfolded protein molecules with a considerable secondary structure with participation of disulfide cross-links and β-sheet hydrophobic forces.

Surface -SH groups are important in heat-induced gel formation. The gel strength increased with increase surface -SH groups [154]. This effect possibly is related to an increase in gel network formation by forming S-S bonds with other protein molecules. Internal S-S groups can lock the proteins into conformations that decrease the gelling capacity. Promotion of the gelling properties due to an increased concentration of surface -SH groups can be obtained by drying as a result of the structure opening of the egg white proteins. The level of heat-exposed SH groups in egg white is affected by heating temperature and time.

Six basic egg albumin proteins were tested for gel formation and lysozyme, conalbumin, ovalbumin and globulins produced gels after heat coagulation in pure aqueous solutions [155]. Conalbumin and lysozyme from chicken egg are rigid globular proteins with good gelling properties. The conalbumin-rich fraction of egg white was reported as a major thiol-dependent gel-forming component of egg white [156].Hydrophobic interactions were responsible for conalbumin gel formation. There are no free sulfhydryl groups in native conalbumin although it contains 15 disulfide linkages. An increase in surface hydrophobicity and sulfhydryl groups was observed when conalbumin was incubated with mercaptoethanol at 35 °C. An increase in turbidity and gel formation was observed after 1 h.

Heat induced gelation of lysozyme resulted from disrupting the disulfide bonds, causing exposure of the hydrophobic regions with subsequent interchanges between sulfhydryl and disulfide groups. The optimal gelation conditions of lysozyme heated at 80 °C were 46.7 mM NaCl and 7.0 mM dithiothreitol (DDT) [157]. Intermolecular associations of lysozyme after the addition of DDT resulted in an increase in turbidity of lysozyme solutions heated at 80 °C. Lysozyme solution (5%) formed a strong, opaque white gel in the presence of 7.0 mM DDT after heating at 80 °C for 12 min.

The Effect of Ionic Strength and pH

Thermal aggregation and gel formation of ovalbumin is influenced by pH and salt concentration. A 5% ovalbumin solution with 20 mM NaCl formed turbid gels around the isoelectric point of ovalbumin (pH 4.7) [158]. Gel hardness had two maxima at pH 3.5 and 6.5. The gel hardness increased with an increase in NaCl concentration. However, transparent hard gels were obtained in a narrow range of pH and salt concentrations. Transparent gels of ovalbumin were prepared at pH 7.5 and 20–50 mM NaCl [159]. Increases in concentration of NaCl cause at first an increase and then a decrease in gel hardness. It was shown that increasing ionic strength shifts the maxima of gel rigidity away from the pI. The maximum values of gel rigidity were higher on the acid than on the alkaline side of the pI. The maximum gel strength occurs at pH values with limited capacity of ovalbumin to aggregate. The general tendency is that maximal rigidities of the ovalbumin gels are on the acid side of the pI. The balance between attractive and repulsive forces during aggregation is dependant on pH.

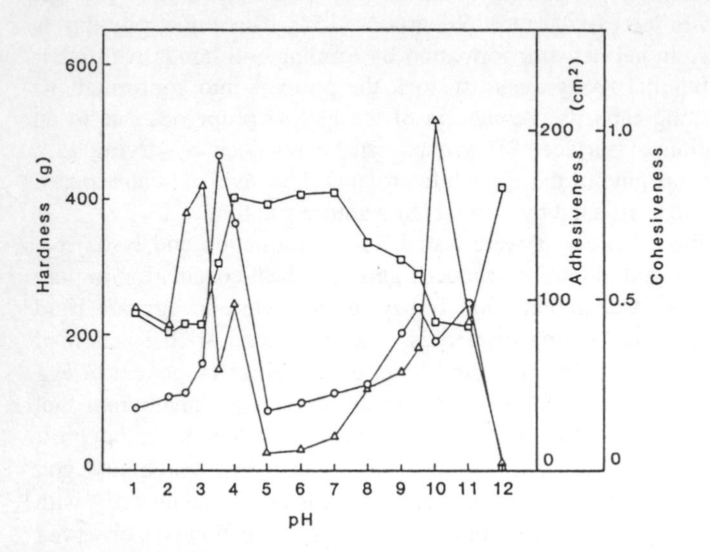

Fig. 6.14. Texture of egg white treated by dialysis, high-speed centrifugation and heat. Hardness (O), adhesiveness (<6,136>), and cohesiveness (□) of heat-treated egg whites were measured

The textural properties of gels are influenced by the pH of the suspension under identical heating conditions. The breaking stress is higher for gels formed at pH 10.0 than for those formed at pH 5 [160].The microstructure of pH 10 gels is uniform and homogeneous, while pH 5 gels are turbid, granular and inhomogeneous. The hardness of egg white gels is greater at pH 5 than at pH 6, 7, and 8 [161]. The effect of pH on gel hardness is influenced by the addition of 0.01 M NaCl and the following relationship was found: pH 9>pH 8>pH 5>pH pH 6.

The gelation temperature of ovalbumin is affected by pH. Gels formed at pH's above 6.0 at a higher temperature are less rigid than gels formed at pH below 4.0 and a lower temperature [162]. Textural characteristics of egg white gels produced by the heating of a supernatant prepared by dialysis and measured at various pH's using a rheometer in the textural mode are presented in Fig. 6.14 [163]. Maximum hardness was found at pH 3.5, with low turbidity, indicating that the gel was transparent. The peak of adhesiveness was found at pH 3.0, slightly acidic from the peak of hardness. Egg-white gels formed at pH's higher than 10 were transparent and showed less syneresis than opaque gels formed near the pI of the main protein components.

Egg white protein modification affected its gelling properties [148]. The egg white was modified with succinic anhydride and oleic acid. A measurable increase in gel rigidity was observed at lower temperatures for oleic acid modified egg white protein. In oleic-acid-modified gels the denatured protein aggregated in a manner favoring the formation of very elastic and rigid gels. Rigidity was influenced by the rate of temperature increase. By using different heating rates and producing different levels of unfolding and aggregation it could be possible to produce gels with similar elasticity but different rigidities. The high final rigidity and elasticity of succinic anhydride modified gels could be related to protein denaturation before aggregates were formed. Succinylation affected the extent of thermal unfolding prior to aggregation and increased gel strength of egg white [164]. Gels prepared by both modification procedures were highly deformable and did not fail in compression testing. The fatty acids and their esters affected gelling properties of spray-dried egg white [108]. Both lauroyl (C_{12}) and myristoyl (C_{14}) derivatives enhanced gelling properties of egg white protein and provided highest increase in gel hardness and storage modulus.

6.5.2 Gelation of Yolk

Gelation of yolk during frozen storage significantly altered the functionality of yolk components. Freezing and subsequent thawing causes yolk gelation which is followed by an increase in viscosity of low-density lipoprotein (LDL) and a marked reduction in solubility. LDL is considered to be responsible for the gelation of egg yolk. The gelation of frozen yolk due to an increase in viscosity is a reflection of the product richness. The gelation of the yolk and its loss of fluidity is difficult to handle in automated processing equipment.

Contacts between particles are developed and strengthened during freezing as the concentration of particles is increased by transformation of the solvent into ice. The solvent conversion into ice increases the solute concentration and mechanical pressure on the yolk particles. The formation of ice crystals may cause the destruction of LDL particles. There is a possibility of disruption of the granules (lipovitellin) during freezing as a result of the high concentration of soluble salts. Nonspecific aggregation occurs due to the concentration of lipoprotein particles by the transformation of solvent water to ice. The increase in concentration of salts in the unfrozen phase is considered as an important gelation factor.

The aggregation of lipoproteins such as lipovitellenin and the decrease in the solubility of the low density fraction during freezing is a possible cause of gelation. The gelation of frozen yolk is apparently caused by the fivefold increase in the concentration of soluble salts that occurs in yolk during freezing to –6 °C [165]. LDL aggregation might be caused not only by the increase in salt concentration but also the pH change in the unfrozen phase. On the other hand, LDL gelation was inhibited by the addition of salts at above the eutectic temperature of the salt. Salt concentration itself is not a key factor in gelation, and the LDL aggregation might be caused by LDL dehydration induced by ice formation. The extent of LDL gelation with NaCl is affected by its concentration and the temperature of frozen storage. Aggregation of lipoproteins is accelerated by reduction of the average distances between them. Non-specific aggregation is caused by cross-linking of peptides and/or phospholipid polar heads of lipovitellinen particles. LDL gelation is induced only by conformational changes accompanied by a decrease in hydrated water [166]. It has been found that low molecular weight material is not liberated from frozen-thawed LDL. However, centrifugation shows a greater amount of precipitate in frozen-thawed LDL than in unfrozen. Yolk gelation is influenced by several processing factors, such as time and temperature of frozen storage, rate of freezing and thawing, level of salt and sugar addition.

Interactions between proteins following the disruption of the lipid-protein complex are involved in gelation and the interacting proteins have an intrinsic property of foaming aggregates. Kurisaki et al. [167] found that LDL on freezing and thawing looses the surface components which stabilize their structure. This change is considered as an initial step of gelation and structural rearrangement and aggregation of LDL with the formation of a mesh-type structure. The rate of freezing and thawing influences the level of gelation. During slow freezing or slow thawing, the aggregation is complete and gelation occurs with the liberation of LDL fragments. However, upon fast freezing and thawing, gelation is not observed, although some of LDL constituents are liberated from the particles. The change of viscosity of thawed yolk is dependent on a storage time-temperature relationship. The rate of viscosity increase is enhanced by a decrease in the storage temperature below –6 °C.

The gelation of egg yolk can be partially reversed by heating after thawing. This treatment improves the functional properties of proteins. The stiffness of the gels obtained after frozen storage can be reduced by more than 50% and become pourable at 21 °C by heating at up to 45 °C for 1 h.

Cryoprotective agents or proteolytic enzymes, are incorporated into yolk to minimize gelation [165]. Cryoprotective agents are: NACl at levels of 1–10% and sugars, such as glucose, sucrose, and galactose at the 10% level. Presumably, the addition of NaCl to the LDL solution increases the unfrozen water content through the formation of an LDL-water-NACl complex. Salt inhibition of egg yolk gelation is also due to an increase in the unfrozen water in the lipovitellin solutions through the formation of a lipoprotein-water-NaCl complex where the water is hardly frozen.

Frozen storage at –20 to –25 °C has almost no effect on the viscosity of LDL solutions with 1~10% NaCl. The viscosity of frozen-thawed LDL solutions increases at 4–10% NaCl concentration and at a storage temperature below –30 °C. LDL is the main component responsible for yolk gelation in the presence of NaCl. Yolk gelation is significantly reduced by the addition of sugar. The degree of viscosity change is lowered as the sucrose concentration increases from 0.001 to 0.014 mol/100 g of yolk.

Hen's egg yolk LDL gives a more stable gel on heating than ovalbumin and BSA [168]. Ovalbumin and BSA give a stable gel within a narrow pH range, LDL gives one at every pH between 4 and 9 due to the special properties of LDL. The gel strength of LDL is affected by added proteins and varies significantly with the pH of the solution. Some differences are found among tested proteins, however a similar effect on the rigidity of LDL gels is found when other proteins are added

[169]. The gel strength of yolk LDL gel increased markedly with the addition of ovalbumin at pH 6.2 and 7.0 or lysozyme at pH 9.0.

A concentrated liquid egg-white product is produced by vacuum evaporation of liquid egg white. Liquid egg white is concentrated with an increase in the solids content from 11.5% to 21.8% [170]. Gels prepared from concentrated liquid egg white exhibit an increased shear stress with an increased solids concentration, but no effect is found on the rue gel strain. The shear stress of gels increases with increasing water-jacket temperature. Egg white gels prepared from concentrates produces fairly elastic gels with the typical cooked egg white gel texture.

The powder of egg white obtained by freeze-drying was rehydrated with distilled water, reheated and the gelling capacity was determined. Freeze-drying did not lessen the ability of egg white to form transparent gels. Nishikawa et al. [171] developed a method of egg white treatment that gave transparent gels. This method included the removal of the coagulum that occurs during foaming and autoclave treatment. During foaming the protein component causing turbidity is removed.

Irradiation treatment at 1.0; 2.5 and 4.0 kGy does not cause significant changes in egg white gel hardness and storage moduli which are related to gel rigidity [172]. The data were in contrast to those observed in shell eggs which show an enhancement in the thermal gelation property of egg white proteins by gamma irradiation.

6.6 Gelling Properties of Soy Proteins

The first soy protein utilization as a food in the Orient was in the form of gel tofu. Soy gels have the capacity to act as a matrix and to hold moisture, lipids, polysaccharides, flavors and other ingredients. The characteristic property of soy protein gels is a considerably higher water holding capacity (WHC) than in milk and other gels. Because of this, soy proteins can be utilized in gel systems in which syneresis is undesirable (yoghurt).

Dispersions of soy proteins with 7% or more protein form gels after heating for 30 min at 100 °C. The increase in viscosity is observed and then gelling of the sol. Soy proteins interact with calcium and the production of soybean curd (tofu) involves Ca-induced and heat-induced gelling mechanisms. Addition of Ca increases the gel strength of soy proteins while decreasing the WHC. A significant attribute of soy proteins is, that by controlling the processing conditions, soy protein products can be produced with different gelling properties. Textural properties of soy gels, such as hardness are correlated with soy protein content and gels are classified from hard to soft.

Extended studies have been carried out to determine the gelation mechanism and the molecular forces participating in the formation and stabilization of the soy protein gel structure [173]. Attention has been focused on the gelling properties of 7 S and 11 S globulins that comprise approximately 50% of the protein in soy isolate [174]. The 2 S fraction does not contribute to the three-dimensional network formation of soy gels [175]. The major gel-forming components of soy isolate are globulins glycinin (11 S) and β-conglycinin (7 S) in term of quantity and functionality. The 7 S and 11 S fractions participate in heat-induced gelation after heating to 88 °C for 42 min [176]. The mechanism of gelation could be different for gels formed from 7 S and 11 S globulins, and soy isolate. Molecular forces involved in the gelation of 11 S globulin may be different at 80 °C compared to 100 °C.

The mechanism of the gelation of soy proteins is determined mainly by the heat-induced association/dissociation behavior of soy proteins. As a result of this behavior, 11 S globulin soluble aggregates with a molecular weight of 8×10^6 are formed when 0.5 and 5% protein solutions are heated for 1 min at 100 °C [177]. Differences in the gelling capacity of soy 7 S and 11 S globulins are caused by association/dissociation reactions and thermal unfolding characteristics of their subunits [178]. It has been demonstrated that mixtures of 7 S and 11 S globulins or soy protein isolate (SPI) show better gelling capacity than either of the constituent protein fractions. The gelling capacity has been related to the thermal interaction between 7 S and 11 S globulin fractions of soy proteins.

In soy gel formation, crosslinking during gelation through hydrophobic and hydrogen interactions and disulfide bonding is responsible for the network formation in the gel [173]. Textural properties of the gel are affected by the intensity of the network formation. The relationship between gel strength and WHC has been found [173]. However, most of the studies have been carried out with native SPI, glycinin and conglycinin. Soy isolates produced commercially due to denaturation and aggregation during processing may have different properties. Heat or chemicals applied during processing would produce a structure similar to that obtained during the sol-progel-gel transition.

Heat-induced interactions between soy 7 S and 11 S have been reported and dissociation of both with formation of subunits has been demonstrated [179]. Dissociated subunits of 7 S and 11 S globulins interact with each other with the formation of high molecular weight soluble macromolecules. These macromolecules contain predominantly subunits of 11 S globulin and the β, α, and α' subunits of 7 S globulin. The electrostatic and disulfide bonds between basic subunits of 7 S and 11 S globulins are involved in the formation of soluble macromolecules. The gelling capacity of the monomers of soybean globulins correlates with the parameter of hydrophobicity. The mechanism for the heat-induced gelation of 11 S globulin includes the association of glycinin molecules through hydrophobic interactions [180]. However, later it was found that dissociation of 11 S into acidic and basic subunits impairs the formation of strong gels.

The gel forming ability of individual soy globulins decreased in the order 2.8 S>basic 11 S>α-7 S=α'-7 S>acidic 11 S [174]. In the gel formation, hydrophobic interactions predominate and the gelling ability of the subunits decreases in the same order as an effective mean hydrophobicity. However, other data [176] showed that the subunits of 7 S globulin participate equally in the formation of gel network by conglycinin. Hardness of the gels formed by 7 S, 11 S, and soy isolate decreases in the order 7 S>>SPI>11 S. The heat-induced gels of 11 S globulin show a higher tensile and shear strength and WHC than those prepared from 7 S globulin or soy isolate. This is mainly due to the different sensitivity of heat denaturation of 11 S and 7 S globulins. Proteins, 7 S and 11 S interact with each other during heating. Electrostatic, hydrophobic interactions and -S-S- bonds participate in the formation of 11 S globulin gels [176]. In 7 S gels, hydrogen bonding is most important and in soy isolate gels. hydrogen bonding and hydrophobic interactions. Three subunits of 7 S participate uniformly in the gel formation of the gel matrix. Preferential interaction is detected between the β-subunit of 7 S and the basic subunits of 11 S in soy isolate gels [181]. Bibkov et al. have proposed an unfolded peptide chain model for heat-set soy protein gels in which the chain undergoes a coil-to-globule transformation as the gel tempera-ture is lowered.

Gelation of the dispersions of soy globulins is influenced by protein concentration and heating conditions, i.e. temperature and time of heating. Soy protein gels exhibit thixotropic properties depending on the preheating temperature. Gelation of 11 S globulin is developed at 100 °C, pH 7.6, and an ionic strength 0.5 M through the aggregation of 11 S molecules in the form of strands (stage 1), followed by interaction between the molecules to form the gel network (stage 2). Lower heating temperatures result in the formation of a weaker gel, because globulins are not sufficiently unfolded and the appropriate three-dimensional network of gel is not formed. Electron microscopy studies showed that the dimensions of the network in glycinin gels would suggest that the strands are formed by association of partially unfolded glycinin molecules [180]. Gel formation in soy protein systems appeared to be promoted by the Maillard reaction. The Maillard gels showed less syneresis, had a higher breaking force and were more elastic.

The basic factors influencing gelation of soy proteins are protein processing history, concentration, heating time and temperature, pH, ionic strength, lipid content and cooling conditions. As the pH of soy protein dispersions increases, the viscosity increases. Increased solubility of the protein at the higher pH values might be responsible for increased viscosity. The gel strength decreases at high acidic and alkaline pH, and the maximum strength is obtained at neutral pH.

The gel strength of soy protein gels with various ratios of 7 S/11 S increased with the addition of 2% NaCl [183]. The addition of 0.2 M NaCl increased protein-protein interactions in the gels of soy proteins and increased the building up of the aggregated structure in whey protein gels. Increased NaCl concentration can enhance hydrophilic or ionic interactions between the polar groups of polypeptides. However, a high NaCl level (10%) caused the failure of soy protein gel formation due to a change in protein conformation, and as a result an increase in hydrophobic interactions between proteins.

Ultrasonic treatment might be utilized to obtain desired functional properties in soybean foods. Ultrasonic treatment improved protein solubility, and promoted the subsequent formation of protein aggregates and mostly 7 S protein was involved in the aggregation process [184]. Separation of protein in the gel showed transformational changes of 7 S proteins with conversion of 7 S into 40–50 aggregates. Ultrasonic-treated soy proteins contained more aggregates than unsonicated samples. This effect is due to the unique sensitivity of 7 S to ultrasound treatment.

Gelling properties of soy proteins are influenced by chemical modification. The thiolation, i.e. introduction of new SH groups into soy isolate improved their gelling properties in terms of elasticity, cohesiveness, and viscosity when measured as the extent of resistance to a compressing force on a rheometer. The exception was hardness which decreased when compared to the control. The increase in bulk density of the thiolated soy protein gels demonstrated the decrease in a space between polypeptide chains or the formation of a more compact protein matrix. The increase in compactness of the protein matrix and gel structure is due to the formation of protein-protein linkages through S-S bonds.

References

1. Kinsella, J. E. (1976). Functional properties of proteins in foods: A survey, *Crit. Rev. Food Sci. Nutr.*, *7*: 219.
2. Paulson, M., Hegg, P., and Castberg, H.B. (1986). Heat-induced gelation of individual whey proteins. A dynamic rheological study, *J. Food Sci.*, *51*: 87.
3. Clark, A.H. and Lee-Tuffnell, C.D. Gelation of Globular Proteins. In Functional Properties of Food Macromolecules. Mitchell J.R. and Ledward D.A. eds. (1986). Elsevier Appli. Sci. Publ., New York, NY, p. 203.
4. Schmidt, R.H. Gelation and coagulation. In: Protein Functionality in Foods. J.P. Cherry (Ed.) (1981). ACS Symposium Series 147, Washington, DC, p. 131.
5. Foegeding, E.A. Molecular Properties and Functionality of Proteins in Food Gels. In: Food Proteins, J.E. Kinsella and W.G. Soucie (Eds.) (1989). The Amer. Oil Chem. Soc., Champaign, IL, p. 185.
6. Hillier, R.M., Lyster, R.L.J. and Cheeseman, G.C. (1980). Gelation of reconstituted whey powders by heat. J. Sci. Food Agric. 31:1152.
7. Schmidt, G.R., Mawson, R.F., and Siegel, D.G. (1981). Functionality of the protein matrix in comminuted meat products. Food Technol. 35 (5):235.
8. Katsuta, K., Rector, D., and Kinsella, J.E. (1990). Viscoelastic properties of whey protein gels: Mechanical model and effects of protein concentration on creep. J. Food Sci. 55:516.
9. Fretheim, K., Egelandsdal, B., Harbitz, O., and Samejima, K. (1985). Slow lowering of pH induces gel formation. Food Chem. 18: 169.
10. Voutsinas, L.P. and Nakai, S. (1983). Relationships between protein hydrophobicity and thermal functional properties of food proteins. Can. Inst. Food Sci. Technol. J. 16(3):185.
11. Nakai, S., Li-Chan, E., and Hayakawa, S. (1986). Contribution of protein hydrophobicity to its functionality. Die Nahrung 30:327.
12. Wang, C.-H., and Damodaran, S. (1990). Thermal gelation of globular proteins: weight-average molecular weight dependence of gel strength. J. Agric. Food Chem. 38:1157.
13. Ferry, J.D. (1948). Protein gels. Adv. Prot. Chem. 4:1.
14. Shimada, K. and Matsushita, S. (1980). Thermal coagulation of egg albumin. J. Agric. Food Chem. 28:409.

15. Samejima, K., Ishioroshi, M., and Yasui, T. (1981). Relative roles of the head and tail portions of the molecule in the heat-induced gelation of myosin. J. Food Sci. 46: 1412.
16. Hermansson, A.M. Water and fatholding. In: Functional Properties of Macromolecules. Mitchell, J.R. and Ledward, D.A. (Eds.), (1986). Elsevier Applied Sci. Publ., London, p. 273.
17. Kuntz, I.D. and Kauzmann, W. Hydration of proteins and polypeptides. In: „Advances in Protein Chemistry", C.B. Anfinsen, J.T. Edsall, and F.M. Richards, eds., (1974). V. 28, p. 239. Academic Press, New York.
18. Hermansson, A.M. and Lucisano, M. (1982). Gel characteristics-water binding properties of blood plasma gels and methodological aspects on the water binding of gel systems. J. Food Sci. 47:1955.
19. Mulvihill, D.M. and Kinsella, J.E. (1987). Gelation characteristics of whey proteins and β-lactoglobulin, Food Technol. 9:102.
20. Hayakawa, S., Ogawa, T. and Sato, Y. (1982). Some functional properties under heating of the globin prepared by carboxymethyl cellulose procedure. J. Food Sci. 47: 1415.
21. Hermansson, A.M., Harbitz, O. and Langton, M. (1986). Formation of two types of gels from bovine myosin. J. Sci. Food Agric. 37:69.
22. Stone, A. P., and Stanley, D. W. (1994). Muscle protein gelation at low ionic strength. Food Res. Internat., 27: 155.
23. Knight, M. K. (1988). Utilization of meat fractions and predictive modeling of meat product cooking losses and texture. Proc. 34th Int. Congr. Meat Sci. Technol., Brisbane, Australia, p. 305.
24. Acton, J.C., and Dick, R.L. Functional roles of heat induced protein gelation in processed meat. In: Food Proteins, Eds. Kinsella, J.E. and Soucie, W.G. (1989). Amer. Oil Chem. Soc. Champaign, IL, p. 195.
25. Ishioroshi, M., Samejima, K., and Yasui, T.G. (1982). Further studies on the roles of the head and tail regions of the myosin molecule in heat-induced gelation. J. Food Sci. 47:114.
26. Hamm, R. Functional properties of the myofibrillar system and their measurement. In „Muscle as Food", ed. P.J. Bechtel, (1986), p.135. Academic Press, New York.
27. Choe, I.S., Morita, J.I., Yamamoto, K., Samejima, K., and Yasui, T. (1991). Heat-induced gelation of myosins/subfragments from chicken leg and breast muscle at high ionic strength and low pH. J. Food Sci. 56:884.
28. Egelandsdal, B., Fretheim, K., and Samejima, K. (1986). Dynamic rheological measurements on heat-induced myosin gels: effect of ionic strength, protein concentration and addition of adenosine triphosphate or pyrophosphate. J. Sci. Food Agric., 37: 915.
29. Asghar, A., Samejima, K., and Yasui, T. (1985). Functionality of muscle proteins in gelation mechanisms of structured meat products. CRC Crit. Rev. Food Sci. Nutr. 22:27.
30. Yamamoto, K., Samejima, K., and Yasui, T. (1988). Heat-induced gelation of myosin filaments. Agric. Biol. Chem 52:1803.
31. Samejima, K., Yamauchi, H., Asghar, A., and Yasui, T. (1984). Role of myosin heavy chains from rabbit skeletal muscle in the heat-induced gelation mechanism. Agric. Biol. Chem. 48:2225.
32. Sano, T., Noguchi, S.F., Matsumoto, J.J. and Tsuchiya, T. (1990). Thermal gelation characteristics of myosin subfragments. J. Food Sci. 55:55.
33. Ishioroshi, M., Samejima, K., and Yasui, T. (1983). Heat-induced gelation of myosin filaments at a low salt concentration. Agric. Biol. Chem. 47:2809.
34. Morita, J.-I., Sugiyama, H., and Kondo, K. (1994). Heat-induced gelation of chicken gizzard myosin. J. Food Sci., 59: 720.
35. Yasui, T., Ishioroshi, M., and Samejima, K. (1980). Heat-induced gelation of myosin in the presence of actin. J. Food Biochem. 4:61.
36. Morita, J.-I., Choe, I., Yamamoto, K., Samejima, K., and Yasui, T. (1987). Heat-induced gelation of myosin from leg and breast muscles of chicken. Agric. Biol. Chem., 51: 2895.

37. Yasui, T., Ishioroshi, M., and Samejima, K. (1982). Effect of actomyosin on heat-induced gelation of myosin. Agric. Biol. Chem. 46:5.
38. Ishioroshi, M., Samejima, K., and Yasui, T. (1983). Heat-induced gelation of myosin filaments at a low salt concentration. Agric. Biol. Chem. 47:25.
39. Acton, J.C., Hanna, M.A., and Satterlee, L.D. (1981). Heat-induced gelation and protein-protein interaction of actomyosin. J. Food Biochem. 5:101.
40. Sano, T., Noguchi, S.F., Tsuchiya, T., and Matsumoto, J. (1989). Paramyosin-myosin-actin interactions in gel formation of invertebrate muscle. J. Food Sci. 54:796.
41. Siegel, D.G., and Schmidt, G.R. (1979). Ionic, pH, and temperature effects on the binding ability of myosin. J. Food Sci. 44:1686.
42. Yasui, T., Ishioroshi, M. and Samejima, K. (1982). Effect of actomyosin on heat-induced gelation of myosin. Agric. Biol. Chem. 46:1049.
43. Samejima, K., Ishioroshi, M., and Yasui, T. (1982). Heat-induced gelling properties of actomyosin. Effect of tropomyosin and troponin. Agric. Biol. Chem. 46:535.
44. Foegeding, E.A. (1987). Functional properties of turkey salt-soluble proteins. J. Food Sci. 52:1495.
45. Kim, S. H., Carpenter, J.A., Lanier, T.C., and Wicker, L. (1992). Polymerization of beef actomyosin induced by transglutaminase. Book of Abstracts, IFT Annual Meeting, New Orleans, p. 73.
46. Schmidt, G.R. Functional behavior of meat components in processing. In: The Science of Meat and Meat Products. J.F. Price, and B.S. Schweigert, Eds. Food and Nutrition Press, Inc. Westport, CN (1987), p. 413.
47. Jimenez-Colmero, F., Careche, J., Corballo, J., and Cofrades, S. (1944). Influence of thermal treatment on gelation of actomyosin from different mysosystems. J. Food Sci., 59: 211.
48. Xiong, Y.L. (1992). Thermally induced interactions and gelation of combined myofibrillar protein from white and red broiler muscles. J. Food Sci. 57:581.
49. Xiong, Y.L. and Brekke, C.J. (1991). Protein extractability and thermally induced gelation properties of myofibrils isolated from pre- and post-rigor chicken muscles. J. Food Sci. 56:210.
50. Fretheim, K., Samejima, K., and Egelandsdal, B. (1986). Myosins from red and white bovine muscles: Part I. Gel strength (elasticity) and water-holding capacity of heat-induced gels. Food Chemistry, 22:107.
51. Wicker, L., Lanier, T.C., Hamann, D.D., and Akahane, T. (1986). Thermal transitions in myosin-ANS fluorescence and gel rigidity. J. Food Sci. 51:1540.
52. Northcutt, J.K., Lavelle, C.L., and Foegeding, E.A. (1993). Gelation of turkey breast and thigh myofibrils: changes during isolation of myofibrils. J. Food Sci. 58:983.
53. Amato, P.M., Hamann, D.D., Ball, H.R., and Foegeding, E.A. (1989). Influence of poultry species, muscle groups, and NaCl level on strength, deformability, and water retention in heat-set muscle gels. J. Food Sci. 54:1136.
54. Samejima, K., Egelandsdal, B., and Fretheim, K. (1985). Heat gelation properties and protein extractability of beef myofibrils. J. Food Sci. 50:1540.
55. Solomon, L.W., and Schmidt, G.R. (1980). Effect of vacuum and mixing time on the extractability and functionality of pre-and post-rigor beef. J. Food Sci. 45:283.
56. Beuschel, B.C. Partridge, J.A., and Smith, D.M. (1992). Insolubilized whey protein concentrate and/or chicken salt-soluble protein gel properties. J. Food Sci. 57:852.
57. Li-Chan, E., Kwan, L., and Nakai, S. (1986). Physicochemical and functional properties of salt-extractable proteins from chicken breast muscle deboned after different post-mortem holding times. Can. Inst. Food Sci. Technol. J. 19(5):241.
58. Xiong, Y.L. and Blanchard, S.P. (1993). Functional properties of myofibrillar proteins from cold-shortened and thaw-rigor bovine muscles. J. Food Sci. 58:720.
59. Smith, D.M. (1987). Functional and biochemical changes in deboned turkey due to frozen storage and lipid oxidation. J. Food Sci. 52:22.

60. Findlay, C.J., and Stanley, D.W. (1984). Differential scanning calorimetry of beef muscle: influence of postmortem conditioning. J. Food Sci. 49:1513.
61. Xiong, Y. L., and Blanchard, S.P. (1994). Myofibrillar protein gelation: viscoelastic changes related to heating procedures. J. Food Sci., 59: 734.
62. Camou, J.P., Sebranek, J.G. and Olson, D.G. (1989). Effect of heating rate and protein concentration on gel strength and water loss of muscle protein gels. J. Food Sci. 54:850.
63. Foegeding, E.A., Allen, C.E., and Dayton, W.R. (1986). Effect of heating rate on thermally formed myosin, fibrinogen and albumin gels. J. Food Sci. 51:104.
64. Lavelle, C.L., and Foegeding, E.A. (1993). Gelation of turkey breast and thigh myofibrils: Effects of pH, salt and temperature. J. Food Sci. 58:727.
65. Sano, T., Noguchi, S.F., Matsumoto, J.J. and Tsuchiya, T. (1990). Effect of ionic strength on dynamic viscoelastic behavior of myosin during thermal gelation. J. Food Sci. 55: 51.
66. Nuckles, R.O., Smith, D.M., and Merkel, R.A. (1991). Properties of heat-induced gels from beef skeletal, heart, lung and spleen protein fractions. J. Food Sci. 56:1165.
67. Miller, A.J., Ackerman, S.A., and Palumbo, S.A. (1980). Effect of frozen storage on functionality of meat for processing. J. Food Sci. 45:1466.
68. O'Neill, E., Mulvihill, D. M., and Morrisey, P. A. (1994). Molecular forces involved in the formation and stabilization of heat-induced actomyosin gels. Meat Sci., 36(3): 407.
69. Foegeding, E.A., Dayton, W.R. and Allen, C.E. (1986). Interaction of myosin-albumin and myosin-fibrinogen to form protein gels. J. Food Sci. 51:109.
70. Shiga, K., Nakamura, Y., and Taki, Y. (1985). Effects of preheating of soybean protein on interaction between meat protein and soybean protein. Jap. J. Zootech. Sci., 56: 897.
71. Bianchi, M. A., Pilosof, A. M. R., and Bartholomai, G. B. (1985). Rheological behavior of comminuted meat systems containing soy protein isolates. J. Text. Stud., 16: 193.
72. Peng, I.C., Dayton, W.R., Quass, D.W. and Allen, C.E. (1982). Studies on the subunits involved in the interaction of soybean 11S protein and myosin. J. Food Sci. 47:1984.
73. Shiga, K., and Nakamura, Y. (1987). Relation between denaturation and some functional properties of soybean protein. J. Food Sci., 52: 681.
74. Egelandsdal, K., Fretheim, K., and Harbitz, O. (1985). Fatty acid salts and analogs reduce thermal stability and improve gel formability of myosin. J. Food Sci, 50:1399.
75. Decker, E. A., Xiong, Y. L., Calvert, J. T., Crum, A. D., and Blanchard, S. P. (1993). Chemical, physical and functional properties of oxidized turkey white muscle myofibrillar proteins. J. Agric. Food Chem., 41: 186.
76. Xiong, Y. L., Decker, E. A., Robe, G. H., and Moody, W. G. (1993). Gelation of crude myofibrillar protein isolated from beef heart under antioxidative conditions. J. Food Sci., 58: 1241.
77. Lanier, T.C., Lin, T.S., Hamann, D.D. and Thomas, F.B. (1980). Gel formation in comminuted fish systems. 3rd National Technical Seminar on Mechanical Recovery and Utilization of Fish Flesh, p. 181. Raleigh, NC.
78. Sano, T., Noguchi, S.F., Matsumoto, J.J., and Tsuchiya, T. (1989). Role of F-actin in thermal gelation of fish actomyosin. J. Food Sci. 54:800.
79. Niwa, E., Nakayama, T., and Hamada, I. (1981). Effect of arylation for setting of muscle proteins. Agric. Biol. Chem. 45:341.
80. Chan, J.K., Gill, T.A., and Paulson, A.T. (1992). Cross-linking ability of myosin heavy chains from cod, herring and silver hake during thermal setting. J. Food Sci. 57:906.
81. Chan, J.K., Gill, T.A., and Paulson, A.T. (1993). Thermal aggregation of myosin subfragments from cod and herring. J. Food Sci. 58:1057.
82. Taguchi, T, Ishizaka H., Tanaka, M., Nagashima, Y., and Amano, K. (1987). Protein-protein interaction of fish myosin fragments. J. Food Sci. 52:1103.
83. Lanier, T.C., Lin, T.S., Liu, Y.M., and Hamann, D.D. (1982). Heat gelation properties of actomyosin and surimi prepared from Atlantic croaker. J. Food Sci. 47:1921.
84. Careche, M., Currall, J., and Mackie, I.M. (1991). A study of the effects of different factors on the heat-induced gelation of cod (Gadus morhua, L.) actomyosin using response surface methodology. Food Chemistry., 42(1): 39.

85. Wu, M.C., Akahane, T., Lanier, T.C. and Hamann, D.D. (1985). Thermal transitions of actomyosin and surimi prepared from Atlantic croaker as studied by differential scanning calorimetry. J. Food Sci. 50:10.
86. MacDonald, G.A., Lelievre, J., and Wilson, N.D.C. (1992). Effect of frozen storage on the gel-forming properties of hoki (Macruronus novaezelandiae). J. Food Sci. 57:69.
87. French, J.S., Kramer, D.E. and Kennish, J.M. (1988). Protein hydrolysis in Coho and Sockeye Salmon during partially frozen storage. J. Food Sci. 53:1014.
88. Kim, B.Y., Hamann, D.D., Lanier, T.C. and Wu, M.C. (1986). Effects of freeze-thaw abuse on the viscosity and gel-forming properties of surimi from two species. J. Food Sci. 51:951.
89. Chung, Y. C., Gebrehiwot, A., Farkas, D. F., and Morrisey, M. T. (1994). Gelation of surimi by high hydrostatic pressure. J. Food Sci., 59: 523.
90. Lee, C.M. and Abdollahi, A. (1981). Effect of hardness of plastic fat on structure and material properties of fish protein gels. J. Food Sci. 46:1755.
91. Miller, A.T., Karmas, E., and Lu, M.F. (1983). Age-related changes in the collagen of bovine corium: Studies on extractability, solubility and molecular size distribution. J. Food Sci. 48:681.
92. Sims, T.J., and Bailey, A.J. Connective tissue. In „Developments in Meat Science"-2 R. Lawrie, Ed. (1981). Applied Science Publishers, (1981), London.
93. Ledward, D.A. Gelation of gelatin. In Functional Properties of Food Macromolecules. Eds. Mitchell, J.R., and Ledward, D.A. Elsevier Appl. Sci. Publ. London, (1986), p. 171.
94. Watanabe, M., Toyokawa, H., Shimada, A. and Arai, S. (1981). Proteinaceous surfactants produced from gelatin by enzymatic modification: evaluation for their functionality. J. Food Sci. 46:1467.
95. Morgan, R.G. Suter, D.A., Carpenter, Z.L. (1980). A comparison of heat-induced gel strengths of bovine plasma and egg albumen proteins. J. Anim. Sci. 51:69.
96. Hickson, D.W., Dill, C.W., Morgan, R.G., Sweat, V.E., Suter, D.A. and Carpenter, Z.L. (1982). Rheological properties of two heat-induced protein gels. J. Food Sci. 47:783.
97. Howell, N.K., and Lawrie, R.A. (1984). Functional aspects of blood plasma proteins. II. Gelling properties. J. Food Technol. 19:289.
98. Howell, N.K. and Lawrie, R.A. (1984). Functional aspects of blood plasma proteins. III. Interaction with other proteins and stabilizers. J. Food Technol. 19:297.
99. Matsudomi, N., Oshita, T., Kobayashi, K., and Kinsella, J. E. (1983). α-lactalbumin enhances the gelation properties of bovine serum albumin. J. Agric. Food Chem., 41: 1053.
100. Singer, N. Simplesee all natural fat substitute and the dairy industry. In Proc. Dairy Prod. Tech. Conf., Am. Dairy Prod. Inst. and Center for Dairy Res., Chicago, (1990), p. 85.
101. Kim, Y. A., Chism, G. W., and Mangino, M. E. (1987). Determination of the β-lactoglobulin, α-lactalbumin and bovine serum albumin of whey protein concentrates and their relationship to protein functionality. *J. Food Sci.*, *52*: 124.
102. Hegg, P.-O. (1982). Conditions for the formation of heat-induced gels of some globular food proteins. *J. Food Sci.*, *47*: 1241.
103. Shimada, K.,and Cheftel, J. C. (1989). Sulfhydryl group/disulfide bond interchange reactions during heat-induced gelation of whey protein isolate. *J. Agric. Food Chem.*, *37*: 161.
104. Hirose, M., Nishizawa, Y., and Lee, J., Y. (1990). Gelation of bovine serum albumin by glutathione. *J. Food Sci.*, *55*: 915.
105. Chatellier, J. Y., Durand, D., and Emery, J. R. (1985). Critical helix content in gelatin gels. *Int. J. Biol., Macromol.*, *7*: 311.
106. Beveridge, T., Jones, L., and Tung, M. (1984). Progel and gel formation and reversibility of whey, soybean and albumen gels. *J. Agric. Food Chem.*, *32*: 307.
107. Richardson, T., and Kester, J. J. (1984). Chemical modifications that affect nutritional and functional properties of proteins. *J.Chem. Educ.*, *61*: 325.

108. Ma, C.-Y., Paquet, A., and McKellar, R.C. (1993). Effect of fatty N-acylamino acids on some functional properties of two food proteins. *J. Agric. Food Chem.*, *41*: 1182.
109. Clark, A. H., Saunderson, D. H. P., and Suggett, A. (1981). Infrared and laser raman spectroscopic studies of thermally-induced globular protein gels. *Int. J. Pept. Protein Res.*, *17*: 353.
110. Mitchell, J. R. (1980). The rheology of gels. *J. Texture Stud.*, *11*: 315.
111. Legowo, A. M., Imade, T., and Hayakawa, S. (1993). Heat-induced gelation of the mixtures of α-lactalbumin and β-lactoglobulin in the presence of glutathione. *Food Research Internat.*, *26*: 103.
112. Pearse, M. J., Linklater, P. M., Hall, R. J., and MacKinlay, A.G. (1985). Effect of heat induced interaction between β-lactoglobulin and k-casein on syneresis. *J. Dairy Res.*, *52*: 159.
113. Mangino, M. E. (1992). Gelation of whey protein concentrates. *Food Technol.*, *46*: 114.
114. Morr, C. V. (1989). Beneficial and adverse effects of water-protein interactions in selected dairy products. *J. Dairy Sci.*, *72*: 575.
115. Stading, M., and Hermansson, A.-M. (1990). Viscoelastic behavior of β-lactoglobulin gel structures. *Food Hydrocolloids.*, *4*: 121.
116. Hashizume, K., and Sato, T. (1987). Gel forming characteristics of milk proteins. 1. Effect of heat treatment. *J. Dairy Sci.*, *71*: 1439.
117. Beuschel, B.,C., Culbertson, J. D., Partridge, J. A., and Smith, D. M. (1992). Gelation and emulsification properties of partially insolubilized whey protein concentrates. *J. Food Sci.*, *57*: 605.
118. Zirbel, F., and Kinsella, J. E. (1988). Factors affecting the rheological properties of gels made from isolated whey protein. *Milchwissenschaft*, *43*: 691.
119. Shimada, K., and Cheftel, J, C. (1988). Texture characteristics, protein solubility, and sulfhydryl group/disulfide bond contents of heat-induced gels of whey protein isolate. *J. Agric. Food Chem.*, *36*: 1018.
120. Kohnhorst, A. L., and Mangino, M. E. (1985). Prediction of the strength of whey protein gels based on composition. *J. Food Sci.*, *50*: 1403.
121. Paulsson, M., Dejmek, P., and Vliet, T. V. (1990). Rheological properties of heat induced β-lactoglobulin gels. *J. Dairy Sci.*, *73*: 45.
122. Matsudomi, N., Rector, D., and Kinsella, J. E. (1991). Gelation of bovine serum albumin and β-lactoglobulin; effects of pH, salts and thiol reagents. *Food Chemistry*, *40*: 55.
123. Rector, D., Matsudomi, N., and Kinsella, J. E. (1991). Changes in gelling behavior of whey protein isolate and β-lactoglobulin during storage: Possible mechanism(s). *J. Food Sci.*, *56*: 782.
124. Rector, D., Kella, N. K., and Kinsella, J. E. (1989). Reversible gelation of whey proteins: Melting, thermodynamics and viscoelastic behavior. *Texture studies.*, *20*: 457.
125. Yasuda, K., Nakamura, R., and Hayakawa, S. (1986). Factors affecting heat-induced gel formation of bovine serum albumin. *J. Food Sci.*, *51*: 1289.
126. Green, M. L.,and Morant, S. V. (1981). Mechanism of aggregation of casein micelles in rennet-treated milk. *J. Dairy Res.*, *48*: 57.
127. Modler, H. W., Larmond, M. E., Lin, C. S., Froehlich, D., and Emmons, D. B. (1983). Physical and sensory properties of yogurt stabilized with milk proteins. *J. Dairy Sci.*, *66*: 422.
128. Jost, R., Baechler, R., and Masson, G. (1986). Heat gelation of oil-in-water emulsions stabilized by whey protein. *J. Food Sci.*, *51*: 440.
129. Yost, R. A.,and Kinsella, J. E. (1993). Properties of acidic whey protein gels containing emulsified butter fat. *J. Food Sci.*, *58*: 158.
130. Schmidt, R.H.,and Morris, H.A. (1984). Gelation properties of milk proteins, soy proteins, and blended protein systems. *Food Technol.*, *38*: 85.
131. Chen, S. X., Swaisgood, H. E., and Foegeding, E. A. (1994) Gelation of β-lactoglobulin treated with limited proteolysis by immobilized trypsin. *J. Agric. Food chem.*, *42*: 234.

132. Foegeding, E. A., Kuhn, P. R., and Hardin, C. C. (1992). Specific divalent cation-induced changes during gelation of β-lactoglobulin. *J. Agric. Food chem.*, *40*: 2092.
133. Johns, J. E. M., and Ennis, B. M. (1981). The effect of the replacement of calcium with sodium ions in acid whey on the functional properties of whey protein concentrates. *N. Z. J. Dairy Sci. Technol.*, *15*: 79.
134. Mulvihill, D. M., and Kinsella, J. E. (1988). Gelation of β-lactoglobulin: Effects of sodium chloride and calcium chloride on the rheological and structural properties of gels. *J. Food Sci.*, *53*: 231.
135. Damodaran, S., and Kinsella, J. E. (1982). Effects of ions on protein confirmation and functionality. *Amer. Chem. Soc. Symp. Ser. 206, ACS*, New york, p.327.
136. Barbut, S., and Foegeding, E. A. (1993). Ca^{2+}-induced gelation of pre-heated whey protein isolate. *J. Food Sci.*, *58*: 867.
137. Harwalker, V. R., and Kalab, M. (1985). Thermal denaturation and aggregation of β-lactoglobulin in solution. Electron microscopic study. *Milchwissenschaft*, *40*: 65.
138. Dunkerley, J.A., and Zadow, J. G. (1981). Rheological studies on heat induced coagula from whey protein concentrates. *N. Z. J. Dairy Sci. Technol.*, *16*: 243.
139. de Wit, J. N., and Klarenbeek, G. (1984). Effects of heat treatments on structure and solubility of whey proteins. *J. Dairy Sci.*, *67*: 2701.
140. Morr, C.V., and Foegeding, E. A. (1990). Composition and functionality of commercial whey and milk protein concentrates and isolates. *Food Technol.*, *4*: 100.
141. de Wit, J. N. (1989). Functional properties of whey proteins. *In: Developments in Dairy Chemistry, v. 4.* Functional milk proteins. *Elsevier Applied Sci. London, Ch. 7, p. 285.*
142. Poole, S., West, S. I., and Fry, J. C. (1987). Effects of basic proteins on the denaturation and heat gelation of acidic proteins. *Food Hydrocoll.*, *1*: 301.
143. Barker, C. M., and Morr, C. V. (1986). Composition and properties of Spherosil-QMA whey protein concentrate. *J. Food Sci.*, *51*: 919.
144. Brandenberg, A. H., Morr, C. V., and Weller, C. L. (1992). Gelation of commercial whey protein concentrates: Effect of removal of low-molecular-weight components. *J. Food Sci.*, *57*: 427.
145. Nakai, S., and Li-Chan, E. (1985). Structure modification and functionality of whey proteins: Quantitative structure-activity relationship approach. *J. Dairy Sci.*, *68*: 2763.
146. Goldsmith, S. M., and Toledo, R. T. (1985). Studies on egg albumin gelation using nuclear magnetic resonance. *J. Food Sci.*, *50*: 59.
147. Hickson, D. W., Alford, E. S., Gardner, F. A., Diehl, K., Sanders, J. O., and Dill, C. W. (1982). Changes in heat-induced rheological properties during cold storage of egg albumen. *J. Food Sci.*, *47*: 1908.
148. Montejano, J. G., Hamann, D. D., Ball, H. R. Jr., and Lanier, T. C. (1984). Thermally induced gelation of native and modified egg white-rheological changes during processing; final strengths and microstructures. *J. Food Sci.*, *49*: 1249.
149. Gossett, P. W., Rizvi, S. S. H., and Baker, R. C. (1984). Quantitative analysis of gelation in egg protein systems. *Food Techol.*, *38(5)*: 67.
150. Kato, A., and Takagi, T. (1988). Formation of intermolecular β-sheet structure during heat denaturation of ovalbumin. *J. Agric. Food Chem.*, *36*: 1156.
151. Shimada, K., Matsushita, S. (1980). Relationship between thermocoagulation of proteins and amino acid compositions. *J. Agric. Food Chem.*, *28*: 413.
152. Doi, E., Kitabatake, N., Hatta, H., and Koseki, T. Relationship of SH groups to functionality of ovalbumin. In Food Proteins, Eds. Kinsella, J. E., and Soucie, W. G. Amer. Oil Chem. Soc. Champaign, Il. (1989), p. 252.
153. Mine, Y., Noutomi, T., and Haga, N. (1990). Thermally induced changes in egg white proteins. *J. Agric. Food Chem.*, *38*: 2122.
154. . B. A. Margoshes (1990). Correlation of protein sulfhydryls with the strength of heat-formed egg white gels. *J. Food Sci.*, *55*: 1753.

155. Johnson, T. M., and Zabik, M. E. (1981). Gelation properties of albumen proteins, singly and in combination. *Poultry Sci.*, *60*: 2071.
156. Hirose, M., Oe, H., Doi, E. (1986). Thiol-dependent gelation of egg white. *Agric. Biol. Chem.*, *50(1)*: 59.
157. Hayakawa, S., and Nakamura, R. (1986). Optimization approaches to thermally induced egg white lysozyme gel. *Agric. Biol. Chem.*, *50*: 2039.
158. Hatta, H., Kitabatake, N., and Doi, E. (1986). Turbidity and hardness of a heat-induced gel of hen egg ovalbumin. *Agric. Biol. Chem.*, *50*: 2083.
159. Kitabatake, N., Shimizu, A., and Doi, E. (1988). Preparation of transparent egg white gel with salt by two-step heating method. *J. Food Sci.*, *53*: 735.
160. Van Kleef, F. S. M. (1986). Thermally induced protein gelation: Gelation and rheological characterization of highly concentrated ovalbumin and soybean protein gels. *Biopolymers*, *25*: 31.
161. Woodard, S. A., and Cotterill, O. J. (1986). Texture and microstructure of heat-formed egg white gels. *J. Food Sci.*, *51*: 333.
162. Egelandsdal, B. (1986). Conformation and structure of mildly heat-treated ovalbumin in dilute solutions and gel formation at higher protein concentrations. *Int. J. Pept. Res.*, *28*: 560.
163. Kitabatake, N., Shimizu, A., and Doi, E. (1988). Preparation of heat-induced transparent gels from egg white by the control of pH and ionic strength of the medium. *J. Food Sci.*, *53*: 1091.
164. Ball, H. R. Jr. (1987). Functional properties of chemically modified egg white proteins. *J. Am. Oil Chem. Soc.*, *64*: 1718.
165. Wakamatu, T., Sato, Y., and Saito, Y. (1983). On sodium chloride action in the gelation process of low density lipoprotein (LDL) from hen egg yolk. *J. Food Sci.*, *48*: 507.
166. Wakamatu, T., and Sato, Y. (1980). Studies on release of components from frozen-thawed low-density lipoprotein (LDL) of egg yolk. *J. Food Sci.*, *45*: 1768.
167. Kurisaki, J.-I., Kaminogawa, S., and Yamauchi, K. (1980). Studies on freeze-thaw gelation of very low density lipoprotein from hen's egg yolk. *J. Food Sci.*, *45*: 463.
168. Nakamura, R., Fukano, T., Taniguchi, M. (1982). Heat-induced gelation of hen's egg yolk low density lipoprotein (LDL) dispersion. *J. Food Sci.*, *47*: 1449.
169. Kojima, E., and Nakamura, R. (1985). Heat gelling properties of hen's egg yolk low density lipoprotein (LDL) in the presence of other protein. *J. Food Sci.*, *50*: 63.
170. Conrad, K. M., Mast, M. G., MacNeil, J. H., and Ball, H. R. Jr. (1993). Composition and gel forming properties of vacuum-evaporated liquid egg white. *J. Food Sci.*, *58*: 1013.
171. Nishikawa, Y., Kawai, F., Mitsuda, H. (1984). Reduction of thermal coagulation of egg white solution by foaming treatment. *J. Jap. Soc. Nutr. Food Sci.*, *37*: 129.
172. Ma, C.-Y., Harwalkar, V. R., Poste, L. M., and Sahasrabudhe, M. R. (1993). Effect of gamma irradiation on the physicochemical and functional properties of frozen liquid egg products. *Food Research Internat.*, *26*: 247.
173. Furukawa, T., Otha, S., and Yomamoto, A. (1979). Texture -structure relationships in heat-induced soy protein gels. *J. Texture Stud.*, *10*: 333.
174. Bibkov, T. M., Grinberg, V. Y., Grinberg, N. V., Varfolomeyeva, E. P., and Likhodzeivskaya, I. B. (1986). Thermoplastic gelation of proteins. *Nahrung.*, *30*: 369.
175. Utsumi, S., and Kinsella, J.E. (1985). Forces involved in soy protein gelation: Effects of various reagents on the formation, hardness and solubility of heat-induced gels made from 7S, 11S, and soy isolate. *J. Food Sci.*, *50*: 1278.
176. Bibkov, T. M., Grinberg, V. Y., Schmandke, H., Chaika, T. S., Vaintraub, I. A., and Tolstoguzov, V. B. (1981). A study on gelation of soybean globulin solutions. 2. Viscoelastic properties and structure of thermotropic gels of soybean globulins. *Colloid Polym. Sci.*, *259*: 536.
177. Mori, T., Nakamura, T., and Utsumi, S. (1981). Gelation mechanism of soybean 11 S globulin: Formation of soluble aggregates as transient intermediates. J. Food Sci., 47: 26.

178. Utsumi, S. Damodaran, S., and Kinsella, J. E. (1984). Heat-induced interactions between soybean proteins: Preferential association of 11 S basic subunits and β-subunits of 7 S. J. Agric. Food Chem., 32: 1406.
179. Nakamura, T., Utsumi, S., and Mori, T. (1984). Network structure formation in thermally induced gelation of glycinin.J. Agric. Food Chem., 32: 349.
180. Utsumi, S., and Kinsella, J.E. (1985). Structure – function relationship in food proteins: subunit interactions in heat-induced gelation of 7 S, 11 S, and soy isolate proteins. J. Agric. Food Chem., 33: 297.
181. Cabodevila, O., Hill, S. E., Armstrong, H. J., De Sousa, I., and Mitchell, J. R. (1994). Gelation enhancement of soy protein isolate using the Maillard reaction and high temperatures. J. Food Sci., 59: 872.
182. Yao, J. J., Tanteeratatarm, K., and Wei, L. S. (1990). Effects of maturation and storage on solubility, emulsion stability, and gelation properties of isolated soy proteins. J.A.O.C.S., 67(12): 974.
183. Wang, L. C. (1981). Soybean protein agglomeration: Promotion by ultrasonic treatment. J. Agric. Food Chem., 29: 177.
184. Sung, S. Y. Improvement of the functionality of soy protein by introduction of new thiol groups to papain-catalyzed acylation. (1983). J. Food Sci., 48: 708.

Subject Index

Printing: Mercedesdruck, Berlin
Binding: Buchbinderei Lüderitz & Bauer, Berlin